Springer Monographs in Mathematics

More information about this series at http://www.springer.com/series/3733

Karl-Peter Hadeler • Johannes Müller

Cellular Automata: Analysis and Applications

Karl-Peter Hadeler
FB Biologie
Universität Tübingen
Tübingen, Germany

Johannes Müller
Centre for Mathematical Sciences
Technical University Munich
Garching, Germany

ISSN 1439-7382 ISSN 2196-9922 (electronic)
Springer Monographs in Mathematics
ISBN 978-3-319-85047-4 ISBN 978-3-319-53043-7 (eBook)
DOI 10.1007/978-3-319-53043-7

Printed on acid-free paper

This Springer imprint is published by Springer Nature
The registered company is Springer International Publishing AG
The registered company address is: Gewerbestrasse 11, 6330 Cham, Switzerland

Contents

List of Symbols

We give a list of some notations with a reference to the first page where the notation appears.

Symbol	Page	Meaning
$[a]$, $[ab]$	162	Concatenation of signs or (finite) sequences
$[\overline{a}]$,$[\overline{0}]$	151	State in $E^{\mathbb{Z}}$ that is identically a (or 0)
$\|z\|_1$	25	ℓ^1-norm of a vector z
$\|z\|_\infty$	27	ℓ^∞-norm of a vector z
$[G : H]$	29	Index of a subgroup H of G
$\|X\|$	33	Size of some set X
0	26	Neutral element of a group, origin of the grid (Abelian case)
$\mathcal{A}$	116	Conley attractor
$\boldsymbol{\mathcal{A}}$	120	Set of all attractors
$A(f)$	55	Set of bijective functions commuting with f
$B_r(x)$	46	Open ball around x with radius r in a metric space
$\mathcal{B}(.)$	117, 131	Basin of attraction (for attractors resp. chain components)
$\mathfrak{B}$	134	Borel algebra
$\mathfrak{C}$	52	Cantor space or metric, compact, and totally disconnected space
$\mathfrak{C}_{1/3}$	39	The mid-third Cantor set
$CR(f)$	125	Recurrent points
$d(u, v)$	45	Cantor distance between two states in E^Γ
$d(n_1, n_2)$	387	Unsigned distance on $\mathbb{Z}_m$
$d_c(g, h)$	25	Distance between two elements on a Cayley graph

Symbol	Page	Meaning
d_H	76	Hamming distance (for two states or pattern)
d_H	302	Hausdorff distance (for two sets)
d_B	76	Besicovitch pseudometric
$d_{\hat{B}}$	76	Shift invariant Besicovitch pseudometric
d_W	76	Weyl pseudometric
d_*	81	One of $d_{\hat{B}}$, d_B, and d_W
D_0, D_g	26	neighborhood of the unit element resp. of $g \in G$
e	26	Neutral element of a group, origin of the grid (general case/non-Abelian case)
E	28	Local states (a finite set)
$E[\Gamma]$, $E[\Gamma]_c$	287	Alternative representation of $E\Gamma$ respectively $(E^\Gamma)_c$
E^{D_0}, $E^{\tilde{\Gamma}}$	30	States/configurations of a subgraph
E^Γ	28	Global states
$(E^\Gamma)_c$	31	Global states with finite support
$Eq(f)$	169	Set of equicontinuous points
f_0	30	Local function of a cellular automaton
$\mathcal{F}_n$	306	Limit set for linear cellular automata
F_n	34	Free group with n elements
$\mathbb{F}_p$	287	Finite field, in our case: $\mathbb{Z}_p$ with p prime
G	20	Finitely generated group
$\mathcal{G}$	180	Grammar
$\Gamma = \Gamma(G, \sigma)$	22	Cayley graph for group $G =< \sigma >$
Γ_m	33	$\Gamma_m = \{g \in G : d_c(g, 0) \leq m\}$, ball of radius m
$\tilde{\Gamma}^-$	160, 255	Interior of a subgraph (w.r.t. a cellular automaton)
$\tilde{\Gamma}^+$	255	Closure of a subgraph (w.r.t. a cellular automaton)
$\partial\Gamma_m$, $\partial\tilde{\Gamma}$	78, 255	Boundary of a subgraph (w.r.t. a cellular automaton)
$\gamma(n)$	33, 69	Growth function, operator on $E^{\mathbb{N}_0 \times \mathbb{Z}}$
$\mathcal{L}(\mathcal{G})$	180	Language generated by the grammar $\mathcal{G}$
$(\mathcal{N}, \mathcal{E})$	265	Directed graph of local pattern that can be elements of a stationary state
μ	63, 132	Distance measure, Bernoulli measure
$\mathbb{N}$	21	Positive integers: 1, 2, 3,...
$\mathbb{N}_0$	32	Non-negative integers: 0, 1, 2, 3,...
$\omega(x)$, $\omega(B)$	114, 116	ω-limit set
PO	266	Set of double-point free periodic orbits within a directed graph

Symbol	Page	Meaning
$\mathcal{P}$	265	Set of all bi-infinite pathes in a directed graph
$\mathcal{P}$	87	Set of all periodic states (in Cantor sense)
$\hat{\mathcal{P}}$	87	Set of all periodic states (in X_*)
$\check{\mathcal{P}}$	87	Set of all weakly periodic states (in X_*)
$\mathfrak{P}(E)$	132	Power set of E
$\mathcal{Q}$	123, 198	Quasi-attractor; set of internal states of a Turing machine
$\mathbb{Q}_+$	225	Non-negative rational numbers (including zero)
$\mathbb{R}_+$	63	Non-negative real numbers (including zero)
$R[X]$	440	Polynomial ring
$R[X,X^{-1}]$	290	Laurent polynomial ring
S^1	156	Unit circle
$\mathcal{S}$	265	Set of stationary states of a cellular automaton
σ_g	25	Shift operator for group element g
$\sigma(n_1,n_2)$	387	Signed distance on $\mathbb{Z}_m$
$\sigma_{\mathfrak{Y}}:\mathfrak{Y}\to\mathfrak{Y}$	66	Shift over an infinite alphabet
Σ,Σ_F	175, 186	Shift over a finite alphabet, shift of finite type
Σ_2, Σ_n	42,182, 66	Set of binary sequences resp. sequences over $\{0,\ldots,n-1\}$; alphabet of a finite automaton or Turing machine
supp(u)	29	Support of state u
$T(M)$	183	Language of accepted words associated with the automaton M
$\mathfrak{X}$	52	Metric or topological space
$(X_{\hat{B}},\ d_{\hat{B}})$, $(X_B,\ d_B)$	76	Besicovitch, shift invariant Besicovitch, Weyl space
$(X_W,\ d_W)$	76	Weyl space
$(X_*,\ d_*)$	81	One of the spaces $(X_{\hat{B}},\ d_{\hat{B}})$, $(X_B,\ d_B)$, and $(X_W,\ d_W)$
$(X,\mathfrak{B},\mu)$	134	Borel probability space
$(X,\mathfrak{T})$	429	Topological space
$W(u_1,\ldots,u_M)$	389	Winding number
$\mathbb{Z}_m$	20	Cyclic group modulo m

Chapter 1
Introduction

1.1 Discreteness

Early mathematics has been either "geometry", "arithmetics", or "algebra", hence essentially discrete. At the times of Newton and Leibnitz it was found that working with continua is, in some aspects, much easier, although at their time neither the continuum nor continuity was well understood. Nevertheless, many problems can be formulated and solved by methods of "calculus" or "analysis" which cannot be even well posed and cannot be solved in a discrete setting. Nowadays scientists try to formulate their mathematical models in the language of continuous time and space, as integral equations or, more often, as ordinary or partial differential equations.

If we look more closely then we see that differential equations are special in the sense that they are localized. This is so because the concept of a derivative is local: The derivative of a function at a point depends only on the values in an arbitrarily small neighborhood of that point. Using partial differential equations as models in physics, engineering, biology, chemistry and even economy (e.g. the Black-Scholes equation) leads to beautiful mathematical theories and powerful applications. But in some sense mathematics takes revenge for abandoning discreteness. Typically solutions of partial differential equations can only be shown to exist in certain spaces. The solutions may be only "weak", i.e., not be defined as classical functions, they may not be unique etc. In fact, the majority of papers on partial differential equations are, in one or the other way, about "regularity" of solutions. In many cases where partial differential equations are used as models for physical phenomena it is unclear how the restrictions imposed by the regularity and existence problem are related to the physical model. So why not go back and abandon locality and continuity?

K.-P. Hadeler, J. Müller, *Cellular Automata: Analysis and Applications*,
Springer Monographs in Mathematics, DOI 10.1007/978-3-319-53043-7_1

One way is to drop locality. For example, we have integral equations. Integral equations have been used, e.g., in modeling epidemic spread, there the integral results from a contact distribution. The theory parallels that for differential equations in many aspects. Locality is replaced by the concept of rapidly decaying kernels like $\exp\{-x^2\}$. The theory of (non-singular) integral equations is much simpler in terms of regularity but otherwise more complicated since the flow does not compactify. Giving up locality with respect to time leads to delay equations or Volterra integral equations. Volterra integral equations (renewal equations) are useful in population dynamics where age is a typical "history" trait. Delay equations, as models, are often used where complete information on intermediate processes such as transport is lacking.

Giving up continuity in time leads to mappings as opposed to continuous time dynamical systems. Typically, if we have a continuum model and a discrete model for the same phenomenon, then the discrete model shows a richer structure. The classical example is exponential growth with saturation: The logistic equation is utterly simple, the discrete logistic equation has a very rich structure (period doubling, period 3, chaos etc.) Sometimes the choice between discrete and continuous models is just by tradition. Geneticists use discrete time models because they think of generations, ecologists use differential equations because they think of mass action kinetics.

Systems which are discrete in time and space like coupled map lattices are still continuous with respect to the state variable. At each point of a discrete grid there is a map, this map is evaluated at discrete time steps and is coupled to similar maps at neighboring grid points. For such systems we can still use analytic tools, e.g., we can linearize at a stationary state and discuss stability in terms of the spectrum of some operator.

What if we abandon the continuum altogether? Then we have a grid with "cells" or "vertices", at each cell a set of finitely many states, and a set of rules that tell how to compute new states from given ones. If we further require that the new value at a given cell depends on the values in a neighborhood (again a locality principle) then we have a cellular automaton. In most established examples there are also symmetries of the grid and of the neighborhood and some translation invariance of the local function. We do not need these properties at this moment.

Thus, a cellular automaton is something like an extreme caricature of a partial differential equation. If the grid is very large then perhaps one can even approximate a cellular automaton by a partial differential equation, or the other way round.

Cellular automata, as compared to continuous systems of any kind, have the great advantage that everything is discrete. Hence there are no problems with existence or regularity of solutions, and no need for any kind of numerical method. As long as the grid is finite and not too large, the cellular automaton can be easily implemented on a computer and, in the typical case of a 2D grid, the behavior can be observed on the screen.

The problems start when we want to describe what we see. The typical notions of stationary state, convergence almost sure (a.s.), do not quite fit, nor do we have any analytical tools at hand. This situation becomes obvious if we look at the many textbooks and monographs on cellular automata. We see graphically that automata can model pattern formation, epidemic spread, and animal coats, but it is difficult to cast these observations into general statements or "theorems".

The goal of this volume is to present some fundamental mathematical tools for discrete dynamical systems in general and cellular automata in particular which exist in the literature since long but have not been well received by the modeling community, to apply them to concrete problems, and to connect the theory of cellular automata to other, partly discrete types of systems such as interacting particle systems or 0,1 neural networks. Before developing a coherent theory of cellular automata, we present some typical examples: (1) the game of life, (2) contact automata for epidemic spread, (3) Wolfram automata and pattern formation, (4) Greenberg-Hastings automata, (5) Langton's ant.

1.2 The Game of Life

The Game of Life has been invented by J.H. Conway around 1970 and it became popular by articles by Martin Gardner [64]. It works on a square grid. Cells are living (occupied, black, 1) or dead (empty, white, 0). A living cell survives at the next time step if it has not too many (death because of crowding) and not too few neighbors (death following loneliness): Let s denote the number of living neighbors in the eight neighboring cells. The cell which is alive will stay alive if $s \in \{2, 3\}$, and a dead cell will become alive if $s = 3$. In all other cases, the cell is dead. This very simple rule generates an astounding variety of patterns depending on the initial data. The Game of Life has become so popular because these patterns appeal to the human brain. First, cellular automata in 2D and also time evolutions of cellular automata in 1D, when presented as 2D graphs, appeal to our optical perception system because the optical system itself is essentially 2D. Second, Conway's automaton produces isolated black objects on a white screen with characteristic, reproducible shapes which can be memorized and which can be given nicknames, like the gun, the glider, the eater. Third, these fictitious objects interact as if they were machines or even living things. The gun shoots (or gives birth to) gliders, gliders move on straight lines and interact in collisions, the eater eats gliders. In some cases these interactions can be formulated in a mathematically rigorous way. Here, we only present a period 3 object (the Cambridge pulsar, Fig. 1.1) which can be easily computed as the limit set of an orbit that starts from a very simple configuration after 21 steps. We refer to the monograph [12] which contains a detailed discussion of the mathematics of the Game of Life together with many graphical examples.

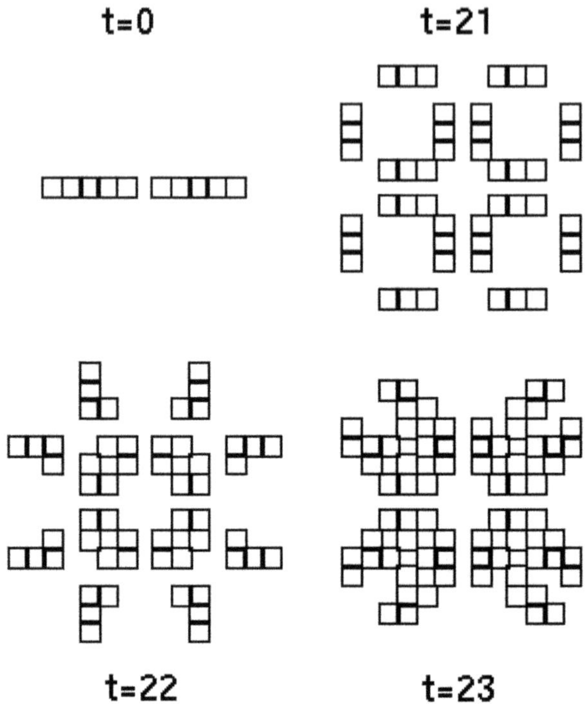

Fig. 1.1 The Cambridge Pulsar, initial state and periodic orbit

1.3 Contact Automata

A contact automaton can be seen as one of the simplest models for the spread of an infectious disease. Imagine a square grid of cells. Assume that all cells are "white" (or "uninfected") except a finite number of "black" (or "infected") cells. For a given cell, define the neighbors as the eight immediately adjacent cells and the cell itself. A deterministic version of a contact process works in discrete time steps as follows. A black cell stays black, a white cell which is a neighbor to a black cell turns black. This system can be called a deterministic contact process.

If we start with a single black cell, than the number of black cells increases as $1, 9, 25, \ldots$. The overall form stays quadratic. If we start with three black cells in an arbitrary relative position then things get already more complicated, see Fig. 1.2.

The number of black cells is ever increasing, eventually we get a large black patch. The patch does not become an approximate circular disk, the asymptotic shape is closer to a rectangle although its boundary keeps some memory of the initial configuration. Such anisotropy is a typical phenomenon in deterministic automata. However, most real world problems we want to describe do not show this kind of anisotropy. To overcome this discrepancy we can leave the domain of

Fig. 1.2 The contact automaton with Moore neighborhood starting from three infected states. The 'age since infection' of a cell is indicated by the size of the block. The total infected area never becomes a rectangle

deterministic systems and introduce probabilistic rules in various ways. We give some examples:

i) A black cell stays black, but a white cell adjacent to a black cell becomes black with some probability $p \in (0, 1)$. Then the resulting set of black cells looks much more irregular and may resemble a snowflake (see Fig. 1.3).

   This stochastic contact automaton works in discrete time. From the viewpoint of stochastic processes it is a disadvantage that several cells are updated at the same time. In the theory of stochastic processes one does not like multiple events.

ii) Assume that there are only finitely many black cells. A white boundary cell is white and has at least one black neighbor. At a given time step a white boundary cell is chosen (with equal probability from all such cells) and then converted into black.

   We have a discrete time process. At each time step exactly one cell changes its state.

iii) A process in continuous time. Each cell has its own Poisson process with rate $\lambda > 0$. These processes are independent. If a cell "fires" then it changes its state if and only if it is a white neighbor of a black cell.

   The process ii) is embedded in the process iii) in the following sense. In the process iii) multiple events occur only with probability zero (we have only countably many cells). If we follow one realization and number the successive changes then we get the process ii).

Fig. 1.3 The random contact automaton starting from a single infected cell. Infection occurs only with probability 0.05. The square grid structure is largely destroyed

In the present book, we will not discuss stochastic particle systems; we restrict ourselves to deterministic cellular automata. However, we will find that a rich variety of phenomena, also epidemiological models, can be formulated and analyzed in this deterministic setup, and yield useful results. Furthermore, even if the dynamics is deterministic, the initial value may be chosen in a stochastic way. This "backdoor" allows to use stochastic tools for the analysis of cellular automata (see Chap. 11).

1.4 Some Wolfram Automata

Cellular automata with a 1D grid, a neighborhood of three cells and the states 0 and 1 are called Wolfram automata, because in 1983 S. Wolfram [181] defined a useful code to enumerate these automata (see below) and, more importantly, he tried to classify the graphical representations of the dynamical behaviors. In the standard graphical representation the state at a given time is printed horizontally and the time axis points downward. According to the long term behavior observed, a cellular automaton is assigned to one of the four groups: (1) tends to a homogeneous stationary state (2) tends to periodic structures (3) tends to aperiodic, chaotic pattern, (4) tends to persistent, complex and localized pattern.

Wolfram's classification has stimulated research on cellular automata (see, e.g., the review article of Sutner [159]). However, at the present level of knowledge, the phenomenological classification cannot be reformulated in terms of formal criteria. Some of these automata have existed, as Pascal's triangle or as a Sierpinski gasket, much earlier in other fields of mathematics. Think of Pascal's triangle for the binomial coefficients. The local rule is: add the two neighbors to the right and to the left and complete the row on either side with a "1".

If we call the entries "cells" then in the standard representation of Pascal's triangle we do not have a square grid since only every second row fits. The rows in between are shifted by half a cell. This problem can be easily solved by putting some zeros in between. Next we can take all numbers mod 2. What we get is the time evolution of the cellular automaton with the local rule: Add the two neighbors and take the sum mod 2. If we start with an infinite array and a single 1 then we find the pattern displayed in Fig. 1.4. Assume we continue the above scheme to an ever increasing number of rows and at the same time reduce the scale of the representation. Then we see an apparently random structure with repeated triangular shapes. This structure is called a Sierpinski gasket. Of course, by this procedure the picture gets an ever finer structure, and we may guess that the picture becomes eventually self-similar. Indeed, Sierpinski gaskets have been among the earliest examples of what we call

Fig. 1.4 Time evolution of the Wolfram automaton rule 18 starting from a single occupied cell

now fractal sets[1] (see also Sect. 3.1 on Cantor sets and Sect. 10.3.3 on fractal sets). But there is more to that, these things do occur in nature. Indeed, if we look at the shell of the marine snail *Oliva porphyria* then we see structures very close to Sierpinski gaskets. In an early paper Ermentrout et al. [55] have designed a type of cellular automaton (they called it a neural net) as a model for molluscan shell patterns ([170] was still earlier). Later Meinhardt [125] has used reaction diffusion systems to imitate many naturally occurring patterns on the shells of snails and mussels. If we type "Oliva porphyria" in Google Images then we get a vast array of color photographs of this beautiful shell.

For better understanding we explain the difference in coat pattern of vertebrates, say fish, and the shell patterns of mollusks. The outer surface of a fish is "topologically" a sphere (neglecting mouth and digestive tract, gills etc., that play no or only a minor role for pattern formation) and it grows by expansion. Cells are dividing and grow, and they manage to do this while staying in contact with their neighbors (no gaps) and keep a smooth surface. If this process goes wrong then there is a malformation or a tumor. While the fish is growing, its skin pattern changes. In a striped fish, the stripes may get wider, but also the number of stripes may increase. Hence the skin pattern of a fish is a truly 2D phenomenon. On the other hand think of a snail. The body of the snail also grows "everywhere", but the existing parts of the shell do not grow. The shell grows only at the outer rim where new material is added by the "mantle". Hence any pattern on the shell cannot be changed. Pattern can only be added. Hence the pattern on the shell of a grown-up snail shows the history of the deposition process at the rim, an essentially 1D structure. That is why Sierpinski gaskets and sea shell patterns are not only intriguing by their similarity but also by analogous production processes.

Wolfram's Nomenclature The following family of one-dimensional cellular automata is easy to define and to simulate. The grid is $G = \mathbb{Z}$ (or an interval in $\mathbb{Z}$), the elementary states are $E = \{0, 1\}$, and the neighborhood is $D_0 = \{-1, 0, 1\}$. There are $2^3 = 8$ local states. A local function is defined if for each local state the resulting elementary state is specified. Thus there are $2^8 = 256$ different local functions. Wolfram has constructed a simple enumeration system for this family. Hence these automata have now established names in terms of decimal numbers. These numbers, unfortunately, do not reflect the structural properties of these automata. The system works as follows:

[1]The original construction of Sierpiński (around 1915/16) went the other way round. Take an equilateral triangle, divide in four, remove the open center triangle, in the remaining four triangles, divide each in four, remove the open center triangle, a.s.o. The resulting compact set, a 2D analogue of the Cantor set, is the same set as that obtained by scaling Pascal's triangle.

First order the local states as three digit binary numbers. There are eight local states. Then associate to each local state the value of the local function. Read these results as an eight digit binary number. Encode this number in decimal notation. An example is given below. The first column contains the eight local states in increasing order, the second the values of the local function, the third the powers of 2 in decimal notation, the last the terms of the sum in decimal notation. The last column gives the sum 18. Hence this automaton is called "rule 18".

φ	$f_0(\varphi)$	2^k	$2^k f_0(\varphi)$
000	0	1	0
001	1	2	2
010	0	4	0
011	0	8	0
100	1	16	16
101	0	32	0
110	0	64	0
111	0	128	0

The local function of "rule 18" can also be represented as

$$f_0(\varphi) = \begin{cases} 1 & \text{for } (0,0,1) \text{ and } (1,0,0) \\ 0 & \text{otherwise.} \end{cases}$$

Let us look at a second example: rule 50. If we write $50 = 32 + 16 + 2$, we find that the local function is defined by

$$f_0(\varphi) = \begin{cases} 1 & \text{for } (0,0,1), \ (1,0,0), \text{ and } (1,0,1) \\ 0 & \text{otherwise.} \end{cases}$$

The space-time pattern for an initial state starting with a single occupied cell is depicted in Fig. 1.5.

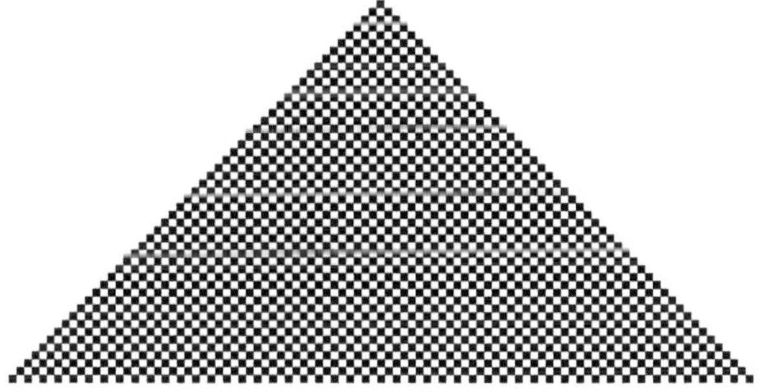

Fig. 1.5 Time evolution of the Wolfram automaton rule 50 starting from a single occupied cell

1.5 Greenberg-Hastings Automata

In 1946 Norbert Wiener and Arturo Rosenblueth [175] introduced a model for the propagation of excitation in the cardiac muscle. They were interested in heart flutter and fibrillation, i.e., in situations when the regular physiological heart beat is replaced by a continuing excitation wave traveling around the heart. They assumed that the cardiac muscle is homogeneous, i.e., "that the fibers which compose it have similar properties ..." [175]. According to their model a fiber can be in one of three states which are described by 'epoch' numbers: active $= 0$ (excited), refractory $= u \in (0,1)$ (not excitable) or resting $= 1$ (excitable). The dynamics is defined by the following rules: Any number less than 1 grows at a constant rate. If a fiber in state 1 has a neighbor in state 0 then it becomes excited, i.e., 0, otherwise it stays in state 1. Neighbors are fibers within a certain distance to the fiber considered. The Wiener-Rosenblueth model has typical features of a cellular automaton. Discrete cells (fibers) form a grid (tissue). There is a set of possible states of the fibers, a neighborhood is defined, and the interaction between fibers is local, i.e., restricted to neighbors. In this model time does not proceed in discrete steps although this is implicitly assumed by the authors. Wiener and Rosenblueth, at their time, could not simulate their model but they drew several conclusions, e.g., that on a one-dimensional torus one external stimulus will produce traveling waves which cancel when they meet while a persisting wave can only be produced by two stimuli, with proper spatial distance and time delay.

In 1978 Greenberg and Hastings [74], see also [75, 76], designed a class of cellular automata which are close to the Wiener-Rosenblueth system, the main difference being a discrete array of delay states rather than a continuous delay variable. These automata are constructed as follows.

The grid is the square lattice, and neighbors are "North, East, West, South" = "NEWS" (this will later be called the von Neumann neighborhood). The possible states of a cell are the natural numbers $0, 1, 2, \ldots, g, g+1, \ldots, g+a$ whereby 0 is interpreted as a resting state, states between $g+1$ and $g+a$ as "excited" and the states between 1 and g as "refractory". The local function is defined as follows. For some cell let s the number of neighbors (including the cell itself) with states z with $g < z \leq g+a$. Let $\bar{s}$ be some threshold. If the cell has state 0 and $s \geq \bar{s}$ then the cell is excited to the state $g+a$. If the cell is in state 0 and $s < \bar{s}$ then it remains in the state 0. In all other cases the state is reduced by 1. A cell which has been excited to the state $g+a$ "runs down" through a number of excited states and becomes eventually resting; in the resting state it can be excited again.

Greenberg and Hastings observed wave patterns evolving from simple initial conditions, e.g. rotating spirals emerging from simple initial states. They gave also sufficient conditions for the persistence of patterns. We shall consider these automata in more detail in Sect. 12.3.

1.6 Langton's Ant and Life Without Death

Langton's ant (see [114]) is described in Math World as a 4 state 2-dimensional Turing machine, see also [61, 62]. We describe it in plain words as follows. We have the standard grid $\mathbb{Z}^2$ and the von Neumann neighborhood. A cell can be black or white. In addition a cell can carry the ant (there is only one ant on the grid). The ant has four directions which we can conveniently describe as NEWS (or up, right, left, down). The ant moves on the grid and leaves a trace according to the following rules.

i) If the ant is on a black cell then it turns 90° to the right.
ii) If the ant is on a white cell then it turns 90° to the left.
iii) If the ant leaves a cell then the cell changes color.

The description seems a little bit ambiguous as to what it means "is in a cell". We guess it means "enter the cell with the previous direction and then turn."

The system is started with a single ant on a white grid. The ant moves around in an erratic fashion and then, after thousand steps or so, its motion becomes periodic with period 104. Then the black part of the path grows in a periodic fashion to infinity.

Langton's ant can be realized as a cellular automaton with 18 (possibly less) states. But there is a less known cellular automaton with only two states, obtained by slightly modifying the Game of Life, which shows very similar behavior. This automaton "Life without Death" has been presented in [77] without reference to Langton, and it has been shown to produce "ladders" which are quite similar to the path of the ant, see Fig. 1.6 and also [73].

The examples show a variety of phenomena and suggests various lines of research, for example investigating asymptotic behavior, stationary states and periodic phenomena, classification of behaviors, complexity, realization of a given dynamics by a cellular automaton, modeling physical phenomena, isotropy, etc.

Fig. 1.6 A ladder for Langton's ant

1.7 A Nice Little Automaton

and a cute trick to find its behavior

Consider the following automaton on $\mathbb{Z}$ or $\mathbb{Z}_n$. The neighborhood is $\{-1, 0, 1\}$, the set of elementary states is $\{0, 2, 1\}$, and the local rule is given as follows.

"0" stays "0" if both neighbors are "0".
"0" becomes "2" if at least one neighbor is different from "0".
"2" becomes "1"
"1" becomes "0". Hereby "neighbor" denotes a neighbor except the cell itself.

The automaton is motivated by an SIS epidemic: "0" is susceptible, "2" is recently infected and "1" has been infected for some time. It turns out that every orbit becomes eventually periodic with period three (or it becomes stationary).

Although this claim can be easily verified for special cases, it is not so clear how to do this in general. Here we use the following trick. There are 27 triads 000, 001, 002, ..., 221, 222. Applying the rule, one finds that each triad has, depending on the environment, either one or two possible successors. Connecting each triad to its possible successors, one gets a directed graph. This graph has four connected components. Then the result follows by inspection of the flow in these components (Fig. 1.7). In the smallest component (three elements) everything becomes stationary. In the large symmetric component (12 elements) and in the two

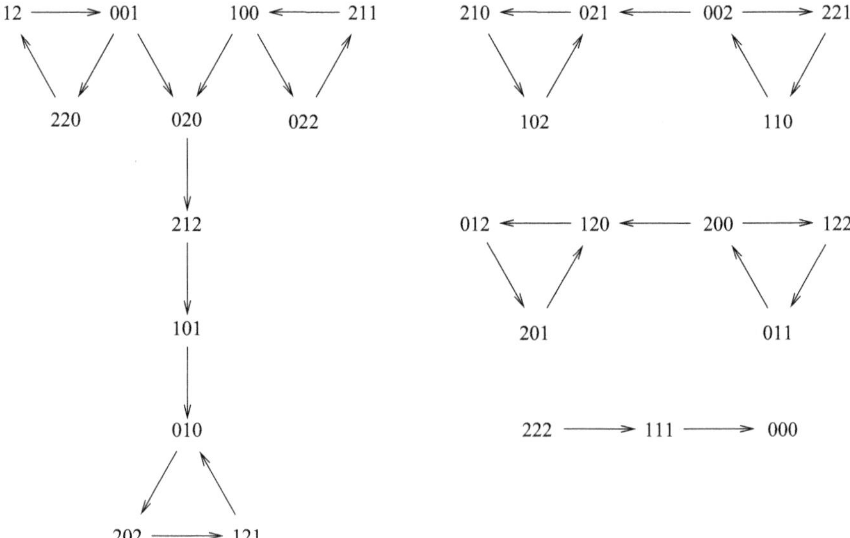

Fig. 1.7 The four components of the triad graph. In the large component there are three cycles of order three. Either the state stays always in the upper left cycle or it stays always in the *upper right cycle* or, after finitely many steps in the *upper left* or the *upper right*, it ends up in the *lower cycle*. Similarly for the other two cyclic components

other components (six elements each), symmetric to each other, everything runs into a period three cycle.

Hence in any finite grid $\mathbb{Z}_n$ every trajectory becomes eventually periodic with period three. In an infinite grid $\mathbb{Z}$ finite segments of trajectories with finite support become eventually periodic with period three.

1.8 History and Applications

One of the earliest papers on cellular automata is J. von Neumann's work on self-replicating automata [168]. His goal was to construct a cellular automaton on a finite portion of $\mathbb{Z}^2$ which would construct an identical copy of itself in another portion of $\mathbb{Z}^2$ and also start the copy working. Von Neumann's construction was complicated (he needed 29 elementary states) but intuitive: different parts of the automaton could be understood as pliers, handles etc. Codd [32] gave a mathematically simpler but less intuitive construction with fewer elementary states.

Later the Game of Life has stimulated research on cellular automata although it has been said "... which has little biological meaning other than a metaphor" [54].

There are some monographs on special classes of cellular automata like [66] and [180]. Most "books" on this topic are actually proceedings of conferences. An exception is the nice book of Ceccherini-Silberstein and Coornaert [27], who focuses on the connection between properties of the grid and properties of the cellular automaton (like surjectivity and injectivity). Cellular automata can be investigated from very different points of view: A more formal localization will put cellular automata in the field between dynamical systems and discrete mathematics: in symbolic dynamics [106, 112, 115]. For many classes of cellular automata, however, often the connection to applications in e.g. physics may lead to deeper insight, as e.g. Wolfram explains [183].

There is a wide range of applications of cellular automata (in a broad sense) as modeling tools in biology. The early paper [54] gives a variety of examples with graphical illustrations: prey-predator models, reaction and diffusion, shell patterns, fibroblast aggregation, immunology, ocular dominance, ant trails. They are also valuable tools in image processing [145, 149]. Physical applications are first of all fluid and gas dynamics [121, 180]. There are also applications in the simulation of the Ising model [166], and we find even an approach to superstring theory in terms of deterministic cellular automata [161].

1.9 Outline of This Work

In the examples above we already observe the richness of cellular automata. The main theme of the present work is the quest for efficient approaches that allow for an insight beyond pure simulations. Chapter 2 sets the scene and introduces cellular automata in a rather algebraic and combinatorial manner. As we already know from the examples above, cellular automata are maps ("global functions") acting on a state space consisting of a grid (e.g., $\mathbb{Z}^2$) where each grid point assume one of a finite number of possible local states (e.g., $\{0, 1\}$). This map incorporates certain properties, in particular locality and translational invariance.

It is by no means clear how complex cellular automata really are. This question is of central importance for our task: No one would have the idea that there is a useful theory covering all operators on $L^2(\mathbb{R})$, say. This set is clearly much too large and inhomogeneous. There are certain classes of operators (Hilbert-Schmidt operators, compact operators, etc.) that are small and homogeneous enough such that powerful results can be developed. Therefore, our first aim (Chaps. 3 and 4) is to estimate the degree of generality that is inherent to cellular automata and to become acquainted with some fundamental properties. We approach this task from the topological dynamical systems point of view. There are two different sensible ways to define a topology on the state space: The Cantor metric, and the Besicovitch resp. Weyl topology.

The Cantor topology is a well known and widely used tool in discrete mathematics. It is based on the idea to focus on finite subsets of the grid, that is, balls around zero. The maximal radius of such balls at which two states agree is determined. If this radius becomes infinite, the states are identical; if the radius is small, their difference is large. The reciprocal value of this maximal radius (plus one) is a metric. The resulting metric space incorporates all kind of nice properties, where the most important one is compactness. However, this topology only gives us control over some region around zero, and possibly no idea about the differences of two states far away from zero. Moreover, cellular automata are shift invariant, while the Cantor topology focus at the region around a special site in the grid, the origin; these two facts seem not to fit.

In contrast, the idea of Besicovitch resp. Weyl topology is to measure the probability that two states differ at a randomly selected vertex. Therefore these topologies incorporate translational invariance in their construction (though it turns out that the Besicovitch pseudodistance is not always translational invariant). However, the resulting topological spaces are rather difficult to handle; the Weyl space is even not complete.

In both cases (Cantor and Besicovitch resp. Weyl), cellular automata can be well characterized. The fundamental Curtis-Hedlund-Lyndon theorem, introduced in Chap. 3 for Cantor topology and Chap. 4 for Besicovitch and Weyl topology, shows that cellular automata are exactly those functions that are continuous and translational invariant.

The degree of generality of cellular automata is indicated by the fact (discussed in Chap. 3) that any continuous function on a metrizable Cantor space can be represented by a cellular automaton. This result implies that cellular automata are by far too general to allow for many useful theorems valid for all of them. The fundamental Curtis-Hedlund-Lyndon theorem is one of the very few examples for such a statement.

It is thus necessary to define small, homogeneous classes of cellular automata if we aim at deeper results. There are two strategies for the construction of such classes. The algebraic, top-down approach selects a certain aspect, and classifies cellular automata with respect to this aspect. This approach yields a partition of cellular automata in disjoint sets; any given example can be localized in one of these classes. The second, bottom-up approach starts with the investigation of certain cellular automata, often driven by applications. These applications are useful as guiding tools since they force the cellular automaton to behave in a rather special way to resemble some real-world phenomenon. If we aim to produce self-similar patterns, for example, then also the dynamics inherits some self-similarity that can be used to analyze the automaton. The advantage of this approach is that we find homogeneous classes of cellular automata that can be readily analyzed.

Chapters 5–7 focus on the top-down approach. Each of these chapters we concentrate on one particular aspect, and develop accordingly a classification scheme for the set of cellular automata.

We proceed with the idea already used in Chaps. 3 and 4 that a cellular automaton is a dynamical system (Chap. 5). For the complexity of a dynamical system the long term behavior, represented by the complexity of the attractor, is decisive. Hurley proposed a classification along these lines, by investigating the set of all Conley-attractors for a given cellular automaton.

In the next chapter we go away from the dynamical systems idea, and focus on properties of the global function. The Curtis-Hedlund-Lyndon theorem indicates that this global function is continuous. Gilman used this observation as the starting point for a classification scheme (Chap. 6). A continuous function may be merely continuous at a certain point, or equicontinuous under iteration. That is, Gilman identifies Lyapunov-continuous states. If a cellular automaton is chaotic, a tiny difference between two states will grow exponentially fast. At Lyapunov-continuous states this exponential fast dispersion does not happen. The more Lyapunov-continuous states a cellular automaton has, the more uniform is the long term behavior.

Another idea, strongly driven by Kůrka, is to take the term “automaton” in “cellular automaton” serious (Chap. 7). There is a long tradition to classify automata according to their complexity. Certain classes of automata and languages (sets of words of finite length over a finite alphabet) are the two sides of the same coin: The classifications of languages and of automata are strongly intertwined. Kůrka constructs a language for a given cellular automaton: Start with one state, and iterate the cellular automaton for a finite number of times (arbitrarily, but only finitely often). Code the states by a finite alphabet (that is, decompose the state space in a finite number of sets). This procedure yields a set of finite words over a finite alphabet. The complexity of the grammar necessary to generate this language will tell us something about the complexity of the cellular automaton. And indeed, the class with lowest complexity is precisely the Gilman class of Lyapunov-continuous cellular automata.

These considerations culminate in the question if all properties of cellular automata are decidable. The inner core of undecidable problems is the ancient paradox of Epimenides: Epimenides, the Cretan, states: “Cretans, always liars”. Clearly, we cannot decide if this statement is true or false. This idea, however, has to be reformulated in a mathematical language (halting problem) and to be adapted to our application (domino problem). Therefore, we discuss in Chap. 8 Turing machines and tessellations. Of special interest is the proof that it is in general undecidable if a given tile set allows for the tessellation of $\mathbb{Z}^2$. We present a nice recent version of this proof due to Kari. Tessellations are close to cellular automata—they have also been named “tessellation automata”—and so we have some tool at hand to investigate the decidability of certain properties in Chap. 9: In particular existence of stationary states, injectivity and surjectivity are interesting. At this point, we also discuss the relation of surjectivity, injectivity, and the Garden-of-Eden theorems: For a map on a finite set, injectivity and surjectivity are clearly coupled. If the grid is not too complex, also for a cellular automaton injectivity and surjectivity are not independent, but injectivity implies surjectivity and hence bijectivity. Concerning decision problems, we find that for essentially one-dimensional structures (free groups) we are able to decide many properties. In particular for $\mathbb{Z}^2$, it is not possible any more to always decide even very basic properties as injectivity, surjectivity or the existence of stationary states.

Chapters 10–13 change the point of view and use a bottom-up approach. The implications of a particular property, only given for a (small) class of cellular automata is investigated. This property can be mathematically motivated or driven by an application.

We begin in Chap. 10 with a mathematical structure that is often useful: linearity. If we superimpose an algebraic structure to the local states, this algebraic structure can be lifted to the state space. It is suggestive to consider cellular automata that respect this algebraic structure. And indeed, this class of cellular automata allows for many nice results. In particular, many of these automata inherit some self-similarity in the dynamics that allows to connect their long term behavior with fractal, self-similar sets. The Wolfram automaton with rule 18 discussed above is an example.

Next, in Chap. 11, we move on to a physical application: motion of particles. Traditional models for particle motion consist of hyperbolic or parabolic partial differential equations (e.g. Fokker-Planck equation for Brownian motion, the heat equation). A naive approach based on discretization of time, space and state for these classical models is investigated. It turns out that this route to cellular automata is rather not worthwhile to follow. The result of a naive discretization is still too close to partial differential equations to lead to many useful ideas about the discrete structure represented by cellular automata. There are two different approaches that are better suited. The ultradiscrete limit is based on a qualitative instead of a quantitative discretization of the local state: Instead of a linear discretization of the local state, the ultradiscretization distinguishes between small, large and huge, using some ε-expansion, and the limiting behavior for $\varepsilon \to 0$. The second approach offers a microscopic model for particle motion, where single particles are simulated (the HPP model). It is intriguing that in some cases the ultradiscretization yields a cellular automaton that can be well interpreted as a microscopic model for particle motion. These microscopic models can be related (at least by formal arguments) with partial differential equations, in the same way as the Boltzmann (gas) kinetics can be related to the Euler- and Navier-Stokes equation. As the analysis of partial differential equations is well developed, we are able to investigate the dynamics of these cellular automata by analyzing the related partial differential equations.

A topic prevalent in chemistry and in particular in biology is pattern formation (Chap. 12). Based on ideas of Bar-Yam, we first discuss a cellular-automaton version of the Turing instability. The Turing mechanism is able to explain many skin pattern quite accurate. Afterward, we turn to excitable media, and present the seminal Greenberg-Hastings automata. The idea of the analysis is unexpectedly based on structures originating in complex analysis: winding number and potential. As the class of Greenberg-Hastings automata is rather small and homogeneous, specialized methods allow for a profound understanding of the dynamics.

In the last chapter, we browse through a variety of applications: self-organized criticality, epidemiology, and evolution. All the cellular automata models formulated for these applications yield useful results and insights for the applications as well as tools to infer information about the behavior of the cellular automata at hand. We are led to the idea that there are much more cellular automata out there that deserve our attention.

Karl-Peter Hadeler was not permitted to hold this book in his hands; he passed away shortly before the proofs of the manuscript had been prepared. Karl was a key person in the field of biomathematics, hard-working and committed, with a tremendous knowledge in biology as well as in mathematics. In his very own way, he taught and inspired a whole generation of biomathematicians. But above all Karl was a friend, who was always interested in and caring towards his collaborators, colleagues and students. The contribution Karl-Peter Hadeler made to biomathematics is outstanding.

Chapter 2
Cellular Automata: Basic Definitions

What is a cellular automaton? This question is not too easy to answer. In the present approach, we define cellular automata in a very narrow sense: the automaton is deterministic and it has a high degree of symmetry. This narrow definition allows us to develop a relatively rich theory. In applications, however, these assumptions are quite often relaxed. Many mathematical systems describe the behavior of single particles, molecules or cells. Such small entities, described on a small spatial scale, do not follow strict deterministic laws but are subject to stochastic variation. Hence it is quite natural to generalize the concept of a cellular automaton in such a way that the local rule depends on random variables.

We have learned from the introduction that a cellular automaton has four components:

(1) the grid
(2) the neighborhood
(3) the states of a single cell
(4) the local function.

We introduce these notions step by step, and then assemble a cellular automaton from these components.

2.1 The Grid

The grid of a cellular automaton can be introduced as a graph. Most applications are based on a square grid in the plane, but also triangular and hexagonal grids and even irregular grids have been used. We define the grid as a graph with symmetries which can be described in terms of a group.

K.-P. Hadeler, J. Müller, *Cellular Automata: Analysis and Applications*,
Springer Monographs in Mathematics, DOI 10.1007/978-3-319-53043-7_2

2.1.1 Abelian or Regular Grids

The simplest grid starts from a row of points. If this row is infinite in both directions then we can represent the row by the entire numbers. Here we consider the entire numbers as a point set $\hat{\mathbb{Z}}$. If the row has only a finite number of points, say $m+1$, then we can identify the first and the last point and obtain a circular point set which we may call $\hat{\mathbb{Z}}_m$.

An important property of these rows of points is invariance under a shift. Define $\sigma : \hat{\mathbb{Z}} \to \hat{\mathbb{Z}},\ z \mapsto z+1$, then $\hat{\mathbb{Z}}$ is invariant under the map σ. We can even use σ to generate $\hat{\mathbb{Z}}$, more exactly, to define $\hat{\mathbb{Z}}$ as being generated by σ. We start with a point $p \in \hat{\mathbb{Z}}$ and apply all (positive and negative) powers of σ. Apparently here we use $\mathbb{Z}$ as the additive group of entire numbers. Hence we recover $\hat{\mathbb{Z}}$ as

$$\hat{\mathbb{Z}} = \{\sigma^n(p)\ :\ n \in \mathbb{Z}\}.$$

We can also introduce the group G generated by the shift σ,

$$G = \{g\ :\ g = \sigma^n,\ n \in \mathbb{Z}\}$$

and write

$$\hat{\mathbb{Z}} = \{g(p) : g \in G\}.$$

Furthermore, we may embed $\hat{\mathbb{Z}}$ in the real numbers: choose any $p \in \mathbb{R}$ and define $\hat{\mathbb{Z}}$ as above. Notice that this construction is very similar to looking at $\mathbb{R}^d$ as an affine space (a manifold) and also use $\mathbb{R}^d$ as a vector space of translations of the affine space (tangent space). We define a grid Γ as a graph with $\hat{\mathbb{Z}}$ as the set of vertices. Two vertices x and y are connected by an edge if and only if $x = \sigma y$ or $y = \sigma x$.

This introductory example can be generalized to square grids in the plane or "cubic" grids in $\mathbb{R}^d$ (see Fig. 2.1a). We start with one point $p \in \mathbb{R}^d$ and apply shifts $\sigma_1, \ldots, \sigma_d$ that generate a group G isomorphic to the additive group $\mathbb{Z}^d$. If

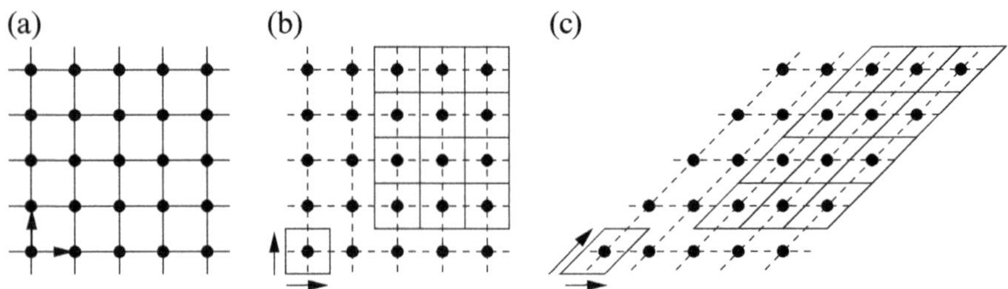

Fig. 2.1 Examples of lattice grids: (**a**) square grid produced by shifting points, (**b**) square grid and (**c**) rhombic grid produced by shifting a unit cell. The two translation vectors (corresponding to σ_1 and σ_2) are indicated

we represent the shifts as vectors in the vector space $\mathbb{R}^d$ then they are independent. Again the grid is the graph with vertex set $\hat{\mathbb{Z}}^d = \{g(p) : g \in G\}$ and the property that there is an edge from x to y if there is a shift σ_i such that $x = \sigma_i y$ or $y = \sigma_i x$ for some $i \in \{1, \ldots, d\}$.

We have constructed the square grid in three steps, first a set of grid points and then the edges and then the embedding in $\mathbb{R}^2$. We can generate it in a quite different way. We define a "unit cell" or "elementary tile" as a subset of $\mathbb{R}^2$, for example a square of edge length 1 centered at the origin (see Fig. 2.1b) and use the same generators σ_1 and σ_2 as before. We see that the plane is tessellated with copies of the elementary cell. A rhombic pattern can be produced by the first approach using a different embedding of the graph into $\mathbb{R}^2$ or by using a rhombic unit cell and appropriate shift operators, see Fig. 2.1c.

We know that by definition every planar periodic pattern is generated by a "unit cell" and two shift operators. The differences between the two approaches become obvious if we look at a triangular pattern as in Fig 2.1b. With the first approach we need three generators x, y, z as in Fig. 2.2a. If we use only two of these we get the same set of grid points but only a rhombic grid. On the other hand we can choose a unit cell, for example a rectangle centered at the origin (see Fig. 2.2b), and use two shift operators. Finally we look at the hexagonal grid in Fig. 2.2c. There is no way to produce the set of grid points as the images of a single point under the action of some shifts. But we can define a rhombic elementary cell as in Fig. 2.2b and produce the grid with two generators.

We will later use the graph structure to indicate the neighborhood of a point. Points within a neighborhood can, but do not have to, interact. However, formally we may choose the neighborhood larger than the sets of points that indeed interact locally. The construction in Fig. 2.2c only allows to distinguish copies of the unit cell, not single points. This may force us to choose a large neighborhood, but is allowed in the framework for cellular automata we develop here.

Definition 2.1.1 A regular or Abelian grid Γ consists of an elementary tile c_0 and commuting shifts $\sigma = \{\sigma_1, \ldots, \sigma_d\}$ generating a group G. The grid is determined

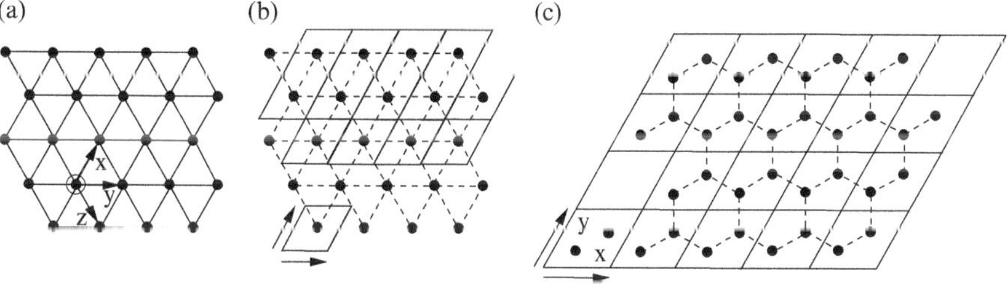

Fig. 2.2 A triangular grid is generated (**a**) by three shift operators or (**b**) using a special unit cell and two shift operators. (**c**) Also a hexagonal grid can be generated by an appropriate unit cell and two shift operators. The generated hexagonal structure is indicated by the *dashed lines*

by G and the set of generators σ, $\Gamma = \Gamma(G, \sigma)$. The tiles are given as

$$gc_0, \quad g \in G$$

or as

$$\Pi_{i=1}^{d} \sigma_i^{z_i} c_0, \qquad z_i \in \mathbb{Z}.$$

Either $\sigma_i^z \neq \sigma_i^{z'}$ for $z \neq z'$, or there is $n_i \in \mathbb{N}$ such that $\sigma_i^0 = \sigma_i^{n_i}$. In the first case, we call the direction corresponding to the shift operator σ_i unbounded, in the second case bounded.

Remark 2.1.2

(1) Regular grids play a central role in crystallography, where one aims at classifying all regular patterns. In general, the elementary tiles exhibit a certain symmetry. One can show that the symmetry group of the resulting grid in the plane is one of seventeen possible groups, the so-called crystallographic groups of the plane. Also in arts, e.g. in the classification of mosaics, these groups play a role [79]. Similar results hold true for regular grids embedded in spaces of higher dimension.

(2) We do not need to know the nature of the object c_0. Although it may be quite useful to have points in $\mathbb{R}^d$ in mind, this is by no means necessary. Think of the pictures of M.C. Escher who fills his tiles with birds and fishes.

2.1.2 Non-Abelian Grids

We started with a regular grid and we obtained an Abelian group which is in some sense equivalent to the grid. Can we also do the converse? Given a group, can we construct a grid, or—more general—a graph? Already in 1878, Arthur Cayley proposed a "graphical representation" of a finitely generated group (find in Appendix A.2 the definition of a finitely generated group) which we now call a Cayley graph.

Definition 2.1.3 Consider a (not necessarily Abelian) finitely generated group G with a set of generators $\sigma = \{\sigma_1, \ldots, \sigma_d\}$. The vertices of the Cayley graph $\Gamma = \Gamma(G, \sigma)$ are the elements of G. The vertices $g_1, g_2 \in G$ are connected by an edge if and only if there is a generator $\sigma_i \in \sigma$ such that

$$g_1 = g_2 \, \sigma_i \qquad \text{or} \qquad g_2 = g_1 \, \sigma_i.$$

We call the neutral element $e \in G$ the origin of the Cayley graph. The vertices are called sites or cells of the grid.

Remark 2.1.4 The Cayley graph is constructed from a group. In general, the Cayley graph depends on the choice of the generators. If different choices are made, then the corresponding Cayley graphs may be not isomorphic. For example, if $G =< \sigma_1, \ldots, \sigma_m >$, then we can add any other element $\tau \in G$ to the set of generators, $G =< \sigma_1, \ldots, \sigma_m, \tau >$. The Cayley graphs corresponding to $\sigma_1,\ldots,\sigma_m$ and to $\sigma_1,\ldots,\sigma_m$, τ are different since the grid contains more edges. However, for a given group, the number of vertices in a Cayley graph is always the same.

Example 2.1.5 The Cayley graph of a finitely generated Abelian group is a grid (possibly finite).

Example 2.1.6 Can we visualize the Cayley graph of a non-Abelian group? Yes, we can, and the visualization is used in group theory to find underlying symmetries of a given group. However, the resulting graph need not be planar. Consider the symmetry group of an equilateral triangle, the dihedral group D_3. Let σ be the rotation by $2\,\pi/3$ and let τ be one of the reflections (left-right in a standard picture). Then

$$D_3 =< \sigma, \tau\ :\ \sigma^3 = e,\ \tau^2 = e,\ \tau\sigma\tau\sigma = e > = \ \{e, \sigma^1, \sigma^2, \tau, \tau\sigma, \tau\sigma^2\}.$$

The Cayley graph has six vertices. The table of multiplications from the right can be found in Table 2.1.
The Cayley graph is shown in Fig 2.3. This finite graph shares some features with a torus: there is an origin (the unit element), and two "shift operators" σ and τ generating the graph. The origin can be shifted into any point and any point can be shifted into the origin.

If we are honest, we did not gain much by this example: the topology of this graph resembles that of the regular grid described by the Abelian group $\mathbb{Z}_3 \times \mathbb{Z}_2$. In general: A Cayley graph for D_n resembles a Cayley graph for $\mathbb{Z}_n \times \mathbb{Z}_2$. There is only one difference: If we apply the generator σ with $\sigma^n = e$ then in the Abelian case we visit the vertices in the inner and in the outer circle in the same direction while we visit them in the opposite direction in the case of D_n. Here we consider undirected graphs, so this difference has no consequences for the theory of cellular automata.

Table 2.1 Multiplication table (from the right) for the symmetry group of an equilateral triangle

Vertex	Multiply with σ	Multiply with σ^2	Multiply with τ
e	σ	σ^2	τ
σ	σ^2	e	$\sigma\tau = \tau\sigma^2$
σ^2	e	σ	$\sigma^2\tau = \sigma\tau\sigma^2 = \tau\sigma$
τ	$\tau\sigma$	$\tau\sigma^2$	e
$\tau\sigma$	$\tau\sigma^2$	τ	$\tau\sigma\tau = \sigma^2$
$\tau\sigma^2$	τ	$\tau\sigma$	$\tau\sigma^2\tau = \tau\sigma\tau\sigma^2 = \sigma$

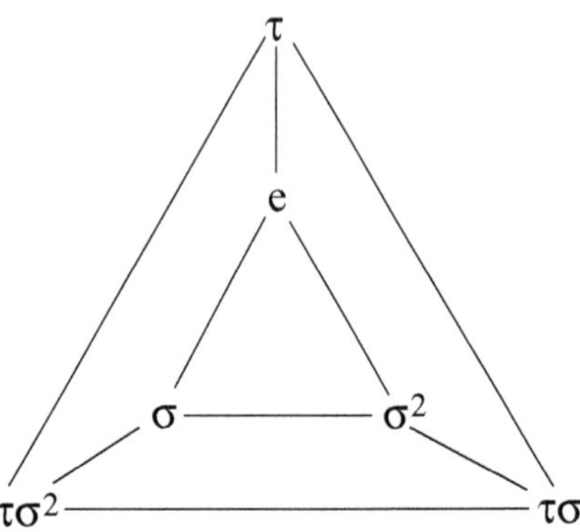

Fig. 2.3 Cayley graph for the group D_3

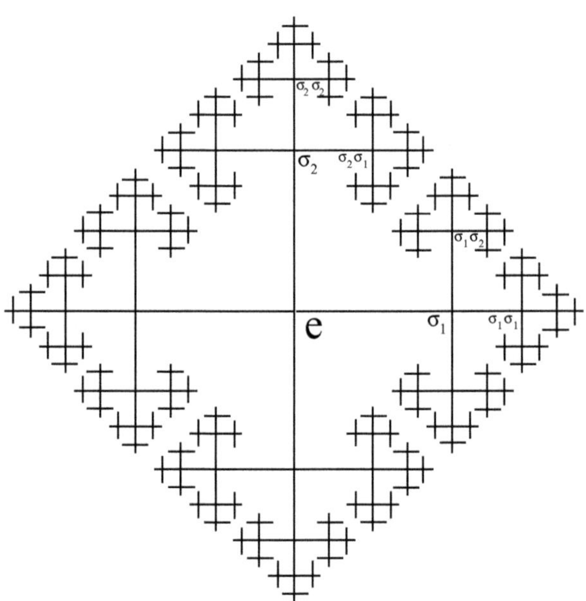

Fig. 2.4 First levels of the Cayley graph for the free group $< \sigma_1, \sigma_2 >$. The points $e, \sigma_1, \sigma_2, \sigma_1\sigma_2, \sigma_2\sigma_1, \sigma_1\sigma_1$ and $\sigma_2\sigma_2$ are indicated. For the complete graph, the branching proceeds to ever finer levels

Example 2.1.7 Consider the Cayley graph of the free group generated by two symbols,

$$G =< \sigma_1, \sigma_2 > .$$

The groups $< \sigma_1 >$ and $< \sigma_2 >$ are isomorphic to $(\mathbb{Z}, +)$. At each vertex there are two crossing branches (or four arms) corresponding to the application of $< \sigma_1 >$ and $\sigma_2 >$ separately. In order to obtain a nice picture we let the scale shrink with distance from the unit element. We obtain a self-similar graph with a D_4 symmetry. The unit element is not distinguished. We can choose any element as the starting point, see Fig. 2.4.

Some properties of regular grids extend to general Cayley graphs.

Proposition 2.1.8 *Let G be a finitely generated group. Any Cayley graph of G is connected. Furthermore, the group action of a group element from the left,*

$$\sigma_g : \Gamma \to \Gamma, \quad h \mapsto g \cdot h$$

is an automorphism of the Cayley graph.

Proof Connectedness: Let $g \in G =< \sigma_1, \ldots, \sigma_d >$. The element g can be represented as

$$g = \prod_{k=1}^{l} \tau_k$$

with $\tau_k \in \{\sigma_1, \sigma_1^{-1}, \ldots, \sigma_d, \sigma_d^{-1}\}$. Let $g_j = \prod_{i=1}^{j} \tau_i$ with g_0 being the neutral element and $g_l = g$. Since $g_{i+1} = g_i \tau_{i+1}$, there is a path in the Cayley graph $e \to g_1 \to \cdots \to g_l = g$. There is a path from any point to the unit element and hence a Cayley graph is connected.
Automorphism: First of all, multiplication of G with a group element is a permutation of G. Furthermore, since $g_1\tau = g_2$ is equivalent with $h g_1 \tau = h g_2$, multiplication from the left with a group element neither destroys nor creates edges. Thus, multiplication from the left is an automorphism of the Cayley graph. □

Remark 2.1.9

(1) For an Abelian grid, e.g. $\mathbb{Z}^d$, the operator σ_g corresponds directly to a shift of the complete grid by the vector g. Also in non-Abelian grids the operators σ_g play the role of shift operators.
(2) An immediate consequence of Proposition 2.1.8 is that any element g of the Cayley graph may be moved into the origin (resp. the origin into any given element) via $\sigma_{g^{-1}} : G \to G, h \mapsto g^{-1}h$ (resp. σ_g). This observation corresponds to the fact that in a regular grid there is a tagged origin, but it does not matter where in the grid this origin is located. Via translation this origin may be moved to any point.
(3) A further immediate consequence is that any vertex (group element) has the same degree, i.e., the same number of adjacent edges. Also this fact is obvious for a regular grid.

Definition 2.1.10 Since a Cayley graph is connected, a finite distance measure can be introduced. Define the length of any edge as 1. Then the distance $d_c(g_1, g_2)$ of two elements g_1 and g_2 in the Cayley graph is defined as the length of a shortest path between these two elements. The metric d_c on the Cayley graph is called the Cayley metric.

In the case of $\mathbb{Z}^d$, the Cayley metric coincides with the l^1-metric for coordinate vectors,

$$d_c(x, y) = \|x - y\|_1 \qquad \text{for } x, y \in \mathbb{Z}^d.$$

Notation In the following, a grid Γ denotes the Cayley graph of a finitely generated group G. We call the grid regular or Abelian if this group is Abelian. We will denote the neutral element by e, in the Abelian case we also use 0. Accordingly, the group operation is either denoted by multiplication $\cdot$ or, if we wish to emphasize the Abelian case, by addition $+$.

Note that the finite generated groups are by no means the most general structures that allow to define cellular automata. Cellular automata are e.g. also discussed on free monoids [29, 30]. However, as we will see, in some sense finitely generated groups are sufficient to understand cellular automata completely, at least in a topological sense: any continuous map on a metric Cantor space can be embedded in a cellular automaton on an finitely generated group (see Chap. 3 below). This generality is the main reason, why finitely generated groups as spatial structure are of special interest and discussed in depth, see e.g. the book by Ceccherini-Silberstein and Coornaert [27].

2.2 The Neighborhood

One essential property of a cellular automaton is locality: each cell interacts only with a fixed number of neighboring cells. Let a Cayley graph Γ be given.

First define the neighborhood D_0 of the origin. D_0 can be any finite set of vertices. The elements of the neighborhood need not be connected to the origin by edges, and the origin itself need not be in its neighborhood. In this generality, the concept of a neighborhood bears little relation to a topological neighborhood.

In order to define the neighborhood of other vertices we use the homogeneity of the Cayley graph: The neighborhood of the point g is obtained by shifting D_0 to this point.

Definition 2.2.1 Let G be a finitely generated group and $\Gamma = \Gamma(G, \sigma)$ a Cayley graph. Let D_0 be a neighborhood of e (i.e., any finite set of vertices of Γ). The neighborhood of a point $g \in \Gamma$ is given by

$$D_g = \sigma_g D_0 = g\, D_0.$$

In regular grids we can define certain types of neighborhoods using the distance (Fig. 2.5).

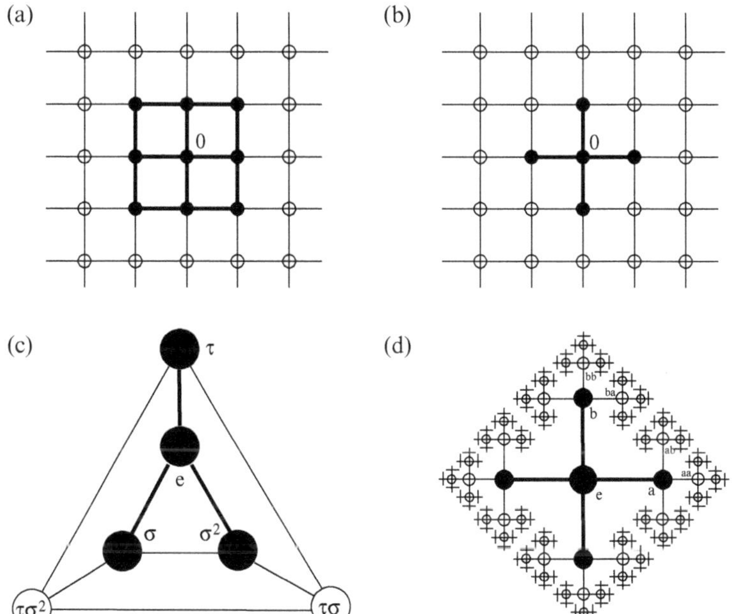

Fig. 2.5 (**a**) Moore neighborhood of $0 \in \mathbb{Z}^2$. (**b**) von Neumann neighborhood of $0 \in \mathbb{Z}^2$. (**c**) and (**d**) the von Neumann neighborhood of the unit element for the group D_3 and the free group $< a, b >$, respectively. The neighborhood is indicated by *bullets* and *fat edges*

Definition 2.2.2 For the regular grid $\mathbb{Z}^d$, the von Neumann neighborhood of a point x is defined as the set

$$D_x = \{y \in \mathbb{Z}^d \; : \; \|y - x\|_1 \leq 1\}$$

and the Moore neighborhood is defined as

$$D_x = \{y \in \mathbb{Z}^d \; : \; \|x - y\|_\infty \leq 1\}.$$

In this case translation invariance is evident.

The concept of a von Neumann neighborhood can be extended to arbitrary Cayley graphs.

Definition 2.2.3 Let G be a finitely generated group and $\Gamma = \Gamma(G, \sigma)$ a Cayley graph with set of generators σ. The von Neumann neighborhood of the origin is defined as

$$D_0 = \{g \in \Gamma \; : \; d_c(g, e) \leq 1\}$$

i.e., it consists of the immediate neighbors of the origin in the Cayley graph and the origin itself.

By definition, the von Neumann neighborhood of any element g is obtained by shifting the von Neumann neighborhood to that element. We can also obtain the neighborhood by evaluating distances near g.

Remark 2.2.4 The von Neumann neighborhood of an element g is given by

$$D_g = \{h \in \Gamma \, : \, d_c(h, g) \leq 1\}.$$

The Moore neighborhood on the grid $\mathbb{Z}^2$ cannot be defined by using the distance alone. We can however introduce two more generators and get an extended grid with an increased number of edges. Then the Moore neighborhood of $\mathbb{Z}^2$ becomes the von Neumann neighborhood of the extended grid. But, unfortunately, the extended grid is not planar. Draw it!

2.3 Elementary State and the Global State

Next we introduce the local and global state.

Definition 2.3.1 Let $\Gamma(G, \sigma)$ be a Cayley graph. We define functions on Γ with values in a finite set E which we call the set of elementary or local states, or the "alphabet". A function

$$u : \Gamma \to E$$

is called a (global) state. The set $E^\Gamma = \{u : \Gamma \to E\}$ is the state space. If g is a vertex (cell) then $u(g) \in E$ is called the state of the cell.

We study transformations of E^Γ. Any permutation π of the elements of E induces an invertible transformation of the set E^Γ by

$$\hat{\pi} : E^\Gamma \to E^\Gamma, \qquad u \mapsto \pi \circ u.$$

In this way we get a group acting on E^Γ which is isomorphic to the group of permutations of E. A simple example is $E = \{\text{white, black}\}$. Then the non-trivial permutation is the flip which changes black cells into white and conversely white cells into black.

These transformations act on the image of u. Now consider transformations of the pre-image. An element of the group $g \in G$ induces a shift isomorphism of the Cayley graph, $\sigma_g : \Gamma \to \Gamma, h \mapsto g \cdot h$. Also the shift operators can be lifted to E^Γ by

$$\hat{\sigma}_g : E^\Gamma \to E^\Gamma, \qquad u \mapsto u \circ \sigma_g.$$

In this case, the states of cells are shifted by the action of the operator σ_g. In the Abelian case the state of a cell is moved by a constant shift vector to another cell on the grid.

The lift $\hat{\pi}$ commutes with the shift operators $\hat{\sigma}_g$,

$$\hat{\pi} \circ \hat{\sigma}_g = \hat{\sigma}_g \circ \hat{\pi}.$$

If you think about it, this is a basic property of maps: transformations of the image commute with transformations of the pre-image. Since no confusion of σ_g and $\hat{\sigma}_g$ is possible, we drop the hat again.

Notation We write $\sigma_g : E^\Gamma \to E^\Gamma$ instead of $\hat{\sigma}_g : E^\Gamma \to E^\Gamma$, as it is always possible to clearly decide if σ_g acts on Γ or on E^Γ.

Sometimes it makes sense to abandon the homogeneous structure of the set E and introduce a neutral state or resting state r. In many applications we have $E = \{0, 1\}$ and the neutral state is 0.

Definition 2.3.2 E is called a set of states with a resting state, if there is exactly one marked element $r \in E$ which is called the resting, quiescent or background state.

Let Γ be a grid. For $u \in E^\Gamma$ the support is defined as the set of cells $g \in \Gamma$ with a state different from the resting state,

$$\text{supp}(u) = \{g : u(g) \neq r\}.$$

A state has finite support, if and only if $|\text{supp}(u)| < \infty$. A state with finite support is also called a configuration. The set of all configurations is denoted by $(E^\Gamma)_c$.

Later, we will, in a slightly abuse of this definition, also call the state of a finite (sub-)graph a configuration. A pattern always indicates the state of a finite (sub-)graph, and a word is a pattern on a one-dimensional graph ($\mathbb{Z}$, $\mathbb{Z}_m$).

Another important subset of E^Γ are the periodic states. For a regular grid $\mathbb{Z}^d$ generated by shift operators $\sigma_1,\ldots,\sigma_n$, a state u is periodic if for each i there is some $k_i \in \mathbb{N}$ such that u is invariant under $\sigma_i^{k_i}$. It is not immediately clear how to generalize this concept to arbitrary Cayley graphs. Yukita [186] proposes the following approach. Given a state u on a Cayley graph $\Gamma(G, \sigma)$, we first determine all group elements $g \in G$ that correspond to transformations $\sigma_g : E^\Gamma \to E^\Gamma$ that leave u invariant. This set is non-empty as the unit element always has this property. Moreover, these group elements form a subgroup $H = H(u)$. In the Abelian case considered before, this subgroup has a finite index which is given by $\Pi_{i=1}^n k_i$ if we choose the numbers k_i to be minimal. We turn this observation into a definition.

Definition 2.3.3 Let G be a finitely generated group, $\Gamma = \Gamma(G, \sigma)$ a Cayley graph for this group, and $u \in E^\Gamma$ a state. Define a subgroup of G by

$$H(u) = \{g \in G : \sigma_g(u) = u\}.$$

The state u is called (H-)periodic if the index $[G : H(u)]$ is finite. The set of all periodic states is denoted by $(E^\Gamma)_p$.

Note that we need to know an H-periodic state only for one representative out of each right-coset to know the state completely, as it is constant on Hg for all $g \in G$. As the index of H in G is finite, u is determined by its value at a finite number of sites.

Another useful concept is to equip the set E with an algebraic structure, e.g. to consider alphabets that have group or ring structure. If the dynamics respects this algebraic structure, it is possible to understand the dynamics much better than in the general case. An example is again $E = \{0, 1\}$ which can be seen as $\mathbb{F}_2$, the field with two elements. We will return to this aspect in Sect. 10.

2.4 The Local and the Global Function

By now we have a grid Γ (a Cayley graph), the elementary states E, the state space E^Γ and the neighborhood D_0. Next we introduce the local state space as the set of all functions (local states) from the neighborhood D_0 to the set E, i.e.,

$$E^{D_0}.$$

For any point $g \in \Gamma$ we define E^{D_g} as the local state space at g.

The local function f_0 is a map from E^{D_0} to E,

$$f_0 : E^{D_0} \to E.$$

The local function is the central "machine" of a cellular automaton. Knowing the local function, we construct the global function which maps a global state into a global state. The new state of the cell g is computed as follows: Shift D_0 by g, then restrict the state u to $g(D_0)$, then apply f_0.

We define a cellular automaton as follows.

Definition 2.4.1 Let G be a finitely generated group, $\Gamma = \Gamma(G, \sigma)$ a Cayley graph. Let D_0 be a finite subset of Γ, called the neighborhood of the unit element $e \in G$ and E a finite set of local states. Let furthermore $f_0 : E^{D_0} \to E$. We define

$$f : E^\Gamma \to E^\Gamma, \qquad u \mapsto f(u)$$

with

$$f(u)(g) = f_0((u \circ \sigma_g)|_{D_0}) \qquad \text{for } g \in \Gamma.$$

The tuple (Γ, D_0, E, f_0) is called a cellular automaton. The time course of a state $u_0 \in E^\Gamma$ (the orbit or trajectory starting with u_0) is given by the sequence $(u_0, u_1, u_2, \ldots)$ with $u_i = f(u_{i-1})$, $i = 1, 2, \ldots$.

An alternative description, slightly imprecise, reads

$$f(u)(g) = f_0(u|_{g(D_0)}) \qquad \text{for } g \in G.$$

The class of cellular automata is very rich and exhibits all kind of behaviors, from very simple, predictable behavior to complex and chaotic dynamics. Only very few theorems apply to the whole class—often enough one considers subclasses. At present we only mention a few classes.

Definition 2.4.2 A cellular automaton (Γ, D_0, E, f_0) is called a cellular automaton with resting state, if the set E has a resting state r and if for $u_0 \in E^{D_0}$, $u_0(g) = r$ for all $g \in D_0$, we have

$$f_0(u_0) = r.$$

A cellular automaton with resting state maps any state with finite support into a state with finite support.

Proposition 2.4.3 *A cellular automaton with resting state leaves the set* $(E^\Gamma)_c$ *(forward) invariant,*

$$f : (E^\Gamma)_c \to (E^\Gamma)_c.$$

As $(E^\Gamma)_c$ is invariant, the restriction of the cellular automaton to this set is well defined.

Definition 2.4.4 Let (Γ, D_0, E, f_0) denote a cellular automaton, and $X \subset E^\Gamma$ such that X is closed w.r.t. all shift operators, $\sigma_g(X) \subset X$ for all $g \in G$. If $f(X) \subset X$, we call the restriction of f to X a (cellular) sub-automaton.

Corollary 2.4.5 *A cellular automaton with resting state* (Γ, D_0, E, f_0) *has the sub-automaton* $f|_{(E^\Gamma)_c}$.

In general, a subautomaton is not an automaton.

We want to show that the set of periodic states is forward invariant. We first prove another simple but important property of the global function: it commutes with the lifts of all shift operators.

Proposition 2.4.6 *Let* (Γ, D_0, E, f_0) *be a cellular automaton with global function* f. *Then,*

(a) f *commutes with the lifts of all shift operators.*
(b) *The set of periodic states is invariant under* f.

Proof

(a) For $h, g \in G$, $u \in E^{\Gamma}$, we find

$$\sigma_g \circ f(u)(h) = f(u)(g\,h) = f_0(u|_{\sigma_{g\,h}(D_0)}) = f_0(\sigma_g(u)|_{\sigma_h(D_0)}) = f \circ \sigma_g(u)(h).$$

(b) Let $g \in H(u)$. Then, $\sigma_g \circ f(u) = f \circ \sigma_g(u) = f(u)$, i.e., $H(u) \subset H(f(u))$ and hence for the index

$$[G : H(f(u))] \leq [G : H(u)].$$

Thus, the image of a periodic state under a cellular automaton is again periodic. □

Corollary 2.4.7 *Any cellular automaton* (Γ, D_0, E, f) *has the sub-automaton* $f|_{(E^{\Gamma})_p}$.

Certain classes of cellular automata have very simple local functions and can be easily analyzed. We introduce some of the most common types.

Definition 2.4.8 The automaton (Γ, D_0, E, f_0) with $E = \{0, 1\}$ is called totalistic if $f_0(u)$ depends only on the number of cells occupied, i.e., on the number of "1",

$$f(u) = f^*(s), \quad s = \sum_{y \in D_0} u(y)$$

with some function $f^* : \mathbb{N}_0 \to E$. The automaton (Γ, D_0, E, f_0) is called semitotalistic if $f_0(u)$ depends only on the value $u(0)$ and on the sum of the values on $D_0' = D_0 \setminus \{0\}$. The automaton is semi-totalistic if there is a function of two variables $f^* : E \times \mathbb{N}_0 \to E$ such that

$$f_0(u) = f^*(u(0), s'), \quad s' = \sum_{y \in D_0'} u(y).$$

Given the automaton (Γ, D_0, E, f_0) with $E = \{0, 1\}$, there is a partial order between local states $u_i \in E^{D_0}$, $i = 1, 2$, defined by

$$u_1 \leq u_2 \quad \Leftrightarrow \quad \forall g \in D_0 : u_1(g) \leq u_2(g).$$

We introduce the class of automata that respect this order.

Definition 2.4.9 The automaton (Γ, D_0, E, f_0) with $E = \{0, 1\}$ is called monotone if

$$u_1 \leq u_1 \Rightarrow f_0(u_1) \leq f_0(u_2).$$

2.5 Excursion: The Growth Function of a Cayley Graph

In later sections we will use the concept of the growth function of a Cayley graph.

Definition 2.5.1 Given a Cayley graph Γ, let $\Gamma_n = \{g \in \Gamma \mid d_c(g,e) \leq n\}$ for $n \in \mathbb{N}_0$. The growth function of a Cayley graph Γ is defined as

$$\gamma(n) = |\Gamma_n|.$$

The asymptotic growth rate is defined as the limit

$$\rho = \lim_{n\to\infty} \gamma(n)^{1/n}.$$

Proposition 2.5.2 *The limit ρ exists and $\rho \geq 1$. If the group is finite then $\rho = 1$. The following dichotomy holds. If two Cayley graphs of the same group are given, then the corresponding two limits are both equal to 1 or both strictly greater than 1.*

Proof ([127])

(a) We first show that the limit exists. Let Γ be a Cayley graph for some group $G =< \tau_1, \ldots, \tau_m >$, and $\Gamma_l = |\{g : d(g,e) \leq l\}$. Since

$$\gamma(k+l) = |\{g \in \Gamma : d_c(g,e) \leq k+l\}| = \left|\cup_{g\in\Gamma_k}\Gamma_l\right| \leq \sum_{g\in\Gamma_k}\gamma(l) \leq \gamma(k)\gamma(l),$$

we find

$$\gamma(k) \leq \gamma(1)^k,$$

and thus

$$\limsup_{n\to\infty} \gamma(n)^{1/n} \leq \gamma(1).$$

We now fix $l \in \mathbb{N}$, and define $k = \lfloor n/l \rfloor + 1$. Then we have $kl \geq n$. Hence

$$\gamma(n) \leq \gamma(kl) \leq \gamma(l)^k = \gamma(l)^{\lfloor n/l\rfloor+1},$$

and hence for any $l \in \mathbb{N}$

$$\gamma(n)^{1/n} \leq \gamma(l)^{(\lfloor n/l\rfloor+1)/n} \leq \gamma(l)^{(n/l+1)/n} \leq \gamma(l)^{1/l+1/n}.$$

Since this is true for any $l \in \mathbb{N}$, we obtain

$$\limsup_{n\to\infty} \gamma(n)^{1/n} \leq \gamma(l)^{1/l}$$

and hence

$$\limsup_{n\to\infty} \gamma(n)^{1/n} \leq \liminf_{l\to\infty} \gamma(l)^{1/l}.$$

Hence the limit exists and is not less than $\gamma(1)$.

(b) Now we investigate the dependence of ρ on the Cayley graph for a given group. If the group is infinite, then $\gamma(n+1) > \gamma(n)$. Hence $\gamma(n) \geq n$ and

$$\rho \geq \lim_{n\to\infty} n^{1/n} = 1.$$

Let G be generated by two sets of generators, $G = < \tau_1, \ldots, \tau_m > = < \zeta_1, \ldots, \zeta_l >$ and let $\Gamma^\tau, \Gamma^\zeta$ be the corresponding Cayley graphs. Let γ_τ, γ_ζ be the growth functions of Γ^τ, Γ^ζ, respectively.

Let Γ_n^τ be the ball of radius n with center e with respect to the Cayley graph B^τ, and ρ_τ the corresponding limit, similarly define γ_ζ, Γ_n^ζ, and ρ_ζ for Γ^ζ.

Then there is $k_0 \in \mathbb{N}$, such that $\zeta_i \in \Gamma_{k_0}^\tau$, $i = 1, \ldots, l$, i.e., each of the generators ζ_i can be expressed as a word of generators τ_i of length k_0 or less. Thus, a word of length less or equal n (with respect to Γ^ζ) can be expressed as a word of length $k_0\, n$ or less in terms of τ_i. Hence $\gamma_\zeta(n) \leq \gamma_\tau(k_0 n)$ and

$$\rho_\zeta \leq \rho_\tau^{k_0}.$$

Similarly, there is a k_1 such that $\rho_\tau \leq \rho_\zeta^{k_1}$.

□

Notice that the proof indicates that the same group, if it is exponentially growing, may have—in dependence on the choice of the generators—different asymptotic growth rates. All these asymptotic growth rates are larger than one. This proposition gives rise to the following definition.

Definition 2.5.3 Let G be an infinite, finitely generated group for which $\rho > 1$ holds for one Cayley graph. Then G is called a group of exponential growth. If there is a polynomial $p(n)$ such that $\gamma(n) \leq p(n)$, then it is called a group of polynomial growth. If the group is not of exponential nor of polynomial growth, then it is called of intermediate growth.

Remark 2.5.4 There are groups of all three growth types. The existence of groups of intermediate growth has been proved as late as 1984 [78].

Example 2.5.5 Let $\gamma^{(n)}(l)$ be the growth function for F_n, the free group over n symbols, $n > 1$. First of all, $\gamma^{(n)}(l)$ for $l = 0, 1$ is given by

$$\gamma^{(n)}(0) = 1, \qquad \gamma^{(n)}(1) = 2n + 1.$$

Γ_{l+1} is the disjoint union of Γ_l and $\Gamma_{l+1} \setminus \Gamma_l$. Each vertex in $\Gamma_l \setminus \Gamma_{l-1}$ has exactly $2n-1$ neighbors in $\Gamma_{l+1} \setminus \Gamma_l$. As two different vertices in $\Gamma_l \setminus \Gamma_{l-1}$ only have common neighbors in Γ_{l-1} we obtain $|\Gamma_{l+1} \setminus \Gamma_l| = (2n-1)\ |\Gamma_l \setminus \Gamma_{l-1}|$. Therefrom we find the recursion

$$\begin{aligned}\gamma^{(n)}(l+1) &= |\Gamma_l| + |\Gamma_{l+1} \setminus \Gamma_l| = \gamma^{(n)}(l) + (2n-1)|\Gamma_l \setminus \Gamma_{l-1}| \\ &= \gamma^{(n)}(l) + (2n-1)(\gamma^{(n)}(l) - \gamma^{(n)}(l-1)) \\ &= 2n\,\gamma^{(n)}(l) - (2n-1)\ \gamma^{(n)}(l-1).\end{aligned}$$

Hence,

$$\begin{pmatrix} \gamma^{(n)}(l+1) \\ \gamma^{(n)}(l) \end{pmatrix} = \begin{pmatrix} 2n & -(2n-1) \\ 1 & 0 \end{pmatrix} \begin{pmatrix} \gamma^{(n)}(l) \\ \gamma^{(n)}(l-1) \end{pmatrix}.$$

The eigenvalues of this matrix are $\lambda_- = 1$ resp. $\lambda_+ = 2n-1$, and the corresponding eigenvectors read

$$x_- = \begin{pmatrix} 1 \\ 1 \end{pmatrix}, \qquad x_+ = \begin{pmatrix} 2n-1 \\ 1 \end{pmatrix}.$$

As

$$\begin{pmatrix} \gamma^{(n)}(1) \\ \gamma^{(n)}(0) \end{pmatrix} = \begin{pmatrix} 2n+1 \\ 1 \end{pmatrix} = \frac{n}{n-1} x_+ - \frac{1}{n-1} x_-$$

we find

$$A^l \begin{pmatrix} \gamma^{(n)}(1) \\ \gamma^{(n)}(0) \end{pmatrix} = \frac{n}{n-1}(2n-1)^l x_+ - \frac{1}{n-1} x_-.$$

This equation implies for $l \in \mathbb{N}_0$

$$\gamma^{(n)}(l) = [n(2n-1)^l - 1]/(n-1).$$

The growth function $\gamma(l)$ behaves asymptotically like $\exp(\ln(2n-1)l)$, i.e., the free groups over more than one symbol are groups of exponential growth with asymptotic growth rate $2n-1$.

Chapter 3
Cantor Topology of Cellular Automata

A cellular automaton *per se* is a discrete structure. Concepts like "continuous" or "arbitrarily close" seem not to be suited for the finite set of states a cell may assume. But if we consider infinite grids, then these concepts suddenly make sense. Imagine an observer placed at the origin of the grid. Although this point is arbitrary, it is tagged and may be used as a reference point. The observer looks at two different states of the cellular automaton. If theses states differ only in sites far away from the tagged location, then the two states look similar as seen from the observer and should be considered "close" to each other (see Fig. 3.1).

In the following we cast this idea into the definition of a metric—the Cantor metric. If we have a metric, then we have a topology. If we have a topology, then we have continuous functions. So we may ask whether the global function of a cellular automata is continuous in some sense.

We investigate the topology of E^Γ, and we find intriguing properties like perfectness or disconnectedness. A prototype for a topological structure with these features is the Cantor set, and it is a good idea to discuss the Cantor set first in order to get used to topological spaces with strange properties. Once we have an idea about the topology, we study functions on the topological space E^Γ. It turns out—even more intriguing—that these considerations lead to a second definition of cellular automata as continuous functions that commute with the shift operators. The equivalence of these definitions is the statement of the Curtis-Hedlund-Lyndon Theorem. In this alternative definition, we do not find any hint about the discrete structure of our objects. The discrete structure is completely hidden in the topology. Using this approach, we may utilize the theory of topological dynamical systems in order to obtain deeper insights into the properties of cellular automata.

Historically, the Curtis-Hedlund-Lyndon theorem has been developed in the late sixties (see the famous article by Hedlund [85]). However, the importance of this theorem became clear only 20 years later. Meanwhile, some publications on the dynamics of cellular automata start with this alternative definition of cellular

K.-P. Hadeler, J. Müller, *Cellular Automata: Analysis and Applications*,
Springer Monographs in Mathematics, DOI 10.1007/978-3-319-53043-7_3

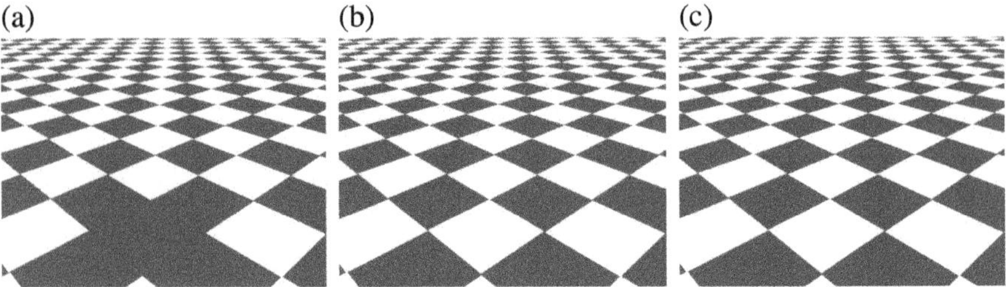

Fig. 3.1 Idea of the Cantor metric: the two checkerboards (**a**) and (**b**) as well as (**b**) and (**c**) differ in one site only. But (**a**) and (**b**) appear less similar than (**b**) and (**c**)

automata as continuous functions, only mentioning in passing that this definition is equivalent to the classical one.

The Cantor metric focusses attention on the origin (the tagged element) of the Cayley graph and it is not invariant under shifts. In Sect. 4 we discuss other approaches to equip the state space of a cellular automaton with a topology in which one tries to overcome this "deficiency" of the Cantor metric.

3.1 Prelude: Cantor Sets and Cantor Spaces

In order to introduce the kind of topology we shall encounter later on, we investigate Cantor sets and Cantor spaces.

3.1.1 The Classical Mid-Third Cantor Set

The classical Cantor set is constructed in the following way: take the closed interval $[0, 1]$. Delete the mid third, i.e., the open interval $(1/3, 2/3)$ (see Fig. 3.2). The set $[0, 1/3] \cup [2/3, 1]$ remains. Now, take the first connected component and again delete the mid third, the same for the second one. We obtain $[0, 1/9] \cup [2/9, 3/9] \cup [6/9, 7/9] \cup [8/9, 1]$. We delete the mid thirds of the connected components of this set, and so on. After n steps the set

$$I_n = \left\{ \sum_{i=1}^{n} \frac{a_i}{3^i} + x \,\middle|\, a_i \in \{0, 2\} \text{ and } 0 \leq x \leq 1/3^n \right\}$$

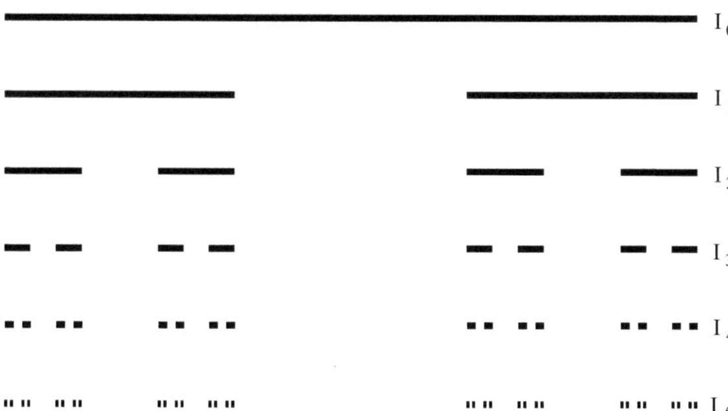

Fig. 3.2 Construction of the Cantor mid third set. In every step, the mid third of every connected component is deleted

remains. Step by step the connected components become smaller and smaller. In the limit, we obtain the mid third Cantor set

$$\mathfrak{C}_{1/3} = \bigcap_{n=0}^{\infty} I_n.$$

Since the I_n are nested, $I_0 \supset I_1 \supset I_2 \cdots$, non-empty and compact, the Cantor-set $\mathfrak{C}_{1/3}$ is non-empty and compact. Which points remain in the Cantor set? A point that will be an end-point of a connected component in I_n for some n will also be an end-point of a connected component for all $I_{n'}$ with $n' > n$ and hence remains. All points that never become end-points of connected components are eventually deleted. Thus, the points in the Cantor set have the form

$$\frac{a_1}{3} + \frac{a_2}{3^2} + \frac{a_3}{3^3} + \cdots = \sum_{i=1}^{\infty} \frac{a_i}{3^i} \qquad \text{with } a_i \in \{0, 2\}.$$

We may use this observation to define the mid-third Cantor set in a more formal way.

Definition 3.1.1 The (mid-third) Cantor set $\mathfrak{C}_{1/3}$ is the set

$$\mathfrak{C}_{1/3} = \left\{ x \in [0, 1] \, : \, x = \sum_{i=1}^{\infty} \frac{a_i}{3^i} \text{ with } a_i \in \{0, 2\} \right\}.$$

The topology of this set is defined as the trace topology originating from $\mathbb{R}$.

This set seems to be strange. What properties does this set have? We state the obvious properties as a proposition.

Proposition 3.1.2 *The Cantor set $\mathfrak{C}_{1/3}$ is non-empty and compact.*

Next, we investigate the connected components.

Proposition 3.1.3 *No open interval is a subset of $\mathfrak{C}_{1/3}$; all connected components consist of one point.*

Proof Assume that an open interval (a, b) is a subset of $\mathfrak{C}_{1/3}$. Then, $(a, b) \subset I_n$, where I_n are defined as above, for all $n \in \mathbb{N}$. However, for n large enough (e.g. $(b-a) > 3^{-n}$) this is not possible. □

This property is so nice and intriguing that we assign a name to it.

Definition 3.1.4 A set $X \subset \mathbb{R}$ that does not contain any (in $\mathbb{R}$) open interval is called totally disconnected.

We want to use the notion of disconnectedness in relation to cellular automata. Therefore we reformulate the definition in terms of open and closed sets.

Assume $X \subset \mathbb{R}$ is totally disconnected, and $x \in X$. Let $U \subset X$ be (with respect to the trace topology) an open set that covers x. Since U is open in the trace topology, there is an open set $\tilde{U} \subset \mathbb{R}$ such that $U = \tilde{U} \cap X$. In particular, there is an interval $(a, b) \subset \tilde{U}$ such that $a < x < b$. Since there are no intervals contained in X, there are points $\hat{a}$ and $\hat{b}$ with $\hat{a}, \hat{b} \notin X$ and $\hat{a} \in (a, x)$, $\hat{b} \in (x, b)$. Hence,

$$V = [\hat{a}, \hat{b}] \cap X = (\hat{a}, \hat{b}) \cap X$$

and $x \in V \subset U$. In other words: In every open set that covers x there is a clopen (closed and open) subset that covers this point. Thus, we have found a statement in topological language that does not explicitly refer to $\mathbb{R}$.

Now, let us try the reverse direction. Consider a set $Y \subset \mathbb{R}$ with the property stated. Is it possible that the set Y contains an open interval (a, b)? No, it is not, since this open interval (a, b) covers a smaller open interval $(\hat{a}, \hat{b})$ with $a < \hat{a} < \hat{b} < b$. And this smaller interval is not clopen nor does it cover any clopen set. Hence we have the following lemma.

Lemma 3.1.5 *Let $X \subset \mathbb{R}$ be equipped with the trace topology. X is totally disconnected if and only if for every $x \in X$ and open $U \ni x$ there is a clopen subset V such that $x \in V \subset U$.*

In view of this lemma we define "totally disconnected" in entirely topological terms.

Definition 3.1.6 A topological space $(X, \mathfrak{T})$ is totally disconnected if for any x and every U with $x \in U \in \mathfrak{T}$ there is a clopen (closed and open) set V with

$$x \in V \subset U.$$

This is the first time we meet clopen sets. They will appear over and over again, since these sets are something like the Swiss army knives for Cantor sets and Cantor spaces. Especially, it is possible to construct disjoint decompositions of Cantor spaces that are quite useful.

A topological Hausdorff space (and hence in particular any metric space) that is totally disconnected does not have a connected component consisting of more than one point: If we consider any two different points $x_1, x_2 \in X$, we find two disjoint open sets U_1 and U_2 with $x_i \in U_i$. Thus, there are clopen sets V_i with $x_i \in V_i \subset U_i$, $i = 1, 2$. We may decompose X in the two clopen sets V_1 and $X \setminus V_1$. The first set covers x_1 and the second x_2, such that any two different points x_1 and x_2 belong to different connected components.

If we look at the picture of the Cantor set (Fig. 3.2), the points seem to accumulate. They seem not to be isolated. We prove a precise statement.

Proposition 3.1.7 *No point in the Cantor set $\mathfrak{C}_{1/3}$ is isolated.*

Proof Let $x \in \mathfrak{C}_{1/3}$,

$$x = \sum_{i=1}^{\infty} \frac{a_i}{3^i}, \quad a_i \in \{0, 2\}.$$

We construct a sequence $\{y_n\}_{n \in \mathbb{N}} \subset \mathfrak{C}_{1/3} \setminus \{x\}$ which converges to x. Define

$$y_n = \sum_{i=1}^{\infty} \frac{a_i^{(n)}}{3^i}$$

with

$$a_i^{(n)} = \begin{cases} a_i & \text{for } i \neq n \\ 2 - a_i & \text{for } i = n. \end{cases}$$

Then,

$$|y_n - x| = \frac{2}{3^n} \to 0 \quad \text{for } n \to \infty.$$

□

This property, together with disconnectedness, makes the Cantor set a non-trivial structure. Therefore it deserves a definition.

Definition 3.1.8 A topological space that does not contain isolated points is called perfect.[1]

[1] In some texts it is required that a perfect space is compact.

We may remove one point from a perfect topological space without loosing important information. If we close the remaining set with respect to its given topology, we get the deleted point back. We have shown

Corollary 3.1.9 *The Cantor set $\mathfrak{C}_{1/3}$ is*

(1) *non-empty and compact*
(2) *totally disconnected*
(3) *perfect.*

It turns out that these three properties characterize the Cantor set $\mathfrak{C}_{1/3}$ very well. Moreover, we have formulated all three properties in terms of topology only. This observation leads to the following, more general definition.

Definition 3.1.10 Let $(X, \mathfrak{T})$ be a topological space. If this space is

(1) a subset of $\mathbb{R}$
(2) non-empty and compact
(3) totally disconnected and
(4) perfect

then it is called a Cantor set. If X is homeomorphic to a Cantor set but not necessarily a subset of $\mathbb{R}$, then we call X a Cantor space.

3.1.2 Cantor Spaces

Obviously, the Cantor set $\mathfrak{C}_{1/3}$ is a Cantor space in the sense of Definition 3.1.10. Can we find more examples? Of course, we may repeat the construction of the Cantor set, not removing the mid third, but larger or smaller pieces. However, this is boring since all these sets look alike. Can we construct something like a Cantor set that is not a subset of $\mathbb{R}$? A hint how to do this is given by the representation of the points $x \in \mathfrak{C}_{1/3}$ as sequences $\{a_i\}_{i\in\mathbb{N}} \in \{0, 2\}^{\mathbb{N}}$, $x = \sum_{i=1}^{\infty} a_i/3^i$. Perhaps we should have a look at binary sequences.

Proposition 3.1.11 *Let the set $\{0, 1\}$ be equipped with the discrete topology, and the set of binary sequences (or vectors) $\Sigma_2 = \{0, 1\}^{\mathbb{N}}$ with the product topology. Then the topological space Σ_2 is a Cantor space.*

Proof We prove this proposition by constructing a homeomorphism between the topological spaces Σ_2 and $\mathfrak{C}_{1/3}$, i.e., we construct a bijective function $\psi : \Sigma_2 \to \mathfrak{C}_{1/3}$ such that this function and its inverse are continuous. The definition of the function ψ is quite obvious,

$$\psi : \Sigma_2 \to \mathfrak{C}_{1/3}, \quad \{a_i\}_{i\in\mathbb{N}} \mapsto \sum_{i=1}^{\infty} \frac{2a_i}{3^i}.$$

The function ψ is bijective, because every point in $\mathfrak{C}_{1/3}$ can be represented in a unique way as $x = \sum_{i=1}^{\infty} 2a_i/3^i$.

Next we show that ψ is continuous. Let $\{a^{(n)}\}$, $a^{(n)} \in \Sigma_2$ be a sequence which converges to some $b \in \Sigma_2$ in the product topology. Let $a^{(n)} = \{a_i^{(n)}\}_{i\in\mathbb{N}}$ and $b = \{b_i\}_{i\in\mathbb{N}}$. Convergence in the product topology means that for every number k there is some $n_0 = n_0(k)$ such that $a_i^{(n)} = b_i$ for $i \le k$ and $n > n_0(k)$. Hence

$$|\psi(a^{(n)}) - \psi(b)| \le \sum_{j=i+1}^{\infty} \frac{2}{3^j} = \frac{1}{3^i} \quad \text{for } n > n_0(k)$$

and thus $\psi(a^{(n)}) \to \psi(b)$.

We show that the inverse of ψ is also continuous. We could recur to the known theorem that a bijective, continuous function from a compact set to a compact set has a continuous inverse (Proposition A.1.12). Instead we show directly that ψ^{-1} is continuous.

Assume $x^{(n)}, y \in \mathfrak{C}_{1/3}$ with $x^{(n)} \to y$ for $n \to \infty$. Thus, the distance $|x^{(n)} - y|$ (in $\mathbb{R}$) becomes arbitrarily small. We represent $x^{(n)}$ and y as

$$x^{(n)} = \sum_{i=1}^{\infty} \frac{2a_i^{(n)}}{3^i}, \qquad y = \sum_{i=1}^{\infty} \frac{2b_i}{3^i}.$$

Now choose $k > 0$. Let $\varepsilon(k) = 3^{-k-1}$. Then, there is $n_0(k) = n_0(\varepsilon(k))$ such that

$$|x^{(n)} - y| < \varepsilon(k) \quad \text{for } n > n_0(k).$$

Therefore $a_j^{(n)} = a_j$ for $j \le k$ and $n > n_0(k)$. The sequence $a^{(n)} = \{a_i^{(n)}\}_{i\in\mathbb{N}}$ converges in Σ_2 to $b = \{b_i\}_{n\in\mathbb{N}}$, i.e., ψ^{-1} is continuous. □

The key property of a Cantor set is not really the fact that it is a subset of $\mathbb{R}$, but that it is compact, totally disconnected and perfect. Are there sets with these properties that are not Cantor spaces? A partial answer is given by the following theorem on *metric* spaces.

Theorem 3.1.12 *Any metric, compact topological space X that is perfect and totally disconnected is homeomorphic to $\mathfrak{C}_{1/3}$.*

We give here only a sketch of the proof, and defer the proof itself to Appendix A.1—the details are rather technical. It is sufficient to find a homeomorphism $\psi : \Sigma_2 \to X$ as $\mathfrak{C}_{1/3}$ and Σ_2 are known to be homeomorphic. The homeomorphism is constructed by recursively splitting the set into smaller and smaller parts (see Fig. 3.3). First, we decompose the set in two parts, named part (0) and part (1). Each of these two parts is split again into two parts such that we get four parts, the two parts from (0) are named (00) and (01), and the two parts

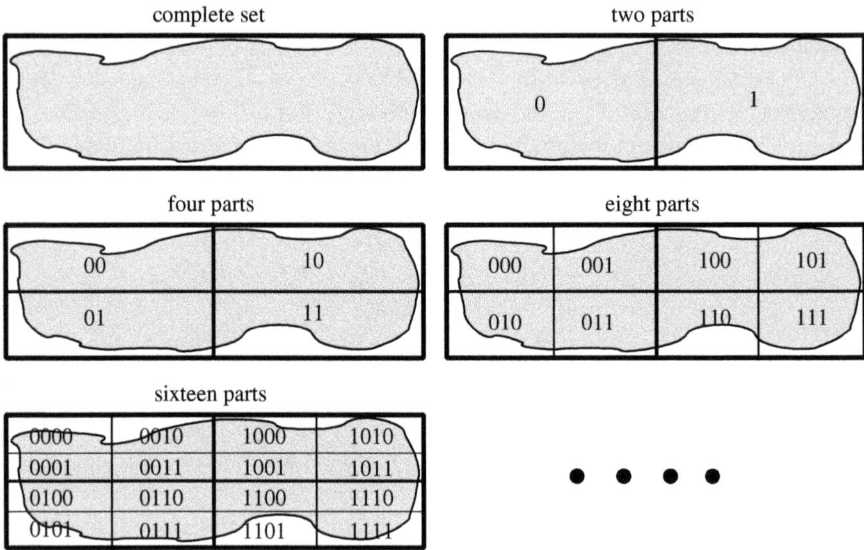

Fig. 3.3 The coding of points within a set by a binary sequence. The set is decomposed recursively again and again. In each step the binary sequence tells us in which part to go

from (1) are named (10) and (11). We then divide these four sets into eight, sixteen, thirty-two, etc. parts.

During every division we control the size of the parts, such that after n splits the diameter of each part is at most const $\cdot\, 2^{-n}$. At this point, we use that X is a compact *metric* space. Taking n to infinity, we assign to each point in X a sequence in Σ_2.

This construction can be performed for any compact metric space X. The example of the interval $X = [0, 1] \subset \mathbb{R}$ shows that in general there will be different sequences corresponding to the same point, e.g. $0.100000\ldots$ and $0.01111\ldots$ (binary numbers) both correspond to the mid-point. Hence the mapping from the binary sequences will be surjective but not injective.

In order to ensure the bijective correspondence between X and Σ_2, we use the other two properties of X, i.e., that X is perfect and totally disconnected: we want the subsets resulting from splitting to be open since then the compactness of X tells us that we only need a finite number of subsets to cover X completely. Also, in every nested sequence of subsets (corresponding to one sequence in Σ_2) there should be at least one point. Thus, these sets should also be closed. Consequently, the construction of successive splits uses appropriately defined clopen sets. After the proof has been completed, it is evident that the conditions (compact, perfect, totally disconnected) are also necessary.

Remark in passing: Remember the construction of the Peano curve by successive subdivisions of a triangle? To each point in the interval $[0, 1]$ given by a binary sequence like $0.00101001\ldots$ we construct a point of the triangle, and this mapping is well-defined—although for some points there are two sequences ($1 = 0.11111..$), surjective and continuous, hence it is a "curve". But obviously there are points in the triangle which have pre-images wide apart. The inverse mapping does not exist (if it would, dimension would be meaningless).

3.2 Cantor Metric for Cellular Automata

In this section we define a metric for cellular automata—the so-called Cantor metric (Cantor did not know cellular automata, presumably). This metric is a strong tool in the context of unbounded grids Γ. For finite grids the results are trivial.

Intuitively, the distance of two states u_1 and u_2 should be considered small if they agree in a large region of the grid centered around the tagged point zero. To be more formal, we consider a Cayley graph $\Gamma(G)$, where G is a group with neutral element e, and define "centered balls" Γ_m in terms of the Cayley metric as

$$\Gamma_m = \{g \in \Gamma \, : \, d_c(g, e) \leq m\}.$$

In the case of $\mathbb{Z}^2$, these are all sites within a "diamond" with side length $(2m+1)\sqrt{2}$ centered around zero.

For two states u_1 and u_2 we determine the smallest m such that they disagree somewhere on Γ_m,

$$k(u_1, u_2) = \inf\{m \, : \, u_1|_{\Gamma_m} \neq u_2|_{\Gamma_m}\}$$

(the infimum over the empty set is formally $+\infty$). The distance is defined as the inverse of $k + 1$,

Definition 3.2.1 The Cantor metric on the state space is defined as

$$d(u_1, u_2) = \begin{cases} \frac{1}{k+1} & \text{if } k = k(u_1, u_2) \text{ is finite} \\ 0 & \text{if } k = k(u_1, u_2) \text{ is infinite.} \end{cases}$$

Note that d_c is the graph metric on Γ (path length) while d is the metric on the state space E^Γ. The distance of two elements of E^Γ assumes only discrete values, $\{0, 1, 1/2, 1/3, \ldots\}$, and cannot exceed 1.

It is easy to see that d defines a metric (prove the triangle inequality!). (E^Γ, d), i.e., E^Γ supplied with the metric d, is a metric space. The metric is not invariant against translations (shifts). Intuitively, we would think that a distance measure that is translational invariant would be better suited to study cellular automata. However, approaches to define invariant metrics and quasi-metrics, e.g. the Weyl topology and the Besicovitch topology (we return to these later) yield weaker results as compared to the Cantor metric.

We have started with a purely combinatorial approach to cellular automata without any connection to topology. Now we have a topology and we can discuss continuity of functions. The definition of the Cantor metric and the corresponding topology opens the world of topological dynamical systems and yields many results, especially results related to existence and structure of attractors.

What are the properties of this topology? The basis of the topology is formed by the balls of radius $\varepsilon > 0$,

$$B_\varepsilon(u) = \{\tilde{u} : d(\tilde{u}, u) < \varepsilon\}.$$

If $\varepsilon \geq 1$, we find $B_\varepsilon(u) = E^\Gamma$, since the distance of any two states is less or equal to one.

Now, let $\varepsilon \in (0, 1)$. If we determine $m \in \mathbb{N}$ by $1/(m+1) < \varepsilon \leq 1/m$, i.e., by $m + 1 > 1/\varepsilon \geq m$, then $B_\varepsilon(u)$ consists of all states which agree on Γ_{m-1}. No information is gained about cells outside of Γ_{m-1}. These "balls" are called the cylinder sets. (Think, why did they get this name?)

We can interpret the cylinder sets in a different way: Equip the alphabet E with the discrete topology, and E^Γ by the product topology. The cylinder sets form the basis of this topology.

Remark 3.2.2 Recall what the product topology is (see also appendix A.1). Suppose we have a sequence (only the case of an infinite sequence is interesting) of topological spaces X_i and we form the product space $X = \prod X_i$ (which can be seen as the space of sequences (x_i) with $x_i \in X_i$). Then the product topology or Tychonov topology is the coarsest topology which makes the "coordinate projections" $P_i : X \to X_i$, $P_i(x) = x_i$, continuous functions. (The finer the topology is, the more open sets it has, and the more likely is it that a function *from* the space to a given space is continuous, since continuous means "the pre-image of an open set is open").

The product topology can be generated from the sub-basis of the cylinder sets. A cylinder set is a set of the form $A = \prod_i A_i$ with $A_i \subset X_i$ an open set such that $A_i = X_i$ for all i except finitely many.

With the metric topology we control the states in a region around the unit element e or 0 of the grid. We do not care what happens far away from 0. If we fix a region, i.e., consider Γ_m for a fixed m, then there is a finite number of elements in E^{Γ_m}. If Γ is finite, this is true anyway. Now assume Γ is not finite. Consider a sequence of states $\{u_n\}_{n\in\mathbb{N}}$. Since this sequence consists of infinitely many elements and E^{Γ_m} is finite, there is at least one element $\hat{u} \in E^{\Gamma_m}$ which is visited by $u_i|_{\Gamma_m}$ infinitely often. Hence we have found the following proposition.

Proposition 3.2.3 *Let $\{u_n\}_{n\in\mathbb{N}} \subset E^\Gamma$ be a sequence, and $m \in \mathbb{N}_0$ any non-negative integer. Then there is a subsequence $\{u_{n_l}\}_{l\in\mathbb{N}}$ such that u_{n_l} is constant on Γ_m,*

$$u_{n_l}|_{\Gamma_m} = u_{n_{l'}}|_{\Gamma_m} \qquad \text{with } l, l' \in \mathbb{N}_0.$$

From this proposition it is immediately clear that for bounded grids we always find a converging subsequence (even a constant subsequence), i.e., for bounded grids the metric space (E^Γ, d) is compact and thus also complete. If we consider unbounded grids, does the picture change? The answer is given in the following theorem.

Theorem 3.2.4 *For any grid Γ (finite or infinite) the metric space (E^Γ, d) is complete and compact.*

Proof Consider a sequence $u_n \in E^\Gamma$. We show that there is a converging subsequence.

Step 1: Construction of subsequences that are constant on larger and larger regions.
Consider $m = 1$. We know, that there is a subsequence of $\{u_n\}_{n\in\mathbb{N}_0}$ that is constant on Γ_1. Denote this sequence by $\{u_n^{(1)}\}_{n\in\mathbb{N}_0}$. For $m = 2$, choose a subsequence of $\{u_n^{(1)}\}_{n\in\mathbb{N}_0}$ that is constant on Γ_2. Denote this subsequence by $\{u_n^{(2)}\}_{n\in\mathbb{N}_0}$. Proceed recursively in the same way, i.e., construct nested subsequences $\{u_n^{(m)}\}_{n\in\mathbb{N}_0}$ that are constant on Γ_m.

Step 2: Convergence of the diagonal sequence.
Define $\tilde{u}_n = u_n^{(n)}$. Then $\{\tilde{u}_n\}$ is a subsequence of u_n. Furthermore, for any $m \in \mathbb{N}_0$, the sequence becomes eventually constant on Γ_m. Define the state $\tilde{u}_\infty$ point wise by

$$\tilde{u}_\infty(g) = \lim_{n\to\infty} \tilde{u}_n(g), \quad g \in \Gamma.$$

Since, for a given $g \in \Gamma$, the sequence $\{\tilde{u}_n(g)\}_{n\in\mathbb{N}_0} \subset E$ eventually becomes constant, the limit is well defined in the discrete topology on E.
We claim that $\tilde{u}_n \to \tilde{u}_\infty$, i.e., $d(\tilde{u}_n, \tilde{u}_\infty) \to 0$. Indeed, let $\varepsilon > 0$. Choose $m \in \mathbb{N}$, $m > 1 + 1/\varepsilon$. Since $\tilde{u}_n$ eventually becomes constant on Γ_m, there is $n_0 = n_0(\varepsilon) > 0$ such that $\tilde{u}_n|_{\Gamma_m} = \tilde{u}_\infty|_{\Gamma_m}$ for $n > n_0$. Hence

$$d(\tilde{u}_n, \tilde{u}_\infty) < \frac{1}{m+1} < \varepsilon \quad \text{for} \quad n > n_0(\varepsilon).$$

Since ε can be arbitrarily small, $d(\tilde{u}_n, \tilde{u}_\infty) \to 0$. A compact space is necessarily complete. □

Assume Γ is not finite and E has at least two elements. Choose any state $u \in E^\Gamma$. We claim that we can find states, different from u but arbitrarily close to u. Define states u_i as follows. For $i \in \mathbb{N}$ choose $g_i \in \Gamma$ with $d_c(g_i, 0) = i$. Then let $u_i(g) = u(g)$ for $g \neq g_i$ and $u_i(g_i) = e_i$ where $e_i \in E$, $e_i \neq u(g_i)$ but otherwise arbitrary. Then $u_i \neq u$, $d(u, u_i) = 1/(i+1)$, and the sequence u_i converges to u with respect to the metric d. Hence, $\overline{E^\Gamma \setminus \{\hat{u}\}} = E^\Gamma$. E^Γ is perfect.

Proposition 3.2.5 *Let Γ be infinite and $|E| > 1$. Then E^Γ, equipped with the Cantor metric, is perfect.*

Next we investigate the topology with respect to connected components.

Proposition 3.2.6 *The metric space (E^Γ, d) is totally disconnected.*

Proof Let $u \in E^\Gamma$ and $U \subset E^\Gamma$ an open set with $u \in U$. Since U is open, we find an $m \in \mathbb{N}$ such that the open ball with center u and radius $\varepsilon = 1/(m+1)$ is a subset of U, $B_\varepsilon(u) \subset U$. In view of the definitions, the distance of two states cannot assume values in the open interval $(1/(m+2), 1/(m+1))$. We find

$$\begin{aligned} B_{1/(m+1)}(u) &= \{v \in E^\Gamma \, : \, d(u,v) < 1/(m+1)\} \\ &= \{v \in E^\Gamma \, : \, d(u,v) < 1/(m+1+1/2)\} \\ &= \{v \in E^\Gamma \, : \, d(u,v) \le 1/(m+1+1/2)\} = \overline{B_{1/(m+1+1/2)}(u)}. \end{aligned}$$

Thus, the open set $B_{1/(m+1)}(u)$ equals the closed set $\overline{B_{1/(m+1+0.5)}(u)}$ and hence it is a clopen set that we aimed to construct: it contains u and it is contained in U. □

Thus, (E^Γ, d) is topologically homeomorphic to the standard Cantor set $\mathfrak{C}_{1/3}$.

Corollary 3.2.7 *Let Γ be infinite and $|E| > 1$. Then (E^Γ, d) is a Cantor space.*

3.3 The Curtis-Hedlund-Lyndon Theorem

Until now we have looked at the state spaces of cellular automata. In this section we investigate the topological properties of the global function. First, we show that the global function is continuous. Then we prove the Curtis-Hedlund-Lyndon theorem: the global functions of cellular automata are exactly those continuous functions which commute with all shift operators. This theorem is valid for all Cayley graphs of finitely generated groups, $\Gamma = \Gamma(G, \sigma)$.

Proposition 3.3.1 *Let (Γ, D_0, E, f_0) be a cellular automaton. The global function $f : E^\Gamma \to E^\Gamma$ is continuous with respect to the metric d.*

Proof Let $u \in E^\Gamma$. We show that for any $\varepsilon > 0$ there is a $\delta > 0$ such that $d(f(u), f(v)) < \varepsilon$ for all v with $d(u,v) < \delta$. Choose $m \in \mathbb{N}$ such that $D_0 \subset \Gamma_m$. Let $\varepsilon > 0$. Choose $k \in \mathbb{N}$ such that

$$\frac{1}{1+k} < \varepsilon.$$

Choose $\delta > 0$ such that

$$\delta < \frac{1}{1+k+m}.$$

For any v with $d(u,v) < \delta$ we have $v|_{\Gamma_{k+m}} = u|_{\Gamma_{k+m}}$ and hence $f(v)(g) = f(u)(g)$ for all $g \in \Gamma_k$. It follows that $d(f(v), f(u)) < \varepsilon$. □

Recall that G is a group generated by a finite set σ, and $\Gamma = \Gamma(G, \sigma)$ is a Cayley graph. A group element g (not necessarily a generator) induces a shift $\sigma_g : h \to gh$ for $h \in \Gamma$. This shift induces a shift operator $\sigma_g : E^\Gamma \to E^\Gamma$ defined by $\sigma_g u(h) = u(gh)$. Since the image of a global state u under a shift σ_g at a site h only depends on the von Neumann neighborhood of h, we find with Proposition 3.3.1 the following corollary.

Corollary 3.3.2 *All shift operators $\sigma_g : E^\Gamma \to E^\Gamma$ act continuously on E^Γ.*

We already know (Proposition 2.4.6) that the global function $f : E^\Gamma \to E^\Gamma$ commutes with the shift operators $\sigma_g : E^\Gamma \to E^\Gamma$,

$$\sigma_g \circ f = f \circ \sigma_g \text{ for all } g \in G.$$

The theorem of Curtis, Hedlund and Lyndon states that a continuous function on E^Γ that commutes with all shift operators can be represented by a cellular automaton. Later we will need a slightly more general version: if the map is not defined on E^Γ but only on a closed, shift invariant subset X, we obtain a cellular sub-automaton. Therefore we prove this general version and get the theorem as a corollary.

Proposition 3.3.3 *Let $X \subset E^\Gamma$ be closed in the metric topology and invariant under all shift operators. Let $F : X \to X$ be a function that is continuous in the metric topology and that commutes with all shift operators. Then there is a cellular sub-automaton with a global function f such that $f|_X = F$.*

Proof

Step 1: Split the space X.

Let $0 \in \Gamma$ denote the origin of the grid (also if $\Gamma = \Gamma(G, \sigma)$ for a non-Abelian group G). For any $\eta \in E$ define

$$U_\eta = \{u \in X : u(0) = \eta\}.$$

Then U_η has the following properties:

1. U_η is closed because limits of sequences u_n with $u_n(0) = \eta$ for all $n \in \mathbb{N}$ are again in U_η due to the properties of the metric topology on E^Γ.
2. U_η is open. Indeed, choose $a \in (1/2, 1)$. Then the statements $d(u, v) < a$ and $u(0) = v(0)$ are equivalent. Thus, for a $u_0 \in X$ with $u_0(0) = \eta$ we find

$$U_\eta = \{u \in X : d(u, u_0) < a\}.$$

Since the map $u \mapsto d(u, u_0)$ is continuous, U_η is open.

3. For elements $\eta,\ \eta' \in E$ holds

$$U_\eta \cap U_{\eta'} = \emptyset \quad \text{if} \quad \eta \neq \eta'.$$

4. The union of all U_η, $\eta \in E$, yields the whole space,

$$X = \dot{\bigcup_{\eta \in E}} U_\eta.$$

Now define, for $\eta \in E$,

$$V_\eta = F^{-1}(U_\eta)$$

and use the continuity of F. The properties of U_η carry over to V_η: the sets V_η are clopen, disjoint and their union yields the whole space X.

Step 2: The minimal distance between V_η and $V_{\eta'}$, $\eta \neq \eta'$, is strictly positive. Assume that for $\eta, \eta' \in E$, $\eta \neq \eta'$ we have

$$\inf_{u \in V_\eta, v \in V_{\eta'}} d(u, v) = 0.$$

There are two sequences $\{u_n\}_{n \in \mathbb{N}} \subset V_\eta$ and $\{v_n\}_{n \in \mathbb{N}} \subset V_{\eta'}$ with

$$d(u_n, v_n) \to 0 \qquad \text{for } n \to \infty.$$

Since (X, d) is compact, there is a sequence $n_k \to \infty$ such that $u_{n_k} \to u^*$ and $v_{n_k} \to v^*$ converge simultaneously. From the triangle-inequality we obtain

$$d(u^*, v^*) \leq d(u^*, u_{n_k}) + d(u_{n_k}, v_{n_k}) + d(v_{n_k}, v^*) \to 0 \text{ for } k \to \infty.$$

Hence $u^* = v^*$. Now $u^* \in V_\eta$ and $v^* \in V_{\eta'}$. Therefore, $F(u^*)(0) = \eta$ and $F(v^*)(0) = \eta'$, contradicting $u^* = v^*$. Hence, the distance between V_η and $V_{\eta'}$ is strictly positive for $\eta \neq \eta'$. Thus, we can find $k \in \mathbb{N}$ such that

$$\inf_{u \in V_\eta, v \in V_{\eta'}} d(u, v) > 1/(k+1).$$

Since E is finite, there is $k \in \mathbb{N}$ such that this inequality is true for all pairs $\eta, \eta' \in E$, $\eta \neq \eta'$.

With this k we define the set $D_0 = \Gamma_k$ as the neighborhood for the automaton to be constructed. Later it could turn out that Γ_k is perhaps larger than it need to be (to fulfill the distance requirement we can choose k very large) but this Γ_k will certainly suffice.

Step 3: Construction of the cellular sub-automaton. Fix an element $\eta_0 \in E$. This element can be seen as a background state. Let $u \in E^\Gamma$, not necessarily $u \in X$. From step 2 we know that $u \in V_\eta$, $v \in V_{\eta'}$, $\eta \neq \eta'$ implies

$$u|_{\Gamma_k} \neq v|_{\Gamma_k}.$$

Now we have two cases. If there is a $v \in X$ which agrees with u on Γ_k then we have no choice. This v is in some V_η, and hence we cannot do other then define $f_0(u|_{\Gamma_k}) = f_0(v|_{\Gamma_k}) = \eta$. If there is no such v then we put $f_0(u|_{\Gamma_k}) = \eta_0$. This procedure is formalized as

$$f_0 : E^{\Gamma_k} \to E$$

by

$$f_0(\varphi) = \begin{cases} \eta & \text{there is a } v \in V_\eta : v|_{\Gamma_k} = \varphi \\ \eta_0 & \text{otherwise.} \end{cases} \qquad (*)$$

We have chosen the grid Γ, the set of elementary states E and the neighborhood $D_0 = \Gamma_k$. With the local function f_0 given in (*), we have constructed a cellular automaton (Γ, D_0, E, f_0) with the global function

$$f(u)(0) = F(u)(0) \quad \text{for} \quad u \in X$$

at the origin. It remains to show that such equation holds everywhere. We use the fact that F and f commute with all translation operators σ_g, $g \in \Gamma$, and find

$$\begin{aligned} &(\sigma_{g^{-1}} \circ f \circ \sigma_g)(u)(0) = (\sigma_{g^{-1}} \circ F \circ \sigma_g)(u)(0) \\ \Rightarrow \quad & (f \circ \sigma_g)(u)(0) = (F \circ \sigma_g)(u)(0) \\ \Rightarrow \quad & f(u)(g) = F(u)(g), \end{aligned}$$

and hence $f|_X = F$. □

If $X \neq E^\Gamma$ then we have some freedom in choosing η_0 and the neighborhood Γ_k. If $X = E^\Gamma$, we do not have any degree of freedom except in the choice of the neighborhood. Indeed, we could have chosen the number k in the proof too large in the sense that a smaller k would also have worked. Even if we had chosen k minimal (with respect to the desired distance between the V_η) there could be some smaller neighborhood that would also work. In all other aspects the cellular automaton is uniquely defined. We can conclude the classical Theorem of Curtis, Hedlund und Lyndon.

Corollary 3.3.4 ***(Theorem of Curtis, Hedlund and Lyndon)*** *The global functions of cellular automata are exactly the continuous functions on (E^Γ, d) that commute with all shift operators.*

In view of this theorem, one can characterize a cellular automaton as a continuous function on the Cantor space (E^Γ, d) which commutes with all shift operators. This alternative definition is often used when one wants to investigate the dynamical structure of cellular automata.

Note that this theorem can be generalized beyond our definition of a cellular automaton: it is not necessary to restrict the set of local states to finite sets. Ceccherini-Silberstein and Coornaert proved a version of the CHL-theorem for infinite alphabets [26, 27].

3.4 Spatial Structure and Simplifications

The Curtis-Hedlund-Lyndon (CHL) theorem leads us from the original cellular automaton, introduced by von Neumann, to a topological structure, the Cantor space, and continuous functions on this space. In this section we investigate the reverse direction. Given a Cantor space and functions operating on this space, under which conditions do we find a cellular automaton that can serve as a model for this structure? Are there continuous functions on metric Cantor spaces that cannot be represented by the global function of a cellular automaton? In other words, are cellular automata rather special and artificial constructions or are they generic objects?

Following [135, 136], we approach out goal in two steps. In the present section, we generalize the CHL theorem, which is a statement about a function f on a Cantor space E^Γ, such that the grid Γ is already given. Here we take a different view. We start from a function f and some further functions $\sigma = \{\sigma_1, \ldots, \sigma_m\}$ on a Cantor set $\mathfrak{C}$. We derive sufficient and necessary conditions such that σ generates a group G, and f can be simulated by a cellular automaton on E^Γ, $\Gamma = \Gamma(G, \sigma)$. This construction can lead to a better understanding of the roles of the different algebraic and topological conditions.

In Sect. 3.5, we proceed with the second step towards our ultimate goal: we start from an arbitrary continuous function on an arbitrary metric Cantor space, and construct a cellular automaton that is able to simulate this function. In that step we have no indication at all how the grid looks like but it is constructed from scratch.

Definition 3.4.1 Let $\mathfrak{X}$ be a metric space with metric d.

(1) A group $(G, \cdot)$ acts on $\mathfrak{X}$, if every $g \in G$ is associated with a continuous map $\varphi_g : \mathfrak{X} \to \mathfrak{X}$ such that ($e \in G$ neutral element, $g_1, g_2 \in G$)

$$\varphi_e = id_{\mathfrak{X}} \quad \text{and} \quad \varphi_{g_1} \circ \varphi_{g_2} = \varphi_{g_1 \cdot g_2}.$$

(2) Let σ be a bijective mapping of $\mathfrak{X}$ to itself. The mapping is called expansive (with constant δ) if there is a real number $\delta > 0$ such that for any $x, y \in \mathfrak{X}$, $x \neq y$, there is a number $k = k(x, y) \in \mathbb{Z}$ with

$$d(\sigma^k(x), \sigma^k(y)) > \delta.$$

(3) Let $\sigma_1, \ldots, \sigma_n$ be bijective mappings of $\mathfrak{X}$ to itself. Let $G =< \sigma_1, \ldots, \sigma_n >$ be the group generated by the σ_i.

Suppose there is a real number $\delta > 0$ such that for every $x, y \in \mathfrak{X}$, $x \neq y$, there is a $g \in G$ (the choice of g may depend on x, y) with

$$d(g(x), g(y)) > \delta.$$

Then $\sigma_1, \ldots, \sigma_n$ are called jointly expansive (with constant δ) and the group G is said to act expansively on $\mathfrak{X}$.

We will often identify the group action φ_g and the group element g and write $g(x)$ instead of $\varphi_g(x)$. Note that φ_g are already bijective and continuous, and the inverse function $\varphi_{g^{-1}}$ is also continuous. The group elements are associated with topological automorphisms.

The following theorem yields the construction of E and Γ. The theorem has been proved by Hedlund [85] for the one-dimensional case. For the construction we do not need all properties of a Cantor space; the space needs not be perfect. But certainly we have a metric Cantor space $\mathfrak{C}$ in mind as the most important case. First we show a lemma which is rather obvious but nevertheless needs a proof: if we have an expansive group on some Cantor space then this group can be carried to any other Cantor space by the homeomorphism between these two spaces.

Lemma 3.4.2 *Let $(\mathfrak{C}_1, d_1)$ and $(\mathfrak{C}_2, d_2)$ be two metric and compact spaces, homeomorphic with the homeomorphism $\varphi : \mathfrak{C}_1 \to \mathfrak{C}_2$. Let $G = <\sigma_1, \ldots, \sigma_m>$ be a group that acts continuously and expansively on $\mathfrak{C}_1$. Let furthermore $\zeta_i = \varphi \circ \sigma_i \circ \varphi^{-1}$. Then, $H = <\zeta_1, \ldots, \zeta_m>$ is a group that acts continuously and expansively on $\mathfrak{C}_2$.*

Proof The functions ζ_i are continuous and bijective on $\mathfrak{C}_2$. It remains to show that H acts expansively on $\mathfrak{C}_2$. Suppose this were not so. Then there are sequences $x_n, y_n \in \mathfrak{C}_2$, such that $x_n \neq y_n$ and

$$\lim_{n\to\infty} \sup_{\eta\in H} d_2(\eta(x_n), \eta(y_n)) = 0.$$

Then

$$\lim_{n\to\infty} \sup_{\sigma\in G} d_2(\varphi \circ \sigma \circ \varphi^{-1}(x_n), \varphi \circ \sigma \circ \varphi^{-1}(y_n)) = 0.$$

This equation can be written as

$$\lim_{n\to\infty} \sup_{\sigma\in G} d_2(\varphi \circ \sigma(u_n), \varphi \circ \sigma(v_n)) = 0,$$

whereby $u_n = \varphi^{-1}(x_n)$ and $v_n = \varphi^{-1}(y_n)$ are elements of $\mathfrak{C}_1$, $u_n \neq v_n$. There are elements $\gamma_n \in G$, such that $d_1(\gamma_n(u_n), \gamma_n(v_n)) \geq \delta$. Let $\hat{u}_n = \gamma_n(u_n)$, $\hat{v}_n = \gamma_n(v_n)$. These two sequences satisfy $d_1(\hat{u}_n, \hat{v}_n) \geq \delta$ and

$$\lim_{n\to\infty} d_2(\varphi(\hat{u}_n), \varphi(\hat{v}_n)) = 0$$

which is impossible since $\mathfrak{C}_1$ is compact and φ is continuous. □

The next theorem shows that a space with expansive bijections is homeomorphic to a subspace of some E^Γ.

Theorem 3.4.3 *Let $\mathfrak{C}$ be a metric, compact and totally disconnected space with metric $d_{\mathfrak{C}}$. Let $\sigma = \{\sigma_1, \dots, \sigma_m\}$ be a set of bijections from $\mathfrak{C}$ to itself. Let $G =< \sigma_1, \dots, \sigma_m >$ be the group generated by σ and let $\Gamma = \Gamma(G, \sigma)$ the Cayley graph corresponding to the generators σ_i. The following two statements are equivalent:*

(1) $\sigma_1, \dots, \sigma_m$ are continuous and jointly expansive on $\mathfrak{C}$.
(2) There is a finite alphabet E, a set $X \subset E^\Gamma$, and a homeomorphism $\varphi : \mathfrak{C} \to X$ with the following properties: X is topologically closed and invariant under σ_g for all $g \in G$, and $\varphi \circ g(x) = \sigma_{g^{-1}} \circ \varphi(x)$ for all $g \in G$.

Proof (1)⇒(2)

Step 1: Find a finite alphabet E and a homeomorphism $\varphi : \mathfrak{C} \to E^\Gamma$.
Let $\delta > 0$ be the constant for the joint expansiveness of the σ_i. Since $\mathfrak{C}$ is compact and totally disconnected, $\mathfrak{C}$ can be covered by finitely many disjoint clopen sets $U_0, \dots, U_K$ with diameter less than δ (see the first step in the proof of Theorem 3.1.12, see Appendix A.1, page 431). We define $E = \{0, \dots, K\}$. Let $G =< \sigma_1, \dots, \sigma_m >$ and let Γ be the Cayley graph of G. We denote the origin of this graph by 0, also in the case that G is non-Abelian. Let $x \in \mathfrak{C}$. With x we associate a mapping $u : G \to E$, i.e., an element $u \in E^\Gamma$, as follows. For a given $g \in G$ the element $g(x)$ is contained in one and only one of the U_i, say, U_{i_0}. Then we define $u(g^{-1}) = i_0$; the reason why we work with g^{-1} instead of g becomes clear in step 3. In this way, we have an element $u \in E^\Gamma$ which depends on x as a parameter. Then we define a mapping $\varphi : \mathfrak{C} \to E^\Gamma$ by $\varphi(x) = u$ where $u \in E^\Gamma$ is the element we have constructed just before. Consequently, we have the equivalence

$$\varphi(x) = u \qquad \Leftrightarrow \qquad g(x) \in U_{u(g^{-1})} \text{ for all } g \in G.$$

Step 2: $\varphi : \mathfrak{C} \to E^\Gamma$ is continuous.
Recall that $d_{\mathfrak{C}}$ is the metric on $\mathfrak{C}$ and d_c is the Cayley metric on Γ while d is the metric on E^Γ. We show that for all $\varepsilon > 0$ there is $\hat{\delta} > 0$ such that $d_{\mathfrak{C}}(x, y) < \hat{\delta}$ implies $d(\varphi(x), \varphi(y)) < \varepsilon$. Choose $x \in \mathfrak{C}$. Choose $\varepsilon > 0$ and $k > 1/\varepsilon$. The functions σ_i are continuous and the sets U_i are open. Hence a finite number of the $g \in G$ are jointly continuous. We can find a $\hat{\delta} > 0$ such that

$$g(y) \in U_{\varphi(x)(g^{-1})}$$

for all y with $d_{\mathfrak{C}}(x, y) < \hat{\delta}$ and all g with $d_c(g, 0) \le k$. Recall that $d_c(g, 0) \le k$ is equivalent with $g \in \Gamma_k$, and that $g \in \Gamma_k$ implies $g^{-1} \in \Gamma_k$. Hence $\varphi(x)(g) = \varphi(y)(g)$ for all $g \in \Gamma_k$ and, by the definition of the metric d, we have $d(\varphi(x), \varphi(y)) \le 1/(k+1) < 1/(1 + 1/\varepsilon) < \varepsilon$. Since $\mathfrak{C}$ is compact, we can choose $\hat{\delta}$ independently of x.

Step 3: $\varphi : \mathfrak{C} \to E^\Gamma$ *is injective; construction of* X.

Assume $x, y \in \mathfrak{C}$, $x \neq y$. Then there is $g \in G$ with

$$d_{\mathfrak{C}}(g(x), g(y)) > \delta \Rightarrow \varphi(g(x))(0) \neq \varphi(g(y))(0) \Rightarrow \varphi(x) \neq \varphi(y).$$

Hence, $\varphi : \mathfrak{C} \to E^\Gamma$ is an injection. The set X is topologically closed because it is the image of a compact set under a continuous mapping.

Let $a \in G$. Then,

$$\varphi(a(x))(g) = k \Leftrightarrow g^{-1}(a(x)) \in U_k \Leftrightarrow k = \varphi(x)((g^{-1}a)^{-1}) = \sigma_{a^{-1}} \circ \varphi(x)(g),$$

and hence $\varphi \circ a(x) = \sigma_{a^{-1}} \circ \varphi(x)$. The invariance of X under σ_g is a consequence of this equation together with the invariance of $\mathfrak{C}$ under the group action.

(2)⇒(1) Let $X \subset E^\Gamma$, $u, v \in X$ and $u \neq v$. There is $g \in G$ such that $u(g) \neq v(g) \Rightarrow \sigma_g(u)(0) \neq \sigma_g(v)(0)$ and $d(\sigma_g(u), \sigma_g(v)) = 1$. G acts jointly expansive on X. X and $\mathfrak{C}$ are homeomorphic with homeomorphism φ, and $\sigma_l = \varphi^{-1} \circ \sigma_{\sigma_i^{-1}} \circ \varphi$. Lemma 3.4.2 implies that G acts jointly expansive on $\mathfrak{C}$. □

Note that the cardinality of the alphabet E is given by the number of sets necessary to construct the decomposition of $\mathfrak{C}$. We will use this fact in Corollary 3.4.8 below.

In the following we determine necessary conditions such that a given function can be interpreted as a cellular (sub-)automaton on an appropriate topological space.

Definition 3.4.4 Let $\mathfrak{X}$ be a topological space and $f : \mathfrak{X} \to \mathfrak{X}$ a continuous function.

(1) The function f defines a discrete time topological dynamical system $\{\mathfrak{X}, f\}$ on $\mathfrak{X}$. The trajectory (semi-orbit) of a point $u \in \mathfrak{X}$ is given by the sequence $\{u_i\}_{i \in \mathbb{N}_0}$ with $u_0 = u$ and $u_{i+1} = f(u_i)$ for $i \in \mathbb{N}_0$.
(2) The group $A(f)$ is defined by

$$A(f) = \{g : \mathfrak{X} \to \mathfrak{X} \; : \; g \text{ homeomorphism and } f \circ g = g \circ f\}.$$

(3) We say that a topological dynamical system $\{\mathfrak{X}, f\}$ can be embedded in a topological dynamical system $\{\mathfrak{Y}, g\}$, if there is a continuous, injective mapping $\varphi : \mathfrak{X} \to \mathfrak{Y}$ such that $\varphi \circ f = g \circ \varphi$.
(4) We say that a topological dynamical system $\{\mathfrak{X}, f\}$ can be (weakly) embedded in a cellular automaton (Γ, D_0, E, g_0) with global function g, if $\{\mathfrak{X}, f\}$ can be embedded in the topological dynamical system $\{E^\Gamma, g\}$, i.e., if there is a continuous, injective function $\varphi : \mathfrak{X} \to E^\Gamma$ such that $\varphi \circ f = g \circ \varphi$. If $\varphi(\mathfrak{X}) \subset E^\Gamma$ is invariant with respect to the shift operators, we call this embedding strong.
(5) The topological dynamical systems $\{\mathfrak{X}, f\}$ and $\{\mathfrak{Y}, g\}$ are conjugated if there is a continuous bijective mapping $\phi : \mathfrak{X} \to \mathfrak{Y}$ such that $\phi \circ f = g \circ \phi$.

Remark 3.4.5 If f bijective, then $A(f)$ is the centralizer of f in the group of all bijective, continuous functions on $\mathfrak{X}$.

The following theorem is essentially a reformulation of the ideas of Hedlund [85].

Theorem 3.4.6 *Let $\mathfrak{C}$ be a metric, compact, and totally disconnected space and let $g : \mathfrak{C} \to \mathfrak{C}$ be a continuous function. The following two statements are equivalent:*

(1) *$\{\mathfrak{C}, g\}$ can be strongly embedded in a cellular automaton (Γ, D_0, E, f_0).*
(2) *The group $A(g)$ contains a finitely generated subgroup that acts expansively on $\mathfrak{C}$.*

Proof (1) $\Rightarrow$ (2): Assume that there is an embedding mediated by $\varphi : \mathfrak{C} \to E^\Gamma$. Let the Cayley graph Γ be generated by $\tau_1, \ldots, \tau_m$. Let f denote the global function of the cellular automaton. Then $\sigma_{\tau_i} \in A(f)$ for all $i = 1, \ldots, m$. Furthermore, $G = < \sigma_{\tau_1}, \ldots, \sigma_{\tau_m} >$ acts expansively on E^Γ (since we can move every site to the origin and achieve that the distance of any two states equals 1). Since $\varphi : \mathfrak{C} \to E^\Gamma$ is injective and since $\varphi(\mathfrak{C})$ is invariant under all shift operators, the inverse of φ is well defined on $\sigma_{\tau_i}(\varphi(\mathfrak{C}))$, i.e., also $\varphi^{-1}(\sigma_{\tau_i}(\varphi(x)))$ is well defined. Define the function $\zeta_i : \mathfrak{C} \to \mathfrak{C}$ by

$$\zeta_i : \mathfrak{C} \to \mathfrak{C}, \quad x \mapsto \varphi^{-1} \circ \sigma_{\tau_i} \circ \varphi(x).$$

Then the ζ_i are continuous, bijective and commute with g, i.e., $\zeta_i \in A(g)$. The group $H =< \zeta_1, .., \zeta_m >$ acts expansively on $\mathfrak{C}$ (Lemma 3.4.2).

(2) $\Rightarrow$ (1): Now assume that $A(g)$ contains a finitely generated subgroup $H = < \zeta_1, \ldots, \zeta_m >$ that acts expansively on $\mathfrak{C}$ with constant $\delta > 0$. Theorem 3.4.3 shows that there is a finite alphabet E such that $\mathfrak{C}$ is homeomorphic to $X \subset E^{\Gamma(H,\zeta)}$ where $\Gamma(H, \zeta)$ denotes the Cayley graph of the group H generated by $\zeta = \{\zeta_1, \ldots, \zeta_m\}$. Let $\varphi : \mathfrak{C} \to X$ be this homeomorphism. The proof of the Curtis-Hedlund-Lyndon theorem shows that the function

$$\tilde{f} : X = \varphi(\mathfrak{C}) \to X = \varphi(\mathfrak{C}), \quad u \mapsto \varphi \circ g \circ \varphi^{-1}(u)$$

can be represented as the global function of a cellular automaton (Γ, D_0, E, f_0) on E^Γ. In general, this automaton is only uniquely defined on $\varphi(\mathfrak{C})$, but can be extended (in a non-unique way) to E^Γ. Thus, $(\mathfrak{C}, g)$ is embedded in (Γ, D_0, E, f_0). □

If $\{\mathfrak{X}, g\}$ can only be weakly embedded in a cellular automaton, then $A(g)$ does not contain a finitely generated group that acts expansively on $\mathfrak{X}$. However, the cellular automaton has this property and hence can be seen as an extension of $\{\mathfrak{X}, g\}$ for which such group exists. We formulate this observation as a corollary.

Corollary 3.4.7 *If $\{\mathfrak{X}, f\}$ can be weakly embedded in a cellular automaton, then there is a continuous extension $\{\tilde{\mathfrak{X}}, g\}$ with a metric Cantor space $\tilde{\mathfrak{X}}$, a continuous function $g \in C(\tilde{\mathfrak{X}}, \tilde{\mathfrak{X}})$ and a continuous, injective function $\varphi : \mathfrak{X} \to \tilde{\mathfrak{X}}$ with*

$$\varphi \circ f = g \circ \varphi$$

such that $A(g)$ contains a finitely generated expansive group that acts expansively on $\tilde{\mathfrak{X}}$.

Corollary 3.4.8 *Let $\mathfrak{C}$ be a set, and let functions $f, \sigma_1,\ldots,\sigma_m : \mathfrak{C} \to \mathfrak{C}$ be given. Assume*

(1) *$\mathfrak{C}$ is metric, compact and totally disconnected*
(2) *$\sigma_1,\ldots,\sigma_m$ are bijective*
(3) *f and $\sigma_1,\ldots,\sigma_m$ are continuous*
(4) *f commutes with $\sigma_1,\ldots,\sigma_m$*
(5) *$G = <\sigma_1,\ldots,\sigma_m>$ acts expansively on $\mathfrak{C}$ (with constant δ).*

Then, these functions can be interpreted in terms of a cellular subautomaton: There exists $n \in \mathbb{N}$ such that $\mathfrak{C}$ can be decomposed in n clopen, disjoint subsets with diameter less than δ, and $(\mathfrak{C},f)$ can be strongly embedded into a cellular automaton (Γ, D_0, E, f_0) with $E = \{0,\ldots,n-1\}$, $\Gamma = \Gamma(G, \{\sigma_1,\ldots,\sigma_m\})$.

3.4.1 Examples: Structures That Are Not Cellular Automata

In order to better understand the roles of the five conditions in Corollary 3.4.8 in the topological characterization of cellular automata, we discuss structures that lack one or several of these properties. The examples are chosen as close to classical cellular automata as possible, but—of course—they are not cellular automata.

(1, a) *$\mathfrak{C}$ is not compact*

Let $E = \mathbb{N}$ and $\Gamma = \mathbb{Z}$. We use the discrete topology on E and the product topology for E^Γ (which is induced by the Cantor metric). The lift of the shift operator $\sigma : \Gamma \to \Gamma, z \mapsto z + 1$ to E^Γ is continuous and expansive.

The property "compact" forces the elementary state space E to be finite (or, at least, compact).

(1, b) *$\mathfrak{C}$ is not totally disconnected*

As a perfect, compact set that is not totally disconnected we could choose the closed unit interval. Since some effort is needed to construct our equivalent of a cellular automaton on the unit interval (especially an expansive function σ), we use the circle S^1 instead. The distance of two points on S^1 is defined as the length of the shorter arc connecting the points. The properties (2)–(5) can be satisfied with the choice $\mathfrak{C} = S^1$ and $\sigma : S^1 \to S^1$ defined by

$$\varphi \mapsto 2\,\varphi \bmod 2\pi.$$

Let the function $f : S^1 \to S^1$ be defined by

$$\varphi \mapsto (\pi\varphi) \bmod 2\pi.$$

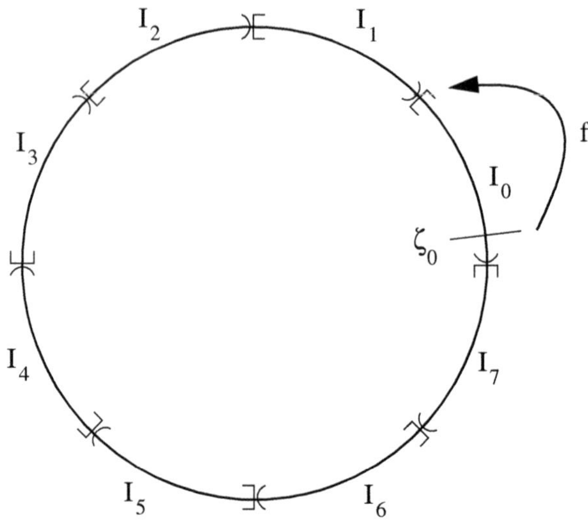

Fig. 3.4 The circle S^1 is decomposed into eight arcs. The point φ_0 is mapped by f to the boundary between I_0 and I_1

The function σ is bijective, f and σ are continuous and commute. The function σ is expansive, since angles which are small enough are doubled. If we start an iteration with σ at two different points then the distance between the iterates is bounded below by $\pi/8$ at a certain number of iterations.

We can decompose S^1 into a finite number of arcs. However, not all of these arcs can be open. A possible choice is $I_i = [i\pi/8, (i+1)\pi/8)$, $i = 0, \ldots, 7$. Let $E = \{0, \ldots, 7\}$ (see Fig. 3.4). In this case, any point $\varphi \in S^1$ can be identified with a point in $u \in E^{\mathbb{Z}}$. The map $\psi : S^1 \to E^{\mathbb{Z}}$ (as in the proof of Theorem 3.1.12) will not be surjective. For example, if the sequence $u = (\ldots, 2, 2, 2, \ldots)$ would occur as an image then the original point x would stay, under multiplication by 2, forever in the interval $I_2 = [\pi/4, 3\pi/8)$, which is not the case for any element in I_2.

The function φ is not continuous at the boundaries of the intervals I_i and at all points that are eventually mapped to one of the boundary points. The topologies of $\mathfrak{C}$ and of $E^{\mathbb{Z}}$ do not agree.

The property "totally disconnected" enables us to construct a representation of $\mathfrak{C}$ by E^Γ using a finite alphabet E and the Cayley graph Γ derived from a finitely generated group. The discrete character of the state space E^Γ is a consequence of this property.

(2) *σ_i is not bijective*

If one of the functions σ_i is not bijective, then the σ_i do not generate a group. The translation symmetry is broken, e.g., instead of a grid $\mathbb{Z}$ we obtain the only $\mathbb{N}$. Points close to the boundary require special treatment.

From a modeling point of view, regions "with boundary" represent boundaries in spatial structures. Most topological concepts will not be changed by such a generalization of a cellular automaton.

(3, a) *f is not continuous*

According to the Curtis-Hedlund-Lyndon theorem, continuity ensures that the resulting global function is localized. Let us consider a cellular structure with $E = \{0, 1\}$ and $\Gamma = \mathbb{Z}$. Let $u_1 = (\ldots 1, 1, 1, \ldots)$ be the state where all cells assume "1", and $u_0 = (\ldots, 0, 0, 0, \ldots)$ the state where all cells assume zero. We define

$$f : E^\Gamma \to E^\Gamma, \quad u \mapsto f(u)$$

where $f(u) = u_1$ if and only if $u = u_1$, and $f(u) = u_0$ otherwise. Formally, the function can be written as

$$f(u) = \prod_{i \in \mathbb{Z}} u(i).$$

This function is not continuous.

(3, b) *σ_{i_0} is not continuous for an $i_0 \in \{1, \ldots, m\}$*

A grid that is not created by continuous generators σ_i has "another idea" of locality. If we lift f to this grid, then the representation of f will not be local in the proper sense.

The simultaneous continuity of f and σ_i ensures that f and σ_i agree in the definition of "local" and create the locality of the cellular automaton. Here it becomes clear that we need the topology as a kind of reference standard such that σ_i and f respect the same concept of locality.

(4) *f does not commute with σ_i*

If f does not commute with some σ_i, we still obtain a local representation, i.e., for a given grid point $g \in G$, there is a finite neighborhood $U(g)$ and a function f_g such that

$$f(u)(g) = f_g(u|_{D(g)}).$$

However, the size of $D(g)$ may be not uniformly bounded in g, such that the concept of "local" may be weakened. Even if $D(g) = \sigma_g D_0$ for all $g \in G$, the functions f_g may be different.

The commutativity properties of f and $< \sigma_1, \ldots, \sigma_m >$ ensure the translational symmetry of the cellular automaton.

(5) *σ_i are not jointly expansive*

Consider a cellular automaton with $E = \{0, 1\}$ and $\Gamma = \mathbb{Z}^2$. If we restrict to only one of the shift operators in the construction above, $G =< \sigma_1 >$, then we obtain a set of local states that resembles Σ_2. Conditions (1), (2), (3) and (4) are satisfied, but not (5).

The joint expansiveness ensures that there are enough generators to span the spatial structure.

3.4.2 Simplification of the State Space

Information about the state can be reformulated in terms of spatial structure (see e.g. [167]).

Proposition 3.4.9 *Let (Γ, D_0, E, f_0) be a cellular automaton with $|E| = n$. Let $\Gamma = \Gamma(G, \sigma)$ the Cayley graph of a finitely generated group G where σ indicates the set of generators. There is a cellular automaton $(\hat{\Gamma}, \hat{E}, \hat{D}_0, \hat{f}_0)$ with $\hat{E} = \{0, 1\}$, and $\hat{\Gamma}$ a Cayley graph on $G \times \mathbb{Z}_{n+1}$, such that the dynamics of the first cellular automaton can be embedded into the dynamics of the latter.*

Proof Without restriction, $E = \{0, \ldots, n-1\}$, and $\Gamma = \Gamma(G, \sigma)$ some set of generators σ. Let $\Gamma = \Gamma(G, \sigma)$ be the Cayley graph of G. Define $\hat{G} = G \times \mathbb{Z}_{n+1}$ (it becomes clear below why we do not work with $\mathbb{Z}_n$ but with $\mathbb{Z}_{n+1}$) as the direct product of G and $\mathbb{Z}_{n+1}$. Let $\hat{\Gamma} = \hat{\Gamma}(\hat{G})$ be the Cayley graph of $\hat{G}$.

Step 1: *Encoding and decoding of E.*
Code a number $m \in E$ by a state $\tilde{u} \in \hat{E}^{\mathbb{Z}_{n+1}}$,

$$\eta : E \to \hat{E}^{\mathbb{Z}_{n+1}}, \qquad \eta(m)(i) = 1 \text{ for } i = 0, \ldots, m, \quad \eta(m)(i) = 0 \text{ else.}$$

As $|Z_{n+1}| = n + 1$ and $n = |E|$, $\eta(m)$ contains for all $m \in E$ at least one site with 1, and one site with 0. To be more precise, $\eta(m)$ consists of a block 1 (length $m + 1$) preceded by a block 0. Any state in $\eta(E)$ possesses a unique transition 0 to 1, $\eta(m)(-1) = 0$, and $\eta(m)(0) = 1$. Let $\ell \in \mathbb{Z}_{n+1}$, then also $\sigma_\ell \circ \eta(m)$ has a unique transition 0 to 1, $\eta(m)(\ell - 1) = 0$, and $\eta(m)(\ell) = 1$. We call ℓ the phase of $\tilde{u} \in \cup_{\ell \in \mathbb{Z}_{n+1}} \varphi(E)$.
Furthermore, define the function ζ assigning to any $\tilde{u} \in \hat{E}^{\mathbb{Z}_{n+1}}$ an element in E,

$$\zeta : \hat{E}^{\mathbb{Z}_{n+1}} \to E, \quad \zeta(\tilde{u}) = \min\{|\{i \in \mathbb{Z}_{n+1} \, : \, \tilde{u}(i) = 1\}| - 1, n - 1\}.$$

Note that $\zeta \circ \eta = id_E$, and $(\eta \circ \zeta)^2 = \eta \circ \zeta$. That is, ζ acts as a decoding function, and $\eta \circ \zeta$ is a projection to $\eta(E) \subset \hat{E}^{\mathbb{Z}_{n+1}}$. Moreover, if $\sigma_\ell : \hat{E}^{\mathbb{Z}_{n+1}} \to \hat{E}^{\mathbb{Z}_{n+1}}$ is any shift, then $\zeta(\tilde{u}) = \zeta \circ \sigma_\ell(\tilde{u})$. The decoding is shift invariant, any phase information in $\tilde{u}$ is lost.

Step 2: *Encoding and decoding of global states.*
Based on this coding, we define

$$\varphi : E^\Gamma \to \hat{E}^{\hat{\Gamma}}, \; u \mapsto \hat{u}, \; \hat{u}(g, i) = \eta(u(g))(i) \quad \text{for } (g, i) \in G \times \mathbb{Z}_{n+1}$$

$$\psi : \hat{E}^{\hat{\Gamma}} \to E^\Gamma, \; \hat{u} \mapsto u, \; u(g) = \zeta(\hat{u}(g, .))$$

where we identified $\hat{u}(g, .)$ with a state in $\hat{E}^{\mathbb{Z}_{n+1}}$. The properties of the function φ and ψ resemble that of η and ζ, that is, $\psi \circ \varphi = id_{E^\Gamma}$, and $(\varphi \circ \psi)^2 = \varphi \circ \psi$.

Naively, one is tempted to define

$$\tilde{f} : \hat{E}^{\hat{\Gamma}} \to \hat{E}^{\hat{\Gamma}}, \quad \tilde{f}(\hat{u}) = \varphi \circ f \circ \psi(\hat{u}).$$

in order to embed (E^{Γ}, f) the in a cellular automaton on $\hat{E}^{\hat{\Gamma}}$. Indeed, we have

$$\tilde{f} \circ \varphi = \varphi \circ f \circ \psi \circ \varphi = \varphi \circ f$$

such that (E^{Γ}, f) is embedded in $(\hat{E}^{\hat{\Gamma}}, \tilde{f})$. Note that $\tilde{f}$ commutes with a shift in $G \times \{e\}$, but is invariant w.r.t. shifts in $\{e\} \times \mathbb{Z}_{n+1}$, $\tilde{f} \circ \sigma_{(e,i)} = \tilde{f}$, as the decoding function ζ (and hence also ψ) dismisses the phase information. That is, $\tilde{f}$ is not a global function of a cellular automaton. A refinement of the construction is required.

Step 3: Shift-invariant global function.

Let $X = \varphi(E^{\Gamma})$, and $X_\ell = \sigma_{(e,\ell)} X$ for $\ell \in \mathbb{Z}_{n+1}$. These sets are disjoint, as from any $\hat{u} \in X_\ell$, and even from $\tilde{u} = \hat{u}(g, .) \in \hat{E}^{\mathbb{Z}_{n+1}}$ ($g \in G$ arbitrarily chosen), we are able to read off the phase information ℓ (see above, step 1).

It is $\tilde{f}(X_\ell) \subset X_0$ for all $\ell \in \mathbb{Z}_{n+1}$, as the phase information is dismissed by the decoding function ζ. In order to incorporate the phase information in the global function, we define $Y = \cup_{\ell \in \mathbb{Z}_{n+1}} X_\ell$, and $\hat{f} : Y \to Y$ by

$$\hat{f}(\hat{u}) = \sigma_{(e,\ell)} \circ \tilde{f}(\hat{u}) \quad \text{if} \quad \hat{u} \in X_\ell.$$

Then, $\hat{f}|_{X_0} = \tilde{f}|_{X_0}$, and (E^{Γ}, f) is embedded in $(\hat{E}^{\hat{\Gamma}}, \hat{f})$. Furthermore, $\hat{f}$ is not only shift invariant w.r.t. $G \times \{0\}$, but also w.r.t. $\{e\} \times \mathbb{Z}_{n+1}$, and therefore w.r.t. $\hat{G}$. As Y is topologically closed as well as closed under the group action of $\hat{G}$, Proposition 3.3.3 guarantees that $(\hat{f}, Y)$ is a cellular subautomaton. □

3.4.3 Simplification of the Neighborhood

In case of a one-dimensional grid, it is possible to replace an arbitrary neighbourhood by $D_0 = \{-1, 0, 1\}$ at the expense of enlarging E. Given a cellular automaton $(\mathbb{Z}, D_0, E, f_0)$ with $D_0 = \Gamma_{d_0}$ and global function f, we define $f_n : E^{\Gamma_n} \to E^{\Gamma_{n-d_0}}$. Therefore, we first note that $f(u)|_{\Gamma_{n-d_0}}$ is determined by $u|_{\Gamma_n}$ for $n \geq d_0$. Given $v \in E^{\Gamma_n}$, we use some arbitrary extension $u \in E^{\Gamma}$ (s.t. $u|_{\Gamma_n} = v$) and then define $f_n(v) = f(u)|_{\Gamma_{n-d_0}}$. That is, we compute the image of $v \in E^{\Gamma_n}$ at all sites where the image already is fully determined.

Proposition 3.4.10 *For every one-dimensional cellular automaton $(\mathbb{Z}, D_0, E, f_0)$ there is another cellular automaton $(\mathbb{Z}, \hat{D}_0, \hat{E}, \hat{f}_0)$ with $\hat{D}_0 = \{-1, 0, 1\}$ such that $(\mathbb{Z}, D_0, E, f_0)$ can be embedded in $(\mathbb{Z}, \hat{D}_0, \hat{E}, \hat{f}_0)$.*

Proof We can assume $D_0 = \Gamma_{d_0}$, and define $x_0 = 2\,d_0 + 1$. Then, $\max(\Gamma_{d_0}) + 1 = \min(\sigma_{x_0}(\Gamma_{d_0}))$. That is, Γ_{d_0} "touches" $\sigma_{x_0}(\Gamma_{d_0})$. Note that $\sigma_{-x_0}(\Gamma_{d_0}) \cup \Gamma_{d_0} \cup \sigma_{x_0}(\Gamma_{d_0}) = \Gamma_{3\,d_0+1}$. Now we define $\hat{E} = E^{\Gamma_{d_0}}$, and a map $\varphi : E^{\mathbb{Z}} \to \hat{E}^{\mathbb{Z}}$ by $u \mapsto v$ with

$$v(z) = u|_{\sigma^z_{x_0}(\Gamma_{d_0})} \qquad z \in \mathbb{Z}.$$

Obviously, φ is continuous and injective but not surjective. Now we construct the local function f_0. We use the notation introduced above, $f_{3d_0+1} : E^{\Gamma_{3d_0+1}} \to E^{\Gamma_{2d_0-1}}$. Given $(v(-1), v(0), v(1)) \in (\hat{E})^3 = \hat{E}^{\Gamma_1}$, we first reconstruct $u \in E^{\Gamma_{3d_0+1}}$ by concatenating the patterns $v(-1)$, $v(0)$ and $v(1)$,

$$u = [v(-1)v(0)v(1)].$$

Next we use the block map f_{3d_0+1} generated by the cellular automaton, and restrict the result to Γ_{d_0} in order to obtain a state in $\hat{E}$. This procedure yields $\hat{f}_0$:

$$\hat{f}_0(v(-1), v(0), v(1)) = f_{3d_0+1}([v(-1)v(0)v(1)])|_{\Gamma_{d_0}}.$$

In this way we construct a cellular automaton $(\mathbb{Z}, \{-1, 0, 1\}, \hat{E}, \hat{f}_0)$ that allows the automaton $(\mathbb{Z}, D_0, E, f_0)$ to be embedded: due to the construction, we have $\hat{f} \circ \varphi = \varphi \circ f$. □

Remark 3.4.11 A similar reduction of a given neighborhood to the Neumann neighborhood with an increased number of elementary states can be performed on any Cayley graph $\Gamma = \Gamma(G, \sigma)$.

3.4.4 Simplification of the Grid

In this section we investigate the influence of the spatial structure given by $\sigma_1, \dots, \sigma_m$. To be more precise, we want to know if we can find criteria for a minimal subset of $\{\sigma_1, \dots, \sigma_m\}$ that still guarantees a cellular automaton. For example, if $\sigma_1, \dots, \sigma_m$ are jointly expansive and $\sigma_m \in < \sigma_1, \dots, \sigma_{m-1} >$, then also $\sigma_1, \dots, \sigma_{m-1}$ are jointly expansive. From a topological point of view, it is not necessary to include σ_m in the spatial structure. Of course, the Cayley graph Γ will be different if we only use a subset of generators—the neighborhood of a point will be smaller. Perhaps even a proper subgroup of $< \sigma_1, \dots, \sigma_m >$ is sufficient to ensure expansivity. We ask under which circumstances a subset of generators still generates an expansive group.

Before we formulate the statement about expansivity, we recall that the distance between two points in a Cayley graph $\Gamma(G,\sigma)$ is the length of the shortest path. That is, if we determine the minimal number n of elements $\tau_i \in \sigma \cup \sigma^{-1}$ such that $g_1 = g_2 \prod_{i=1}^n \tau_i$, then $d_c(g_1, g_2) = n$. This is equivalent with $g_2^{-1} g_1 = \prod_{i=1}^n \tau_i$, which can be reformulated as $d_c(g_2^{-1} g_1, e) = n$. Equivalently, $g_2^{-1} g_2 \in \Gamma_n \setminus \Gamma_{n-1}$. We will use this fact below.

Proposition 3.4.12 *Let $\mathfrak{C}$ be a metric, compact and totally disconnected set. Let G be a finitely generated group acting expansively on $\mathfrak{C}$. Let H be a subgroup of G and let*

$$\mu : G \to \mathbb{R}_+, \quad g \mapsto \mu(g) = \inf_{h\in H} d_c(g, h).$$

The following statements are equivalent:

(1) *H acts expansively on $\mathfrak{C}$.*
(2) *$\mu(\cdot)$ is uniformly bounded on G.*
(3) *The index $[G : H]$ is finite.*

Proof Let $\Gamma = \Gamma(G,\sigma)$ be the Cayley graph of G. There is a finite alphabet $E = \{0, \ldots, k\}$ such that $\mathfrak{C}$ is homeomorphic to a shift-invariant subset of E^Γ. Since a homeomorphism conserves the property "expansive", we may consider E^Γ instead of $\mathfrak{C}$.

(1) ⇒ (2): We show that (1) is not true if (2) is not true. If $\mu(\cdot)$ is not uniformly bounded on G, then there are $g_i \in G$, $i \in \mathbb{N}$, such that

$$\inf_{h\in H} d_c(g_i, h^{-1}) \geq i.$$

Hence, $h^{-1}g \notin \Gamma_{i-1}$ for all $h \in H$ (as $d_c(hg_i, e) \geq i$ for all $h \in H$). Let $u_i,\ v \in E^\Gamma$ be defined by $v(g) = 0$ for all $g \in G$, and $u_i(g) = 0$ if $g \neq g_i$, $u_i(g_i) = 1$. Then,

$$(\sigma_h \circ v)(g) = (\sigma_h \circ u_i)(g) \quad \text{for all } g \in \Gamma_{i-1}$$

and therefrom we conclude $d(\sigma_h \circ v, \sigma_h \circ u_i) \leq 1/i$ for all $h \in H$ and

$$\sup_{h\in H} d(\sigma_h \circ v, \sigma_h \circ u_i) \leq 1/i,$$

H is not expansive.

(2) ⇒ (3): Let $L \in \mathbb{N}$ such that $\mu(g) \leq L$ for all $g \in G$. For every $g \in G$ there is an $h \in H$ and a $\hat{g} \in \Gamma_L$ such that $g = h\hat{g}$. Hence,

$$G = \cup_{\hat{g}\in\Gamma_L} H\hat{g}$$

and $[G : H] \leq |\Gamma_L| < \infty$.

(3) $\Rightarrow$ **(1)**: Let $u, v \in E^\Gamma$, $u \neq v$. Then there is a $g \in G$ such that $u(g) \neq v(g)$. As $[G : H] = m < \infty$, there are $g_1, \ldots, g_m$ such that

$$G = \cup_{i=1}^m H g_i.$$

There is an $L \in \mathbb{N}$ such that $g_1, \ldots, g_m \in \Gamma_L$. Hence,

$$g = h_0 \hat{g}$$

for $h_0 \in H$, $\hat{g} \in \Gamma_L$. We conclude that

$$d(\sigma_{h_0} \circ u,\ \sigma_{h_0} \circ v) \geq \frac{1}{1 + d_c(e, \hat{g})} \geq \frac{1}{1+L}.$$

H acts expansively with constant $\delta = 1/(1 + L)$. □

Remark 3.4.13

(a) If G is an Abelian group and $H \neq G$ is the subgroup generated by σ_{i_0}, then the index of H in G equals the order of σ_{i_0}. We are allowed to remove all cyclic generators of finite order.
(b) If G is finite, then we do not need to take into account any spatial structure. The index of the trivial group $\{e\}$ is always finite in a finite group.
(c) If G is not Abelian, then a generator with a finite order, $< \sigma_{i_0} >= \mathbb{Z}_k$, cannot be removed in general. A cyclic generator does not imply that the index of H in G is finite. An example is the group $G =< a, b\ :\ b^2 = e >$. The Cayley graph of this group is sketched in Fig. 3.5. Here $H =< b >= \mathbb{Z}_2$, but $[G : H] = \infty$.

Remark 3.4.14 If we dispense with the spatial structure, then we can find a homeomorphism between the state space of any cellular automaton on an infinite grid and the standard Cantor set $\mathfrak{C}_{1/3}$. Thus, any cellular automaton can be identified with a continuous function

$$f \in C(\mathfrak{C}_{1/3}, \mathfrak{C}_{1/3}).$$

The shift operators σ_i can be identified with continuous functions $\tilde{\sigma}_i \in C(\mathfrak{C}_{1/3}, \mathfrak{C}_{1/3})$. However, it would be very hard to recognize this one-dimensional structure in an d-dimensional setting for $d > 1$.

Conclusion The world is one-dimensional. But complicated.

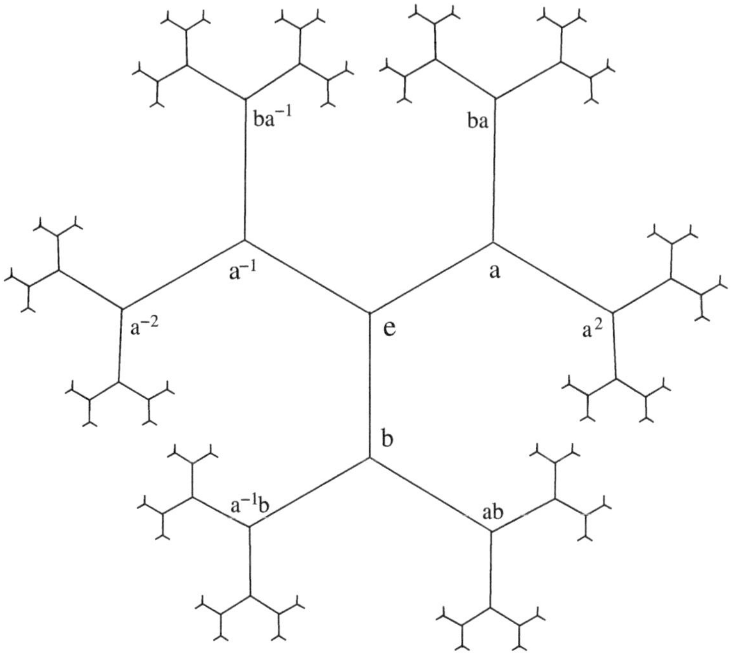

Fig. 3.5 Sketch of the first few levels of the Cayley graph of the group $< a, b \, : \, b^2 = e >$. If b is removed, only the *horizontal zig-zag line* representing $< a >$ remains

3.5 Cellular Automata and Continuous Maps on Cantor Spaces

In this section we decide if the set of continuous functions on metric (metrizable) Cantor spaces is larger than the set of cellular automata or not. We know that the global function of any any cellular automaton is a continuous function on a metric Cantor space. Given an arbitrary continuous function f on an arbitrary metric Cantor space, do we find a cellular automaton that can simulate f? In the present section we indeed construct such an automaton (see also [135, 136]). This finding indicates that the set of cellular automata forms a rather general class of objects. One cannot expected to prove deep theorems about all cellular automata; the Curtis-Hedlund-Lyndon Theorem is one of the few—in the context of Cantor topology perhaps the only one—that applies to all of them. In a similar way as the theory of partial differential equations introduces parabolic, hyperbolic and elliptic equations with rather different properties, the generality of cellular automata urges us to introduce meaningful subclasses. Until today, this problem has not been solved in a satisfactory way.

3.5.1 Bijective Maps

We first consider a topological dynamical system $(\mathfrak{C},f)$ where $\mathfrak{C}$ is a metric Cantor space and f is a continuous and bijective function. In this case, it is possible to construct directly a cellular automaton that allows an embedding of the dynamical system. In the present and the next section, we will use the following definitions.

Definition 3.5.1

(1) Let $E_n = \{0,\dots,n-1\}$, equipped with the discrete topology.
(2) Let Σ_n be the set of all bi-sequences over E_n

$$\Sigma_n = \{x = (x_i)_{i\in\mathbb{Z}} \,:\, x_i \in E_n\}.$$

The metric d_n on Σ_n is the standard Cantor metric, i.e., $d_n(x,y) = 1/(k+1)$ where $k = \inf\{|i| \,:\, x_i \neq y_i\}$ and $d_n(x,y) = 0$ if $x = y$.
(3) Let $\mathfrak{Y}_n = (E_n)^{\mathbb{N}_0}$ be the one-sided (left) shift over the alphabet Σ_n. In the following we identify $(\Sigma_n)^{\mathbb{N}_0}$ with $E^{\mathbb{N}_0\times\mathbb{Z}}$ and call the latter $\mathfrak{Y}$. We equip this set with the usual Cantor metric as follows. For $u, v \in E^{\mathbb{N}_0\times\mathbb{Z}}$ define $d_{\mathfrak{Y}}(u,v) = 0$ if $u = v$. If $u \neq v$, determine $k = \inf\{i \in \mathbb{N}_0 \,:\, \exists j \in \mathbb{Z} : |j| \leq i$ a $u(i,j) \neq v(i,j)\}$ and put $d_{\mathfrak{Y}}(u,v) = 1/(k+1)$.
(4) Let $\sigma_{\mathfrak{Y}}$ be the left shift on $\mathfrak{Y}_n$, i.e., for $u \in \mathfrak{Y}_n$ define $\sigma_{\mathfrak{Y}}(u)(i,j) = u(i+1,j)$ for $i \in \mathbb{N}_0, j \in \mathbb{Z}$.

We will later refine the definition of shifts (Definitions 6.3.13 and 7.4.1). For now, we stay with this first definition.

Proposition 3.5.2 *Let $(\mathfrak{C},f)$ be a topological dynamical system, $\mathfrak{C}$ a metric Cantor space and f a bijective, continuous function. Let furthermore (Γ, D_0, E, g_0) be a cellular automaton with $\Gamma = \mathbb{Z}^2$, $E = E_2 = \{0,1\}$ and D_0 the von Neumann neighborhood. Let the global function g of this cellular automaton be the shift operator in the x_1-direction of the grid. Then, $(\mathfrak{C},f)$ can be embedded into the cellular automaton.*

Proof Since $\mathfrak{C}$ is a metric Cantor space, it is homeomorphic to Σ_2. Thus, there is a bijective function $\varphi_1 \in C(\mathfrak{C},\Sigma_2)$. Let $f_1 = \varphi_1 \circ f \circ \varphi_1^{-1} : \Sigma_2 \to \Sigma_2$. Note that f_1 is bijective. We find that $\varphi_1 \circ f = f_1 \circ \varphi_1$ holds, i.e., $(\mathfrak{C},f)$ and (Σ_2,f_1) are conjugated.

Next we identify the function f_1 with its forward-backward trajectory. Let $\varphi_2 : \Sigma_2 \to E^{\mathbb{Z}^2}$ be defined by

$$\varphi_2(x)(i,j) = f_1^i(x)(j), \qquad i,j \in \mathbb{Z}^2.$$

Define the shift operator

$$\sigma_1 : E_2^{\mathbb{Z}^2} \to E_2^{\mathbb{Z}^2}, \quad u \mapsto v, \qquad \text{with } v(i,j) = u(i+1,j).$$

Then,

$$\varphi_2(f_1(x)) = f_1^{i+1}(x)(j) = \sigma_1(\varphi_2(x)).$$

The function φ_2 is continuous and injective. Thus, $(\mathfrak{C},f)$ is embedded in the cellular automaton that represents the shift operator on $E_2^{\mathbb{Z}^2}$. □

This proposition allows to characterize all continuous bijective functions on a metric Cantor space.

Lemma 3.5.3 *The bijective continuous functions on a metric Cantor space are conjugated to a shift operator acting on a closed subset of* $\{0,1\}^{\mathbb{Z}^2}$ *(topologically closed and closed with respect to all shift operators). Subsets of* $\{0,1\}^{\mathbb{Z}^2}$ *that are topologically closed and closed under the shift operator can be identified with the bijective, continuous functions on a metric Cantor space.*

If we drop the assumption that f is bijective and consider a general continuous function on a Cantor space then we can perform the construction as above, using the positive orbit $(f^i(x))_{i\in\mathbb{N}}$ instead of the complete orbit. If we extend this sequence for $i < 0$ by zero, then we get a cellular automaton, this time the shift on $\{0,1\}^{\mathbb{Z}}$. However, the dynamical system $(\mathfrak{C},f)$ is not exactly embedded in the cellular automaton: If we first apply f and then map to $\{0,1\}^{\mathbb{Z}}$, the support of the result is contained in the sites with $i \geq 0$. If we first map a point to $\{0,1\}^{\mathbb{Z}}$ and shift it then to the left, the support of the result is contained in $i \geq -1$. The two states agree on $i \geq 0$, but we are forced to use the restriction operator to non-negative indices in order to establish a connection. For this reason, we use another idea for non-bijective functions in the next section.

3.5.2 General Maps: The Universal Cellular Automaton

For a continuous function f on a metric Cantor space $\mathfrak{C}$ we want to construct a cellular automaton such that $(\mathfrak{C},f)$ can be weakly embedded into this cellular automaton. The first step is a representation of $(\mathfrak{C},f)$ in a suitable way as in the next theorem.

Theorem 3.5.4 *Let* $\mathfrak{C}$ *be a metric Cantor space with metric* $d_{\mathfrak{C}}$ *and let* $f : \mathfrak{C} \to \mathfrak{C}$ *be a continuous function. Then,* $(\mathfrak{C},f)$ *can be embedded in the full shift over the alphabet* Σ_3*, i.e., in* $(\mathfrak{Y}_3, \sigma_{\mathfrak{Y}_3})$.

Proof

Step 1: Representation of $\mathfrak{C}$.
Since a metric Cantor space is homeomorphic to any other metric Cantor space, we find a topological homeomorphism $\varphi_1 : \mathfrak{C} \to \Sigma_3$. Define $h : \Sigma_3 \to \Sigma_3$ by $h(x) = \varphi_1 \circ f \circ \varphi_1^{-1}(x)$. Then, $(\mathfrak{C},f)$ and (Σ_3, h) are topologically conjugated.

Step 2: Representation of f.

We represent f as a one-sided shift over the alphabet Σ_3 by defining $\varphi_2 : \Sigma_3 \to \mathfrak{Y}_3$ by

$$\varphi_2(x)(i,j) = f^i(x)(j), \qquad i \in \mathbb{N}_0,\ j \in \mathbb{Z}.$$

We now state and prove three properties of φ_2 and $\sigma_{\mathfrak{Y}}$.

Property (a): $\varphi_2 \circ h = \sigma_{\mathfrak{Y}} \circ \varphi_2$

This property is a direct consequence of the construction of φ_2.

Property (b): $\varphi_2 \in C(\Sigma_3, \mathfrak{Y})$

Consider $x \in \Sigma_3$, and chose $\varepsilon > 0$, arbitrarily fixed. We show that there is a constant $\delta > 0$ such that $d_3(x,y) < \delta$ implies $d_{\mathfrak{Y}}(\varphi_2(x), \varphi_2(y)) < \varepsilon$. Let $n = \lceil 1/\varepsilon \rceil + 1$. Since f is continuous, there is a $\delta > 0$ such that $h^i(x)(j) = h^i(y)(j)$ for $|j| \leq n$ and $i = 0, \ldots, n$ if $d_3(x,y) < \delta$. Then, also $d_{\mathfrak{Y}}(\varphi_2(x), \varphi_2(y)) < \varepsilon$ is a consequence of $d_3(x,y) < \delta$.

Property (c): φ_2 is injective

If $x \neq y$, then $\varphi(x)(0,\cdot) = x \neq y = \varphi(y)(0,\cdot)$. Hence this function is injective.

Properties (a)–(c) imply that $(\mathfrak{C}, f)$ is embedded in the shift $(\mathfrak{Y}_3, \sigma_{\mathfrak{Y}_3})$. □

Remark 3.5.5

(1) Let $\hat{\mathfrak{Y}} \subset \mathfrak{Y}_3$ denote the subshift generated by $(\mathfrak{C}, f)$. Then the set of the first components of all elements in $\hat{\mathfrak{Y}}$ is the complete set Σ_3, i.e.,

$$\{x : \exists u \in \varphi_2 \circ \varphi_1(\mathfrak{C}) : u(0,\cdot) = x(\cdot)\} = \Sigma_3.$$

(2) Since $\mathfrak{C}$ is a metric Cantor space and $\varphi_2 \circ \varphi_1$ is continuous and injective, also $\hat{\mathfrak{Y}} = \varphi_2 \circ \varphi_1(\mathfrak{C})$ is a metric Cantor space.

Now we construct elements of $A(\sigma_{\mathfrak{Y}_3})$ (recall that $A(f)$ is the set of all homeomorphisms that commute with f, see page 55). We must find finitely many continuous bijections that generate an expansive group in order to construct a cellular automaton as in Corollary 3.4.8.

Definition 3.5.6

(1) Define $\tau : \mathfrak{Y}_3 \to \mathfrak{Y}_3$ by

$$\tau(u)(i,j) = u(i, j+1) \qquad \text{for } i \in \mathbb{N}_0,\ j \in \mathbb{Z}.$$

(2) Let $\zeta : E_3 \times E_3 \to E_3$ be defined as

$$\begin{aligned} &\zeta(0,0) = 0,\ \zeta(0,1) = 2,\ \zeta(0,2) = 0,\\ &\zeta(1,0) = 1,\ \zeta(1,1) = 1,\ \zeta(1,2) = 1,\\ &\zeta(2,0) = 2,\ \zeta(2,1) = 0,\ \zeta(2,2) = 2 \end{aligned}$$

and define $\gamma : \mathfrak{Y}_3 \to \mathfrak{Y}_3$ as

$$\gamma(u)(i,j) = \zeta(u(i,j), u(i+1,j)) \qquad \text{for } i \in \mathbb{N}_0,\ j \in \mathbb{Z}.$$

(3) Let $\hat{\omega}$ be the cyclic permutation of E_3 given by $\hat{\omega}(0) = 1, \hat{\omega}(1) = 2, \hat{\omega}(2) = 3$. The lift ω of $\hat{\omega}$ to $\mathfrak{Y}_3$ is defined by

$$\omega : \mathfrak{Y}_3 \to \mathfrak{Y}_3, \quad u \mapsto \omega(u) \quad \text{with } \omega(u)(i,j) = \hat{\omega}(u(i,j)).$$

Then, for $i \in \{0,1,2\}$ define $\gamma_i : \mathfrak{Y}_3 \to \mathfrak{Y}_3$, $u \mapsto v$ by $\gamma_i = \omega^{-i} \circ \gamma \circ \omega^i$.

Proposition 3.5.7 *The function* $\tau : \mathfrak{Y}_3 \to \mathfrak{Y}_3$ *is continuous, bijective and commutes with the map* $\sigma_{\mathfrak{Y}} : \mathfrak{Y}_3 \to \mathfrak{Y}_3$.

Proof Continuity is ensured by the local definition of the shift τ. The inverse of τ is given by

$$u \mapsto v, \qquad v(i,j) = v(i,j-1) \qquad \text{for } i \in \mathbb{N}_0,\ j \in \mathbb{Z},$$

hence τ is bijective. The shifts τ and $\sigma_{\mathfrak{Y}_3}$ act on different indices, and hence commute. □

Proposition 3.5.8 *The function* $\gamma : \mathfrak{Y}_3 \to \mathfrak{Y}_3$ *is continuous, bijective and commutes with the maps* $\sigma_{\mathfrak{Y}_3} : \mathfrak{Y}_3 \to \mathfrak{Y}_3$ *and* $\tau : \mathfrak{Y}_3 \to \mathfrak{Y}_3$.

Proof This proof follows closely [85, p. 335] (although the definitions of the functions are somewhat different).

(i) Continuity follows as usual from the locality of the map $\sigma_{\mathfrak{Y}_3}$.
(ii) Bijectivity of γ requires more attention. We first show injectivity.

Assume that $u, v \in \mathfrak{Y}_3$, $u \neq v$ and $\gamma(u) = \gamma(v) =: w$. Then there is an index $(i_0, j_0) \in \mathbb{N}_0 \times \mathbb{Z}$ such that $u(i_0, j_0) \neq v(i_0, j_0)$ and $\gamma(u)(i, j_0) = \gamma(v)(i, j_0)$ for all $i \in \mathbb{N}_0$. Since $\zeta(a, b) = 1$ if and only if $a = 1$, we have $w(i_0, j_0) \neq 1$. Hence $w(i_0, j_0) \in \{0, 2\}$ and therefore also $u(i_0, j_0)$, $v(i_0, j_0) \in \{0, 2\}$. Without restriction we may assume $u(i_0, j_0) = 0$ and $v(i_0, j_0) = 2$.

Assume $w(i_0, j_0) = 0$. Then necessarily $v(i_0 + 1, j_0) = 1$ which in turn implies $w(i_0 + 1, j_0) = 1$. If $w(i_0 + 1, j_0) = 1$ then we can conclude that $u(i_0 + 1, j_0) = 1$. However, in this case we find $\zeta(u(i_0, j_0), u(i_0 + 1, j_0)) = \zeta(0, 1) = 2 \neq 0 = w(i_0, j_0)$ such that this case cannot occur.

Now consider the second case, $w(i_0, j_0) = 2$. The argument parallels the last one: If $w(i_0, j_0) = 2$, necessarily $u(i_0 + 1, j_0) = 1$ which in turn implies $w(i_0 + 1, j_0) = 1$ and hence also $v(i_0 + 1, j_0) = 1$. However, $\zeta(v(i_0, j_0), v(i_0 + 1, j_0)) = \zeta(2, 1) = 0 \neq 2 = w(i_0, j_0)$ such that this case cannot occur.

All in all we conclude that γ is injective.

We show that γ is surjective. First note, that the map $\zeta(\cdot,\cdot)$ acts like a permutation with respect to the first entry, i.e., for given $b \in E_3$, the map $a \mapsto \zeta(a, b)$ is a permutation of E_3. Consider $v \in \mathfrak{Y}_3$. We construct a sequence of states $\{u_n\}_{n\in\mathbb{N}}$, $u_n \in \mathfrak{Y}_3$ such that $\gamma(u_n) \to v$: Let $u_n(i,j) = 0$ for $i > n$. Due to the indicated permutation property, there is a unique state $u_3(n,j) \in E_3$ such that $\zeta(u_3(n,j), u_3(n+1,j)) = \zeta(u_3(n,j), 0) = v(n,j)$. Recursively, we can obtain values $u_n(n-2,j), \ldots, u_n(0,j)$. Thus, $\gamma(u_n)(i,j) = v(i,j)$ for $i = 0, \ldots, n$ and $j \in \mathbb{Z}$. Since $\mathfrak{Y}_3$ is compact, there is a converging subsequence, $u_n \to u$. Since γ is continuous, $\gamma(u) = v$.

(iii) Finally we show that γ commutes with τ and $\sigma_{\mathfrak{Y}_3}$. We find

$$\sigma_{\mathfrak{Y}_3} \circ \gamma(u)(i,j) = \zeta(u(i+1,j), u(i+2,j)) \quad = \gamma \circ \sigma_{\mathfrak{Y}_3}(u)(i,j),$$
$$\tau \circ \gamma(u)(i,j) = \zeta(u(i,j+1), u(i+1,j+1)) = \gamma \circ \tau(u)(i,j).$$

□

The function γ_i is essentially the function γ, up to a local permutation of the states of a cell, and the properties of γ carry over to γ_i. Hence we have the following corollary.

Corollary 3.5.9 *The functions $\gamma_i : \mathfrak{Y}_3 \to \mathfrak{Y}_3$, $i = 0, 1, 2$, are continuous, bijective and commute with the maps $\sigma_{\mathfrak{Y}_3} : \mathfrak{Y}_3 \to \mathfrak{Y}_3$ and $\tau : \mathfrak{Y}_3 \to \mathfrak{Y}_3$.*

The functions τ, γ_i, and ω are in $A(\sigma_{\mathfrak{Y}_3})$. Let $\mathcal{G}$ be the group generated by τ, γ_0, γ_1, γ_2. We ask whether this group is expansive. We recall the concept of "expansiveness". If we have two states u, v and they differ in some coordinate $(i,j) \in \mathbb{N}_0 \times \mathbb{Z}$ then it should be possible to transport the information about this disagreement to the coordinate $(0,0)$ (or any other fixed coordinate) by some group element. If this is the case, then the images of u and v under a suitable sequence of transformations (induced by group elements) have a fixed distance equal to 1.

The function τ transports information in the direction of the second coordinate by a simple shift. Transport of information in the direction of the first coordinate is less simple, since the canonical shift in this direction, $\sigma_{\mathfrak{Y}_3}$, is not bijective. We need a more refined mechanism. The following proposition shows, that the functions γ_i jointly perform the required transport of information.

Proposition 3.5.10 *Let $u, v \in \mathfrak{Y}_3$, $u(i,j) = v(i,j)$ for $i = 0, \ldots, n-1$ and $j \in \mathbb{Z}$, and let $u(n,j_0) \neq v(n,j_0)$ for $n \in \mathbb{N}_0$, $j_0 \in \mathbb{Z}$. Then there is $\gamma \in \{\gamma_0, \gamma_1, \gamma_2\}$ such that $\gamma(u)(n-1,j_0) \neq \gamma(v)(n-1,j_0)$.*

Proof Let $x = u(n-1,j_0) = v(n-1,j_0)$, $y = u(n,j_0) \neq v(n,j_0) = z$. Then there are nine combinations for (x,y), and—given (x,y)—there are two possibilities for z, hence there are 18 cases for (x,y,z) for which we check whether there is an $i \in \{0,1,2\}$ such that

$$\hat{\omega}^{-i} \circ \zeta(\hat{\omega}^i(x)\hat{\omega}^i(y)) \neq \hat{\omega}^{-i} \circ \zeta(\hat{\omega}^i(x)\hat{\omega}^i(z)),$$

or, equivalently,

$$\zeta(\hat{\omega}^i(x)\hat{\omega}^i(y)) \neq \zeta(\hat{\omega}^i(x)\hat{\omega}^i(z)).$$

Table 3.1 All cases for the proof of Proposition 3.5.10

x	y	z	A_1	B_1	$\hat{\omega}(x)$	$\hat{\omega}(y)$	$\hat{\omega}(z)$	A_2	B_2	$\hat{\omega}^2(x)$	$\hat{\omega}^2(y)$	$\hat{\omega}^2(z)$	A_3	B_3
0	0	1	**0**	**2**	1	1	2	1	1	2	2	0	2	2
0	0	2	0	0	1	1	0	1	1	2	2	1	**2**	**0**
0	1	0	**2**	**0**	1	2	1	1	1	2	0	2	2	2
0	1	2	**2**	**0**	1	2	0	1	1	2	0	1	**2**	**0**
0	2	0	0	0	1	0	1	1	1	2	1	2	**0**	**2**
0	2	1	**0**	**2**	1	0	2	1	1	2	1	0	**0**	**2**
1	0	1	1	1	2	1	2	**0**	**2**	0	2	0	0	0
1	0	2	1	1	2	1	0	**0**	**2**	0	2	1	**0**	**2**
1	1	0	1	1	2	2	1	**2**	**0**	0	0	2	0	0
1	1	2	1	1	2	2	0	2	2	0	0	1	**0**	**2**
1	2	0	1	1	2	0	1	**2**	**0**	0	1	2	**2**	**0**
1	2	1	1	1	2	0	2	2	2	0	1	0	**2**	**0**
2	0	1	**2**	**0**	0	1	2	**2**	**0**	1	2	0	1	1
2	0	2	2	2	0	1	0	**2**	**0**	1	2	1	1	1
2	1	0	**0**	**2**	0	2	1	**0**	**2**	1	0	2	1	1
2	1	2	**0**	**2**	0	2	0	0	0	1	0	1	1	1
2	2	0	2	2	0	0	1	**0**	**2**	1	1	2	1	1
2	2	1	**2**	**0**	0	0	2	0	0	1	1	0	1	1

$A_1 = \zeta(x, y)$, $B_1 = \delta(x, z)$, $A_2 = \zeta(\hat{\omega}(x), \hat{\omega}(y))$, $B_2 = \zeta(\hat{\omega}(x), \hat{\omega}(z))$, $A_3 = \zeta(\hat{\omega}^2(x), \hat{\omega}^2(y))$, $B_3 = \zeta(\hat{\omega}^2(x), \hat{\omega}^2(z))$. Numbers A_i, B_i are in bold face whenever $A_i \neq B_i$

These cases are listed in Table 3.1. If $\gamma_i(x, y) \neq \gamma_i(x, z)$ then the result of the functions are marked in bold face. We find in each line bold numbers, which tells that the claim holds. □

Proposition 3.5.11 *The group $\mathcal{G} =< \tau, \gamma_0, \gamma_1, \gamma_2 >$ acts expansively on $\mathfrak{Y}_3$.*

Proof Let $u \neq v$. Then there is at least one cell $(i_0, j_0) \in \mathbb{N}_0 \times \mathbb{Z}$ such that $u(i_0, j_0) \neq u(i_0, j_0)$. Without restriction, let i_0 be minimal, i.e., $u(i, j) = v(i, j)$ for $i = 0, \ldots, i_0 - 1$, $j \in \mathbb{Z}$. According to Proposition 3.5.10, we find an element $\eta \in< \gamma_0, \gamma_1, \gamma_2 >$, such that $\eta(u)(0, j_0) \neq \eta(v)(0, j_0)$. Applying τ often enough (τ^{-j_0} suffices), we find

$$\tau^{-j_0} \circ \eta(u)(0, 0) \neq \tau^{-j_0} \circ \eta(v)(0, 0).$$

Thus,

$$d(\tau^{-j_0} \circ \eta(u), \tau^{-j_0} \circ \eta(v)) = 1;$$

the group $\mathcal{G}$ acts expansively on $\mathfrak{Y}_3$. □

At this point, it is useful to review our construction done so far:

- We started from $f \in C(\mathfrak{C}, \mathfrak{C})$, where $\mathfrak{C}$ is a metric Cantor space.
- We represented $\mathfrak{C}$ by Σ_3, i.e., we used a homeomorphism $\varphi_1 : \mathfrak{C} \to \Sigma_3$ and defined $f_1 : \Sigma_3 \to \Sigma_3$ by $f_1 = \varphi_1 \circ f \circ \varphi_1^{-1}$. Then, $\varphi_1 \circ f = f_1 \circ \varphi_1$.
- We represented f_1 by a subshift, i.e., we defined $\varphi_2 : \Sigma_3 \to \Sigma_3^{\mathbb{N}_0} = \mathfrak{Y}_3$, $u \mapsto (f^i(u))_{i\in\mathbb{N}_0}$. Then, $\varphi_2 \circ f_1 = \sigma_{\mathfrak{Y}_3} \circ \varphi_2$.
- We found a finitely generated group acting continuously and expansively on $\mathfrak{Y}_3$. The group commutes with $\sigma_{\mathfrak{Y}_3}$.

We are now in the situation to conclude by Corollary 3.4.8 that there is a cellular automaton with alphabet $E = \{0, 1, 2\}$ and grid $\Gamma = \Gamma(G, \{\tau, \gamma_0, \gamma_1, \gamma_2\})$ that is able to simulate $(\sigma_{\mathfrak{Y}_3}.\mathfrak{Y}_3)$, i.e., $(\sigma_{\mathfrak{Y}_3}.\mathfrak{Y}_3)$ is weakly embedded into this cellular automaton. We obtain the central theorem of this section.

Theorem 3.5.12 *Consider* $(\mathfrak{C},f)$ *where* $\mathfrak{C}$ *is a metric Cantor space and* f *a continuous function on* $\mathfrak{C}$. *Then there is a cellular automaton* (Γ, D_0, E, g_0) *with group*

$$G = < \tau, \omega, \gamma \, : \, \tau \circ \omega = \omega \circ \tau, \;\; \tau \circ \gamma = \gamma \circ \tau, \;\; \omega^3 = e \} >,$$

Cayley graph $\Gamma = \Gamma(G, \{\tau, \omega, \gamma\})$, *and elementary states* $E = \{0, 1, 2\}$, *such that* $(\mathfrak{C},f)$ *can be weakly embedded into the cellular automaton.*

Remark 3.5.13

(1) Figure 3.6 represents the Cayley graph of the non-Abelian part of the group.
(2) Note that in general $\mathfrak{Y}$ is not invariant under the translation operators of the grid Γ.
(3) The function g_0 does not depend on f. The function f is coded in the state. The cellular automaton constructed here is a "universal cellular automaton" in the sense that it is able to simulate any cellular automaton. In order to simulate a given cellular automaton, the initial state has to be chosen appropriately. With respect to this property, it is what a "universal Turing machine" is for the class of all Turing machines. An universal cellular automaton is also computational universal (i.e., is able to simulate any Turing machine), since there are computational universal cellular automaton which can be simulated by the universal cellular automaton. Conversely, no Turing machine is not able to simulate the universal cellular automaton (in the sense that the Turing machine computes the next global state of the cellular automaton within a finite number of steps) as the grid of the cellular automaton is unbounded.
(4) The structure of the Cayley graph indicates several obstacles to further mathematical exploration: the graph $\mathbb{Z}^2$ is a subgraph of Γ. In general, injectivity and surjectivity of the global function are not decidable (see Sect. 9.3). Furthermore, the grid has an exponential growth function; in general the claims of Garden-of-Eden theorems need not hold (see Sect. 9.1).

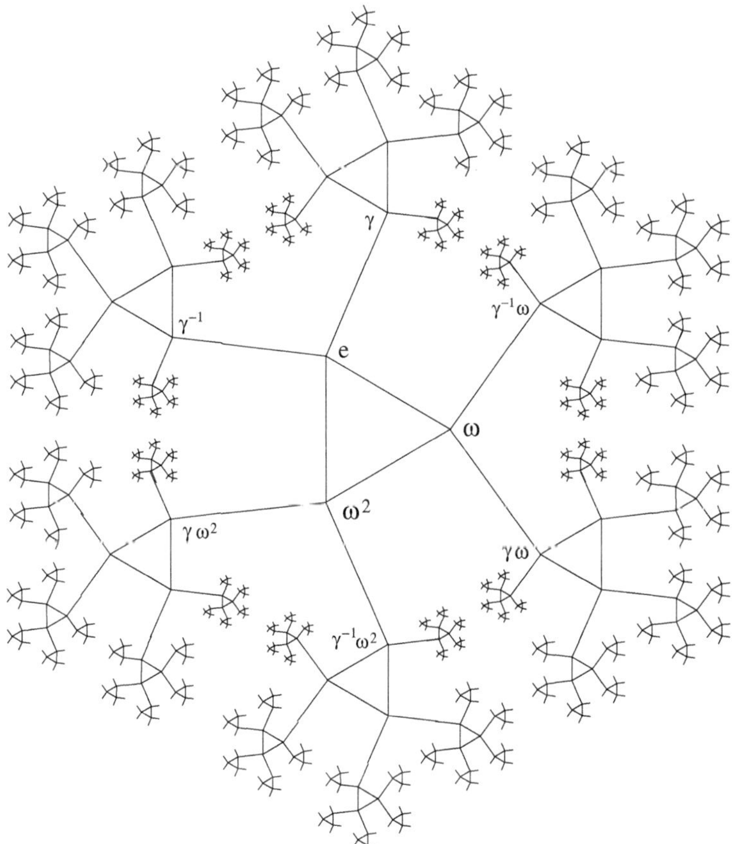

Fig. 3.6 Cayley graph of $< \gamma, \omega \ : \ \omega^3 = e >$

Chapter 4
Besicovitch and Weyl Topologies

In the last sections we realized that the Cantor metric and cellular automata fit very well together: any cellular automaton can be viewed as a continuous map of a metric Cantor space and any continuous function on a metric Cantor space can be embedded into a cellular automaton. However, there is one major drawback: the Cantor metric is not translational invariant. The metric is focused on a region near the origin and everything far away is neglected. The amount of information that is neglected may be tremendously large, as "everything else" is an infinite region, while the "near region" is finite. A perturbation of a state is considered small if it agrees with the state on a (large, but finite) region around the origin. The states at cells in the remaining, infinite part of the grid may be arbitrary. The intuitive idea of a "small perturbation" is not met by the concept of the Cantor metric.

The Besicovitch and the Weyl topologies for cellular automata assume a more global point of view. These topologies have been introduced for spaces of quasi-periodic functions [13]. Only in 1997 it became clear that it is possible to use them for one-dimensional cellular automata [24], later also for cellular automata on Abelian groups [23] and finitely generated groups [22]. The underlying concept aims at the probability that the local state of a randomly chosen cell will be different for two given states.

4.1 Definition of the Besicovitch and Weyl Space

The basic problem with the idea of Besicovitch and Weyl arises from the fact that there is no way to choose one site randomly (in the sense of a uniform distribution) from an infinite number of cells. Besicovitch and Weyl proposed slightly different solutions to this problem. In both approaches one uses restrictions to finite subgraphs. On these, the probability can be estimated by the relative proportion of numbers of differently occupied cells, i.e., by the Hamming distance

K.-P. Hadeler, J. Müller, *Cellular Automata: Analysis and Applications*,
Springer Monographs in Mathematics, DOI 10.1007/978-3-319-53043-7_4

between the patterns scaled with the inverse of the size of the subgraph. In order to extend the distance to the whole grid, two limits are considered: the supremum of the probability is determined when the subgraph is shifted to all locations of the grid, and the size of the subgraph is taken to infinity. The Besicovitch and Weyl approaches differ in the order in which these limits are taken.

Definition 4.1.1 Let $\Gamma = \Gamma(G, \sigma)$ be a Cayley graph, E a set of elementary states and $u_1, u_2 \in E^\Gamma$. The Hamming distance on a finite set $M \subset \Gamma$ is defined as

$$d_H(u_1, u_2; M) = |\{g \in M : u_1(g) \neq u_2(g)\}|.$$

The Besicovitch pseudometric is defined as

$$d_{\hat{B}}(u_1, u_2) = \limsup_{m \to \infty} \frac{d_H(u_1, u_2; \Gamma_m)}{|\Gamma_m|}$$

and the shift invariant Besicovitch pseudometric by

$$d_B(u_1, u_2) = \sup_{g \in G} \limsup_{m \to \infty} \frac{d_H(\sigma_g \circ u_1, \sigma_g \circ u_2; \Gamma_m)}{|\Gamma_m|}.$$

The Weyl pseudometric is given by

$$d_W(u_1, u_2) = \limsup_{m \to \infty} \max_{g \in G} \frac{d_H(\sigma_g \circ u_1, \sigma_g \circ u_2; \Gamma_m)}{|\Gamma_m|}.$$

Remark 4.1.2

(1) The Weyl pseudometric and the shift invariant Besicovitch pseudometric have the shift invariance "incorporated" by definition. The Besicovitch pseudometric is still focussed at the origin, similar to the Cantor metric. Nevertheless, it is $d_{\hat{B}}(\cdot, \cdot)$ which is usually called "Besicovitch pseudometric". We will show that this pseudometric is also shift invariant if we restrict attention to well-behaved groups, e.g. Abelian grids. However, studying the shift invariant Besicovitch pseudometric is more natural, as the difference between the two approaches reduces to the order in which the limits are taken.

(2) Any finite or infinite but—in a certain sense—"small" set of cells can be neglected in the computation of the distance. The following example shows a rather surprising consequence of this fact: Let $\Gamma = \mathbb{Z}$, $E = \{0, 1\}$ and the sequence of states $\{u^i\}_{i \in \mathbb{N}}$,

$$u^i(j) = \begin{cases} 1 & \text{if} \quad |j| < i \\ 0 & \text{otherwise.} \end{cases}$$

Let furthermore $\overline{0}$, $\overline{1}$ be the states that are identically '0' and '1', respectively. Then, $d_B(\overline{0}, u^i) = 0$, and therefore

$$\lim_{i\to\infty} u^i = \overline{0} \qquad \text{w.r.t. the Besicovitch pseudometric,}$$

whereas $d(u^i, \overline{1}) = 1/(1+i)$ such that

$$\lim_{i\to\infty} u^i = \overline{1} \qquad \text{in the Cantor topology.}$$

This observation, that finite sets do not play a role, applies also to the Weyl pseudometric.

(3) We find for any $u_1, u_2 \in E^\Gamma$ that

$$d_{\hat{B}}(u_1, u_2) \leq d_B(u_1, u_2) \leq d_W(u_1, u_2).$$

The first inequality is an immediate consequence of the definition; the second can be shown as follows. Let $c = d_B(u_1, u_2)$. For every $\varepsilon > 0$ there is a g such that

$$\limsup_{m\to\infty} \frac{d_H(\sigma_g \circ u_1, \sigma_g \circ u_2; \Gamma_m)}{|\Gamma_m|} \geq c - \varepsilon/2.$$

Hence there is a sequence $\{m_i\}_{i\in\mathbb{N}}$, $m_i \to \infty$ for $i \to \infty$ such that

$$\frac{d_H(\sigma_g \circ u_1, \sigma_g \circ u_2; \Gamma_{m_i})}{|\Gamma_{m_i}|} \geq c - \varepsilon.$$

Hence, $d_W(u_1, u_2) \geq c - \varepsilon = d_B(u_1, u_2) - \varepsilon$. As ε is arbitrary, we have $d_W(u_1, u_2) \geq d_B(u_1, u_2)$.

Before we proceed, we show that the Besicovitch pseudometric is shift invariant if the underlying group is of finite or intermediate growth. We will demonstrate by an example that shift invariance does not hold in general for Cayley graphs with exponential growth.

Proposition 4.1.3 *Let the group G be of finite or intermediate growth and let τ be a generator of the group. Then, for $u, v \in E^\Gamma$,*

$$d_{\hat{B}}(\sigma_\tau \circ u, \sigma_\tau \circ v) \leq a \; d_{\hat{B}}(u, v)$$

where a is defined by $a = \limsup_{m\to\infty} \gamma(m+1)/\gamma(m)$, and $\gamma(\cdot)$ denotes the growth function of the group.

Proof As $\sigma_\tau(\Gamma_m) \subset \Gamma_{m+1}$, the Hamming distance of $\sigma_\tau \circ u$ and $\sigma_\tau \circ v$ can be estimated by that of u and v in Γ_{m+1}. Hence

$$\begin{aligned} d_{\hat{B}}(\sigma_\tau \circ u, \sigma_\tau \circ v) &= \limsup_{m\to\infty} \frac{d_H(\sigma_\tau \circ u, \sigma_\tau \circ v; \Gamma_m)}{|\Gamma_m|} \\ &\leq \limsup_{m\to\infty} \frac{|\Gamma_{m+1}|}{|\Gamma_m|} \frac{d_H(u, v; \Gamma_{m+1})}{|\Gamma_{m+1}|} \leq a\ d_{\hat{B}}(u, v). \end{aligned}$$

□

Since $d_{\hat{B}}(\sigma_\tau \circ u, \sigma_\tau \circ v) \leq a d_{\hat{B}}(u, v)$ and $d_{\hat{B}}(u, v) \leq a d_{\hat{B}}(\sigma_\tau \circ u, \sigma_\tau \circ v)$, an immediate consequence of the last proposition is the shift invariance in the case of $a = 1$. If there is a constant C and $n \in \mathbb{N}$, such that $\gamma(m)/m^n \to C$ for $m \to \infty$, then clearly we find $\gamma(m+1)/\gamma(m) \sim C(m+1)^n/m^n \to 1 = a$ for $m \to \infty$. Polynomial groups are good candidates for spatial structures on which the pseudodistance of $d_{\hat{B}}(\cdot, \cdot)$ is shift invariant. It is straightforward to check that this is indeed true for $\mathbb{Z}^d$, or for every Abelian group (use the structure theorem for finitely generated Abelian groups). Gromov's theorem [119, Theorem 4.2,Theorem 8.6] does not only characterize the groups of polynomial growth but gives a quantitative estimate. There are positive constants c_1, c_2 and an exponent $n > 0$, such that $c_1 m^n \leq \gamma(m) \leq c_2 m^n$. The exponent n is also called the growth order [119]. This is slightly less than we need to show shift invariance for the Besicovitch pseudometric. In a rather technical paper, Pansu [144] improved this result. He was able to show that $\gamma(m)/m^n$ asymptotically tends to a constant for $m \to \infty$ [119, p. 49] together with [119, p. 86], and [144]. From Proposition 4.1.3 and Pansu's result, we obtain the following corollary.

Corollary 4.1.4 *The Besicovitch pseudometric $d_{\hat{B}}(\cdot, \cdot)$ is shift invariant on groups of polynomial growth.*

The following example indicates that the Besicovitch distance is in general not invariant on graphs with exponential growth.

Example 4.1.5 Consider the free group over two symbols, $G = \langle a, b \rangle$ and let $E = \{0, 1\}$. We define three different states $u_i \in E^\Gamma$ for $i = 0, 1, 2$ by (see Fig. 4.1)

$$u_i(g) = \begin{cases} 1 & \text{if 3 divides } d_c(g, e) - i \\ 0 & \text{otherwise.} \end{cases}$$

If we define $\partial\Gamma_m = \Gamma_m \setminus \Gamma_{m-1}$, then $u_i(g) = 1$ if and only if $g \in \partial\Gamma_{3m-i}$ for some $m \in \mathbb{N}$.

Let $v(g) = 0$ for all $g \in \Gamma$. We will show that $d_{\hat{B}}(u_0, v) \neq d_{\hat{B}}(\sigma_{a^{-1}} \circ u_0, \sigma_{a^{-1}} \circ v)$. It is possible to relate the state of $\sigma_{a^{-1}} \circ u_0$ on the four branches (starting with $a^\pm$ and $b^\pm$) to u_1 and u_2: Let g_1 be a (reduced) word that does not start with a^{-1} and g_2 a word that does not start with a. If we use Fig. 4.1 as a reference, then ag_1 is in the right branch, and g_2 in one of the three other (left, upper and lower) branch of the

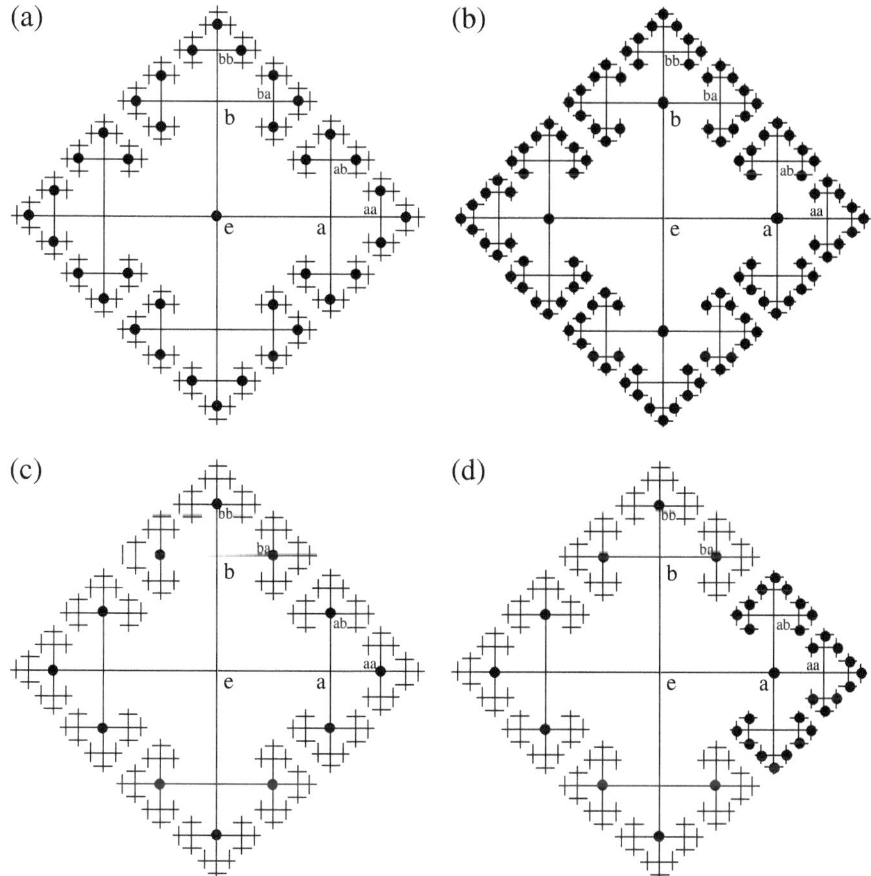

Fig. 4.1 Example 4.1.5 shows that, in general, the Besicovitch metric is not translational invariant on exponentially growing groups (in this case the free group over two symbols a and b). (**a**) u_0, (**b**) u_1, (**c**) u_2, (**d**) $\sigma_{a^{-1}} \circ u_0$. *Black dots* indicate the cells with state '1', elsewhere the state is '0'

graph. Then

$$(\sigma_{a^{-1}} \circ u_0)(ag_1) = u_0(a^{-1}ag_1) = u_0(g_1) = \begin{cases} 1 \text{ if 3 divides } d_c(g_1, e) \\ 0 \qquad \text{otherwise} \end{cases}$$

$$= \begin{cases} 1 \text{ if 3 divides } d_c(ag_1, e) - 1) \\ 0 \qquad \text{otherwise} \end{cases}$$

$$= u_1(ag_1)$$

$$(\sigma_{a^{-1}} \circ u_0)(g_2) = u_0(a^{-1}g_2) = \begin{cases} 1 \text{ if 3 divides } d_c(a^{-1}g_2, e) \\ 0 \qquad \text{otherwise} \end{cases}$$

$$= \begin{cases} 1 \text{ if 3 divides } d_c(g_2, e) + 1) \\ 0 \qquad\qquad \text{otherwise} \end{cases}$$

$$= \begin{cases} 1 \text{ if 3 divides } d_c(g_2, e)) - 2) \\ 0 \qquad\qquad \text{otherwise} \end{cases}$$

$$= u_2(g_2).$$

Thus, $\sigma_{a^{-1}} \circ u_0$ coincides with u_1 on the "right" branch, while this state coincides with u_2 on the three other branches.

In order to obtain the Besicovitch distances of u_0 and $\sigma_{a^{-1}} \circ u_0$ to v, we first derive $d_H(v, u_i; \Gamma_{3m+j})$ for $i, j \in \{0, 1, 2\}$. Let us start with $i = j = 0$. Since $\gamma(l) = 2 \cdot 3^l - 1$, we obtain

$$\begin{aligned} d_H(v, u_0; \Gamma_{3m}) &= 1 + |\partial\Gamma_3| + |\partial\Gamma_6| + \cdots + |\partial\Gamma_{3m}| \\ &= 1 + \sum_{i=1}^{m} (\gamma(3i) - \gamma(3i-1)) \\ &= 1 + \frac{4}{3} \sum_{i=0}^{m-1} 3^{3(i+1)} = 1 + \frac{18}{13}(3^{3m} - 1). \end{aligned}$$

Hence the subsequence $d_H(v, u_0; \Gamma_{3m})/|\Gamma_{3m}|$ converges for $m \to \infty$, and

$$\lim_{m\to\infty} \frac{d_H(v, u_0; \Gamma_{3m})}{\gamma(3m)} = \lim_{m\to\infty} \frac{1 + \frac{18}{13}(3^{3m} - 1)}{2 \cdot 3^{3m} - 1} = \frac{9}{13}.$$

Note, that $d_H(v, u_0; \Gamma_{3m}) = d_H(v, u_0; \Gamma_{3m+1}) = d_H(v, u_0; \Gamma_{3m+2})$. Hence,

$$\lim_{m\to\infty} \frac{d_H(v, u_0; \Gamma_{3m+1})}{\gamma(3m+1)} = \lim_{m\to\infty} \frac{d_H(v, u_0; \Gamma_{3m})}{\gamma(3m+1)} = \frac{3}{13},$$

$$\lim_{m\to\infty} \frac{d_H(v, u_0; \Gamma_{3m+2})}{\gamma(3m+2)} = \lim_{m\to\infty} \frac{d_H(v, u_0; \Gamma_{3m})}{\gamma(3m+2)} = \frac{1}{13}.$$

Similar reasoning leads to Table 4.1. Note, that rows as well as columns add up to one which reflects the fact that $u_0(g) + u_1(g) + u_2(g) = 1$ for all $g \in \Gamma$.

Now we can determine the corresponding limits for $\sigma_{a^{-1}} \circ u_0$: since three branches of this state agree with u_2 and one branch agrees with u_1, we obtain

$$\lim_{m\to\infty} \frac{d_H(v, \sigma_{a^{-1}} \circ u_0; \Gamma_{3m})}{\gamma(3m)} = \frac{3}{4}\frac{1}{13} + \frac{1}{4}\frac{3}{13} = \frac{3}{26}$$

$$\lim_{m\to\infty} \frac{d_H(v, \sigma_{a^{-1}} \circ u_0; \Gamma_{3m+1})}{\gamma(3m+1)} = \frac{3}{4}\frac{9}{13} + \frac{1}{4}\frac{1}{13} = \frac{14}{26}$$

Table 4.1 The three limits of the relative Hamming distances in Example 4.1.5

	$i=0$	$i=1$	$i=2$
$\lim_{m\to\infty}\frac{d_H(v,u_i;\Gamma_{3m})}{\gamma(3m)}$	9/13	3/13	1/13
$\lim_{m\to\infty}\frac{d_H(v,u_i;\Gamma_{3m+1})}{\gamma(3m+1)}$	3/13	1/13	9/13
$\lim_{m\to\infty}\frac{d_H(v,u_i;\Gamma_{3m+2})}{\gamma(3m+2)}$	1/13	9/13	3/13

$$\lim_{m\to\infty}\frac{d_H(v,\sigma_{a^{-1}}\circ u_0;\Gamma_{3m+2})}{\gamma(3m+2)}=\frac{3}{4}\frac{3}{13}+\frac{1}{4}\frac{9}{13}=\frac{9}{26}.$$

As the Besicovitch distance uses the limes superior, we find

$$d_{\hat{B}}(u_0,v)=\frac{9}{13}=\frac{18}{26}>\frac{14}{26}=d_{\hat{B}}(\sigma_{a^{-1}}\circ u_0,\sigma_{a^{-1}}\circ v).$$

4.2 Topological Properties

Neither the Weyl nor the Besicovitch pseudometric is a metric: if two states differ only in a relatively small number of points (a finite number, say), then $d_W(\cdot,\cdot)$ as well as $d_B(\cdot,\cdot)$ and $d_{\hat{B}}(\cdot,\cdot)$ become zero. We introduce equivalence classes and consider the quotient spaces, i.e., we identify all states with distance zero [14, 24].

Definition 4.2.1 Let $u, v\in E^\Gamma$. Define the three equivalence relations

$$u\sim_{\hat{B}} v \Leftrightarrow d_{\hat{B}}(u,v)=0,\quad u\sim_B v\Leftrightarrow d_B(u,v)=0,\quad u\sim_W v\Leftrightarrow d_W(u,v)=0.$$

Denote spaces of equivalence classes by $X_{\hat{B}} := E^\Gamma/\sim_{\hat{B}}$, $X_B := E^\Gamma/\sim_B$, and $X_W := E^\Gamma/\sim_W$.

Then $(X_{\hat{B}},\ d_{\hat{B}})$, $(X_B,\ d_B)$, and $(X_W,\ d_W)$ are metric spaces. We will show below that cellular automata harmonize well with these metric spaces.

Notation In order to clarify the notation, we write $\hat{u}$ for an equivalence class, and u for a state in E^Γ. Accordingly, $u\in\hat{u}$ indicates a state $u\subset E^\Gamma$ that is element of the equivalence class $\hat{u}$. If we write X_* or $d_*(\cdot,\cdot)$, then $*\in\{\hat{B},B,W\}$. The statement can be applied to $X_{\hat{B}}$, X_B and X_W. We apply the (pseudo)-distance $d_*(\cdot,\cdot)$ to members of X_* as well as to members of E^Γ. A function on a Besicovitch or Weyl space will be denoted by $\hat{f}$, and a function on E^Γ by f. The sign "hat" is sometimes even used as an embedding operator: If $u\in E^\Gamma$ then the equivalence class that contains u is $\hat{u}\in X_*$. Similarly, the lift of a function $f: E^\Gamma\to E^\Gamma$ to X_* is denoted by $\hat{f}$ (if the function f allows a lift).

The metric of X_W dominates that of X_B which in turn dominates that of $X_{\hat{B}}$ as

$$d_{\hat{B}}(\hat{u},\hat{v}) \leq d_B(\hat{u},\hat{v}) \leq d_W(\hat{u},\hat{v}).$$

Especially, each equivalence class in X_W is a subset of one equivalence class in X_B. The maps $I : X_W \to X_B$, $\hat{u} \mapsto \hat{u}$, and $I : X_B \to X_{\hat{B}}$, $\hat{u} \mapsto \hat{u}$, are thus well defined.

Proposition 4.2.2 *The injections*

$$I : X_W \to X_B,\ \hat{u} \mapsto \hat{u},\ \textit{and}\quad I : X_B \to X_{\hat{B}},\ \hat{u} \mapsto \hat{u},$$

given by the identity, are continuous maps.

In a way, taking equivalence classes with respect to the Besicovitch or the Weyl topology parallels the transition from pointwise defined functions to Lebesgue classes. Differences on sets of measure zero can be neglected. Also a state in the Besicovitch or Weyl space is only determined up to a set with a relatively small number of cells. But beware: this "small" number may be infinite, if only the relative density of points is zero. Basically, the density of disagreeing cells should decrease sufficiently fast (resp. the number of disagreeing cells should not increase too fast) with the distance from the origin.

4.2.1 Besicovitch Spaces

In this section, we investigate the topological properties of these metric spaces. We start with completeness, connectedness and perfectness of the Besicovitch space. The following proofs have been adapted from [14].

Proposition 4.2.3 *The Besicovitch space $X_{\hat{B}}$ is complete.*

Proof For a finite grid, the Besicovitch topology is the same as the Cantor topology. We may thus assume an infinite grid.

Let $\{\hat{u}^n\}_{n\in\mathbb{N}}$ be a Cauchy sequence in $X_{\hat{B}}$. For each equivalence class $\hat{u}^n$ fix one element $u^n \in \hat{u}^n$. We can find a subsequence for which

$$d_{\hat{B}}(u^{n_i}, u^{n_{i+1}}) = \limsup_{l\to\infty} \frac{d_H(u^{n_i}, u^{n_{i+1}}; \Gamma_l)}{|\Gamma_l|} \leq 2^{-i-1}.$$

We split the graph Γ into different regions. We choose a sequence $\{l_i\}_{i\in\mathbb{N}}$, $l_i \in \mathbb{N}$, such that

(1) $l_{i+1} \geq 2l_i$
(2) $\forall l \geq l_i :\ |\{g \in \Gamma_l\ :\ u^{n_i}(g) \neq u^{n_{i+1}}(g)\}| \leq 2|\Gamma_l|2^{-i-1}$.

We define $u^{\infty} \in E^{\Gamma}$ by

$$u^{\infty}(g) = \begin{cases} u^{1}(g) & \text{for } g \in \Gamma_{l_1} \\ u^{n_i}(g) & \text{for } g \in \Gamma_{l_{i+1}} \setminus \Gamma_{l_i} \end{cases}$$

and show that $d_{\hat{B}}(u^{\infty}, u^{n_i}) \to 0$. Consider, for some $m > i$, the states u^{n_m} and u^{n_i}. For $l > l_m$ we derive an estimate for the number of disagreeing sites in Γ_l,

$$\begin{aligned} |\{g \in \Gamma_l \, : \, u^{n_i}(g) \neq u^{n_m}(g)\}| &\leq \sum_{j=i}^{m-1} |\{g \in \Gamma_l \, : \, u^{n_j}(g) \neq u^{n_{j+1}}(g)\}| \\ &\leq \sum_{j=i}^{m-1} 2\,|\Gamma_l| 2^{-j-1} < 2\,|\Gamma_l| 2^{-i}. \end{aligned}$$

According to its definition, the state u^{∞} agrees with u^{n_i} on $\Gamma_{l_{i+1}} \setminus \Gamma_{l_i}$. We use this fact in estimating the difference of u^{∞} and u_{n_i}. Assume $l > l_i$. Then there is an integer $k \in \mathbb{N}$, such that $l_k \leq l < l_{k+1}$. Then

$$\begin{aligned} &|\{g \in \Gamma_l \, : \, u^{\infty}(g) \neq u^{n_i}(g)\}| \\ &= |\{g \in \Gamma_{l_i - 1} \, : \, u^{\infty}(g) \neq u^{n_i}(g)\}| + \sum_{j=i}^{k-1} |\{g \in \Gamma_{l_{j+1}} \setminus \Gamma_{l_j} \, : \, u^{\infty}(g) \neq u^{n_i}(g)\}| \\ &\qquad + |\{g \in \Gamma_l \setminus \Gamma_{l_k} \, : \, u^{\infty}(g) \neq u^{n_i}(g)\}| \\ &= |\{g \in \Gamma_{l_i - 1} \, : \, u^{\infty}(g) \neq u^{n_i}(g)\}| + \sum_{j=i}^{k-1} |\{g \in \Gamma_{l_{j+1}} \setminus \Gamma_{l_j} \, : \, u^{n_j}(g) \neq u^{n_i}(g)\}| \\ &\qquad + |\{g \in \Gamma_l \setminus \Gamma_{l_k} \, : \, u^{n_k}(g) \neq u^{n_i}(g)\}| \\ &\leq |\Gamma_{l_i - 1}| + \sum_{j=i}^{k-1} |\{g \in \Gamma_l \, : \, u^{n_j}(g) \neq u^{n_i}(g)\}| + |\{g \in \Gamma_l \, : \, u^{n_k}(g) \neq u^{n_i}(g)\}| \\ &\leq |\Gamma_{l_i - 1}| + \sum_{j=i}^{k} |\{g \subset \Gamma_l \, : \, u^{n_i}(g) \neq u^{n_i}(g)\}| \\ &\leq |\Gamma_{l_i - 1}| + \sum_{j=i}^{k+1} 2\,|\Gamma_l| 2^{-j} \leq |\Gamma_{l_i - 1}| + |\Gamma_l| 2^{-i+2}. \end{aligned}$$

This inequality depends on l only via the factor $|\Gamma_l|$ (note that i is fixed). Thus,

$$d_{\hat{B}}(u^{\infty}, u^{n_i}) = \limsup_{l \to \infty} \frac{d_H(u^{\infty}, u^{n_i}; \Gamma_l)}{|\Gamma_l|} \leq \limsup_{l \to \infty} \left(\frac{|\Gamma_{l_i - 1}|}{|\Gamma_l|} + 2^{-i+2} \right) = 2^{-i+2}$$

which implies $d_{\hat{B}}(\hat{u}^\infty, \hat{u}^{n_i}) \leq 2^{2-i}$. The subsequence $\hat{u}^{n_i}$ converges to $\hat{u}^\infty$. As the sequence $\hat{u}^{n_i}$ is a Cauchy sequence, the convergence of a subsequence implies the convergence of the sequence itself. □

At this stage of our investigation it is not clear whether the shift invariant Besicovitch space is complete. On groups of polynomial growth, the Besicovitch and the shift invariant Besicovitch topology coincide. In this case, the shift invariant Besicovitch spaces are complete, indeed. Below we will prove that the Weyl space is not complete; as the shift invariant version of the Besicovitch distance is structural close to the Weyl distance, we may conjecture that the completeness is lost in case of exponentially growing groups.

We proceed with the study of topological properties by discussing connectivity. In case of $\Gamma = \mathbb{Z}$, a well known proof for this property relies on Toeplitz states [14] (for the definition of Toeplitz states, see below). We follow a different line of reasoning, which was made popular by Erdős in number theory: in order to prove that an object with a given property exists, one proves that the probability to pick an object with this property out of a larger set of objects is positive. This so-called probabilistic method for non-probabilistic theorems is easy-going in our case, as the underlying idea of the Besicovitch metric is close to measuring probability. However, this method may be slightly unfamiliar; we formulate the argument at length.

Proposition 4.2.4 *If the grid is infinite, $|\Gamma| = \infty$, and E has at least two different elements, then the Besicovitch spaces X_B and $X_{\hat{B}}$ are pathwise connected.*

Proof Let $a \in E$ be any elementary state and $\overline{a}$ the state with all entries equal to a. Let $\hat{u} \in X_{\hat{B}}$ be arbitrary. We show, that there is a continuous path from $\hat{u}$ to $\widehat{\overline{a}}$. First we construct a path between $u \in \hat{u}$ and $\overline{a}$.

Step 1: Construction of a path.

Let $\{U_g\}_{g\in\Gamma}$ be a family of independent and identically distributed random variables defined on a probability space $(\Omega, \mathcal{F}, P)$. Let U_g be uniformly distributed between zero and one, i.e., $P(U_g \leq \alpha) = \alpha$ for $\alpha \in [0, 1]$ and $g \in \Gamma$. For our argument below it is useful to recall that this is a short-hand notation for $U_g : \Omega \to [0, 1]$, and $P(\{\omega \in \Omega : U_g(\omega) \leq \alpha\}) = \alpha$ for $\alpha \in [0, 1]$.

Let $\tilde{u}_g = U_g(\omega)$ for a given $\omega \in \Omega$ be one realization of these random variables. By means of $\tilde{u}_g(\omega)$ we assign a number between zero and one to each grid point. Now we define a family of (random) states parametrized by $\alpha \in [0, 1]$,

$$v^\alpha(g) = v^\alpha(g, \omega) = \begin{cases} a & \text{for} \quad \tilde{u}_g(\omega) \leq \alpha \\ u(g) & \text{otherwise.} \end{cases}$$

Our candidate for a continuous path between u and $\bar{a}$ is a realization of the random map (depending on $\omega \in \Omega$)

$$\tau : [0, 1] \to E^\Gamma, \quad \alpha \mapsto v^\alpha = v^\alpha(\omega).$$

Step 2: $d_{\hat{B}}(v^{\alpha_1}(\omega), v^{\alpha_2}(\omega)) \leq |\alpha_2 - \alpha_1|$ for $\alpha_1, \alpha_2 \in [0,1] \cap \mathbb{Q}$ and $\omega \in \tilde{\Omega} \subset \Omega$, $P(\tilde{\Omega}) = 1$.

Consider $0 \leq \alpha_1 < \alpha_2 \leq 1$ and any $h \in \Gamma$. As the random variables U_g are i.i.d., and the grid is infinite, we find

$$d_{\hat{B}}(v^{\alpha_1}, v^{\alpha_2}) = \limsup_{l\to\infty} \frac{1}{|\Gamma_l|} d_H(v^{\alpha_1}, v^{\alpha_2}; \Gamma_l) \leq P(U_h \in [\alpha_1, \alpha_2)) = \alpha_2 - \alpha_1 \quad \text{a.s.}$$

Furthermore, v^0 is a.s. the state u and v^1 the state $\bar{a}$.

Note at this point that the states u^α are spatially homogeneous; we also find for any $g \in G$ that $d_{\hat{B}}(\sigma_g \circ v^{\alpha_1}, \sigma_g \circ v^{\alpha_2}) \leq P(U_h \in [\alpha_1, \alpha_2)) = \alpha_2 - \alpha_1$ a.s., and therefore

$$d_B(v^{\alpha_1}, v^{\alpha_2}) \leq P(U_h \in [\alpha_1, \alpha_2)) = \alpha_2 - \alpha_1 \quad \text{a.s.}$$

Since $d_B(\cdot,\cdot)$ is stronger than $d_{\hat{B}}(\cdot,\cdot)$, we focus in the following on X_B. We construct a set $\tilde{\Omega} \subset \Omega$ such that the conditions, which are only "a.s." so far in Ω, are given for all $\omega \in \tilde{\Omega}$. However, we need to be careful: we must not impose uncountably many conditions. Fix $\alpha_1, \alpha_2 \in \mathbb{Q} \cap [0,1]$. Then, there is a set $\tilde{\Omega}_{\alpha_1,\alpha_2} \subset \Omega$ of full measure such that $d_B(v^{\alpha_1}, v^{\alpha_2}) \leq |\alpha_1 - \alpha_2|$ is true. This inequality is simultaneously true for all rational α_1, α_2 if $\omega \in \bigcap_{\alpha_1,\alpha_2\in\mathbb{Q}} \tilde{\Omega}_{\alpha_1,\alpha_2}$. The countable intersection of sets of full measure (see Lemma A.3.19), $\bigcap_{\alpha_1,\alpha_2\in\mathbb{Q}} \tilde{\Omega}_{\alpha_1,\alpha_2}$ is a set of full measure. Of course, we may also include the conditions $v^0 = u$ and $v^1 = \bar{a}$ (that are also true a.s.). All in all, we find a set $\tilde{\Omega} \subset \Omega$ of full measure, such that

$$v^0(\omega) = \hat{a}, \quad v^1(\omega) = u, \quad \forall \alpha_1, \alpha_2 \in [0,1] \cap \mathbb{Q}: \; d_B(v^{\alpha_1}, v^{\alpha_2}) \leq |\alpha_1 - \alpha_2|$$

holds. In particular, $\tilde{\Omega} \neq \emptyset$.

Step 3: *Continuity of* $\hat{\tau} : [0,1] \to X_B, \alpha \mapsto \widehat{v^\alpha(\omega)}$.

We show that for realizations $\omega \in \tilde{\Omega}$, the inequality $d_B(v^{\alpha_1}, v^{\alpha_2}) \leq |\alpha_1 - \alpha_2|$ is true not only for rational values in α_1, α_2, but for all values. Assume that this inequality does not hold for $\omega_0 \in \tilde{\Omega}$, and $\alpha_1, \alpha_2 \in [0,1]$. Then there is $\varepsilon > 0$ such that

$$d_B(v^{\alpha_1}(\omega_0), v^{\alpha_2}(\omega_0)) \geq |\alpha_1 - \alpha_2| + \varepsilon.$$

Without loss of generality $\alpha_1 < \alpha_2$. We choose $\tilde{\alpha}^1, \tilde{\alpha}^2$ in $[0,1] \cap \mathbb{Q}$ such that

$$0 \leq \tilde{\alpha}_1 \leq \alpha_1 < \alpha_2 < \tilde{\alpha}_2 < 1, \qquad |\tilde{\alpha}_2 - \tilde{\alpha}_1| \leq |\alpha_2 - \alpha_1| + \varepsilon/2$$

The monotonicity of the construction implies that

$$d_B(v^{\tilde{\alpha}_1}(\omega_0), v^{\tilde{\alpha}_1}(\omega_0)) \geq d_B(v^{\alpha_1}(\omega_0), v^{\alpha_1}(\omega_0))$$

and hence

$$|\alpha_2 - \alpha_1| + \varepsilon/2 \geq |\tilde{\alpha}_2 - \tilde{\alpha}_1| \geq d_B(v^{\tilde{\alpha}_1}(\omega_0), v^{\tilde{\alpha}_1}(\omega_0))$$
$$\geq d_B(v^{\alpha_1}(\omega_0), v^{\alpha_1}(\omega_0)) \geq |\alpha_1 - \alpha_2| + \varepsilon.$$

This is a contradiction. The inequality is true on $[0, 1]$ and thus the map $\hat{\tau}$ is a Lipschitz-continuous path from $\widehat{\overline{a}}$ to $\hat{u}$ for all $\omega \in \tilde{\Omega}$.
Since the topology of X_B is stronger than that of $X_{\hat{B}}$, this result implies that there is also a continuous path in $X_{\hat{B}}$.

□

It is possible to approximate all elements $\hat{u} \in X_B$ by other elements $\hat{\tau}(\alpha)$ arbitrarily well for $\alpha \to 1$. Hence, $\overline{X_{\hat{B}} \setminus \{\hat{u}\}}^{d_{\hat{B}}} = X_{\hat{B}}$ and $X_{\hat{B}}$ is perfect. Also all elements in X_B can be approximated by other elements in X_B.

Proposition 4.2.5 *If the grid is infinite, $|\Gamma| = \infty$, and E consists of at least two different elements, then the Besicovitch spaces $X_{\hat{B}}$ and X_B are perfect.*

Next we consider the topological dimension. We have seen that the Cantor space is totally disconnected. Hence there is no injective, continuous map $[0, 1] \to (E^\Gamma, d)$. For Besicovitch spaces, we know that there is such a map, a one-dimensional interval can be embedded continuously into X_B. This embedding leads to the topological concept of dimension.

Definition 4.2.6 Let $(\mathfrak{X}, \mathcal{T})$ be a topological space. The topological dimension of $(\mathfrak{X}, \mathcal{T})$ is $n \in \mathbb{N}$, if there is a continuous, injective map from $[0, 1]^n$ into this topological space, but not from $[0, 1]^{n+1}$. If there is no such map for $n = 1$, then we call the topological space zero dimensional. If there is such a map for all $n \in \mathbb{N}$, we call the space infinite dimensional.

Remark 4.2.7 The topological dimension of the Cantor space is zero. Furthermore, we know that the topological dimension of the Besicovitch space is at least one.

Proposition 4.2.8 *Let the grid be infinite, $|\Gamma| = \infty$, and let E consist of at least two elements. Then the topological dimension of the Besicovitch as well as that of the shift invariant Besicovitch space is infinite.*

Proof As the arguments for the shift invariant Besicovitch space X_B and the Besicovitch space $X_{\hat{B}}$ are parallel, it is sufficient to consider $X_{\hat{B}}$. Choose $n \in \mathbb{N}$. We first split the grid points of Γ into n different, uniformly distributed sets: Let $\{W_g\}_{g\in\Gamma}$ a family of i.i.d. distributed random variables that assume all values in $\{1, \ldots, n\}$ with equal probability $1/n$. We construct a map from $[0, 1]^n$ into X_B. The grid sites with $W_g = i$ will code the ith component of the n-dimensional interval.

We now repeat the proof for the pathwise connectivity: In each cell $g \in \Gamma$, we place a random variable U_g. These random variables are i.i.d. and uniformly distributed on the interval $[0, 1]$. Let $e_1, e_2 \in E$. Consider the random map

$$\hat{\tau} : [0, 1]^n \to X_{\hat{B}}, \quad \alpha \mapsto \widehat{u^\alpha}$$

where $\alpha = (\alpha_1, \ldots, \alpha_n)$ is a multi-index and $\widehat{u^\alpha}$ is the equivalence class of

$$u^\alpha(g) = \begin{cases} e_1 & \text{for} \quad u(g) \leq \alpha_{W_g} \\ e_2 & \text{otherwise.} \end{cases}$$

Then, for $\alpha^i = (\alpha^i_1, \ldots, \alpha^i_n)$, $i = 1, 2$, we find

$$d_{\hat{B}}(\widehat{u^{\alpha^1}}, \widehat{u^{\alpha^2}}) \leq \frac{1}{n} \sum_{i=1}^{n} |\alpha^1_i - \alpha^2_i|$$

a.s. With similar arguments as above find that injective and continuous maps from $[0,1]^n$ in $X_{\hat{B}}$ exist. □

X_B is neither separable (i.e., there is no countable basis for the topology) nor locally compact [14]. We do not prove these facts.

Next we wish to find a small subset of states that is dense in X_B. One could think of the spatially periodic states. However, this set is too small.

Remark 4.2.9 There is a lift $\hat{f} : X_* \to X_*$ of a function $f : E^\Gamma \to E^\Gamma$ if and only if $d_*(u, v) = 0$ implies $d_*(f(u), f(v)) = 0$, i.e., if the function f respects the equivalence classes. As $d_H(\sigma_g \circ u, \sigma_g \circ v; \Gamma_m) \leq d_H(u, v; \Gamma_{m+d_c(e,g)})$, we find $d_*(\sigma_g u, \sigma_g v) \leq C d_*(u, v)$ (see also Proposition 4.3.1), and the shift operator possesses a continuous lift $\hat{\sigma}_g$ to X_*.

In Definition 2.3.3 we had defined periodic states on E^Γ. Now we introduce a similar notion for X_*.

Definition 4.2.10 Let G be an infinite, finitely generated group. Periodic states are related to subgroups H of G that have a finite index $[G : H] < \infty$. Let $\mathcal{H}$ be the set of all subgroups with finite index. The set $\mathcal{P}$ consist of the periodic states in the Cantor sense

$$\mathcal{P} = \{\hat{u} \in E^\Gamma : \exists H \in \mathcal{H} : \forall h \in H : d(\sigma_h \circ u, u) = 0\},$$

and the set $\hat{\mathcal{P}}$ of the periodic states in the X_* sense

$$\hat{\mathcal{P}} = \{\hat{u} \in X_* : \exists u \in \hat{u} : \ u \in \mathcal{P}\}.$$

Let furthermore $\check{\mathcal{P}}$ be the set of weakly periodic states in the X_* sense

$$\check{\mathcal{P}} = \{\hat{u} \in X_* : \exists H \in \mathcal{H} : \forall h \in H : d_*(\hat{\sigma}_h \circ \hat{u}, \hat{u}) = 0\}.$$

Remark 4.2.11 In general, the set $\hat{\mathcal{P}}$ is a proper subset of $\check{\mathcal{P}}$: take $\Gamma = \mathbb{Z}$, $E = \{0, 1\}$ and define $u(g) = 0$ for $g < 0$ and $u(g) = 1$ for $g \geq 1$. Then, $\hat{u} \in \check{\mathcal{P}} \setminus \hat{\mathcal{P}}$.

Proposition 4.2.12 *Let Γ be a Cayley graph of a finitely generated group, $|\Gamma| = \infty$, and E a finite alphabet of at least two elements. The set of periodic states $\check{\mathcal{P}} = \{\hat{u} \in X_B : \exists g \in \Gamma \setminus \{e\} : d_B(\hat{u}, \hat{\sigma}_g \circ \hat{u}) = 0\}$ is not dense in X_B.*

Proof Let $e_1, e_2 \in E$, $e_1 \neq e_2$, and let u be the state defined with i.i.d. Bernoulli random variables, placed in all grid points $g \in \Gamma$: These random variables assume e_1 with probability $1/2$, and e_2 with the same probability. For any $g_1, g \in \Gamma, g \neq e$, we find

$$d_{\hat{B}}(\sigma_{g_1} \circ u, \sigma_{g_1} \circ \sigma_g \circ u) = \frac{1}{2} \qquad \text{a.s.}$$

As Γ is countable, there is a set of realizations of full measure such that

$$d_{\hat{B}}(\sigma_{g_1} \circ u, \sigma_{g_1} \circ \sigma_g \circ u) = \frac{1}{2}$$

for all $g_1 \in \Gamma$ and thus

$$d_B(u, \sigma_g \circ u) = \frac{1}{2}.$$

Let u_0 be one of the states that we just constructed, i.e., one generic realization of the random state. Let $v \in \check{P}$. Using $d_B(v, \sigma_g \circ v) = 0$ for $g \in \Gamma \setminus \{e\}$ appropriately chosen, we find

$$\begin{aligned} \frac{1}{2} &= d_B(u_0, \sigma_g \circ u_0) \leq d_B(u_0, v) + d_B(v, \sigma_g \circ u_0) \\ &\leq d_B(u_0, v) + d_B(v, \sigma_g \circ v) + d_B(\sigma_g \circ v, \sigma_g \circ u_0) = 2 d_B(u_0, v). \end{aligned}$$

Hence,

$$\inf_{v \in \check{\mathcal{P}}} d_B(u_0, v) \geq \frac{1}{4}$$

such that u_0 is not an element of the closure of the periodic states. □

For one dimensional automata, $\Gamma = \mathbb{Z}$, the notion of a Toeplitz state generalizes the notion of a periodic state. Furthermore, Toeplitz states are dense in X_B.

Definition 4.2.13 Let $\Gamma = \mathbb{Z}$ and let E be a finite alphabet. A state $u \in E^\Gamma$ is called a Toeplitz state, if for every $i \in \mathbb{Z}$ there is a $p_i \in \mathbb{N}$ such that

$$u(i + p_i k) = u(i) \qquad \forall k \in \mathbb{Z}.$$

Let $T \subset E^\Gamma$ denote the set of all Toeplitz states, and $\hat{T} \subset X_*$ the corresponding set of equivalence classes,

$$\hat{T} = \{\hat{t} : \exists t \in T : t \in \hat{t}\}.$$

Proposition 4.2.14 *Let $\Gamma = \mathbb{Z}$, E a finite alphabet of at least two elements, and let $\hat{T} \subset X_B$ be the set of equivalence classes that contain at least one Toeplitz state. Then,*

(1) each equivalence class $\hat{t}$ in $\hat{T} \subset X_*$ contains exactly one Toeplitz state.
(2) $\hat{T}$ is dense in X_B and $X_{\hat{B}}$.

Proof

1) Let $u, v \in T$ and $u \neq v$. There is a cell $i \in \mathbb{Z}$ such that $u(i) \neq v(i)$. The cell $i \in \mathbb{Z}$ has a period p_1 in state u, and period p_2 in state v, i.e., $u(i + p_1 k) = u(i)$ and $v(i + p_2 k) = v(i)$ for all $k \in \mathbb{Z}$. Let $p = p_1 p_2$ be the product of the two numbers, then

$$u(i + pk) \neq v(i + pk) \qquad \forall k \in \mathbb{Z}$$

for all $k \in \mathbb{Z}$. Hence, $d_W(u, v) \geq 1/p$, and thus also $d_B(u, v) \geq 1/p$ (and also $d_{\hat{B}}(u, v) \geq 1/p$). Therefore, u and v cannot belong to the same equivalence class in X_*, where X_* may be any of the spaces B, $\hat{B}$, or W.

2) Since $\mathbb{Z}$ is a group of polynomial growth, we have $d_{\hat{B}}(u, v) = d_B(u, v)$. It is sufficient to prove the desired inequality for $d_{\hat{B}}(u, v)$.

Let $u \in E^\Gamma$. For a given $\varepsilon > 0$, we construct $v \in T$ such that $d_B(u, v) < \varepsilon$. As a prerequisite we choose a sequence $\{n_k\}_{k \in \mathbb{N}} \subset \mathbb{N}$, such that n_k is a divisor of n_l if $k < l$, and

$$\sum_{i=1}^{\infty} \frac{1}{n_i} < \varepsilon.$$

We can choose e.g. a subsequence of $n_i = i!$.

Step 1: Construction of $\quad v \in T$
We construct v in a recursive way. First, define for $l = 1$ that

$$v(k\, n_l) = u(0) \qquad \forall k \in \mathbb{Z}.$$

Next consider $l = 2, 3, \ldots$.

(a) if $v(l)$ is not already defined, define $v(l + k\, n_{2l}) = u(l)$ for all $k \in \mathbb{Z}$, i.e., especially $v(l) = u(l)$.
(b) if $v(-l)$ is not already defined (even not by (a)), define $v(-l + k\, n_{2l+1}) = u(-l)$ for all $k \in \mathbb{Z}$, especially $v(-l) = u(-l)$.
This construction is well defined, i.e., no site is assigned to a value twice. Furthermore, due to its construction, the state is Toeplitz.

Step 2: $\quad d_B(u, v) < \varepsilon$
We aim at an upper bound for $d_H(u, v; \Gamma_m)$. In which cells do u and v *not* differ? If we consider step l, then we know that $u(\pm l) = v(\pm l)$, if $v(\pm l)$ is defined in

this step. In step one, at most $\lfloor (2m+1)/n_1 \rfloor + 1$ cells have been defined, and at least one of these correctly ($v(0) = u(0)$). Thus, the first step contributes to the Hamming distance with at most $\lfloor (2m+1)/n_1 \rfloor$ incorrect matches. Similarly, step two contributes with at most $\lfloor (2m+1)/n_2 \rfloor$ incorrect matches. In this way, we find that

$$d_H(u, v; \Gamma_m) \leq \sum_{l=1}^{\infty} \left\lfloor \frac{2m+1}{n_l} \right\rfloor \leq \sum_{l=1}^{\infty} \frac{2m+1}{n_l} = (2m+1) \sum_{l=1}^{\infty} \frac{1}{n_l} \leq (2m+1)\varepsilon.$$

Hence,

$$d_{\hat{B}}(u, v) \leq \varepsilon.$$

□

4.2.2 Weyl Spaces

In this section, we focus on Weyl spaces. Many topological properties that are true for Besicovitch spaces, hold also for Weyl spaces. This fact is a consequence of the similar construction principle for both spaces. We find the following properties:

Proposition 4.2.15 *Let $\Gamma = \mathbb{Z}$, and let E be a finite alphabet with at least two elements. Then, X_W is*

(1) pathwise connected
(2) topologically infinite dimensional
(3) perfect
(4) not separable
(5) not locally compact.

Furthermore, one can show that Toeplitz sequences are *not* dense in X_W. The proof for this statement can be found in [14].

There is one striking difference between Besicovitch spaces and Weyl spaces: in general, Weyl spaces are not complete. This fact can be explained as a sequence converging e.g. to zero w.r.t. the Weyl topology has to be spatially much more uniform than a sequence converging to zero w.r.t. the Besicovitch topology. This incompleteness is the only property of Weyl spaces that we discuss here.

The idea of the proof is the following [46]: We construct a certain function on E^Γ, the number of msi-sets (these are defined below) in the closure of the orbit. First we show that this function is lower semi-continuous. Then we construct a Cauchy sequence and show, that its convergence would contradict semi-continuity. As a consequence the Weyl space is not complete.

Definition 4.2.16 Let $\Gamma = \mathbb{Z}$ and $U \subset E^\Gamma$. We call U shift-invariant, if $U \neq \emptyset$, U is closed (w.r.t. the Cantor topology) and if $\sigma(U) \subset U$. A shift-invariant set that does not contain a proper subset that is also shift-invariant is called a minimal shift-invariant set (msi-set).

Let $u \in E^\Gamma$ and

$$\overline{O(u)}^d = \overline{\{\sigma^i(u) \,:\, i \in \mathbb{Z}\}}^d$$

the closure of the trajectory of u under the shift operator with respect to the Cantor topology. Let $m = m(u)$ the number of msi-sets contained in $\overline{O(u)}^d$ if this number is finite, and $m(u) = \infty$ otherwise. The relation $u \mapsto m(u)$ defines a map $m : E^\Gamma \to \mathbb{N} \cup \{\infty\}$.

Remark 4.2.17 Recall Zorn's lemma: Consider a partially ordered set Ω. If any linear chain, that is, a totally ordered subset of Ω, has a lower bound in Ω, then Ω contains at least one minimal element.

Using Zorn's lemma we show that any shift invariant set U contains an msi. Let $\mathcal{M}(U)$ be the family of all shift invariant subsets of U. As $U \in \mathcal{M}(U)$, $\mathcal{M}(U) \neq \emptyset$. The inclusion $\subset$ defines a semi-order in $\mathcal{M}(U)$. Any minimal element with respect to this semi-order is an msi. Let $\mathcal{V}$ be a linear chain in $\mathcal{M}(U)$. As U is closed and $E^\mathbb{Z}$ is compact (in the Cantor topology), all sets in $\mathcal{V}$ are compact, and $V = \cap_{W \in \mathcal{V}} W \neq \emptyset$ is compact and invariant and a lower bound for $\mathcal{V}$ in $\mathcal{M}(U)$. By Zorn's lemma there is a minimal element in $\mathcal{M}(U)$. Therefore, $m(u) \geq 1$.

The next proposition gives a useful criterion to determine $m(u)$. Our construction is based on the idea that an orbit will stay most of the time in a set of "islands". The number of islands is an estimate for the number $m(u)$.

Definition 4.2.18 A set $J \subset \mathbb{Z}$ is called syndetic, if there is a constant $L \in \mathbb{N}$ such that

$$\forall k \in \mathbb{Z} : \{k, k+1, \ldots, k+L\} \cap J \neq \emptyset.$$

Proposition 4.2.19 *Let $u \in E^\mathbb{Z}$. Then, $m(u) \leq m_0$ if and only if there are points $z_1, \ldots, z_{m_0} \in E^\mathbb{Z}$, such that for all $\varepsilon > 0$ the set*

$$J^\varepsilon(z_1, \ldots, z_{m_0}) = \{j \in \mathbb{Z} \,:\, \min_{i=1,\ldots,m_0} \{d(\sigma^j(u), z_i)\} < \varepsilon\}$$

is syndetic.

Proof "$\Rightarrow$" Let $m(u) \leq m_0$. We show that there are $z_1, \ldots, z_{m_0}$ such that $J^\varepsilon(z_1, \ldots, z_{m_0})$ is syndetic.

There are precisely $m = m(u)$ msi-sets $\mathcal{S}_1, \ldots, \mathcal{S}_m$ contained in $\overline{O(u)}^d$. Let $z_i \in \mathcal{S}_i$, arbitrary, fixed. For all $\varepsilon > 0$, the set $J^\varepsilon(z_1, \ldots, z_m)$ is syndetic: assume that this is not the case. Then, there is $\varepsilon > 0$ such that for every $n \in \mathbb{N}$, we find a number $k_n \in \mathbb{Z}$ such that

$$J^\varepsilon(z_1, \ldots, z_m) \cap [k_n - n, k_n + n] = \emptyset.$$

Since E^Γ is compact in the Cantor topology, there is a convergent subsequence of $\{\sigma^{k_n}(u)\}_{n\in\mathbb{N}}$. The limit of this subsequence is denoted by v. As $d(\sigma^{k_n}(u), z_i) \geq \varepsilon$, also $d(v, z_i) \geq \varepsilon$ for $i = 1, \ldots, m$. We show that this inequality holds also for shifted states, $d(\sigma^l(v), z_i) \geq \varepsilon$ for $i = 1, \ldots, m$ and $l \in \mathbb{Z}$. Assume that this is not the case. Then, there is $l_0 \in \mathbb{Z}$ and $i_0 \in \{1, \ldots, m\}$ such that

$$d(\sigma^{l_0}(v), z_{i_0}) < \varepsilon.$$

Thus, the ball V with radius $\delta = (\varepsilon - d(\sigma^{l_0}(v), z_{i_0}))/2$ centered at $\sigma^{l_0}(v)$ is completely contained in the ball with radius ε, centered at z_{i_0}. The set $\sigma^{-l_0}(V)$ is a neighborhood of v, and hence contains an infinite number of elements of the sequence $\{\sigma^{k_n}(u)\}_{n\in\mathbb{N}}$. For every $N > 0$, there is $n > N$ such that

$$\sigma^{l_0}(\sigma^{k_n}(u)) \in V,$$

i.e., $k_n + l_0 \in J^\varepsilon(z_1, \ldots, z_m)$. If $n > |l_0|$, this finding contradicts the assumption $J^\varepsilon(z_1, \ldots, z_m) \cap [k_n - n, k_n + n] = \emptyset$.

Hence, $d(\sigma^l(v), z_i) \geq \varepsilon$ and therefore

$$d(\overline{O(v)}^d, z_i) \geq \varepsilon.$$

As $\overline{O(v)}^d$ contains at least one msi-set that is different from $\mathcal{S}_1, \ldots, \mathcal{S}_m$, we find a contradiction to the assumption that there is no further msi-set other than $\mathcal{S}_1, \ldots, \mathcal{S}_m$. Thus, $J^\varepsilon(z_1, \ldots, z_m)$ is syndetic. Extending the points $z_1, \ldots, z_m$ by arbitrarily chosen points $z_{m+1}, \ldots, z_{m_0}$, we find a syndetic set $J^\varepsilon(z_1, \ldots, z_{m_0})$.

"$\Leftarrow$" Let $J^\varepsilon(z_1, \ldots, z_{m_0})$ be syndetic. We show that $m(u) \leq m_0$.

Let $m(u) > m_0$. Then there are least $m_0 + 1$ msi-sets $\mathcal{S}_1, \ldots, \mathcal{S}_{m_0+1}$. Let

$$\varepsilon = \min\{d(\mathcal{S}_i, \mathcal{S}_j) \,:\, i \neq j\}/4.$$

Thus, there is at least one msi-set $\mathcal{S}_{i_0}$ that has an empty intersection with the ε-balls centered at $z_1, \ldots, z_{m_0}$. As $\mathcal{S}_{i_0}$ is contained in the closed hull of the trajectory $\overline{O(u)}^d$, the iterated $\sigma^l(u)$ come arbitrary close to this set. And since this set is invariant, the continuity of the shift forces the trajectory to stay very long close to $\mathcal{S}_{i_0}$ once it comes close to $\mathcal{S}_{i_0}$. Thus, $J^\varepsilon(z_1, \ldots, z_{m_0})$ is not syndetic, in contradiction to the assumption. □

The next lemma on our way towards the incompleteness result is rather technical: we show that the Cantor distance can be controlled by the Weyl metric in a certain sense: the Weyl metric controls the density of disagreeing sites for two states in a (spatially) uniform manner. If this density is very low, then there are large regions on which the states agree. Within these large regions, the Cantor distance (with respect to some "central point" in each region) is small. The following proposition gives a lower bound for the density of these "good central points".

Lemma 4.2.20 *For every $\varepsilon > 0$ there is a $\delta > 0$, such that for every two elements $u, v \in E^{\mathbb{Z}}$ with $d_W(u, v) < \delta$ there is an integer $m > 0$ that satisfies*

$$\forall k \in \mathbb{Z}: \quad |\{j \in [0, \ldots, m] \, : \, d(\sigma^{k+j}(u), \sigma^{k+j}(v)) < \varepsilon\}| > 2m/3.$$

Proof The logical structure of this statement is

$$\forall \varepsilon \, \exists \delta \, \forall (u_1 u_2) \, \exists m \, \forall k : \cdots .$$

First of all, from $u, v \in E^{\mathbb{Z}}$ and $d_W(u, v) < \delta$ we conclude

$$\limsup_{m \to \infty} \max_{k \in \mathbb{Z}} d_H(u, v; [k, \ldots, k+m])/m \leq \delta.$$

Thus, there is $m_0 > 0$ such that for all $m \geq m_0$ the inequality

$$\max_{k \in \mathbb{Z}} d_H(u, v; [k, \ldots, k+m])/m \leq 2\delta$$

holds, and hence

$$d_H(u, v; [k, \ldots, k+m]) \leq 2\delta m \quad \text{for all } k \in \mathbb{Z}.$$

Now we turn to the Cantor distance. There is a number $L > 0$, such that $d(\sigma^j(u), \sigma^j(v)) < \varepsilon$ is guaranteed by

$$\sigma^j(u)(i) = \sigma^j(v)(i) \qquad \text{for } |i| < L.$$

If the states disagree at site j_0, $u(j_0) \neq v(j_0)$, then there is a region of size $2L + 1$ at which $d(\sigma^j(u), \sigma^j(v)) < \varepsilon$ cannot be true

$$u(j_0) \neq v(j_0) \quad \Rightarrow \quad d(\sigma^j(u), \sigma^j(v)) > \varepsilon \quad \text{for } |j_0 - j| < L.$$

Last, we prove our statement. Consider an arbitrary but fixed region $[k, \ldots, k+m]$ of length $m+1$. Let l denote the number of disagreeing sites within this region. We do not know something about the states outside of this region. Thus, the first L sites and the last L sites may not satisfy $d(\sigma^j(u), \sigma^j(v)) < \varepsilon$ due to sites on which u and v disagree, outside of the region under consideration. Within the region under consideration, each disagreeing site forces at most the sites in a symmetric neighborhood of length $2L + 1$ to satisfy $d(\sigma^j(u), \sigma^j(v)) \geq \varepsilon$. Hence, at least

$$m - 2L - l(2L+1) \geq m - (l+1)(2L+1)$$

sites within our region of interest satisfy $d(\sigma^j(u), \sigma^j(v)) < \varepsilon$. We choose δ small enough such that $2(2L+1)\delta < 1/6$, and subsequently $m > \max\{6, \; m_0\}$. With

$$d_H(u, v; [k, \ldots, k+m]) \leq 2\delta m$$

we find

$$m-(l+1)(2L+1) \geq m-(2\delta m+1)(2L+1) = m(1-2(2L+1)\delta-1/m) \geq 2m/3.$$

□

This lemma provides a tool in proving the semi-continuity of the function m from Definition 4.2.16.

Proposition 4.2.21 *The function* $m : E^\Gamma \to \mathbb{N}\cup\{\infty\}$ *is lower semi-continuous with respect to the Weyl topology.*

Proof

Step 1: Setup.
Let $u \in E^\Gamma$, and $m(u) > m$, $m \in \mathbb{N}$. We will show that there is an X_W-ball around u such that $m(v) > m$ for all v within this ball. We provide four prerequisites:

(1) As $m(u) > m$, there are at least $m+1$ different msi sets $\mathcal{S}_1, \ldots, \mathcal{S}_{m+1}$ contained in $\overline{O(u)}^d$. Let

$$\varepsilon_1 = \min_{s \neq t} d(\mathcal{S}_s, \mathcal{S}_t).$$

(2) We apply Lemma 4.2.20 and find $\varepsilon_2 > 0$ and $\overline{m} \in \mathbb{N}$ such that

$$|\{j \in [0, \ldots, \overline{m}] \; : \; d(\sigma^{k+j}(u), \sigma^{k+j}(v)) < \varepsilon_2/8\}| > 2\overline{m}/3 \qquad \forall k \in \mathbb{Z}.$$

(3) As the shift operator is continuous in the Cantor topology, there is a $\delta > 0$ such that $d(u_1, u_2) < 2\delta$ implies $d(\sigma^j(u_1), \sigma^j(u_2)) < \varepsilon_1/4$ for all $j = 0, \ldots, \overline{m}$.
We will show that all elements of the ε_2-ball around u satisfy $m(v) \geq m$.
(4) Let $v \in \{v' \in X_W \; : \; d_W(u, v') < \varepsilon_2\}$, arbitrary, fixed. If $m(v) = \infty$ we are done. If $m(v) < \infty$, there are points $z_1, \ldots, z_{m(v)}$ for which $J^\delta(z_1, \ldots, z_{m(v)})$ is syndetic with some constant L.

Step 2: Definition of a function $h : \{1, \ldots, m+1\} \to \{1, \ldots, m(v)\}$.
We define a function $h : \{1, \ldots, m+1\} \to \{1, \ldots, m(v)\}$ in a slightly arbitrary manner. Later on, we will show that this function is injective, and thus $m(v) \geq m+1$. The idea is to find a point $z_{h(s)} \in E^\Gamma$ close to each of the sets $\mathcal{S}_s$.
Fix $s \in \{1, \ldots, m+1\}$. As $\mathcal{S}_s$ is an invariant subset of $\overline{O(u)}^d$, the trajectory of u stays arbitrarily long in the vicinity of $\mathcal{S}_s$. We find an integer k locating an interval $[k, \ldots, k+\overline{m}+L]$ such that

$$d(\sigma^j(u), \mathcal{S}_s) < \varepsilon_1/8 \qquad \text{for} \quad j \in [k, \ldots, k+\overline{m}+L].$$

Next we focus on the orbit of the element v we selected above, in prerequisite (4). At this moment, we do not use the inequality $d_W(u, v) < \varepsilon_2$, but we only use the integer k we did just construct: $J^\delta(z_1, \ldots, z_{m(v)})$ is syndetic with constant L. We obtain integers $j_0 \in [k, k+L]$ and $\zeta \in \{1, \ldots, m(v)\}$ such that

$$d(\sigma^{j_0}(v), z_\zeta) < \delta.$$

We define

$$h(s) := \zeta.$$

Of course, there may be more than one element of $\{z_1, \ldots, z_{m(v)}\}$ in the δ-neighborhood of $\{\sigma^k(v), \ldots, \sigma^{k+L}(v)\}$. If so, we arbitrarily fix one of the points z_i.

Note that we did not use the inequality $d_W(u, v) < \varepsilon_2$ up to now. In the next step we use this inequality to estimate the distance of $\mathcal{S}_s$ and iterates of v.

Step 3: The trajectory of v stays close to $\mathcal{S}_s$ for a certain time.

We strengthen the link between the invariant sets $\mathcal{S}_s$ (defined via the orbit of u) and the orbit of v (see Fig. 4.2). Due to the choice of ε_2, we know that

$$d(\sigma^j(v), \sigma^j(u)) < \varepsilon_1/8$$

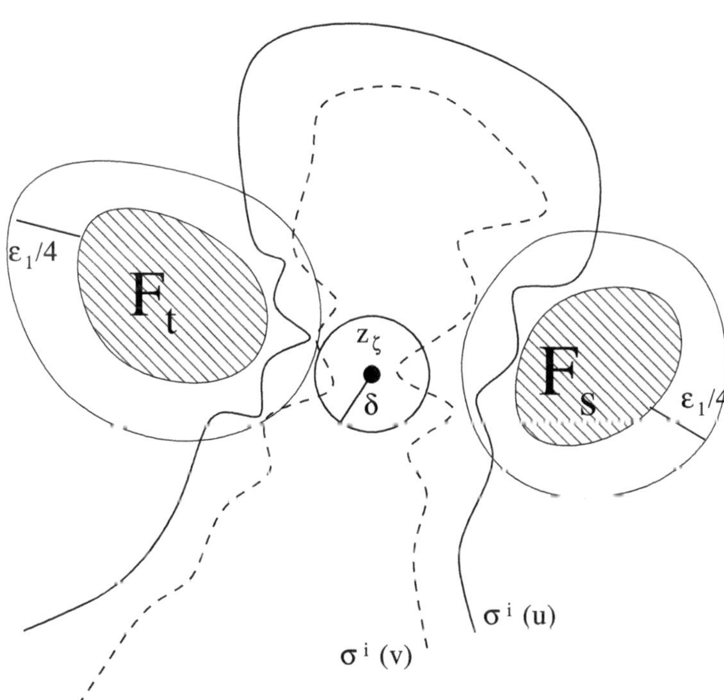

Fig. 4.2 Sketch of the situation in steps 3 and 4

holds for more than two thirds of the integers j in $[j_0, \ldots, j_0 + \overline{m}]$ (where j_0 has been determined in step 2). And we know $d(\sigma^j(u), \mathcal{S}_s) < \varepsilon_1/8$ for all integers j in $[j_0, \ldots, j_0 + \overline{m}]$. Hence, from $\zeta = h(s)$ we conclude that for $\hat{v} := \sigma^{j_0}(v)$ we have $\hat{v} \in B_\delta(z_\zeta)$, and (note that $\sigma^i(\hat{v}) = \sigma^{j_0+i}(v)$)

$$d(\sigma^i(\hat{v}), \mathcal{S}_s) \le d(\sigma^i(\hat{v}), \sigma^{j_0+i}(u)) + d(\sigma^{j_0+i}(u), \mathcal{S}_s) < \varepsilon_1/4$$

for more than two thirds of the integers i in $[0, \ldots, \overline{m}]$.

Step 4: The function h is injective.

Assume $\zeta = h(s) = h(t)$ for $s \neq t$, $t, s \in \{1, \ldots, m+1\}$. According to step 3, we find $\hat{v}_1, \hat{v}_2 \in B_\delta(z_\zeta)$ such that

$$d(\sigma^j(\hat{v}_1), \mathcal{S}_s) < \varepsilon_1/4, \quad d(\sigma^j(\hat{v}_2), \mathcal{S}_t) < \varepsilon_1/4$$

for more than two thirds of the integers j in $[0, \ldots, \overline{m}]$ (see Fig. 4.2). Thus, there is at least one integer $\tilde{k} \in [0, \ldots, \overline{m}]$ such that both inequalities

$$d(\sigma^k(\hat{v}_1)\mathcal{S}_s) < \varepsilon_1/4, \quad d(\sigma^k(\hat{v}_2), \mathcal{S}_t) < \varepsilon_1/4$$

hold. Furthermore, $d(\hat{v}_1, \hat{v}_2) < 2\delta$ as $\hat{v}_1, \hat{v}_2 \in B_\delta(z_\zeta)$. The constant δ has been chosen in such a way that $d(\hat{v}_1, \hat{v}_2) < 2\delta$ implies $d(\sigma^j(\hat{v}_1), \sigma^j(\hat{v}_2)) < \varepsilon_1/4$ for $j = 0, \ldots, \overline{m}$. Thus,

$$d(\mathcal{S}_s, \mathcal{S}_t) \le d(\mathcal{S}_s, \sigma^k(\hat{v}_1)) + d(\sigma^k(\hat{v}_1), \sigma^k(\hat{v}_2)) + d(\sigma^k(\hat{v}_2), \mathcal{S}_t) < 3\varepsilon_1/4.$$

But the minimal distance between different sets $\mathcal{S}_s$ and $\mathcal{S}_t$ is at least ε_1. Hence we found a contradiction, and $h(s) \neq h(t)$ for $s \neq t$, and finally $m(v) > m$.

□

Remark 4.2.22 The function $m : X_W \to \mathbb{N} \cup \{\infty\}$ is not continuous. Let a_n, b_n be blocks of length n given by

$$a_n = \underbrace{1000\cdots0}_{n}, \qquad b_n = \underbrace{0000\cdots0}_{n},$$

and let the states u_n defined by $u_n = \cdots a_n a_n a_n b_n b_n b_n \cdots$. The set $\overline{O(u_n)}^d$ contains at least the periodic states $\cdots a_n a_n a_n \cdots$ and the state that is identically zero, $\overline{0}$. As both states generate different (finite) invariant sets under the action of the shift operator, we find $m(u_n) \geq 2$. However, $d_W(x_n, \overline{0}) \leq 1/n$, i.e., the sequence $(u_n)_{n\in\mathbb{N}}$ tends to zero with respect to the Weyl topology. As $m(\overline{0}) = 1$, the function m is not continuous with respect to the Weyl topology.

Now we are ready to state the central theorem of this section.

Theorem 4.2.23 *The Weyl space X_W is not complete.*

Proof We construct a Cauchy sequence in X_W that would contradict the lower semi-continuity of $m(\cdot)$ in case that it has a limit.

Step 1: Construction of the sequence.
Let $\{v_i\}_{i\in\mathbb{N}} \subset E^\Gamma$ be a sequence of periodic states,

$$v_i = \cdots b_i b_i b_i b_i \cdots .$$

The blocks b_i are defined recursively. Let $b_0 = 0$ be a block of length one. Then, v_0 is the state that is identically zero, $v_0 = \overline{0}$. Let $|b_i|$ denote the length of block i. The block b_i is defined as

$$b_i = \underbrace{b_{i-1}b_{i-1}\cdots b_{i-1}b_{i-1}}_{r_i\text{blocks}}\ \underbrace{ss\cdots ss}_{\text{length } |b_{i-1}|}$$

where "s" is either "1" if i is odd, or "0" if i is even. That is, $|b_i| = |b_{i-1}|(r_i + 1)$. We estimate the Weyl pseudodistance between v_i and v_{i-1}. This bound is based on the periodicity of v_i and v_{i-1}. Within one period of the state v_i, r_i blocks in v_i (of size $|b_{i-1}|$) agree with v_{i-1}, and one block in v_i (of the same size) disagrees with v_{i-1} at some sites. Hence, $d_W(v_i, v_{i-1})$ is less or equal $1/(|b_{i-1}|\,(r_i + 1))$. If we choose r_i large enough, we have

$$d_W(v_i, v_{i+1}) < 2^{-i}.$$

This choice guarantees that $\{v_i\}$ is a Cauchy sequence.
Step 2: The Cauchy sequence $\{v_i\}_{i\in\mathbb{N}}$ does not converge in X_W.
As v_i are spatially periodic, the orbit under the shift operator consists of a finite number of points. Therefore its closure consists of a finite number of points, and there is only one invariant set, $m(v_i) = 1$.
Let us assume that $u = \lim_{i\to\infty} v_i$ in the Weyl topology. We show that $\overline{0}$ is in the closure (w.r.t. the Cantor topology) of the orbit of u. Let $\varepsilon > 0$ and $L > 1 + 1/\varepsilon$, such that $d(u, v) < \varepsilon$ if $u(j) = v(j)$ for $|j| < L$. Now let us choose i_1 large enough that $|b_{i_1}| > 2L + 1$, and let i_2 be an even integer larger than i_1. Then, b_{i_2} contains a block of $s \cdots s = 0 \cdots 0$ of length $|b_{i_2-1}| \geq |b_{i_1}| > 2L + 1$. If we shift the center point of this block of zeros to the origin of the grid, we find $d(v_{i_2}, \overline{0}) < \varepsilon$. Therefore, $\overline{0} \in \overline{O(u)}^d$.
A parallel argument shows that $\overline{1} \in \overline{O(u)}^d$.
Both states, $\overline{0}$ and $\overline{1}$, are minimal invariant sets. Hence, $m(u) \geq 2$. As $m(v_i) = 1$, this finding contradicts the lower semi-continuity of m with respect to the topology of X_W. The Cauchy sequence constructed in step 1 cannot converge. X_W is not complete.

□

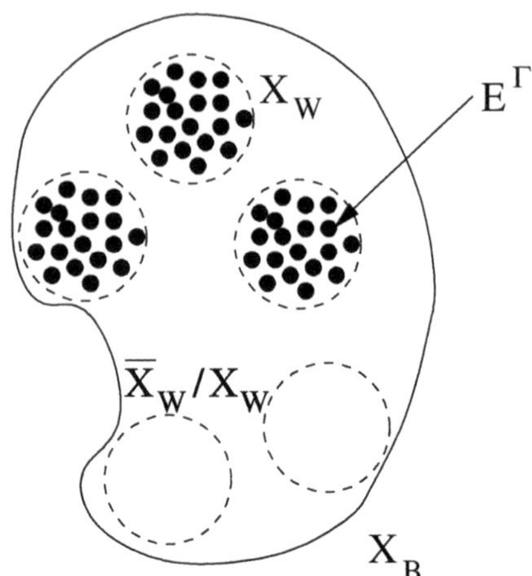

Fig. 4.3 Sketch of the set theoretical connection between the Cantor space $E^{\mathbb{Z}}$, X_W and X_B. The *outer line* symbolizes an equivalence set in $X_{\hat{B}}$, the *dashed circles* equivalent classes in $\overline{X_W}$, and the *small bold dots* elements of the Cantor space $E^{\mathbb{Z}}$

Remark 4.2.24

(a) The closure of X_W w.r.t. $d_W(\cdot,\cdot)$ yields a closed, metric space $\overline{X_W}$. We show that the global function of a cellular automaton can be defined on X_W. This function on X_W is continuous and has a continuous extension to $\overline{X_W}$. Thus, it is possible to define the global function of cellular automata on elements that are not states (elements in E^Γ) any more.

(b) A Cauchy sequence in X_W is also a Cauchy sequence in $X_{\hat{B}}$. The requirements for convergence are less strict in $X_{\hat{B}}$. This fact allows $X_{\hat{B}}$ to be complete. A Cauchy sequence that fails to converge in X_W still has a limit in $X_{\hat{B}}$. The Weyl topology incorporates a higher degree of uniformity in the spatial structure.

On the other hand, one equivalence class of $X_{\hat{B}}$ consists generically of several equivalence classes in X_W, together with some of the new limit points in $\overline{X_W} \setminus X_W$. From a set theoretical point of view the situation is visualized in Fig. 4.3: The largest set are equivalence classes in (elements of) $X_{\hat{B}}$. These equivalent classes are unions of equivalence classes in X_W. In some sense, X_W allows to "zoom" into points of $X_{\hat{B}}$. Some of the points in $\overline{X_W}$ are subsets of E^Γ, others are empty (in the sense that they are limits of Cauchy sequences that do not converge in $\overline{X_W}$, and thus not subsets of E^Γ).

4.3 Cellular Automata on Besicovitch and Weyl Spaces

A cellular automaton defines a map $E^\Gamma \to E^\Gamma$ by means of its global function. The spaces X_*, $* \in \{\hat{B}, B, W\}$, consist of equivalence classes. In general, a map from E^Γ into E^Γ does not respect the equivalence relations. In the special case of cellular automata, the structure can be lifted to X_*.

Proposition 4.3.1 *Let* (Γ, D_0, E, f). *Let* $u, v \in E^\Gamma$ *and* $D_0 \subset \Gamma_n$. *Furthermore, let* $a = \limsup_{l\to\infty} \gamma(l+1)/\gamma(l)$. *Then,*

$$d_{\hat{B}}(f(u),f(v)) \leq a^n \; |D_0| \; d_{\hat{B}}(u,v),$$
$$d_B(f(u),f(v)) \leq a^n \; |D_0| \; d_B(u,v),$$
$$d_W(f(u),f(v)) \leq a^n \; |D_0| \; d_W(u,v).$$

Proof The Hamming distance satisfies

$$\begin{aligned} d_H(f(u),f(v);\Gamma_m) &= |\{g \in \Gamma_m \,:\, f(u)(g) \neq f(v)(g)\}| \\ &\leq |\{g \in \Gamma_m \,:\, u \circ \sigma_g|_{D_0} \neq v \circ \sigma_g|_{D_0}\}| \\ &\leq |D_0| \; |\{g \in \Gamma_{m+n} \,:\, u(g) \neq v(g)\}| \\ &= |D_0| \; d_H(u,v;\Gamma_{m+n}). \end{aligned}$$

Thus,

$$\begin{aligned} d_{\hat{B}}(f(u),f(v)) &= \limsup_{m\to\infty} \frac{d_H(f(u),f(v);\Gamma_m)}{|\Gamma_m|} \leq |D_0| \; \limsup_{m\to\infty} \frac{d_H(u,v;\Gamma_{m+n})}{|\Gamma_m|} \\ &\leq |D_0| \; \limsup_{m\to\infty} \frac{|\Gamma_{m+n}|}{|\Gamma_m|} \; \frac{d_H(u,v;\Gamma_{m+n})}{|\Gamma_{m+n}|} = |D_0| \; a^n \; d_{\hat{B}}(u,v). \end{aligned}$$

Furthermore,

$$\begin{aligned} d_B(f(u),f(v)) &= \sup_{g\in\Gamma} d_{\hat{B}}(\sigma_g \circ f(u), \sigma_g \circ f(v)) = \sup_{g\in\Gamma} d_{\hat{B}}(f(\sigma_g \circ u), f(\sigma_g \circ v)) \\ &\leq |D_0| \; a^n \; \sup_{g\in\Gamma} d_{\hat{B}}(\sigma_g \circ u, \sigma_g \circ v) = |D_0| \; a^n \; d_B(u,v). \end{aligned}$$

Then,

$$\begin{aligned} d_W(f(u),f(v)) &= \limsup_{m\to\infty} \max_{g\in\Gamma} \frac{d_H(f(u),f(v);\Gamma_m)}{|\Gamma_m|} \\ &\leq |D_0| \; \limsup_{m\to\infty} \max_{g\in\Gamma} \frac{d_H(u,v;\Gamma_{m+n})}{|\Gamma_m|} \\ &\leq |D_0| \; \limsup_{m\to\infty} \frac{|\Gamma_{m+n}|}{|\Gamma_m|} \; \max_{g\in\Gamma} \frac{d_H(u,v;\Gamma_{m+n})}{|\Gamma_{m+n}|} \\ &= |D_0| \; a^n \limsup_{m\to\infty} \max_{g\in\Gamma} \frac{d_H(u,v;\Gamma_{m+n})}{|\Gamma_{m+n}|} = |D_0| \; a^n \; d_W(u,v). \end{aligned}$$

□

This proposition ensures that if u and v belong to the same equivalence class, also $f(u)$ and $f(v)$ are members of one single equivalence class. This finding allows to lift the global function f from E^Γ to X_*. Recall the notation introduced in Sect. 4.2; the symbol "hat" can be used as an embedding operator, i.e., for $u \in E^\Gamma$, $\hat{u}$ denotes the equivalence class of u w.r.t. a pseudometric $d_*(\ ,\ )$.

Definition 4.3.2 Let f be a global function of a cellular automaton. Define $\hat{f} : X_* \to X_*$ by $\hat{f}(\hat{u}) = \widehat{f(u)}$ for $u \in \hat{u}$.

Remark 4.3.3 The a priori inequalities imply that the function $\hat{f}$ is well defined. Moreover, the inequalities derived in Proposition 4.3.1 are also valid for $\hat{f}$,

$$d_*(\hat{f}(\hat{u},\hat{f}(\hat{v})) \leq Cd_*(\hat{u}, \hat{v})$$

such that the global function of a cellular automaton $\hat{f}$ is Lipschitz continuous on X_*.

Next we study stationary points and periodic orbits. For the remaining part of this section, let $\Gamma = \mathbb{Z}$. The following theorem has been derived in [15, 16].

Proposition 4.3.4 *Let $\Gamma = \mathbb{Z}$. If a one-dimensional cellular automaton on X_B exhibits two different periodic points in $\hat{\mathcal{P}}$ with prime period p_1 and prime period p_2, respectively, then there are uncountably many periodic points in X_B with a prime period $p = lcm(p_1, p_2)$.*

Proof The basic idea of the proof is the following: If the neighborhood $D_0 \subset \Gamma_n$, then $u, v \in E^\Gamma$ and $u|_{\Gamma_m} = v|_{\Gamma_m}$ implies $f(u)|_{\Gamma_{m-n}} = f(v)|_{\Gamma_{m-n}}$. As the Besicovitch metric neglects disagreement of states on sets of relatively small sizes, we can concatenate pieces of both periodic states, if these pieces are long enough, and the resulting state will again be a periodic orbit. As there are uncountably many possible combinations of these pieces, we are able to construct uncountably many different periodic points.

Step 1: Construction of periodic points.

Let $\hat{u}_1 \in X_B$ be the periodic orbit with period p_1, and $\hat{u}_2 \in X_B$ the periodic orbit with period p_2. Let furthermore $u_i \in \hat{u}_i$, $i = 1, 2$. As $d_B(\hat{u}_1, \hat{u}_2) > 0$, also $d_B(u_1, u_2) > 0$, i.e., $u_1 \neq u_2$ in E^Γ. Let $\{\zeta_i\}_{i\in\mathbb{N}_0}$ be a sequence of strictly increasing natural numbers with $\zeta_0 = 0$ and $2\zeta_i < \zeta_{i+1}$. The sequence diverges exponentially fast; the number of elements ζ_i smaller than a given $m \in \mathbb{N}$ can be estimated by

$$|\{\zeta_i < m\}| \leq \log_2(m) + 1.$$

Given a sequence $\alpha \in \{0, 1\}^{\mathbb{N}_0}$, define

$$w(g) = \begin{cases} u_1(g) & \text{if} \quad \zeta_{i-1} \leq g < \zeta_i, & \alpha_i = 0 \\ u_2(g) & \text{if} \quad \zeta_{i-1} \leq g < \zeta_i, & \alpha_i = 1 \\ u_1(g) & \text{if} \ -\zeta_{i+1} \leq g < -\zeta_i, & \alpha_i = 0 \\ u_2(g) & \text{if} \ -\zeta_{i+1} \leq g < -\zeta_i, & \alpha_i = 1 \end{cases}.$$

We know that u_i is a periodic point of period p_i, $i = 1, 2$. As p_1 and p_2 divide p, we have $d_B(f^p(u_i), u_i) = 0$. Furthermore, in the intervals $I_i = \{\zeta_i + np + 1, \zeta_{i+1} - np - 1\}$ and $I_{-i} = \{-\zeta_{i+1} + np + 1, -\zeta_i - np - 1\}$ either

$$f^p(w)(i)|_I = f^p(u)(i)|_I \quad \text{or} \quad f^p(w)(i)|_I = f^p(v)(i)|_I.$$

depending on the value of α_i. Only in intervals around $\pm\zeta_i$, in $\pm\zeta_i + \{-np, \ldots, np\}$, nothing is known about $f(w)$. Thus,

$$\begin{aligned} d_{\hat{B}}(f^p(w), w) &= \limsup_{m\to\infty} \frac{d_H(w, f^p(w); \Gamma_m)}{|\Gamma_m|} \\ &\leq \limsup_{m\to\infty} \frac{d_H(u_1, f^p(u_1); \Gamma_m) + d_H(u_2, f^p(u_2); \Gamma_m) + \sum_{\zeta_i\in\Gamma_m}(2np+1)}{|\Gamma_m|} \\ &\leq d_{\hat{B}}(f^p(u_1), u_1) + d_{\hat{B}}(f^p(u_2), u_2) + (2np+1)\limsup_{m\to\infty} \frac{|\{\zeta_i \in \Gamma_m\}|}{|\Gamma_m|}. \end{aligned}$$

Since the sequence ζ_i increases in an exponential way, the last term can be estimated by

$$\limsup_{m\to\infty} \frac{|\{\zeta_i \in \Gamma_m\}|}{|\Gamma_m|} \leq \frac{2(\log_2(m)+1)}{2m+1} = 0.$$

Hence, $d_{\hat{B}}(f^p(w), w) = 0$ and w is a periodic point of period p. This period is also the prime period if there are arbitrarily long regions where w agrees with u_1 (requires period p_1) and similarly for u_2.

Step 2: Construction of periodic points that are different in X_B.

In step 1 we constructed a family of X_B-periodic points in $E^{\mathbb{Z}}$. It could be that they belong only to few equivalence classes in X_B. However, we show that the family yields uncountably many different periodic points in X_B.

We know that $d_B(u_1, u_2) > 0$, and hence (we have $\Gamma = \mathbb{Z}$) also $d_{\hat{B}}(u_1, u_2) > 0$. Hence, there is a sequence $\{m_i\}_{i\in\mathbb{N}}$ such that

$$\frac{d_H(u_1, u_2; \Gamma_{m_i})}{|\Gamma_{m_i}|} \geq \delta/2$$

for some $\delta \in (0, 1)$, and therefore

$$\begin{aligned} \delta/2 &\leq \frac{d_H(u_1, u_2; \Gamma_{m_i} \setminus \Gamma_{m_{i-1}}) + d_H(u_1, u_2; \Gamma_{m_{i-1}})}{|\Gamma_{m_i}|} \\ &= \frac{d_H(u_1, u_2; \Gamma_{m_i} \setminus \Gamma_{m_{i-1}})}{|\Gamma_{m_i} \setminus \Gamma_{m_{i-1}}|} \frac{|\Gamma_{m_i} \setminus \Gamma_{m_{i-1}}|}{|\Gamma_{m_i}|} + \frac{d_H(u_1, u_2; \Gamma_{m_{i-1}})}{|\Gamma_{m_{i-1}}|} \frac{|\Gamma_{m_{i-1}}|}{|\Gamma_{m_i}|} \end{aligned}$$

which implies

$$\frac{d_H(u_1,u_2;\Gamma_{m_i}\setminus\Gamma_{m_{i-1}})}{|\Gamma_{m_i}\setminus\Gamma_{m_{i-1}}|} \geq \left(\frac{\delta}{2} - \frac{d_H(u_1,u_2;\Gamma_{m_{i-1}})}{|\Gamma_{m_{i-1}}|}\frac{|\Gamma_{m_{i-1}}|}{|\Gamma_{m_i}|}\right)\frac{|\Gamma_{m_i}|}{|\Gamma_{m_i}\setminus\Gamma_{m_{i-1}}|}$$
$$\geq \left(\frac{\delta}{2} - \frac{|\Gamma_{m_{i-1}}|}{|\Gamma_{m_i}|}\right)\frac{|\Gamma_{m_i}|}{|\Gamma_{m_i}|-|\Gamma_{m_{i-1}}|}.$$

We choose a subsequence of $\{m_i\}$ (that we again denote by $\{m_i\}$) such that $|\Gamma_{m_{i-1}}|/|\Gamma_{m_i}| < \delta/4$. Thus,

$$\frac{d_H(u_1,u_2;\Gamma_{m_i}\setminus\Gamma_{m_{i-1}})}{|\Gamma_{m_i}\setminus\Gamma_{m_{i-1}}|} \geq \left(\frac{\delta}{2} - \frac{|\Gamma_{m_{i-1}}|}{|\Gamma_{m_i}|}\right)\frac{|\Gamma_{m_i}|}{|\Gamma_{m_i}|-|\Gamma_{m_{i-1}}|} > \frac{\delta}{4}\frac{1}{1-\delta/4} = \frac{\delta}{4-\delta}.$$

We again choose a subsequence of $\{m_i\}$ (which we again denote by $\{m_i\}$) such that $m_{i+1} > 2m_i$. We may then define $\zeta_i = m_i$. Now we consider w and w' that have been constructed using the binary sequences α_i and α_i'. Assume $\alpha_i \neq \alpha_i'$. Then,

$$\frac{d_H(w,w';\Gamma_{m_i})}{|\Gamma_{m_i}|} \geq \frac{d_H(u_1,u_2;\Gamma_{m_i}\setminus\Gamma_{m_{i-1}})}{|\Gamma_{m_i}|}$$
$$= \frac{d_H(u_1,u_2;\Gamma_{m_i}\setminus\Gamma_{m_{i-1}})}{|\Gamma_{m_i}|-|\Gamma_{m_{i-1}}|}\frac{|\Gamma_{m_i}|-|\Gamma_{m_{i-1}}|}{|\Gamma_{m_i}|}$$
$$\geq \frac{\delta}{4-\delta}\ (1-\delta/4) = \delta/4$$

independently on i. Thus, w and w' have a positive distance in $X_{\hat{B}}$ (and thus also in X_B) if the two sequences $\{\alpha_i\}$ and $\{\alpha_i'\}$ disagree at infinitely many entries. In the last step, we show that this observation implies the existence of an infinite number of periodic points.

Step 3: The set of binary sequences that are dissimilar at an infinite number of entries is not countable.

Let $\alpha, \alpha' \in \{0,1\}^{\mathbb{N}}$, and define an equivalence relation

$$\alpha \sim \alpha' \quad \Leftrightarrow |\{i : \alpha(i) \neq \alpha'(i)\}| < \infty.$$

This is an equivalence relation, as the relation is symmetric and also transitive:

$$|\{i : \alpha(i) \neq \alpha''(i)\}| \leq |\{i : \alpha(i) \neq \alpha'(i)\}| + |\{i : \alpha'(i) \neq \alpha''(i)\}|$$

i.e., $\alpha \sim \alpha'$ and $\alpha' \sim \alpha''$ forces also $\alpha \sim \alpha''$.

We claim that each equivalence class is only countable. Let A denote an equivalence class and let $\alpha_0 \in A$. We may split this class into disjoint subsets using α_0 as a reference sequence,

$$A_i := \{\alpha \in A : \alpha(i) \neq \alpha_0(i), \quad \forall j > i : \alpha(j) = \alpha_0(j)\}, \qquad A = \cup_{i=1}^{\infty} A_i.$$

Each set A_i only consists of a finite number of elements, and thus A is countable. As $\{0, 1\}^{\mathbb{N}}$ is not countable, the number of equivalence classes is not countable. □

An immediate consequence is the following corollary.

Corollary 4.3.5 *Let* $\Gamma = \mathbb{Z}$. *If a cellular automaton has two fixed points, then it has uncountably many fixed points in* X_B.

Proposition 4.3.6 *Only either of the following two statements is true for a cellular automaton on* X_B *and* $X_{\hat{B}}$:

(a) There is one and only one translation invariant fixed point and no periodic point with prime period larger one.
(b) There are uncountably many periodic or stationary points.

Proof First we show that there is at least one stationary point or periodic orbit: The set of states where all sites assume the same elementary state (constant in space) is finite and forward invariant. Thus, within this set there is at least one fixed point or periodic orbit.

Either (a) or the negation of (a) (together with the fact that there is at least one fixed point/periodic orbit) are true. We show, that the second possibility implies (b): If there are two fixed points, we obtain an uncountably many fixed points. If there is a periodic point with a prime period larger than one, then there is a u and a v in this periodic orbit ($v = f(u)$, say) such that $u \neq v$. We can construct an uncountable set of periodic orbits using u and v as starting points. □

4.4 A CHL Theorem for Besicovitch and Weyl Spaces

The CHL theorem says that continuous functions on a Cantor space can be represented as global functions of cellular automata. Here we show, following [134], a similar theorem for Besicovitch and Weyl spaces. The proof of the representation theorem is constructive, we obtain a cellular automaton from a function that satisfies certain conditions. We restrict ourselves to the one-dimensional case, $\Gamma = \mathbb{Z}$. In this case the topologies of X_B and $X_{\hat{B}}$ are equivalent. As before, X_* and $d_*(\cdot, \cdot)$ refer to $X_{\hat{B}}$, X_B or X_W.

One basic ingredient used for the characterization of a cellular automaton are restriction operators. For convenience, we assume that the alphabet E consists of integers $E = \{0, 1, \ldots, n-1\}$, such that multiplication or addition (modulo n) is possible. However, we will only multiply or add two local states, if at least one of them is zero. We will never use the operation modulo n

Definition 4.4.1 Let $\Gamma = \mathbb{Z}$, $E = \{0, .., n-1\}$, and $A \subset \mathbb{Z}$. The restriction operator $\rho_A : E^{\mathbb{Z}} \to E^{\mathbb{Z}}$, $u \mapsto v$ is defined by

$$v(i) = \begin{cases} u(i) & \text{if} \quad i \in A \\ 0 & \text{otherwise.} \end{cases}$$

Proposition 4.4.2

(a) Let $A \subset \mathbb{Z}$. The restriction operator ρ_A satisfies

$$d_*(\rho_A(u), \rho_A(v)) \leq d_*(u, v).$$

It can be lifted to X_*. The lift $\hat{\rho}_A : X_* \to X_*$ is Lipschitz continuous with Lipschitz constant 1.

(b) Let $A_1, \ldots, A_k \subset \mathbb{Z}$, $\cup_{i=1}^k A_i = \mathbb{Z}$. Then, $d_*(\hat{u}, \hat{v}) \leq \sum_{i=1}^k d_*(\hat{\rho}_{A_i}(\hat{u}), \hat{\rho}_{A_i}(\hat{v}))$.

Proof

a) We first note that

$$d_H(\rho_A(u), \rho_A(v); \Gamma_m) \leq d_H(u, v; \Gamma_m).$$

Thus, $d_*(\rho_A(u), \rho_A(v)) \leq d_*(u, v)$. The restriction operator ρ_A can be lifted to X_*, and its Lipschitz constant is 1.

b) In a similar manner, we find for $u, v \in E^{\mathbb{Z}}$

$$d_H(u, v; \Gamma_m) \leq \sum_{i=1}^{k} d_H(\rho_{A_i}(u), \rho_{A_i}(v); \Gamma_m).$$

Hence,

$$\begin{aligned} d_{\hat{B}}(u, v) &= \limsup_{m\to\infty} \frac{d_H(u, v; \Gamma_m)}{|\Gamma_m|} \leq \limsup_{m\to\infty} \frac{\sum_{i=1}^k d_H(\rho_{A_i}(u), \rho_{A_i}(v); \Gamma_m)}{|\Gamma_m|} \\ &\leq \sum_{i=1}^{k} \limsup_{m\to\infty} \frac{d_H(\rho_{A_i}(u), \rho_{A_i}(v); \Gamma_m)}{|\Gamma_m|} = \sum_{i=1}^{k} d_{\hat{B}}(\rho_{A_i}(u), \rho_{A_i}(v)). \end{aligned}$$

Similarly,

$$\begin{aligned} d_W(u, v) &= \limsup_{m\to\infty} \max_{j\in\mathbb{Z}} \frac{d_H(\sigma_j \circ u, \sigma_j \circ v; \Gamma_m)}{|\Gamma_m|} \\ &\leq \limsup_{m\to\infty} \max_{j\in\mathbb{Z}} \frac{\sum_{i=1}^k d_H(\sigma_j \circ \rho_{A_i}(u), \sigma_j \circ \rho_{A_i}(v); \Gamma_m)}{|\Gamma_m|} \end{aligned}$$

$$\leq \sum_{i=1}^{k} \limsup_{m\to\infty} \max_{j\in\mathbb{Z}} \frac{d_H(\sigma_j \circ \rho_{A_i}(u), \sigma_j \circ \rho_{A_i}(v); \Gamma_m)}{|\Gamma_m|}$$

$$= \sum_{i=1}^{k} d_W(\rho_{A_i}(u), \rho_{A_i}(v)).$$

These inequalities for elements in $E^{\mathbb{Z}}$ imply the corresponding inequalities for elements in X_*.

□

Notation For a non-empty subset $A \subset \mathbb{Z}$, we denote by $A^{(n)}$ the set

$$A^{(n)} = \{i + k \,:\, i \in A,\;\; k \in \{-n, \ldots, n\}\}.$$

Proposition 4.4.3 *Let f be the global function of a cellular automaton $(\mathbb{Z}, D_0, E, f_0)$, where $D_0 \subset \Gamma_n$. Let $\hat{f}$ be the lift to X_*. Let $A \subset \mathbb{Z}$, $A \neq \emptyset$. Then*

$$d_*(\hat{\rho}_A \circ \hat{f}(\hat{u}), \hat{\rho}_A \circ \hat{f}(\hat{v})) \leq |D_0| d_*(\hat{\rho}_{A^{(n)}}(\hat{u}), \hat{\rho}_{A^{(n)}}(\hat{v})).$$

Proof It is sufficient to prove the inequality for $u, v \in E^{\mathbb{Z}}$:

$$\begin{aligned} d_H(\rho_A(f(u)), \rho_A(f(v)); \Gamma_m) &= |\{i \in A \cap \Gamma_m \,:\, f(u)(i) \neq f(v)(i)\}| \\ &\leq |\{i \in A \cap \Gamma_m \,:\, u|_{i+D_0} \neq v|_{i+D_0}\}| \\ &\leq |D_0| \;\; |\{i \in A^{(n)} \cap \Gamma_{m+n} \,:\, u(i) \neq v(i)\}| \\ &= |D_0| \;\; d_H(\rho_{A^{(n)}}(u), \rho_{A^{(n)}}(v); \Gamma_{m+n}). \end{aligned}$$

The inequality now follows as in the proof of Proposition 4.3.1. □

In the CHL theorem for Cantor spaces, locality of the function, expressed by continuity w.r.t. the Cantor topology, and shift invariance played the central role. We introduce the counterpart of the two properties in X_*. We define the weak and the strong localization property, and also the weak and the strong shift invariance. Recall that $\hat{\mathcal{P}}$ denotes the periodic points in X_B, see Definition 4.2.10.

Definition 4.4.4 A function $\hat{f} \in C(X_*)$ is called

(1) weakly shift invariant, if $\hat{\sigma} \circ \hat{f} = \hat{f} \circ \hat{\sigma}$ for all shift operators $\hat{\sigma}$.
(2) strongly shift invariant, if it is weakly shift invariant and leaves $\hat{\mathcal{P}}$ invariant, $\hat{f}(\hat{\mathcal{P}}) \subset \hat{\mathcal{P}}$.
(3) weakly localizing (has the weak localization property), if there is a $C > 0$, such that

$$d_*(\hat{f}(\hat{u}), \hat{f}(\hat{v})) \leq C d_*(\hat{u}, \hat{v}).$$

(4) strongly localizing (has the strong localization property) with index n, if there are local maps

$$\tilde{f}_i : E^{\Gamma_n} \to E, \qquad \text{for } i \in \mathbb{Z}$$

such that the function $f : E^{\Gamma} \to E^{\Gamma}, u \mapsto v$ with $v(i) = \tilde{f}_i(u|_{i+\Gamma_n})$ is a member of $\hat{f}$, i.e., $d_*(f(u), \hat{f}(u)) = 0$ for all $u \in E^{\Gamma}$.

Next we give an equivalent formulation of "strongly localizing" using restriction operators.

Proposition 4.4.5 *A function $\hat{f} \in C(X_*)$ has the strong localization property with index n if and only if there is a constant $C > 0$ such that*

$$d_*(\hat{\rho}_A(\hat{f}(\hat{u})), \hat{\rho}_A(\hat{f}(\hat{v}))) \leq C d_*(\hat{\rho}_{A^{(n)}}(\hat{u}), \hat{\rho}_{A^{(n)}}(\hat{v}))$$

holds for all subsets $A \subset \mathbb{Z}$.

Proof "$\Rightarrow$" If the strong localization property is given, the inequality can be proved along the lines of Proposition 4.4.3.

"$\Leftarrow$" Now assume that the inequality is true. We will show that $\hat{f}$ has the strong localization property.

First we enumerate all patterns in E^{Γ_n},

$$E^{\Gamma_n} = \{\tilde{v}_1, \ldots, \tilde{v}_k\}.$$

We concatenate these patterns periodically, and obtain $v_i \in E^{\mathbb{Z}}$,

$$v_i(l) = \tilde{v}_i(l \bmod 2n + 1), \qquad i = 1, \ldots, k.$$

We want to construct local functions $\tilde{g}_i : E^{\Gamma_n} \to E$ that can be used to assemble a global function $g : E^{\mathbb{Z}} \to E^{\mathbb{Z}}$, which equals $\hat{f}$ (in X_*). With this aim we consider a state $w_{i,j} \in E^{\Gamma}$ that is an arbitrary but fixed member of the set $\hat{f}(\widehat{\sigma^i \circ v_j})$. We then define

$$\tilde{g}_i(\tilde{v}_j) = w_{i,j}(i).$$

The global function $g : E^{\mathbb{Z}} \to E^{\mathbb{Z}}$ is assembled from these local functions,

$$g(u) = w \quad \text{with } w(i) = \tilde{g}_i(u|_{i+\Gamma_n}).$$

As g is Lipschitz-continuous with respect to $d_*(\cdot, \cdot)$, we can lift g to X_*, and obtain $\hat{g} \in C(X_*)$. We prove $d_*(\hat{g}(\hat{u}), \hat{f}(\hat{u})) = 0$: Let $u \in E^{\mathbb{Z}}$. We determine locations $i \in \mathbb{Z}$ for which $u|_{i+\Gamma_n} = \tilde{v}_l$,

$$A_l = \{i \in \mathbb{Z} : u|_{i+\Gamma_n} = \tilde{v}_l\}, \qquad l = 1, \ldots, k.$$

We split the sets A_l into subsets that fit into periodic schemes,

$$A_{l,j} = \{i \in A_l \,:\, i \bmod 2n+1 = j\} \qquad \text{for } j = 0, \ldots, 2n, \quad l = 1, \ldots, k.$$

For $i \in A_{l,j}$ we find

$$u|_{i+\Gamma_n} = \sigma_j \circ v_l|_{i+\Gamma_n}$$

which leads to (for $i \in A_{l,j}$)

$$g(u)(i) = g_i(\tilde{v}_j) = w_{i,l}(i) \quad \text{and} \quad d_*(\hat{\rho}_{A_{l,j}^{(n)}}(\widehat{\sigma_j \circ v_l}), \hat{\rho}_{A_{l,j}^{(n)}}(\hat{u})) = 0.$$

We conclude that in X_* the following equations hold,

$$\hat{\rho}_{A_{l,j}} \circ \hat{f}(\widehat{\sigma_j \circ v_l}) = \hat{\rho}_{A_{l,j}}(\hat{w}_{j,l}) = \hat{\rho}_{A_{l,j}} \circ \hat{g}(\hat{u}).$$

These equations and the assumed inequality yield

$$\begin{aligned} d_*(\hat{g}(\hat{u}), \hat{f}(\hat{u})) &\leq \sum_{j=0}^{n} \sum_{l=1}^{k} d_*(\hat{\rho}_{A_{l,j}} \circ \hat{g}(\hat{u}), \hat{\rho}_{A_{l,j}} \circ \hat{f}(\hat{u})) \\ &= \sum_{j=0}^{n} \sum_{l=1}^{k} d_*(\hat{\rho}_{A_{l,j}} \circ \hat{f}(\widehat{\sigma_j \circ v_l}), \hat{\rho}_{A_{l,j}} \circ \hat{f}(\hat{u})) \\ &\leq C \sum_{j=0}^{n} \sum_{l=1}^{k} d_*(\hat{\rho}_{A_{l,j}^{(n)}}(\widehat{\sigma_j \circ v_l}), \hat{\rho}_{A_{l,j}^{(n)}}(\hat{u})) = 0. \end{aligned}$$

□

Remark 4.4.6 Suppose that $\hat{f}$ leaves the periodic states invariant, $\hat{f}(\hat{\mathcal{P}}) \subset \hat{\mathcal{P}}$. Then we know that $\hat{f}(\widehat{\sigma_j \circ u_l})$ includes exactly one periodic state w.r.t. $E^{\mathbb{Z}}$, i.e., the set $\mathcal{P} \cap \hat{f}(\widehat{\sigma_j \circ u_l})$ contains precisely one element. We can choose $w_{i,j} \in \mathcal{P}$, and, in this way, guarantee that there is a period $m > 0$ in the local functions $\tilde{g}_i$,

$$\tilde{g}_i = \tilde{g}_{i+jm} \qquad \forall j \in \mathbb{Z}.$$

We will use this fact in the proof of the following theorem, a Curtis–Hedlund–Lyndon theorem for Besicovitch and Weyl spaces.

Theorem 4.4.7 *Let E be a finite alphabet, $\mathbb{Z}$ the grid, and X_* the corresponding Besicovitch (Weyl) space. A function $\hat{f} : X_* \to X_*$ is the global function of a cellular automaton if and only if*

(1) the function $\hat{f}$ has the strong localization property for an index n,
(2) the function $\hat{f}$ is strongly shift invariant.

In this case, the neighborhood D_0 of the cellular automaton can be chosen as a subset of Γ_n.

Proof If f is the global function of a cellular automaton such that $d(f(\cdot),\hat{f}(\cdot)) = 0$, then the two properties (1) and (2) follow at once. This is the trivial direction.

We show now that (1) and (2) imply, that $\hat{f}$ can be represented by a lift of a cellular automaton from $E^{\mathbb{Z}}$ to X_*. First of all, according to Proposition 4.4.5, the function $\hat{f}$ can be represented by local functions: There are functions

$$\tilde{g}_i : E^{\Gamma_n} \to E$$

such that the lift of the global function $g : E^{\mathbb{Z}} \to E^{\mathbb{Z}}$, $g(u)(i) = \tilde{g}_i(u|_{i+\Gamma_n})$, coincides with $\hat{f}$,

$$d_*(\hat{g}(\hat{u}),\hat{f}(\hat{u})) = 0 \qquad \forall \hat{u} \in X_*.$$

As the strong localization property implies that $\hat{f}(\hat{\mathcal{P}}) \subset \hat{\mathcal{P}}$, the local functions $\tilde{g}_i$ can be constructed in a periodic way. There is an integer $m > 0$ such that

$$\forall j \in \mathbb{Z} : \;\; \tilde{g}_i = \tilde{g}_{i+jm}.$$

The shift invariance in X_* implies the equation $0 = d_*(\hat{\sigma} \circ \hat{f}(u), \hat{f} \circ \hat{\sigma}(u))$, i.e., (recall the notation in the proof of Proposition 4.4.5) the sets

$$B_l = \{i \in \mathbb{Z} \,:\, g_i(\tilde{v}_l) \neq g_{i+1}(\tilde{v}_l)\}, \quad l = 1, \ldots, k$$

have zero density in Besicovitch sense (if we consider $X_{\hat{B}}$ or X_B),

$$\limsup_{m\to\infty} \frac{|B_l \cap \Gamma_m|}{|\Gamma_m|} = 0$$

and in the Weyl sense (if we focus on X_W),

$$\limsup_{m\to\infty} \max_{i\in\mathbb{Z}} \frac{|(B_l + i) \cap \Gamma_m|}{|\Gamma_m|} = 0.$$

If one of the sets B_l is non-empty, i.e., there is an $l_0 \in \{1, \ldots, k\}$ and $i_0 \in \mathbb{Z}$ with

$$i_0 \in B_l,$$

then—due to the m-periodicity of $\tilde{g}_i$—we conclude

$$i_0 + mj \in B_l \quad \forall j \in \mathbb{Z}.$$

In this case, the density of B_l in the Besicovitch as well as in the Weyl sense is larger or equal $1/m$, contradicting $d_*(\hat{\sigma} \circ \hat{f}(u), \hat{f} \circ \hat{\sigma}(u)) = 0$. Hence, $\tilde{g}_i = \tilde{g}_{i+1}$, and thus g is shift invariant (in the Cantor sense). The function g is the global function of a cellular automaton. □

Remark 4.4.8

(1) As in the classical case of the Cantor topology, also on X_* a cellular automaton is well characterized by its effect on periodic functions, although the closure of the set $\hat{\mathcal{P}}$ of periodic functions in X_* is a proper subset of X_*. There are continuous functions that agree on $\hat{\mathcal{P}}$ with cellular automata but not on $X \setminus \hat{\mathcal{P}}$. We have an extension theorem for cellular automata from their action on $\hat{\mathcal{P}}$ to X_*, which is non-trivial in the sense that it is not the usual case of a continuous function that is defined on a dense subset of a metric space. The algebraic structure (localization and shift invariance) allows to extrapolate the effect of the cellular automaton on points in X_* that are far away from $\hat{\mathcal{P}}$.

(2) If we consider $\mathbb{Z}^d$ (or any Cayley graph corresponding to a finitely generated, Abelian group) the theorem can be easily extended, using an appropriate definition of periodic states (periodic w.r.t. all generators).

(3) It can be shown that the strong versions of "shift invariance" and "localization" are necessary for the Curtis–Lyndon–Hedlund theorem [134].

Chapter 5
Attractors

Classification schemes are the holy grail in the theory of cellular automata. One aspect that may serve as an identifier for different categories of cellular automata is asymptotic behavior. Although transient states are interesting, often enough the dynamics settles quite fast on typical structures. The set consisting of these asymptotic states is the attractor. The first part of this section is devoted to the definition of an attractor and the proof that every cellular automaton has attractors. Attractors are not amorphous sets without further features but can be decomposed into smaller parts. We prove the central structure theorem: Conley's decomposition theorem. Essential for the relevance of an attractor is the "size" of the set of trajectories that eventually tend to the attractor. This "size" of the set of attracted states can be described in terms of measure theory. Therefore we define the Bernoulli measure on the set of states. Using Conley's decomposition theorem and measure theory, it is possible to classify cellular automata by the structure of their attractor set: the Hurley classification theorem.

Subsequent sections are devoted to two other classification approaches. We know that every cellular automaton is continuous, but continuity is a local property of the global function. Here we are interested in sensitive dependence on initial data after many time steps. This question leads to the definition of Lyapunov stability and then to a classification that is based on the number of states that are stable in the sense of Lyapunov.

A third approach is based on the theory of languages. It is possible to define languages by their grammatical structures. One can distinguish four different classes of complexity for such languages. The states of one-dimensional cellular automata can be interpreted as words of a language. In this way, it is possible to classify cellular automata according to the complexity of their language. At first glance, this line of thinking is completely different from the previous two classification approaches which are based on topology. But it turns out that the language classification contains much information on the topological structure of

K.-P. Hadeler, J. Müller, *Cellular Automata: Analysis and Applications*,
Springer Monographs in Mathematics, DOI 10.1007/978-3-319-53043-7_5

the global function and is not at all independent from the first two classification schemes.

To prepare for the first and perhaps most natural classification approach, we need some notions that allow to characterize long term and asymptotic behavior. Central notions are ω-limit sets that describe the asymptotic behavior of one trajectory or of all trajectories starting in a certain set. The notion of a limit set is extended to the concept of an attractor which incorporates some convergence and stability properties. While an ω-limit set may be that small that it is non-generic, an attractor is always large enough to allow small perturbations without essentially changing long term behavior.

For finite grids, the concepts of ω-limit sets and attractors coincide, because every state is an isolated point. Furthermore, since there are only finitely many states, long time behavior is always periodic. In the case of a finite grid, the interesting questions are about the number of periodic orbits and their periods. In the infinite case, the attractors exhibit in general a more complex structure.

5.1 Dynamical Systems, ω-Limit Sets and Attractors

5.1.1 Dynamical Systems

A dynamical system is defined by a state space and the dynamics on the state space. In our case, we are concerned with discrete time systems. We had before a brief look on topological dynamical systems (Definition 3.4.4); now we consider the special case of a dynamical system on a compact, metric space.

Definition 5.1.1 Let X be a compact metric space with metric d and let $f : X \to X$ be a continuous function. We call the triple (X, d, f) a dynamical system. For $x \in X$, the sequence $\{f^n(x)\}_{n \in \mathbb{N}_0}$ is called the trajectory or the (semi-) orbit of x.

Remark 5.1.2 If (Γ, D_0, E, f_0) is a cellular automaton with the Cantor topology then the global function defines a dynamical system on a compact metric space.

Often trajectories tend to settle on a small portion of the state space. We try to find subsets of the state space which in some sense contain the asymptotic behavior of the trajectories and exclude as much as possible the transient behavior. In other words, we aim at ω-limit sets.

In ordinary differential equations "singular elements" as stationary points and periodic orbits play an important role, on the 2-sphere they essentially characterize the underlying vector field. But in discrete time systems, even in dimension one (interval maps) knowing stationary points and periodic points in general does not provide much information about the global dynamics. Nevertheless we need these concepts also for discrete systems.

Definition 5.1.3 A periodic point of period n is a state $x \in X$ such that $f^n(x) = x$. The period is called minimal, if n is the smallest positive integer with this property. A point with minimal period 1 is called a stationary point or fixed point. A stationary point x is called stable if for every $\varepsilon > 0$ there is a $\delta > 0$ such that for all y with $d(y, x) < \delta$ holds $d(f^k(y), x) < \varepsilon$ for all $k \in \mathbb{N}$. The point x is called asymptotically stable if, in addition, there is $\varepsilon > 0$ such that for all $y \in B_\varepsilon(x) = \{y \in X \,:\, d(x, y) < \varepsilon\}$

$$\lim_{k\to\infty} f^k(y) = x. \tag{$*$}$$

A periodic point $x \in X$ with period n is called (asymptotically) stable if x is an (asymptotically) stable point of the function $f^n : X \to X$.

Remark In a compact metric space the property ($*$) for some ε implies stability. Stability proper does not play a role in the following. Hence we shall use the word "stable" when we mean "asymptotically stable".

Here we state some first implications for cellular automata. For one-dimensional automata on $\mathbb{Z}$ it is possible to construct all periodic points (see Sect. 9.2.1). But even in this case, assuming the Cantor topology, several asymptotically stable periodic points cannot coexist since small perturbations are changes far away from the origin.

Proposition 5.1.4 *Let* (Γ, D_0, E, f_0) *be a cellular automaton with* $|\Gamma| = \infty$, *and let the state space* E^Γ *be equipped with the Cantor metric d. Then there is at most one asymptotically stable stationary state. Moreover, any asymptotically stable stationary state is spatially constant.*

Proof Assume that $u, v \in E^\Gamma$, $u \neq v$, are two stationary states which are both asymptotically stable, i.e., there is $\varepsilon > 0$ such that all perturbations of u or v with a distance of at most $\varepsilon > 0$ from either u or v will vanish if time goes to infinity. Any state that agrees on a ball $\Gamma_{1/\varepsilon}$ with u (v) will tend to u (v). As $|\Gamma| = \infty$, there is a grid element $\hat{g}$ with $d_c(e, \hat{g}) > 1 + 2/\varepsilon$. Define a state w by

$$w(g) = \begin{cases} u(g) & \text{if} \quad d_c(e, g) < 1 + 1/\varepsilon \\ v(\hat{g}^{-1}\, g) & \text{otherwise.} \end{cases}$$

Then, $f^n(w)$ tends to u, and—at the same time—$f^n(\sigma_{\hat{g}}(w))$ tends to v, which implies that $f^n(w)$ tends to $\sigma_{\hat{g}^{-1}}(v)$. Therefore, $\sigma_{\hat{g}^{-1}}(v) = u$. Since $\hat{g}$ is any group element with $d_c(e, g) > 1 + 2/\varepsilon$, v is shift invariant and so is u. Thus, $u = v$.

If an asymptotically stable stationary state is not shift invariant, there is a grid element g such that $\sigma_g(u) \neq u$. Then, we have two different, asymptotically stable stationary states, which is impossible. □

Corollary 5.1.5 *A cellular automaton with unbounded grid has no asymptotically stable periodic orbit with prime period larger one.*

Proof If we have a period n orbit with prime period $n > 1$, then there is a state u with $u = f^n(u) \neq f(u)$. u as well as $f(u)$ are asymptotically stable fixed points of the function $g = f^n$. As the function g is again the global function of a cellular automaton, this is not possible. □

As there are only $|E|$ states that are constant in space, it is straightforward to find all spatially invariant periodic orbits: Draw a graph with vertices E such that an edge runs from e_1 to e_2 whenever the constant state e_1 is mapped to the constant state e_2. All loops correspond to periodic orbits.

For the preceding arguments the Cantor metric is essential. A small perturbation is an arbitrary perturbation far away from the origin of the grid; only a finite number of cells of the original state need to be preserved, all others may change. In other contexts one would not call this a small perturbation. However, the framework of the Cantor metric is extremely useful to describe the behavior of automata in general.

5.1.2 ω-Limit Sets and Attractors

Now we return to dynamical systems on general compact metric spaces. We start with one single orbit and define the ω-limit set.

Definition 5.1.6 Let (X, d, f) be a dynamical system. The ω-limit set of a point $x \in X$ is defined as

$$\omega(x) = \{y \in X \,:\, \exists\, t_k \to \infty,\ f^{t_k}(x) \to y\}.$$

Proposition 5.1.7 *An equivalent characterization of $\omega(x)$ reads*

$$\omega(x) = \bigcap_{N>0} \overline{\{f^n(x) \,:\, n > N\}}^{d}.$$

Proof Call

$$A = \{y \in X \,:\, \exists\, t_k \to \infty : f^{t_k}(x) \to y\}$$

and

$$B = \bigcap_{N>0} \overline{\{f^n(x) \,:\, n > N\}}^{d}.$$

$A \subset B$: If $y \in A$ then trivially $y \in B$.

$B \subset A$: If $y \in B$, then

$$y \in \overline{\{f^n(x) : n > N\}}^d$$

for all $n \in \mathbb{N}$. Hence there is a sequence $n_k \to \infty$ such that $f^{n_k} \to y$.

□

Definition 5.1.8 A set $Y \subset X$ is called forward invariant, if $f(Y) \subset Y$, and backward invariant if $Y \subset f(Y)$. The set is called invariant, if $f(Y) = Y$.

Remark 5.1.9 The two properties "forward invariant" and "backward invariant" together imply "invariant". Note that the property "invariant" does not exclude that there is a $z \in X \setminus Y$ such that $f(z) \in Y$.

Theorem 5.1.10 *Let $x \in X$. Then*

(1) $\omega(x) \neq \emptyset$,
(2) $\omega(x)$ *is compact,*
(3) $\omega(x)$ *is invariant.*

Proof

1) Since X is compact, the sequence $\{f^n(x)\}$ has an accumulation point y,

$$y \in \overline{\bigcup_{n>N} f^n(x)}^d$$

for any $N \in \mathbb{N}$, and also

$$y \in \bigcap_{N>0} \overline{\bigcup_{n>N} f^n(x)}^d = \omega(x).$$

Therefore $\omega(x) \neq \emptyset$.
2) $\omega(x)$ is closed and X is compact.
3) *Forward invariant:* Let $y \in \omega(x)$. We show $f(y) \in \omega(x)$. Since $y \in \omega(x)$, there is a strictly increasing sequence n_k such that

$$y = \lim_{k\to\infty} f^{n_k}(x).$$

Since f is continuous,

$$f(y) = \lim_{k\to\infty} f^{n_k+1}(x)$$

and therefore $f(y) \in \omega(x)$.

4) *Backward invariant:* Let $m > 0$ and $y \in \omega(x)$ be fixed. We construct $\tilde{y} \in \omega(x)$, such that

$$f^m(\tilde{y}) = y.$$

Let

$$y = \lim_{k\to\infty} f^{n_k}(x).$$

For k large enough we have $n_k - m \geq 0$. If we neglect finitely many terms, we can assume that the sequence has already this property. Then the sequence

$$\{f^{n_k-m}(x)\}_{n\in\mathbb{N}}$$

is well defined. This sequence has a converging subsequence. Hence we can assume that the sequence converges. Define

$$\tilde{y} = \lim_{n\to\infty} f^{n_k-m}(x).$$

It follows that $\tilde{y} \in \omega(x)$, and (with the same argument as before) $f^m(\tilde{y}) = y$. □

Definition 5.1.11 Let $B \subset X$. The ω-limit set of B is defined as

$$\omega(B) = \bigcap_{N>0} \overline{\bigcup_{n>N} f^n(B)}^d.$$

Remark 5.1.12 There is an equivalent definition using sequences:

$$\omega(B) = \{y \in X \,:\, \exists\,(t_k, x_k) \in \mathbb{N} \times B,\ t_k \to \infty \,:\, f^{t_k}(x_k) \to y\}.$$

Remark 5.1.13 In general $\bigcup_{x\in B} \omega(x) \subset \omega(B)$ is a proper subset. As an example, consider $X = [0, 1]$ (with the canonical metric) and $F : X \to X, x \mapsto x^2$. Then

$$\bigcup_{x\in X} \omega(x) = \{0, 1\}.$$

However, since $f(X) = X$, we find $\omega(X) = X$.

Definition 5.1.14 Let (X, d, f) be a dynamical system. An open set $U \subset X$, $U \neq \emptyset$, is called a pre-attractor, if $f(\overline{U}) \subset U$. A subset $\mathcal{A} \subset X$ is called an attractor (or Conley attractor), if there is a pre-attractor U such that

$$\mathcal{A} = \cap_{n\geq 0} f^n(U).$$

The set U is called a pre-attractor for $\mathcal{A}$. The basin of attraction $\mathcal{B}(\mathcal{A})$ is defined as

$$\mathcal{B}(\mathcal{A}) = \bigcup_{n\geq 0} f^{-n}(U).$$

We show below in Remark 5.1.17 that $\mathcal{B}(\mathcal{A})$ is well defined, i.e., is independent of the choice of the pre-attractor.

Proposition 5.1.15 *Let $\mathcal{A}$ be an attractor with pre-attractor U. Then $\mathcal{A} = \omega(U) = \omega(\overline{U})$.*

Proof As $f(U) \subset f(\overline{U}) \subset U$, we obtain $f^{i+1}(U) \subset f^i(U)$ and hence

$$\bigcup_{i\geq n} f^i(U) = f^n(U).$$

Furthermore, as $f(\overline{U}) \subset U$, we conclude that $f(U) \subset U$ and the sets $f^i(U)$ are nested. Therefore

$$\begin{aligned}\omega(U) &= \bigcap_{n\geq 0} \overline{\bigcup_{i\geq n} f^i(U)} = \bigcap_{n\geq 0} \overline{f^n(U)} = \bigcap_{n\geq 1} \overline{f^n(U)}\\ &= \bigcap_{n\geq 1} f^n(\overline{U}) \subset \bigcap_{n\geq 1} f^{n-1}(U) = \mathcal{A}.\end{aligned}$$

On the other hand,

$$\begin{aligned}\omega(U) &= \bigcap_{n\geq 0} \overline{\bigcup_{i\geq n} f^i(U)} = \bigcap_{n\geq 0} \overline{f^n(U)}\\ &= \bigcap_{n\geq 0} f^n(\overline{U}) \supset \bigcap_{n\geq 0} f^n(U) = \mathcal{A},\end{aligned}$$

hence $\omega(U) = \mathcal{A}$. Furthermore,

$$\omega(U) = \bigcap_{n\geq 0} \overline{\bigcup_{i\geq n} f^i(U)} = \bigcap_{n\geq 0} \overline{f^n(U)} = \bigcap_{n\geq 0} \overline{\bigcup_{i\geq n} f^i(\overline{U})} = \omega(\overline{U}).$$

□

Proposition 5.1.16 *Assume an attractor $\mathcal{A}$, an open set $C \subset X$ with $\mathcal{A} \subset C$ and a closed set $K \subset \mathcal{B}(\mathcal{A})$. Then there is an $n \in \mathbb{N}$ such that*

$$f^n(K) \subset C.$$

Proof Since $\mathcal{A}$ is an attractor, there is a U such that $f(\overline{U}) \subset U$, $\mathcal{B}(\mathcal{A}) = \bigcup_{n\geq 0} f^{-n}(U)$ and $\mathcal{A} = \bigcap_{n\geq 0} f^n(U)$.

Step 1: There is an $m \in \mathbb{N}$, such that $K \subset f^{-m}(U)$.

Suppose there in no such m. For every $m \in \mathbb{N}$ there is a $u_m \in K \setminus f^{-m}(U)$. Since

K is compact, there is a subsequence of the u_m that converges to some $u^* \in K$. Since the $f^n(U)$ are nested, we have

$$u^* \notin \bigcup_{m\geq 0} f^{-m}(U) = \mathcal{B}(\mathcal{A})$$

contradicting $u^* \in K \subset \mathcal{B}(\mathcal{A})$.

Step 2: *There is an* $l \in \mathbb{N}$ *such that* $f^l(\overline{U}) \subset C$.

Suppose that there is no such l. For every $l \in \mathbb{N}$ there is a u_l with

$$u_l \in f^l(\overline{U}) \setminus C.$$

We have $u_l \in f(\overline{U})$ which is closed and hence compact in X. There is a convergent subsequence with limit $u_* \in f(\overline{U})$, and for this point follows a contradiction

$$u_* \in \bigcap_{l\geq 0} f^l(\bar{U}) \setminus C = \mathcal{A} \setminus C = \emptyset.$$

Step 3: Immediately it follows that

$$f^{m+l}(K) \subset C.$$

□

Remark 5.1.17 Let $\mathcal{A}$ be an attractor, and U_1, U_2 pre-attractors for $\mathcal{A}$. As $f(\overline{U}_1)$ is closed and subset of the pre-attractor U_1, and U_2 is open, Proposition 5.1.16 indicates that there is n with $f^n(U_1) \subset f^n(\overline{U}_1) \subset U_2$. Hence, $U_1 \subset f^{-n}(U_2)$, and $\bigcup_{n\geq 0} f^{-n}(U_1) \subset \bigcup_{n\geq 0} f^{-n}(U_2)$. For symmetry reasons, also $\bigcup_{n\geq 0} f^{-n}(U_1) \supset \bigcup_{n\geq 0} f^{-n}(U_2)$, and $\mathcal{B}(\mathcal{A})$ only depends on $\mathcal{A}$ and not on the choice of the pre-attractor.

Proposition 5.1.18 *The set* $\mathcal{B}(\mathcal{A})$ *is open and invariant.*

Proof $\mathcal{B}(\mathcal{A})$ is open as it is the union of open sets. Since $f(U) \subset f(\overline{U}) \subset U$, we find

$$\begin{aligned} f(\mathcal{B}(\mathcal{A})) &= f\left(\bigcup_{n\geq 0} f^{-n}(U)\right) = \bigcup_{n\geq 0} f^{-n+1}(U) \\ &= \left(\bigcup_{n\geq 0} f^{-n}(U)\right) \cup f(U) = \bigcup_{n\geq 0} f^{-n}(U) = \mathcal{B}(\mathcal{A}). \end{aligned}$$

□

Proposition 5.1.19 *Let* (X, d, f) *be a dynamical system. Then there is at least one attractor.*

Proof X is compact. The $f^n(X)$ are nested, nonempty and compact sets. Hence, $\bigcap_{n\geq 0} f^n(X) \neq \emptyset$ is an attractor. □

5.2 Structure of Attractors: Finite Grids

In this section we look at finite grids, $|\Gamma| < \infty$. In this case (E^{Γ}, d) is compact and totally disconnected. The structure of attractors becomes very simple.

Remark 5.2.1 Let X be a finite set equipped with the discrete topology. Then every orbit of the dynamical system (X,f) is eventually periodic. Especially, $\omega(x)$ is a periodic orbit for every x.

Proposition 5.2.2 *Let (X,f) be a dynamical system with X a finite set equipped with the discrete topology. Then, for every $x \in X$ the set $\omega(x)$ is an attractor. Conversely, any attractor is a union of (finitely many) ω-limit sets.*

Proof

Step 1: *$\omega(x)$ is an attractor.*
Let $x \in X$. Then, $\omega(x)$ is clopen and invariant. Thus, $U = \omega(x)$ is a pre-attractor; the corresponding attractor is $\mathcal{A} = \omega(x)$.

Step 2: *An attractor is the union of ω-limit sets.*
Let $\mathcal{A}$ be an attractor. Then there is a pre-attractor $U \supset \mathcal{A}$. As $\mathcal{A} \subset \omega(U)$ and U is finite, we have $\omega(U) = \cup_{u \in U} \omega(u)$, i.e.,

$$\mathcal{A} \subset \bigcup_{u \in U} \omega(u).$$

Furthermore $\omega(u)$ consists of a single periodic orbit and $\mathcal{A}$ is invariant. Thus, we can conclude from $\omega(u) \cap \mathcal{A} \neq \emptyset$ that $\omega(u) \subset \mathcal{A}$. Assume that we have n periodic orbits contained in $\mathcal{A}$. Choose from each periodic orbit in $\mathcal{A}$ one element u_i, then

$$\mathcal{A} = \bigcup_{i=1}^{n} \omega(u_i).$$

□

For the finite, one-dimensional case, $\Gamma = \mathbb{Z}_m$, there are results about the number and structure of periodic orbits [96].

5.3 Intersection of Attractors and Quasi-Attractors

The concepts of this section work for attractors in compact metric spaces. In fact they work with much weaker requirements on the topology. Conley's decomposition theorem, the central topic of this and of the next section, tells us something about the topological structure of attractors. The decomposition theorem says little about the structure of the dynamics within an attractor but it tells something about the number of attractors and their topological relation. Using the theorem, a classification scheme of topological dynamical systems on compact, metric spaces can be achieved. This scheme has been developed by Hurley for cellular automata [93] and it has been refined and extended to more general dynamical systems by Kůrka [111]. For general dynamical systems the classification scheme does not yield strong results. Only the special structure of cellular automata, in particular the existence of translation operators, exhibits more specific properties of the different classes.

Define $\mathbf{\mathcal{A}} = \{\mathcal{A} \subset X : \mathcal{A} \text{ is an attractor}\}$ as the set of all attractors.

Proposition 5.3.1 *$\mathbf{\mathcal{A}}$ is a countable set.*

Proof

Step 1: Choose a countable base of neighborhoods of X.
For each $x \in X$ there is a base of neighborhoods $U_{i,x} = \{y \in X : d(x,y) < 1/i\}$. For each i the sets $\{U_{i,x} : x \in X\}$ cover X. From these we can choose a finite covering

$$X \subset \cup_{j=1}^{m(i)} U_{i,x_j}.$$

The family

$$\mathcal{U} = \{U_{i,x_j} : j = 1, \ldots, m(i);\ i = 1, \ldots, \infty\}$$

is a countable base of neighborhoods. Let $\mathcal{V}$ be the set of all finite unions of elements from $\mathcal{U}$, i.e.,

$$\mathcal{V} = \{ V : \exists m \in \mathbb{N},\ \ U_1, \ldots, U_m \in \mathcal{U}; V = \cup_{i=1}^{m} U_i\}.$$

Then $\mathcal{V}$ is also countable.
Step 2: For every $\mathcal{A} \in \mathbf{\mathcal{A}}$ there is a $U^* = U^*(\mathcal{A}) \in \mathcal{V}$ such that $U_0 \supset \overline{U}^* \supset U^* \supset \mathcal{A}$ for some pre-attractor U_0 of $\mathcal{A}$.
Since $\mathcal{A}$ is an attractor, there is an open set $U_0 = U_0(\mathcal{A})$ such that

$$f(\overline{U}_0) \subset U_0, \quad \mathcal{A} \subset U_0.$$

For every $a \in \mathcal{A}$ there is an open set $U(a) \in \mathcal{U}$, with $a \in U(a)$ and $\overline{U}(a) \subset U_0$. Since $\mathcal{A}$ is compact, there is a finite sub-covering

$$\mathcal{A} \subset \cup_{i=1}^{m} U(a_i) =: U^* \in \mathcal{V}.$$

Now it follows that $\mathcal{A} \subset U^* \subset \overline{U}^* \subset U_0$. Of course U^* is not unique. With every attractor $\mathcal{A}$ we associate one set U^* constructed as above. Then we have a mapping

$$\mathbf{\mathcal{A}} \to \mathcal{V}, \qquad \mathcal{A} \mapsto U^*(\mathcal{A}).$$

Step 3: $\mathcal{A} \neq \tilde{\mathcal{A}}$ *implies* $U^*(\mathcal{A}) \neq U^*(\tilde{\mathcal{A}})$.
Consider two attractors $\mathcal{A} \neq \tilde{\mathcal{A}}$. Suppose $U^*(\mathcal{A}) = U^*(\tilde{\mathcal{A}})$. Then there is a pre-attractor U_0 of $\mathcal{A}$ such that $U^*(\mathcal{A}) \subset U_0$. As $\tilde{\mathcal{A}}$ is invariant,

$$\tilde{\mathcal{A}} \subset f^n(U^*(\tilde{\mathcal{A}})) = f^n(U^*(\mathcal{A})) \subset f^n(U_0).$$

The sets $f^n(U_0)$ are nested, and $\mathcal{A} = \cap_{n\in\mathbb{N}} f^n(U_0)$. This implies $\tilde{\mathcal{A}} \subset \mathcal{A}$. Now exchange the roles of $\tilde{\mathcal{A}}$ and $\mathcal{A}$.
Hence the mapping $\mathcal{A} \to \mathcal{V}$ is injective. But $\mathcal{V}$ is countable. Hence the set of attractors is at most countable. □

We investigate the structure of attractors further. We are especially interested in situations where more than one attractor exists. Therefore we need results on the union and intersection of attractors.

Proposition 5.3.2 *The union of two attractors is an attractor. In particular, there is a maximal attractor. Let* $\tilde{\mathcal{A}}$ *and* $\hat{\mathcal{A}}$ *be two attractors with nonempty intersection. Then there is an attractor* $\mathcal{A} \subset \tilde{\mathcal{A}} \cap \hat{\mathcal{A}}$.

Proof Let $\tilde{\mathcal{A}}$ and $\hat{\mathcal{A}}$ be two attractors. Then there are corresponding pre-attractors $\tilde{U}$ and $\hat{U}$. Then the set $U := \tilde{U} \cup \hat{U}$ is a pre-attractor. We find at once that $\hat{\mathcal{A}} \cup \tilde{\mathcal{A}} \subset \omega(U)$. Let $a \in \omega(U)$. Then, there are sequences $x_i \in \tilde{U} \cup \hat{U}$, $n_i \to \infty$, such that $f^{n_i}(x_i) \to a$. There is a subsequence of $\{x_i\}$ completely contained in $\tilde{U}$ (or in $\hat{U}$), and hence $a \in \tilde{\mathcal{A}}$ (or $a \in \hat{\mathcal{A}}$). Therefore,

$$\hat{\mathcal{A}} \cup \tilde{\mathcal{A}} = \omega(U)$$

is an attractor.

This observation implies that the relation "$\subset$" defines a partial order on the set of attractors with maximal element $\omega(X)$.

Assume $\tilde{\mathcal{A}} \cap \hat{\mathcal{A}} \neq \emptyset$. Then it follows from $\hat{\mathcal{A}} \subset \hat{U}$, $\tilde{\mathcal{A}} \subset \tilde{U}$ that $U = \hat{U} \cap \tilde{U} \neq \emptyset$. The set U is a pre-attractor: If $u \in \overline{U}$, then u is in the closure of $\hat{U}$ and in the closure of $\tilde{U}$; therefore $f(u) \in \hat{U} \cap \tilde{U}$, i.e., $f(\overline{U}) \subset U$.

Let $\mathcal{A}$ be the attractor to U, i.e., $\mathcal{A} = \omega(U)$. Then

$$\mathcal{A} = \omega(\tilde{U} \cap \hat{U}) \subset \omega(\tilde{U}) \cap \omega(\hat{U}) = \tilde{\mathcal{A}} \cap \hat{\mathcal{A}}.$$

□

Let $\mathcal{A}_1, \mathcal{A}_2$, be two attractors with nonempty intersection. In general $\mathcal{A}_1 \cap \mathcal{A}_2$ is not an attractor. Here follows an example.

Example 5.3.3 We describe the example only verbally. In Fig. 5.1a, a dynamical system on a branched manifold is depicted. The set $X \subset \mathbb{R}^2$ consist of two lines, $[C_-, C_+] \cup [A, B]$. The two lines joint in point B. With the Euclidean metric d on $\mathbb{R}^2$ restricted to X, the space (X, d) is a compact metric space. We define a map $f : X \to X$ that possesses five stationary points. p_1, p_3 and p_5 are locally asymptotically stable, p_2 and p_4 are unstable. The function f is well behaved

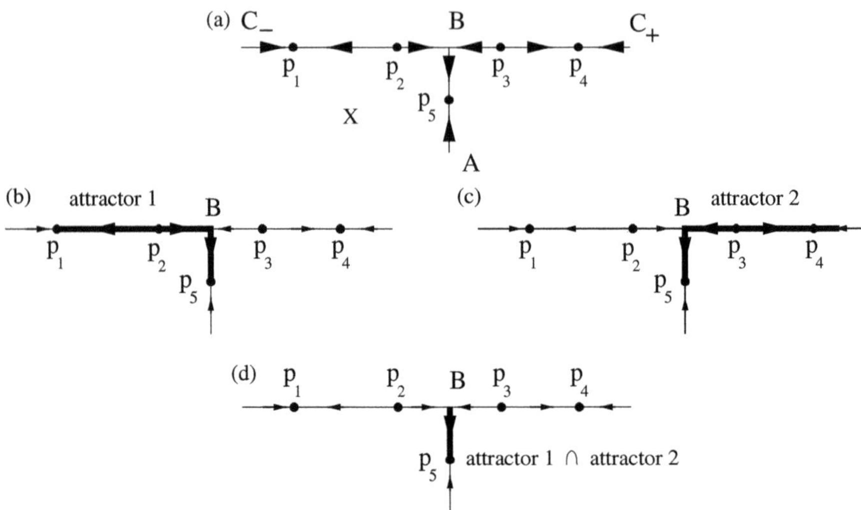

Fig. 5.1 Visualization of Example 5.3.3. In (**b**) attractor 1, (**c**) attractor 2 and in (**d**) their intersection are marked by *bold lines*

in that all trajectories are monotonously converging towards the nearest locally asymptotically stable stationary point; at point B, this implies that trajectories starting in (p_2, B) eventually leave $[C_-, C_+]$, and converge to p_5; the points "take a right turn" at point B. Also trajectories starting in (B, p_3) will (by means of a "left turn" at point B) tend to p_5. The line $[B, A]$ is forward invariant, while the line $[C_-, C_+]$ is not.

The set $[p_1, B] \cup [B, p_5]$ is an attractor (Fig. 5.1b) as well as $[p_4, B] \cup [B, p_5]$ (Fig. 5.1c). The intersection of these two attractors is $[B, p_5]$ (Fig. 5.1d). This set is forward invariant, but not backward invariant. That is, $[B, p_5]$ is not an attractor. However, it contains the set $\{p_5\}$, which again is an attractor.

We have shown that the non-empty intersection of two attractors contains again an attractor. Then the non-empty intersection of finitely many attractors contains again an attractor. But the nonempty intersection of infinitely many attractors may not contain an attractor as is shown by the next example.

Example 5.3.4 (from Hurley [93]) Consider $X = S^1$ (realized as the one-point compactification of $\mathbb{R}$) and a continuously differentiable function $f : X \to X$ with the following properties. It has exactly the fixed points $z_i = 1/i, i \in \mathbb{N}$, and the fixed points 0 and ∞. Furthermore $f'(z_i) > 0$, and $f'(z_i) > 1$ for even i and $f'(z_i) < 1$ for odd i. For $x < 0$ the function satisfies $0 > f(x) > x$. Then

$$x_n \to \frac{1}{2i+1} \qquad \text{for } x_0 \in (\frac{1}{2i}, \frac{1}{2i+2}),$$

$x_n \to 1$ for $x_0 > 1$ and $x_n \to 0$ for $x_0 < 0$.

Then the sets

$$A_i = [0, \frac{1}{2i+1}]$$

are attractors for $i \in \mathbb{N}$, but $\{0\}$ is not. Nevertheless, $\{0\}$ has an interesting property which leads us to the following definition of a quasi-attractor.

Definition 5.3.5 Let (X, d, f) be a dynamical system.

(1) An attractor $\mathcal{A}$ is called minimal if there is no attractor $\tilde{\mathcal{A}} \subset \mathcal{A}$, $\tilde{\mathcal{A}} \neq \mathcal{A}$.
(2) Any nonempty intersection of finitely or countably many attractors $Q = \bigcap_{n \geq 0} \mathcal{A}_n \neq \emptyset$ is called a quasi-attractor.
(3) A quasi-attractor Q is called minimal if no proper (nonempty) subset is a quasi-attractor.

The concept of a quasi-attractor, which is weaker than that of an attractor, is useful in describing the asymptotic behavior of dynamical systems.

Even a dynamical system on the unit interval may have more than countably many quasi-attractors. From Hurley we have the following example.

Example 5.3.6 (from Hurley [93]) Let $R_1 = [0, 1] \times [0, 1]$ be a square in the plane (starting in dimension two makes things easier to understand, see Fig. 5.2). There are two attractors in the squares $R_{(1,0)} = [1/5, 2/5] \times [2/5, 3/5]$ and $R_{(1,1)} = [3/5, 4/5] \times [2/5, 3/5]$, respectively, and their domains of attraction are separated by a saddle point at $(1/2, 0)$. Inside these two squares we have exactly the same flow as in R_1 (after an appropriate affine mapping). Then the set of quasi-attractors is a Cantor set on the x-axis. We can construct the example in such a way that the x-axis is an invariant set. Then we can restrict the system to the x-axis. Since the

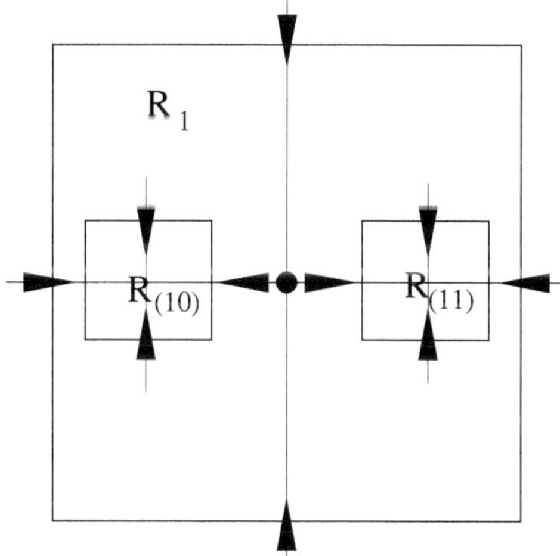

Fig. 5.2 Visualization of Example 5.3.6

Cantor set is not countable we get more than countably many quasi attractors for a dynamical system on [0, 1].

Proposition 5.3.7 *Let (X, d, f) be a dynamical system. Let $\mathbf{Q}$ be the set of quasi-attractors. The set $\mathbf{Q}$ has a unique maximal element $\omega(X)$ with respect to inclusion "$\subset$". If any two attractors have nonempty intersection then the set $\mathbf{Q}$ has also a unique minimal element.*

Proof Let $\mathcal{A}$ be any attractor. There is a pre-attractor $U \subset X$, and hence $\mathcal{A} = \omega(U) \subset \omega(X)$. Hence the global attractor $\omega(X)$ is the maximal element.

There are only countably many attractors, the set $\boldsymbol{\mathcal{A}}$ of all attractors can be written as

$$\boldsymbol{\mathcal{A}} = \{\mathcal{A}_i \,:\, i \in \mathcal{I}\}$$

with some index set $\mathcal{I}$ which is either $\{1, 2, \ldots, m\}$ or $\mathbb{N}$. Define $Y_1 = \mathcal{A}_1$ and successively choose attractors Y_i such that

$$Y_{i+1} \subset Y_i \cap \mathcal{A}_{i+1} \text{ for } i \in \mathcal{I}.$$

For each i the choice of Y_{i+1} may not be unique. The attractors Y_i are nested, hence the set

$$Q = \bigcap_{i \in \mathcal{I}} Y_i$$

is not empty and it is contained in each of the $\mathcal{A} \in \boldsymbol{\mathcal{A}}$. The set Q is an element of $\mathbf{Q}$. Let $\tilde{Q}$ be any element of $\mathbf{Q}$. Then there is an index set $\tilde{\mathcal{I}} \subset \mathcal{I}$ such that

$$\tilde{Q} = \bigcap_{i \in \tilde{\mathcal{I}}} \mathcal{A}_i \supset Q.$$

Hence Q is the minimal element. Of course Q is unique. □

Theorem 5.3.8 (Hurley's Classification of Dynamical Systems) *Let (X, d, f) be a dynamical system. Then it belongs to exactly one of the following classes:*

[Hurley 1] *There are at least two disjoint attractors.*
[Hurley 2] *There is a unique minimal quasi-attractor which is not an attractor.*
[Hurley 3] *There is a unique minimal attractor which is different from $\omega(X)$.*
[Hurley 4] *There is exactly one attractor $\omega(X) \neq X$.*
[Hurley 5] *There is exactly one attractor $\omega(X) = X$.*

Proof Suppose the dynamical system is not in class [Hurley 1]. Then any two attractors have nonempty intersection. Hence there is a unique minimal quasi-attractor. Either this quasi-attractor is not an attractor (class [Hurley 2]), or it is an attractor but a proper subset of the global attractor (class [Hurley 3]). The said quasi-attractor may be the global attractor (in this case the minimal attractor is also the maximal attractor, hence it is the unique attractor). Either this unique attractor is smaller than the whole space [Hurley 4]) or this attractor is the space X (class [Hurley 5]). □

5.4 Conleys Decomposition Theorem, Attractors, and Chains

In order to better understand the structure of quasi-attractors we weaken the concept of a trajectory to that of an ε-chain. An ε-chain is something like a trajectory that allows a small error at each iteration step. With this concept we can find further properties of the Hurley classes. At the end we will have objects that are larger than ω-limit sets of points but smaller than attractors. In a way these objects are the "atoms" of attractors.

Definition 5.4.1 Let (X,d,f) be a dynamical system. A finite sequence of points $\{x_i\}_{i=0,\dots,n}$ is called an ε-chain (from x_0 to x_n), if

$$d(f(x_i), x_{i+1}) < \varepsilon \qquad \text{for } i = 0,\dots,n-1.$$

A point $x \in X$ is called recurrent, if for every $\varepsilon > 0$ there is an ε-chain from x to x. The set of recurrent points is called $\mathrm{CR}(f)$,

$$\mathrm{CR}(f) = \{x \,:\, x \text{ is recurrent}\}.$$

Example 5.4.2 Consider a dynamical system in $\mathbb{R}^2$ given in polar coordinates. The mapping $f : \mathbb{R}^2 \to \mathbb{R}^2$ is given by

$$(\phi, r) \mapsto (f_1(\phi), f_2(r))$$

with $f_1(\phi) = \phi + a$ and

$$f_2(r) = \frac{2r}{1+r}.$$

The function f_2 has two fixed points $r = 0$ and $r = 1$, and the iteration $r \mapsto f_2(r)$ carries all points $r \in (0,\infty)$ to $r = 1$. Hence for the two-dimensional system we have $\omega(0) = 0$ and $\omega(x) \subset \{x \in \mathbb{R}^2 : \|x\| = 1\}$ if $x \neq 0$. Furthermore, every point on the unit circle is recurrent, i.e., it is a periodic point (if $a/(2\pi) \in \mathbb{Q}$), or the trajectory is dense on the unit circle (if $a/(2\pi) \in \mathbb{R} \setminus \mathbb{Q}$). In either case we find two attractors,

$$\mathcal{A}_1 = \{x \in \mathbb{R}^2 : \|x\| \leq 1\}, \qquad \mathcal{A}_2 = \{x \in \mathbb{R}^2 : \|x\| = 1\},$$

and the set of recurrent points is

$$\mathrm{CR}(f) = \{0\} \cup \{x \in \mathbb{R}^2 \,:\, \|x\| = 1\}.$$

In a way the recurrent points appear as the important part of the attractor.

Conley [35] proved a fundamental result on recurrent points which is now called the Conley decomposition theorem.

Theorem 5.4.3 (Conley Decomposition Theorem) *The non-recurrent points are exactly those points that are in the basin of an attractor but not in the attractor itself,*

$$X \setminus CR(f) = \bigcup_{\mathcal{A} \in \mathcal{A}} (\mathcal{B}(\mathcal{A}) \setminus \mathcal{A}).$$

Remark 5.4.4 In a heuristic view the set $\mathrm{CR}(f)$ is formed by the minimal attractors and the boundaries of their domains of attraction.

Conley's decomposition theorem holds also for non-compact spaces [31, 95]. Here we present the theorem only for the compact case. The main part of the proof is preceded by some lemmas. The following two propositions yield a device to inhibit an ε-chain which is contained in the basin of an attractor, but not in the attractor itself, from returning to its origin. The proposition yields a pre-attractor that is contained in a given open set. With this result we show that a recurrent point in a basin is already contained in the attractor. In the last step before the proof of Conley's decomposition theorem we construct pre-attractors from chains.

Proposition 5.4.5 *Let $\mathcal{A}$ be an attractor, let C be open and $\mathcal{A} \subset C$. Then there is an open set $V \subset C$ with the following properties:*

(1) $f(\overline{V}) \subset V$
(2) $\bigcap_{n \geq 0} f^n(V) = \mathcal{A}$.

Proof Let U be a pre-attractor of $\mathcal{A}$. Without restriction $C \subset U$, as otherwise we consider $C \cap U$ instead of C.

Step 1: There is an open set W such that $\mathcal{A} \subset W \subset C \subset U$ and $f^n(W) \subset C$ for all $n \geq 0$.

First we construct an open set W_1 such that

$$\mathcal{A} \subset W_1 \subset \overline{W}_1 \subset U.$$

For each $a \in \mathcal{A}$ choose $\varepsilon(a)$ such that

$$B_{2\varepsilon(a)}(a) := \{y \in X \,:\, d(a, y) < 2\,\varepsilon(a)\} \subset U.$$

Then

$$a \in B_{\varepsilon(a)}(a) \subset \overline{B_{\varepsilon(a)}(a)} \subset B_{2\varepsilon(a)}(a) \subset U.$$

The compact set $\mathcal{A}$ can be covered by finitely many of these balls $B_{\varepsilon(a)}(a)$, there are points $a_1, \ldots, a_m \in \mathcal{A}$ such that

$$\mathcal{A} \subset \bigcup_{i=1}^{m} B_{\varepsilon(a_i)}(a_i) =: W_1.$$

The set W_1 is open, and $\mathcal{A} \subset W_1 \subset \overline{W}_1 \subset U$.
Now apply Proposition 5.1.16 with $C \cap W_1$ instead of C and $\overline{W}_1$ instead of B. There is n_0 such that

$$f^{n+n_0}(\overline{W}_1) \subset C \cap W_1 \qquad \text{for all } n \in \mathbb{N}.$$

We define $W = f^{n_0}(W_1)$ and find

$$\mathcal{A} \subset W \subset \overline{W} \subset C \subset U$$

and

$$f^n(W) \subset C \qquad \text{for all } n \in \mathbb{N}.$$

Step 2: Construction of the set V.
Since $\overline{W}$ is compact and $\mathcal{A} \subset \overline{W} \subset U$, we conclude from Proposition 5.1.16 that there is some $N > 0$ such that $f^N(\overline{W}) \subset W$. Now choose $\eta > 0$. Define $V_1 = W$, and for $i > 1$ define V_i as the η-neighborhood of $f(V_{i-1})$, i.e.,

$$V_i = \{u \in X \,:\, \exists v \in V_{i-1} \,:\, d(u, f(v)) < \eta\} \text{ for } i > 1.$$

For η small enough we have $V_i \subset C$ for $i = 1, \ldots, N+1$, $f(\overline{V}_i) \subset V_{i+1}$ for $i = 1, \ldots, N-1$, and $f(\overline{V}_N) \subset W = V_1$. Next define

$$V = \bigcup_{i=1}^{N} V_i$$

Then V is open, $V \subset C$, and

$$f(\overline{V}) = \bigcup_{i=1}^{N} f(\overline{V}_i) \subset \bigcup_{i=1}^{N-1} f(\overline{V}_i) \,\cup\, f(\overline{V}_N) \subset \bigcup_{i=2}^{N} V_i \,\cup\, V_1 = V.$$

Step 3: Property (2).
The pre-attractor U has the properties (1) and (2) claimed for V, and $C \subset U$. We have $\mathcal{A} = \omega(U)$ and hence $\mathcal{A} = \omega(V)$ for every V with $\mathcal{A} \subset V \subset U$. In particular we have property (2) for all V with property (1) and $\mathcal{A} \subset V \subset C \subset U$.

□

Proposition 5.4.6 *Let $\mathcal{A}$ be an attractor and $x \in CR(f) \cap \mathcal{B}(\mathcal{A})$. Then $x \in \mathcal{A}$.*

Proof Let $x \in \mathcal{B}(\mathcal{A}) \setminus \mathcal{A}$. Choose an open set $C \supset \mathcal{A}$ with $x \notin C$. Apply Proposition 5.4.5 and find an open set V with $\mathcal{A} \subset V \subset C$ such that $x \notin V$, $f(\overline{V}) \subset V$. Since $X, f(\overline{V})$ and $X \setminus V$ are compact, we have

$$\min\{\mathrm{dist}(f(z), X \setminus V) \, : \, z \in \overline{V}\} = \delta > 0.$$

We show that no ε-chain starting from x returns to x if ε is small enough. Since $\mathrm{dist}(f^n(x), \mathcal{A}) \to 0$, there is $n_0 \in \mathbb{N}$ with $f^{n_0}(x) \in V$. The function f is continuous and, since X is compact, even uniformly continuous. Hence we find $\varepsilon_0 > 0$, such that for every ε-chain $\{y_i\}_{i=1,\dots,n_0}$ with $y_1 = x$, $\varepsilon < \varepsilon_0$, we have $y_{n_0} \in V$. For small ε, the y_{k+1} are close to $f^k(x)$. Hence we can find $\varepsilon_1 > 0$ such that for $\varepsilon \in (0, \varepsilon_1)$ no ε-chain with length $n \leq n_0$ connects x to x. For $\varepsilon < \min\{\varepsilon_0, \varepsilon_1\}$, every ε-chain connecting x to x has length greater than n_0 and first enters V and then approaches x. We exclude also this possibility by showing that any ε-chain is trapped in V if $\varepsilon < \delta/2$: Let $y_i \in V$. We have $f(y_i) \in V$, and

$$\begin{aligned} \delta &\leq \mathrm{dist}(f(y_i), X \setminus V) \leq \mathrm{dist}(y_{i+1}, X \setminus V) + d(y_{i+1}, f(y_i)) \\ &\leq \mathrm{dist}(y_{i+1}, X \setminus V) + \varepsilon < \mathrm{dist}(y_{i+1}, X \setminus V) + \delta/2. \end{aligned}$$

Hence, $\mathrm{dist}(y_{i+1}, X \setminus V) > \delta/2$ and $y_{i+1} \in V$. Now choose $\varepsilon < \min\{\varepsilon_0, \varepsilon_1, \delta/2\}$ and consider an ε-chain starting with x. The arguments above show that the chain cannot return to x, in contradiction to the hypothesis $x \in \mathrm{CR}(f)$. □

Proposition 5.4.7 *Let $x \in X$ and $\varepsilon > 0$. The set*

$$U = \{y \, : \, \textit{there is an } \varepsilon\textit{-chain from } x \textit{ to } y\}$$

is a pre-attractor.

Proof The set U is open. We show that $f(\overline{U}) \subset U$. If there is an ε-chain from x to y, then also from x to $f(y)$ (just extend the chain by $f(y)$).

Let $y \in \overline{U}$. For every $\delta > 0$ we have $B_\delta(y) \cap U \neq \emptyset$ and $f(B_\delta(y)) \cap f(U) \neq \emptyset$. Since f is continuous, we find (for the given ε) a $\delta > 0$ such that $f(B_\delta(y)) \subset B_\varepsilon(f(y))$. For this δ choose $\tilde{z} \in U, f(\tilde{z}) \in f(B_\delta(y)) \cap f(U) \subset B_\varepsilon(f(y))$. Then,

$$d(f(y), f(\tilde{z})) < \varepsilon.$$

We have an ε-chain $\{z_i\}_{i=1,\dots,n}$ from x to $\tilde{z}$. Extend the chain by $z_{n+1} = f(y)$. Then $\{z_i\}_{i=1,\dots,n+1}$ is an ε-chain from x to $f(y)$, hence $f(\overline{U}) \subset U$. □

Proof (of Theorem 5.4.3)

Step 1: $X \setminus \mathrm{CR}(f) \supset \cup_{\mathcal{A}\in\mathfrak{A}} \mathcal{B}(\mathcal{A}) \setminus \mathcal{A}$.
From Proposition 5.4.6 we know $\mathrm{CR}(f) \cap \mathcal{B}(\mathcal{A}) \subset \mathcal{A}$. Hence we have the implication

$$(\mathcal{B}(\mathcal{A}) \setminus \mathcal{A}) \cap \mathrm{CR}(f) = \emptyset \quad \Rightarrow \quad (\mathcal{B}(\mathcal{A}) \setminus \mathcal{A}) \subset X \setminus \mathrm{CR}(f).$$

Step 2: $X \setminus \mathrm{CR}(f) \subset \cup_{\mathcal{A}\in\mathfrak{A}} \mathcal{B}(\mathcal{A}) \setminus \mathcal{A}$.
Let $x \notin \mathrm{CR}(f)$. Then there is a (sufficiently small) $\varepsilon > 0$ such that there is no ε-chain connecting x to itself. Let

$$U = \{y \,:\, \text{there is an } \varepsilon\text{-chain from } x \text{ to } y\},$$

Then $x \notin U$, but $f(x) \in U$. Since U is a pre-attractor (Proposition 5.4.7), we find an attractor $\mathcal{A} = \cap_{n\geq 0} f^n(U)$. Therefore $x \in \mathcal{B}(\mathcal{A}) \setminus \mathcal{A}$.

□

Definition 5.4.8 Define an equivalence relation on $\mathrm{CR}(f)$. For $x, y \in \mathrm{CR}(f)$ define $x \sim y$ if for every $\varepsilon > 0$ there is an ε-chain from x to y and from y to x. The equivalence classes $\mathrm{CR}(f)/\sim$ are called chain components.

A chain component is a subset of X which consists of a point in $\mathrm{CR}(f)$ and all equivalent points.

What is the idea behind this definition? Chain components are somewhat similar to attractors. We can show that limit sets $\omega(x)$ are always contained in chain components and that chain components are invariant. If several ω-limit sets are close to each other then they may coalesce if an ε-perturbation is applied, see Example 5.4.9. On the other hand the concept of a chain component is weaker than the concept of an attractor. A chain component needs not to be attracting. We will see that a minimal element within the set of all attractors (or quasi-attractors) consists of exactly one chain component. Hence chain components are the nuclei of attractors. This property makes them interesting. Now we shall prove the statements just mentioned.

Example 5.4.9 Let $f : S^1 \to S^1$, $\phi \mapsto (\phi + 2\pi a) \bmod 2\pi$. We know that for $a \notin \mathbb{Q}$ there are no periodic points, and $\omega(\phi_0) = S^1$ for every $\phi_0 \in S^1$. On the other hand, if $a = m/n \in \mathbb{Q}$ then there are only periodic points. If m and n have common divisor 1 then these have period n. In this case $|\omega(\phi_0)| = n$ is finite. However, for all choices of a, S^1 is a single chain component.

Proposition 5.4.10 *Let $C \in CR(f)/\sim$ be a chain component. Then C is invariant under f.*

Proof

Step 1: If $x \in C$ then any accumulation point of ε-chains connecting x with x is also in C.

Let $x \in C$ and let $(x_i^\varepsilon)_{i=0,\ldots,n(\varepsilon)}$ be an ε-chain from x to x. Let y^* be an accumulation point of the set

$$\bigcup_{\varepsilon=1/l,\, l\in\mathbb{N}} \{x_i^\varepsilon : i = 0, \ldots, n(\varepsilon)\}.$$

We show that for every $\varepsilon > 0$ there is an ε-chain from y^* to x: Since f is continuous, there is $\delta > 0$ such that $d(x, y^*) < \delta$ implies $d(f(x), f(y^*)) < \varepsilon/2$. Since y^* is an accumulation point, there is an η-chain, $\eta > 0$, $\eta < \varepsilon/2$, and an $i^* = i^*(\eta)$ with $d(y^*, x_{i^*(\eta)}^\eta) < \delta$. Define

$$y_0 = y^*, \quad y_i = x_{i+i^*(\eta)-1}^\eta, \quad i = 1, \ldots, n(\eta).$$

Then $\{y_i\}$ is an ε-chain from y^* to x (recall $\eta < \varepsilon$).

We can also connect x to y^* by choosing, for the given ε and η, an η-chain such that there is an $i^*(\eta)$ with $d(x_{i^*(\eta)}^\eta, y^*) < \varepsilon/2$. Then the chain

$$y_i = x_i \quad \text{for } i = 1, \ldots, i^*(\eta) - 1, \quad y_{i^*(\eta)} = y^*$$

is an ε-chain from x to y.

Step 2: C is forward invariant.

Let $x \in C$. Then $f(x)$ is an accumulation point of $(x_2^\varepsilon)_{\varepsilon=1/l}$ (compare part (1) of the proof), and hence $f(x) \in C$.

Step 3: C is backward invariant.

Again let $(x_i^\varepsilon)_{i=1,\cdots,n(\varepsilon)}$ be an ε-chain from x to x. For $\varepsilon = 1/l$, $l \in \mathbb{N}$, consider the sequence $y_l = x_{n(\varepsilon)-1}^\varepsilon$. Since X is compact, there is an accumulation point y^* of $(y_l)_l \in \mathbb{N}$. Then it follows that $y^* \in C, f(y^*) = x$ and hence also $C \subset f(C)$.

□

Proposition 5.4.11 *For every $x \in X$ the set $\omega(x)$ is contained in a single chain component. In particular $\omega(x) \subset CR(f)$ for all $x \in X$.*

Proof Let $y_1, y_2 \in \omega(x)$. For every $i \in \mathbb{N}$ there are numbers n_i^1, n_i^2 such that

$$d(f^{n_i^j}(x), y_j) < 1/i \qquad \text{for } j \in \{1, 2\}.$$

Assume $\varepsilon > 0$. Since f is continuous and X compact, there is $\delta = \delta(\varepsilon)$ such that

$$d(f(x), f(y)) < \varepsilon \quad \text{whenever } d(x, y) < \delta.$$

Now choose $i_1 > 1/\min\{\delta, \varepsilon\}$. Then it follows that

$$d(y_1, f^{n^1_{i_1}}(x)) < \min\{\delta, \varepsilon\}$$

and hence, with $z_0 = y_1$, $z_k = f^{n^j_i+k}(x)$, $k \in \mathbb{N}_0$, that

$$d(z_0, f^{n^1_{i_1}}(x)) < \delta,$$
$$d(f(z_0), z_1) = d(f(z_0), f(f^{n^1_{i_1}}(x))) < \varepsilon.$$

Next choose $i_2 > i_1$ such that $n^2_{i_2} > n^1_{i_1}$. Then

$$d(f^{n^2_{i,2}}(x), y_2) < 1/i_2 < 1/i_1 < \varepsilon.$$

Hence $\{z_k\}_{k=0,\ldots,n^2_{i_2}-n^1_{i_1}}$ is an ε-chain from y_1 to y_2. We can exchange the roles of y_1 and y_2 and obtain, for any $\varepsilon > 0$, an ε-chain from y_2 to y_1.

In particular we can choose $y_1 = y_2$. Then we get $\omega(x) \subset CR(f)$. The argument further shows that any two points of $\omega(x)$ belong to the same chain component. □

Definition 5.4.12 The basin of attraction of a chain component C is the set of all points $x \in X$ for which the ω-limit set is contained in C,

$$\mathcal{B}(C) = \{x \in X : \omega(x) \subset C\}. \qquad (*)$$

Remark 5.4.13 Let $\mathcal{A}$ be an attractor with a pre-attractor U and assume that $\mathcal{A}$ consists of a single chain component C, i.e., $C = \mathcal{A}$. Then we have two definitions of a basin of attraction, one for the attractor,

$$\mathcal{B}(\mathcal{A}) = \bigcup_{n\in\mathbb{N}} f^{-n}(U),$$

and one for the chain component, as in $(*)$. But these sets are the same: We have $\mathcal{B}(C) \subset \mathcal{B}(\mathcal{A})$ since $\omega(x) \subset \mathcal{A}$ implies that the $f^n(x)$ approach the set $\mathcal{A}$ arbitrarily close and are contained in the pre-attractor U after finitely many steps. Conversely, any point $x \in \mathcal{B}(\mathcal{A})$ (the set $\{x\}$ is closed) is attracted by $\mathcal{A}$, i.e., $\text{dist}(f^n(x), \mathcal{A}) \to 0$ for $n \to \infty$. Hence $\omega(x) \subset \mathcal{A}$ and hence $\mathcal{B}(C) \supset \mathcal{B}(\mathcal{A})$ and further $\mathcal{B}(C) = \mathcal{B}(\mathcal{A})$.

Proposition 5.4.14 *Let C, C_1, C_2 be chain components and let $\mathcal{A}, \mathcal{A}_1, \mathcal{A}_2$ be attractors. Then the following statements hold:*

(1) $\mathcal{B}(C_1) \cap \mathcal{B}(C_2) \neq \emptyset \quad \Rightarrow \quad C_1 = C_2$

(2) $\mathcal{B}(\mathcal{A}_1) \cap \mathcal{B}(\mathcal{A}_2) \neq \emptyset \quad \Rightarrow \quad \mathcal{A}_1 \cap \mathcal{A}_2 \neq \emptyset$

(3) $\mathcal{B}(C) \cap \mathcal{B}(\mathcal{A}) \neq \emptyset \quad \Rightarrow \quad C \subset \mathcal{A}.$

Proof

1) For $x \in \mathcal{B}(C_1) \cap \mathcal{B}(C_2)$ we have $\omega(x) \subset C_1$ and also $\omega(x) \subset C_2$, i.e., $C_1 \cap C_2 \neq \emptyset$.
Hence there is an element $y \in C_1 \cap C_2$. From every point $c_1 \in C_1$ there is an ε-chain to y, and another ε-chain to every $c_2 \in C_2$ and back. Hence c_1 and c_2 are in the same chain component, and $C_1 = C_2$.

2) If $x \in \mathcal{B}(\mathcal{A}_1) \cap \mathcal{B}(\mathcal{A}_2)$ then $\omega(x) \subset \mathcal{A}_1$ and $\omega(x) \subset \mathcal{A}_2$, hence $\mathcal{A}_1 \cap \mathcal{A}_2 \neq \emptyset$.

3) For every $x \in \mathcal{B}(C) \cap \mathcal{B}(\mathcal{A})$ we have $\omega(x) \in C \cap \mathcal{A}$. Hence $C \cap \mathcal{A} \neq \emptyset$. Let $x \in C \cap \mathcal{A}$ and let $y \in C$. Then there is an ε-chain from x to y. Now we repeat the construction in the proof of Proposition 5.4.6. Let U be a pre-attractor for $\mathcal{A}$. We have $\min\{\mathrm{dist}(f(x), X \setminus U) \,:\, x \in \overline{U}\} = \delta > 0$. Hence a $\delta/2$-chain cannot leave U. Hence $y \in \mathcal{B}(\mathcal{A})$ and (again because of $y \in \mathrm{CR}(f)$) also $y \in \mathcal{A}$, i.e., $C \subset \mathcal{A}$.

□

Proposition 5.4.15 *A minimal quasi-attractor consists of exactly one chain component.*

Proof Let $x \in Q$ and

$$U^{(\varepsilon)} = \{z \,:\, \text{there is an } \varepsilon\text{-chain from } x \text{ to } z\}.$$

Then $U^{(\varepsilon)}$ is a pre-attractor (see Proposition 5.4.7). Define $\mathcal{A}_i = \omega(U^{(1/i)})$. Then $f(x) \in U^{(1/i)} \subset \mathcal{B}(\mathcal{A}_i)$, i.e., $Q \cap B(\mathcal{A}_i) \neq \emptyset$. As Q is a quasi-attractor, there are attractors $\mathcal{A}_j$ such that $Q = \cap_j \mathcal{A}'_j$. Then, $x \in \mathcal{A}_i$, $x \in \mathcal{A}'_j$ and hence $Q' = (\cap_i \mathcal{A}_i) \cap (\cap_j \mathcal{A}'_j) \subset Q$ is a quasi-attractor. As Q is minimal, $Q' = Q$ and

$$Q \subset \cap_{i \in \mathbb{N}} \mathcal{A}_i.$$

In other words, $Q \subset U^{(\varepsilon)}$ for all $\varepsilon > 0$, i.e., for every point $y \in Q$ and every arbitrarily small ε there is an ε-chain from x to y. By the same argument there is a chain from y to x. Therefore x and y are in $\mathrm{CR}(f)$, and they are in the same chain component C. Since $x, y \in Q$ have been arbitrary, $Q = C$. □

5.5 Bernoulli Measure on Cellular Automata

Here we define measures on the state spaces of cellular automata. We first define measures for a single cell and then we form product measures.

Definition 5.5.1 Assume a probability measure on E. For each $\eta \in E$ there is a probability $p_\eta \in (0, 1]$ such that $\sum_{\eta \in E} p_\eta = 1$. This measure induces a Borel probability measure μ on the power set $\mathfrak{P}(E)$. The corresponding product measure μ on E^Γ is called the Bernoulli measure induced by $\{p_\eta\}_{\eta \in E}$.

In the present and the following section, we exclude the trivial cases $|E| = 1$, and $\exists e \in E : p_e = 0$.

Remark 5.5.2 At this point, it is useful to recall how a product measure is defined, and to relate this concept to the situation at hand. For the precise definitions and proofs consult Appendix A.3 or e.g. [150]. We start from the Borel probability measure on $\mathfrak{P}(E)$. First consider a state over a finite part of the grid, $u \in E^{\Gamma_n}$ for some $n \in \mathbb{N}$. It is natural to assign to u the measure

$$\mu(u) = \prod_{g\in\Gamma_n} p_{u(g)}.$$

And indeed, this definition yields a probability measure on $\mathcal{P}(E^{\Gamma_n})$. For $A \subset E^{\Gamma_n}$, we set

$$\mu(A) = \sum_{u\in A} \mu(u).$$

Next we turn to subsets of E^Γ that only carry information on a finite part of the grid: the cylinder sets. Together with the empty set, these sets form a semi-algebra $\mathfrak{S}$. Let $A \in \mathfrak{S}$. In order to define the (pre)measure of this set, we first determine the part of the grid on which A is non-trivial. For a subset of the grid $\hat{\Gamma} \subset \Gamma$ let $A|_{\hat{\Gamma}} = \{u|_{\hat{\Gamma}} : u \in A\}$ the set of restricted states. For $u \in A_{\Gamma_n}$ let the set of all continuations of u in A be $C(u; \Gamma_n) = \{u' \in A : u'|_{\Gamma_n} = u\}$. By definition of a cylinder set, there is $n_0 \in \mathbb{N}$ such that for all $u \in A|_{\Gamma_{n0}}$ we have $C(u; \Gamma_{n0})|_{\Gamma\setminus\Gamma_{n0}} = E^\Gamma|_{\Gamma\setminus\Gamma_{n0}}$: The state $u \in A|_{\Gamma_{n0}}$ allows for arbitrary extensions in A outside a finite region. This is meant by the statement "A only carries information on a finite part of the grid".

The measure for $A \in \mathfrak{S}$ is $\mu(A) = \mu(A|_{\Gamma_{n0}})$. Note that we can replace n_0 in this formula by any $n \geq n_0$ since the cylinder set A carries no information outside Γ_{n0}. Standard arguments show that $\mu(\cdot)$ defines a premeasure on $\mathfrak{S}$.

In the next step we introduce the algebra $\mathfrak{A}$ that is generated by $\mathfrak{S}$. Any element $A \in \mathfrak{A}$ can be represented by a finite union of sets $S_i \in \mathfrak{S}$ that have pairwise empty intersection,

$$A = \cup_{i=1}^n S_i, \qquad S_i \cap S_j = \emptyset \text{ for } i \neq j.$$

It is clear that $\mu(A)$ is defined as the sum $\sum \mu(S_i)$. It is less clear and deserves a proof that the extension of $\mu(\cdot)$ from $\mathfrak{S}$ to $\mathfrak{A}$ is unique and well defined (Proposition A.3.8).

In the last step, we move from finite intersections to infinite intersections. We define the measure on the Borel algebra, the smallest σ-algebra that covers $\mathfrak{A}$. This step is rather involved (see Definition A.3.9–Proposition A.3.12). At the center is an appropriate approximation of sets by elements of the algebra, and the selection of well-behaved sets in $\mathcal{P}(E^\Gamma)$. Technically, one does not work with the Borel algebra directly, but with the algebra $\mathcal{A}$ and approximation results such as Theorem A.3.14.

Remark 5.5.3

(1) We denote the measure on E and the measure on E^G both by μ since confusion should not arise.
(2) The Bernoulli measure is invariant under shifts.
(3) Let $U \subset E^\Gamma$ be a nonempty open set. Then $\mu(U) > 0$: There is $\psi \in U$ and $\{\phi \in E^G : d(\phi, \psi) < \varepsilon\} \subset U$ for ε small. But this set is a cylinder set. Every cylinder set has positive measure.

We must clarify how shift operators and measures interact. We will find that a measurable set that is invariant under a non-trivial translation is either a set of measure zero or it has full measure. We need some more definitions.

Definition 5.5.4 Let (X, d) be a compact metric space and $f \in C(X)$. Further let $(X, \mathfrak{B}, \mu)$ be a Borel probability space. The mapping f is called

(1) measure preserving if $B \in \mathfrak{B}$ implies $f^{-1}(B) \in \mathfrak{B}$ and

$$\mu(f^{-1}(B)) = \mu(B).$$

(2) topologically transitive if for every pair of open sets A, B there is an n_0 such that

$$f^{n_0}(A) \cap B \neq \emptyset.$$

(3) topologically mixing if for every pair of open sets A, B there is an n_0 such that for all $n > n_0$ holds

$$f^n(A) \cap B \neq \emptyset.$$

(4) strongly mixing with respect to the measure if f is measure preserving and for all $A, B \in \mathfrak{B}$ holds

$$\lim_{n\to\infty} \mu(f^{-n}(A) \cap B) = \mu(A)\,\mu(B).$$

(5) ergodic if f is measure preserving and if $B \in \mathfrak{B}, f^{-1}(B) = B$, implies $\mu(B) \in \{0, 1\}$.

Ergodicity can be considered as a kind of irreducibility: Assume that f is *not* ergodic. There is a measurable set B, $0 < \mu(B) < 1$, with $f^{-1}(B) = B$. Hence, $f(B) = B$ and $f(X \setminus B) = X \setminus B$. We may define the restrictions of f to B respectively $X \setminus B$, $f_1 = f|_B$, $f_2 = f|_{X\setminus B}$. At the same time, we may define probability measures μ_1 (μ_2) on B ($X \setminus B$) as the conditioned probability measures, $\mu_1(A) = \mu(A)/\mu(B)$ for measurable $A \subset B$, $\mu_2(A) = \mu(A)/\mu(X \setminus B)$ for measurable $A \subset X \setminus B$. The measures μ_i are invariant w.r.t. f_i. We obtain two smaller dynamical systems with similar properties as the larger one. Ergodicity prevents us to define the conditioned measures, as either B or $X \setminus B$ has zero measure.

It is useful to know whether a given mapping is ergodic. In given examples it may be difficult to check this property. The following theorem [171, Theorem 1.5, p. 27]) provides some equivalent conditions which sometimes can be checked more easily.

We use the symmetric difference of two sets

$$A \Delta B = (A \cup B) \setminus (A \cap B).$$

Theorem 5.5.5 *Let (X,f) be a dynamical system, X compact metric space, f continuous. Let $(X, \mathfrak{B}, \mu)$ be a Borel probability space. Assume f preserves the measure μ. The following statements are equivalent.*

(1) *f is ergodic.*
(2) *From $B \in \mathfrak{B}$, $\mu(f^{-1}(B)\Delta B) = 0$ it follows that $\mu(B) \in \{0, 1\}$.*
(3) *For all $B \subset \mathfrak{B}$ with $\mu(B) > 0$ holds $\mu(\cup_{n\geq 1} f^{-n}(B)) = 1$.*
(4) *For every pair A, $B \in \mathfrak{B}$ with $\mu(A)$, $\mu(B) > 0$ there is an $n > 0$ such that*

$$\mu(f^{-n}(A) \cap B) > 0.$$

Proof **(1) ⇒ (2)**
Let $B \in \mathfrak{B}$, $\mu(f^{-1}(B)\Delta B) = 0$. We construct a measurable set B_∞ such that $f^{-1}(B_\infty) = B_\infty$, $\mu(B \Delta B_\infty) = 0$ holds.

Step 1: $\mu(f^{-n}(B)\Delta B) = 0$ *for all* $n \in \mathbb{N}$.

For any three sets A, B, C we have (draw Venn diagrams)

$$A\Delta C \subset (A\Delta B) \cup (B\Delta C),$$

successively

$$A\Delta C \subset (A\Delta B_1) \cup (B_1\Delta C) \subset (A\Delta B_1) \cup (B_1\Delta B_2) \cup (B_2\Delta C) \subset \cdots .$$

Furthermore, $f^{-(i+1)}(B) \setminus f^{-i}(B) = f^{-i}(f^{-1}(B) \setminus B)$, as

$$x \in f^{-(i+1)}(B) \setminus f^{-i}(B) \Leftrightarrow f^{i+1}(x) \subset B \text{ and } f^i(x) \notin B$$
$$\Leftrightarrow f^i(x) \in f(B) \text{ and } f^i(x) \notin B \Leftrightarrow x \in f^{-i}(f^{-1}(B) \setminus B).$$

Similarly, $f^{-(i+1)}(B)\Delta f^{-i}(B) = f^{-i}\left(f^{-1}(B)\Delta B\right)$. We find in particular

$$f^{-n}(B)\Delta B \subset \bigcup_{i=0}^{n-1} \left(f^{-(i+1)}(B)\Delta f^{-i}(B)\right) = \bigcup_{i=0}^{n-1} f^{-i}\left(f^{-1}(B)\Delta B\right)$$

and hence

$$0 \leq \mu(f^{-n}(B)\Delta B) \leq n\mu(f^{-1}(B)\Delta B) = 0.$$

Step 2: For the set $B_\infty := \cap_{n=0}^{\infty} \cup_{i=n}^{\infty} f^{-i}(B)$ *we show* $\mu(B_\infty \Delta B) = 0$ *and* $f^{-1}(B_\infty) = B_\infty$.

For any three sets A, B, C we have the inclusions $(A \cap B)\Delta C \subset (A\Delta C) \cup (B\Delta C)$ and $(A \cup B)\Delta C \subset (A\Delta C) \cup (B\Delta C)$. The sets $\cup_{i=n}^{\infty} f^{-i}(B)$ are nested. Hence it follows that

$$\begin{aligned} 0 \leq \mu(B_\infty \Delta B) &= \mu(\left(\bigcap_{n=0}^{\infty} \bigcup_{i=n}^{\infty} f^{-i}(B)\right) \Delta B) \\ &\leq \mu(\bigcup_{n=0}^{\infty} \left((\bigcup_{i=n}^{\infty} f^{-i}(B))\Delta B\right)) \leq \mu(\bigcup_{n=0}^{\infty} \bigcup_{i=n}^{\infty} (f^{-i}(B)\Delta B)) \\ &\leq \mu(\bigcup_{i=0}^{\infty} (f^{-i}(B)\Delta B)) = \lim_{n\to\infty} \mu(\bigcup_{i=0}^{n} f^{-i}(B)\Delta B) = 0. \end{aligned}$$

Furthermore

$$\begin{aligned} f^{-1}(B_\infty) = f^{-1}\left(\bigcap_{n=0}^{\infty} \bigcup_{i=n}^{\infty} f^{-i}(B)\right) &= \bigcap_{n=0}^{\infty} \bigcup_{i=n}^{\infty} f^{-i-1}(B) \\ &= \bigcap_{n=0}^{\infty} \bigcup_{i=n+1}^{\infty} f^{-i}(B) = \bigcap_{n=1}^{\infty} \bigcup_{i=n}^{\infty} f^{-i}(B). \end{aligned}$$

Since $\cup_{i=0}^{\infty} f^{-i}(B) \supset \cup_{i=1}^{\infty} f^{-i}(B)$, we have

$$\begin{aligned} f^{-1}(B_\infty) &= \bigcap_{n=1}^{\infty} \bigcup_{i=n}^{\infty} f^{-i}(B) \\ &= \left(\bigcap_{n=1}^{\infty} \bigcup_{i=n}^{\infty} f^{-i}(B)\right) \cap \left(\bigcup_{i=0}^{\infty} f^{-i}(B)\right) = \bigcap_{n=0}^{\infty} \bigcup_{i=n}^{\infty} f^{-i}(B) = B_\infty. \end{aligned}$$

Because of the ergodicity (1) it follows that $\mu(B_\infty) \in \{0, 1\}$, and with $\mu(B_\infty \Delta B) = 0$ also $\mu(B) \in \{0, 1\}$.

(2) ⇒ (3)

Let $B \in \mathfrak{B}$, $\mu(B) > 0$. Define $A = \cup_{n\geq 1} f^{-n}(B)$. It follows that $f^{-1}(A) \subset A$ and, since f preserves the measure, also $\mu(f^{-1}(A)) = \mu(A)$, and

$$\mu(A\Delta f^{-1}(A)) = \mu(f^{-1}(A)\setminus A) + \mu(A\setminus f^{-1}(A)) = \mu(\emptyset) + \mu(A) - \mu(f^{-1}(A)) = 0.$$

Hence $\mu(A) \in \{0, 1\}$. Since $\mu(B) > 0$ and $\mu(A) \geq \mu(f^{-1}(B)) = \mu(B) > 0$, we have $\mu(A) = 1$.

(3) ⇒ (4)
Let $\mu(A) > 0$, $\mu(B) > 0$. Then $\mu(\cup_{n\geq 0} f^{-n}(A)) = 1$ and hence $\mu(X \setminus \cup_{n\geq 0} f^{-n}(A)) = 0$. Then it follows that

$$0 < \mu(B) = \mu(B \cap (\bigcup_{n\geq 0} f^{-n}(A))) + \mu(B \cap (X \setminus \bigcup_{n\geq 0} f^{-n}(A))) = \mu(\bigcup_{n\geq 0} (B \cap f^{-n}(A))).$$

Hence there is an $n \geq 0$ such that $\mu(B \cap f^{-n}(A)) > 0$.
(4) ⇒ (1)
Let $B \in \mathfrak{B}, f^{-1}(B) = B$. If $0 < \mu(B) < 1$, then $0 < \mu(X \setminus B) < 1$. Since $B \cap (X \setminus B) = \emptyset$, we have also

$$0 = \mu(B \cap (X \setminus B)) = \mu(f^{-n}(B) \cap (X \setminus B))$$

for all n. This equality contradicts (4) since $0 < \mu(B),\ \mu(X \setminus B)$. Hence we have $\mu(B) \in \{0, 1\}$. □

Remark 5.5.6 We see that a measure-preserving mapping which is strongly mixing with respect to the measure, is already ergodic.

We will show that a non-trivial translation operator σ_x, $x \neq 0$, acting on a cellular automaton with $\mathbb{Z}^n$ as a grid is topologically mixing and mixing with respect to the measure, and is also ergodic (see also [171, Theorem 1.12, p. 32, Theorem 1.17, p. 41]). This observation, however, is not sufficient if we also aim to handle non-Abelian, infinite grids. The central question here is, if a finitely generated group of infinite size always has $\mathbb{Z}$ as a subgroup. This problem, raised by Burnside 1902, has been decided 62 years later, by Evgeny Golod and Igor Shafarevich [69, 70]. They constructed a finitely generated group of infinite size, where each element is periodic. Hence, we cannot assume in general that $\mathbb{Z}$ is a subgroup of a finitely generated, infinite group. What we *can* assume is, that there are group elements with arbitrary distance (w.r.t. the Cayley metric) from the unit element. We will find that this observation is sufficient for our purposes. Very often, the next definition is not stated for a family of functions $f_n \in C(X)$, but for the case that this family is generated by iteration of one single function f, $f_n = f^n$. For the reasons mentioned, we require a slightly more general definition for topological mixing, strongly mixing and ergodic.

Definition 5.5.7 Let (X, d) be a compact metric space and $f_n \in C(X)$ for $n \in \mathbb{N}$. Further let $(X, \mathfrak{B}, \mu)$ be a Borel probability space. The family of mappings $(f_n)_{n\in\mathbb{N}}$ is called

(1) measure preserving if, for all $n \in \mathbb{N}$, $B \in \mathfrak{B}$ implies $f_n^{-1}(B) \in \mathfrak{B}$ and

$$\mu(f_n^{-1}(B)) = \mu(B).$$

(2) topologically mixing if for every pair of open sets A, B there is an n_0 such that for all $n > n_0$ holds

$$f_n(A) \cap B \neq \emptyset.$$

(3) strongly mixing with respect to the measure if f is measure preserving and for all $A, B \in \mathfrak{B}$ holds

$$\lim_{n\to\infty} \mu(f_n^{-1}(A) \cap B) = \mu(A)\,\mu(B).$$

(4) ergodic if $(f_n)_{n\in\mathbb{N}}$ is measure preserving and if $B \in \mathfrak{B}$, $f_n^{-1}(B) = B$ for all $n \in \mathbb{N}$, implies $\mu(B) \in \{0, 1\}$.

Theorem 5.5.8 *Let G an infinite group, σ a finite set of generators, $\Gamma = \Gamma(G, \sigma)$ a Cayley graph, E finite, and let E^Γ be endowed with the Cantor metric. Let $g_n \in G$, $n \in \mathbb{N}$, $d_c(e, g_n) \geq n$. Then, the family of shift operators $f_n = \sigma_{g_n}$ has the properties:*

(1) $(f_n)_{n\in\mathbb{N}}$ *is topologically mixing,*
(2) $(f_n)_{n\in\mathbb{N}}$ *is strongly mixing with respect to every Bernoulli measure,*
(3) $(f_n)_{n\in\mathbb{N}}$ *is ergodic with respect to every Bernoulli measure.*

Proof We had already observed that σ_g leaves the measure invariant for all $g \in G$. Obviously $\sigma_g^{-1} = \sigma_{g^{-1}}$ has the same property. The family $(f_n)_{n\in\mathbb{N}}$ is measure preserving.

1) Let $A, B \subset E^\Gamma$ nonempty open sets. Then there are an $\varepsilon > 0$ and elements $u_a \in A$, $u_b \in B$ such that

$$A_1 = \{v \in E^\Gamma \,:\, d(v,\, u_a) < \varepsilon\} \subset A, \quad B_1 = \{v \in E^\Gamma \,:\, d(v,\, u_b) < \varepsilon\} \subset B.$$

Let $k > 1/\varepsilon + 1$. Then the elements in A_1 and B_1 are fixed (by u_a, u_b) on the set Γ_k; on $\Gamma \setminus \Gamma_k$ the elements of A_1, B_1 run through all possible combinations of elementary states. Hence it follows that for all $n > 2\,k$ we have $A_1 \cap f_n(B_1) \neq \emptyset$.

2) Let $\mathfrak{S}$ the semi-algebra of the cylinder sets. The semi-algebra generates the algebra $\mathfrak{A}$ and the σ-algebra $\mathfrak{B}$.
Let $A, B \in \mathfrak{S}$ be fixed. There is a $k > 0$ such that the elements of A and B run through all combinations of elementary states on $\Gamma \setminus \Gamma_k$. For $n \geq k$ we have

$$\mu(f_n^{-1}(A) \cap B) = \mu(A)\,\mu(B). \qquad (*)$$

If $A, B \in \mathfrak{A}$ then there are pairwise disjoint sets $A_1, \ldots, A_m \in \mathfrak{S}$ and pairwise disjoint sets $B_1, \ldots, B_{m'} \in \mathfrak{S}$ such that $A = \cup_i A_i$, $B = \cup_i B_i$. Thus, also in this case, there is an $n_0 > 0$ such that $(*)$ holds for all $n \geq n_0$.

Let $A, B \in \mathfrak{B}$. In view of the approximation Theorem A.3.14 for σ-algebras and probability measures we can find sets $\tilde{A}, \tilde{B} \in \mathfrak{A}$ with

$$\mu(\tilde{A}\Delta A) < \varepsilon, \quad \mu(\tilde{B}\Delta B) < \varepsilon.$$

In particular, we have $|\mu(A) - \mu(\tilde{A})| < \varepsilon$, $|\mu(B) - \mu(\tilde{B})| < \varepsilon$. For $i \geq 0$ we have (draw a Venn diagram!)

$$(f_i(A) \cap B)\Delta(f_i^{-1}(\tilde{A}) \cap \tilde{B}) \subset (f_i(A)\Delta f_i^{-1}(\tilde{A})) \cup (B\Delta\tilde{B})$$

and therefore

$$|\mu(f_i^{-1}(A) \cap B) - \mu(f_i^{-1}(\tilde{A}) \cap \tilde{B})| \leq 2\varepsilon.$$

Now we find

$$\begin{aligned}
&|\mu(f_i^{-1}(A) \cap B) - \mu(A)\mu(B)| \\
&\quad \leq |\mu(f_i^{-1}(A) \cap B) - \mu(f_i^{-1}(\tilde{A}) \cap \tilde{B})| \\
&\qquad +|\mu(f_i^{-1}(\tilde{A}) \cap \tilde{B}) - \mu(\tilde{A})\mu(\tilde{B})| \\
&\qquad +|\mu(\tilde{A})\mu(\tilde{B}) - \mu(A)\mu(\tilde{B})| + |\mu(A)\mu(\tilde{B}) - \mu(A)\mu(B)| \\
&\quad \leq (2 + \mu(\tilde{B}) + \mu(\tilde{A}))\,\varepsilon + |\mu(f_i^{-1}(\tilde{A}) \cap \tilde{B}) - \mu(\tilde{A})\mu(\tilde{B})| \\
&\quad \leq 4\varepsilon + |\mu(f_i^{-1}(\tilde{A}) \cap \tilde{B}) - \mu(\tilde{A})\mu(\tilde{B})|
\end{aligned}$$

whereby we have used, in the last step, the inequalities $\mu(\tilde{A}) \leq 1$, $\mu(\tilde{B}) \leq 1$. For large i the last term vanishes and it follows that

$$\lim_{i\to\infty} |\mu(f_i^{-1}(A) \cap B) - \mu(A)\mu(B)| \leq 4\,\varepsilon,$$

and hence, with $\varepsilon \to 0$,

$$\lim_{i\to\infty} \mu(f_i^{-1}(A) \cap B) = \mu(A)\mu(B).$$

3) From (2) we obtain that $\forall A, B \in \mathfrak{B}$, $\mu(A), \mu(B) > 0$

$$\exists n \in \mathbb{N} : \mu(f_n^{-1}(A) \cap B) > 0.$$

The ergodicity of $(f_n)_{n\in\mathbb{N}}$ can be shown by the same argument we used in Theorem 5.5.5 for (4) $\Rightarrow$ (1). □

In the case of $\mathbb{Z}^n$, we may choose a non-trivial shift operator σ_x as the generator of the family, $f_n = \sigma_x^n$. We obtain the following corollary.

Corollary 5.5.9 *Let $\Gamma = \mathbb{Z}^d$, E finite, let E^Γ be endowed with the Cantor metric and $\sigma_x : E^\Gamma \to E^\Gamma$ a non-trivial translation operator ($x \neq 0$). Then the following statements hold:*

(1) *σ_x is topologically mixing,*
(2) *σ_x is strongly mixing with respect to every Bernoulli measure,*
(3) *σ_x is ergodic with respect to every Bernoulli measure.*

If an attractor $\mathcal{A}$ or a chain component C are shift invariant, $\sigma_g(\mathcal{A}) = \mathcal{A}$ resp. $\sigma_g(C) = C$ for all generators g of the group, then also their basin of attraction $B(\mathcal{A})$ resp. $B(C)$ is invariant. In this case, the basins of attraction are invariant w.r.t. to any shift operator, in particular to the family of shift operators used in Theorem 5.5.8. Hence, the measure of the basins of attraction are either 0 or 1.

Corollary 5.5.10 *Consider a Cayley graph for an infinite, finitely generated group. The basin of attraction B of a shift invariant attractor resp. shift invariant chain component has either measure 0 or 1,*

$$\mu(B) \in \{0, 1\}.$$

5.6 Structure of Attractors—Infinite Grids: Hurley Classification

In this section we consider an infinite grid. The grid may be generated by a non-Abelian group. As before, we assume $|E| > 1$, and exclude Bernoulli measures with $p_e = 0$ for some $e \in E$.

Remark 5.6.1 If there is a unique attractor then it is invariant under translations. Suppose there are two attractors $\mathcal{A}_1, \mathcal{A}_2$ which are both invariant under translations. Suppose they are disjoint. Then the basins of attraction $\mathcal{B}(\mathcal{A}_1)$ and $\mathcal{B}(\mathcal{A}_2)$ are open, invariant under translations, and disjoint. Take a (non-trivial) shift operator σ_g. Then there is an $n \in \mathbb{N}$ such that

$$\emptyset \neq \mathcal{B}(\mathcal{A}_1) \cap \sigma_g^n(\mathcal{B}(\mathcal{A}_2)) = \mathcal{B}(\mathcal{A}_1) \cap \mathcal{B}(\mathcal{A}_2)$$

i.e., $\mathcal{A}_1 \cap \mathcal{A}_2 \neq \emptyset$ which is a contradiction. This observation suggests the following proposition.

Proposition 5.6.2 *Let $(\Gamma(G, \sigma), D_0, E, f_0)$ be a cellular automaton that has two disjoint attractors. Then:*

i) *These attractors are not translation invariant.*
ii) *Each of them contains two disjoint attractors.*
iii) *The cellular automaton has countably many attractors and more than countably many quasi-attractors.*

Proof Let $\mathcal{A}_1$ and $\mathcal{A}_2$ be two disjoint attractors with basins $\mathcal{B}(\mathcal{A}_1), \mathcal{B}(\mathcal{A}_2)$. Let $g_\ell \in G$, $d_c(g_\ell, e) \geq \ell$, $\ell \in \mathbb{N}$. The sets $\mathcal{B}(\mathcal{A}_i)$ are open. Since $(\sigma_{g_\ell})_{\ell\in\mathbb{N}}$ is a topologically mixing family, there is a number n_0 such that

$$\mathcal{B}(\sigma_{g_\ell}(\mathcal{A}_i)) \cap \mathcal{B}(\mathcal{A}_j) \neq \emptyset \qquad \text{for all } \ell > n_0, \text{ and all } i,j \in \{1,2\}.$$

Suppose $\mathcal{A}_1$ is invariant under translations. Then it follows that $\emptyset = \mathcal{B}(\mathcal{A}_1) \cap \mathcal{B}(\mathcal{A}_2) = \sigma_{g_\ell}(\mathcal{B}(\mathcal{A}_1)) \cap \mathcal{B}(\mathcal{A}_2) \neq \emptyset$ for large l. Hence we have a contradiction. Both attractors are not invariant under translations.

The basins $\mathcal{B}(\mathcal{A}_i)$ and $\mathcal{B}(\sigma_{g_\ell}(\mathcal{A}_j))$, for $i,j \in \{1,2\}$, and $l > n_0$ have pairwise nonempty intersection. Hence the same is true for the attractors $\mathcal{A}_i$ and $\sigma_{g_\ell}(\mathcal{A}_j)$. Fix $\ell > n_0$. We find further attractors in the intersections

$$\mathcal{A}_{1,1} \subset \mathcal{A}_1 \cap \sigma_{g_\ell}(\mathcal{A}_1), \quad \mathcal{A}_{1,2} \subset \mathcal{A}_1 \cap \sigma_{g_\ell}(\mathcal{A}_2),$$
$$\mathcal{A}_{2,1} \subset \mathcal{A}_2 \cap \sigma_{g_\ell}(\mathcal{A}_1), \quad \mathcal{A}_{2,2} \subset \mathcal{A}_2 \cap \sigma_{g_\ell}(\mathcal{A}_2).$$

Since $\mathcal{A}_1 \cap \mathcal{A}_2 = \emptyset$, also $\sigma_{g_\ell}(\mathcal{A}_1) \cap \sigma_{g_\ell}(\mathcal{A}_2) = \emptyset$. It follows that

$$\mathcal{A}_{1,1} \cap \mathcal{A}_{1,2} \subset \mathcal{A}_1 \cap \sigma_{g_\ell}(\mathcal{A}_2) \cap \mathcal{A}_1 \cap \sigma_{g_\ell}(\mathcal{A}_1) = \emptyset$$

and similarly $\mathcal{A}_{2,1} \cap \mathcal{A}_{2,2} = \emptyset$.

Recursively we can associate an attractor with any finite sequence $\{a_i\}_{i=1}^n$, $a_i \subset \{1,2\}$, such that it has empty intersection with the other attractors associated with other such sequences. There are infinitely many disjoint attractors.

Now consider infinite sequences in $\{1,2\}^{\mathbb{N}}$. Each of these sequences encodes a quasi-attractor and these quasi-attractors are distinct. Hence the set of quasi-attractors is not countable. □

Proposition 5.6.3 *Let $(\Gamma(G), D_0, E, f_0)$ be a cellular automaton and let C be a chain component. Assume $\mu(\mathcal{B}(C)) > 0$ for some Bernoulli measure μ. Let $g_\ell \in G$ with $d_c(e, g_\ell) \geq \ell$, $\ell \in \mathbb{N}$. Then the following statements hold:*

i) C is invariant with respect to a family of shift operators $(\sigma_{g_\ell})_{\ell\in\mathbb{N}}$.
ii) $\mu(\mathcal{B}(C)) = 1$.
iii) Every attractor contains C and C is a minimal quasi-attractor.

Proof

Step 1: C is invariant under translations.

Assume that C is not shift invariant w.r.t. the family $(\sigma_{g_\ell})_{\ell\in\mathbb{N}}$. Since $\mu(\mathcal{B}(C)) > 0$ and $\mu(\sigma_{g_\ell}(\mathcal{B}(C))) > 0$ and $(\sigma_{g_\ell})_{\ell\in\mathbb{N}}$ is strongly mixing with respect to any Bernoulli measure, there is an n_0 such that

$$\mu(\mathcal{B}(C) \cap \sigma_{g_\ell}(\mathcal{B}(C))) \neq 0 \quad \text{for} \quad \ell > n_0,$$

in particular $\mathcal{B}(C) \cap \sigma_{g_\ell}(\mathcal{B}(C)) \neq \emptyset$ for all $\ell > n_0$, and hence $C = \sigma_{g_\ell}(C)$ by Proposition 5.4.14. As σ_{g_ℓ} is bijective, $C = \sigma_{g_\ell}^{-1}(C)$.

Step 2: $\mu(\mathcal{B}(C)) = 1$.

Since $(\sigma_{g_\ell})_{\ell \in \mathbb{N}}$ is ergodic and $\sigma_{g_\ell}^{-1}(\mathcal{B}(C)) = \mathcal{B}(C)$, we have

$$\mu(\mathcal{B}(C)) \in \{0, 1\}.$$

Now $\mu(\mathcal{B}(C)) > 0$ implies $\mu(\mathcal{B}(C)) = 1$.

Step 3: $C \subset \mathcal{A}$ *for every attractor* $\mathcal{A}$.

Suppose $\mathcal{A}$ is an attractor such that $C \not\subset \mathcal{A}$. In view of Proposition 5.4.14 the basins of attraction are disjoint. Hence $\mathcal{B}(\mathcal{A}) \subset E^\Gamma \setminus \mathcal{B}(C)$. Since $\mathcal{B}(\mathcal{A})$ is open, it follows that $\mu(\mathcal{B}(\mathcal{A})) > 0$. Since $\mu(\mathcal{B}(C)) = 1$, we have $\mu(E^\Gamma \setminus \mathcal{B}(C)) = 0$. Hence such attractor cannot exist.

There is a unique minimal quasi-attractor. As a minimal quasi-attractor is a chain component, this unique minimal quasi-attractor coincides with C. □

These results lead to a better characterization of the Hurley classes from Theorem 5.3.8.

Corollary 5.6.4 *The Hurley classes have the following additional properties.*

Hurley 1: *There are countably many attractors and there are more than countably many minimal quasi-attractors.*
For every chain component the basin of attraction has measure zero.
Hurley 2: *The following statements are mutually exclusive:*

i) For every chain component and every Bernoulli measure the basin of attraction has measure zero.
ii) The basin of attraction of the minimal quasi-attractor has full measure.

Hurley 3, 4, 5: *The minimal attractor is translation invariant and its basin of attraction has measure* 1.

Proof **For Hurley 1:** Propositions 5.6.2 and 5.6.3.

For Hurley 2: Proposition 5.6.3: Either the basin of attraction of every chain component has measure zero, or there is a chain component for which the basin of attraction has positive measure. In that case the measure is 1, and the chain component is the minimal attractor.

Hurley 3, 4, 5: Let $\mathcal{A}$ be the minimal attractor. It is invariant under translations since otherwise translations would produce further attractors. It is also the minimal quasi-attractor, i.e., it consists of a single chain component (Proposition 5.4.15). Let $B_c(\mathcal{A})$ be the basin of attraction of the chain component $\mathcal{A}$ and $B_a(\mathcal{A})$ that of the attractor $\mathcal{A}$. Then we know $B_c(\mathcal{A}) = B_a(\mathcal{A}) = \mathcal{B}(\mathcal{A})$. Again we have a dichotomy: Either $\mu(\mathcal{B}(\mathcal{A})) = 0$ or $\mu(\mathcal{B}(\mathcal{A})) = 1$. Since $\mathcal{B}(\mathcal{A})$ is open and hence $\mu(\mathcal{B}(\mathcal{A})) > 0$, it follows that $\mu(\mathcal{B}(\mathcal{A})) = 1$. □

Remark 5.6.5 The situation of Hurley 3 can be visualized by a discrete time dynamical system on S^1 where the "north pole" is an unstable stationary point, the "south pole" is locally asymptotically stable, and the ω-limit set of every other orbit is the south pole. Then the global attractor is all of S^1, the minimal quasi-attractor is the south pole, and the basin of attraction of the latter is all of S^1 except the north pole and hence has measure 1.

Now we follow [111] in giving examples for the different classes.

Example 5.6.6 (Hurley 1) Consider the Wolfram automaton 204 with global function $f = \text{id}$. Then every set of the form

$$B_\varepsilon(u) = \{v \in E^{\mathbb{Z}} : d(u, v) < \varepsilon\}, \quad \varepsilon^{-1} \notin \mathbb{N},$$

is an attractor: $B_\varepsilon(u)$ is closed and open, i.e., $f(\overline{B_\varepsilon(u)}) \subset B_\varepsilon(u)$ and $\mathcal{A} = \omega(B_\varepsilon(u)) = B_\varepsilon(u)$. Hence there are infinitely many attractors, and every point in $E^{\mathbb{Z}}$ is a quasi-attractor.

Example 5.6.7 (Hurley 2 with $\mu(B(Q)) = 1$) Consider the Wolfram automaton 136 with local function

$$f_0(x_{-1}, x_0, x_1) = x_0 x_1.$$

We determine the stationary points. The outcome of the local function is '1' if and only if the right and the center cell both carry '1'. Hence a step function ...000011111... remains while a step ...111000... moves to the left. Let $[\overline{0}]$, the state with "0" only, $[\overline{1}]$, the state with "1" only, and ϕ_i the step functions ($\phi_i(j) = 0$ for $j < i$ and $\phi_i(j) = 1$ for $j \geq i$). These states are the only stationary points of the cellular automaton.

Claim *There are no disjoint attractors.*

Proof Assume $\mathcal{A}$ is an attractor. The corresponding pre-attractor U is open and hence there is a state $u_0 \in U$ with

$$\forall i \in \mathbb{Z}\ \exists j > i : u_0(j) = 0.$$

Since $\omega(u_0) = \{[\overline{0}]\}$, we have $[\overline{0}] \in \mathcal{A}$ for any attractor $\mathcal{A}$. Hence all attractors contain the zero state. □

Claim $U_i = \{u : u(i - 1) = 0\}$ *are pre-attractors of pairwise different attractors.*

Proof The sets U_i are cylinder sets, and hence open and closed. The cellular automaton never destroys a local state '0' at a cell. Hence, $f(\overline{U_i}) = f(U_i) \subset U_i$; U_i are pre-attractors. Let $\mathcal{A}_i = \omega(U_i)$ the attractors within these pre-attractors. These attractors are different: $\mathcal{A}_i \neq \mathcal{A}_j$ for $i < j$ as $\phi_i \in \mathcal{A}_i$ but $\phi_i \notin \mathcal{A}_j$. □

Claim $Q = \{[\overline{0}]\}$ *is the minimal quasi-attractor and not an attractor.*

Proof Let $B = \cap_{i \in \mathbb{N}} \mathcal{A}_i$. If $u \in B$, then $u \in \mathcal{A}_i \subset U_i$, and hence $u(i) = 0$. Since $i \in \mathbb{Z}$ has been arbitrary, $u = [\overline{0}]$. Therefore, $Q = B = \{[\overline{0}]\}$. $Q = \cap_i \mathcal{A}_i$ is a subset of every attractors and it is minimal since it consists of one point only. Hence, it is the minimal quasi-attractor. Suppose Q were an attractor. Then there would be an open pre-attractor U with $\omega(U) = \{[\bar{0}]\}$. This set contains a ϕ_{i_0} (of the form given above) if i_0 is large enough. Since ϕ_{i_0} is a stationary point, we get $\phi_{i_0} \in \omega(U)$ in contradiction to the assumption. □

Claim $\mu(\mathcal{B}(Q)) = 1$.

Proof A minimal quasi-attractor is a chain component (Theorem 5.4.15), and its basin of attraction as a quasi-attractor is the same as its basin of attraction as a chain component,

$$\mathcal{B}(Q) = \{u : \forall i \in \mathbb{Z}\ \exists j > i \text{ such that } u(j) = 0\}.$$

Define $M_i = \{u : \forall j \geq i \text{ holds } u(j) = 1\}$. Then, $\mu(M_i) = 0$, and $\mathcal{B}(Q) = E^\Gamma \setminus (\cup_{i\in\mathbb{Z}} M_i)$. Hence,

$$\mu(\mathcal{B}(Q)) = \mu(E^\Gamma) - \mu(\bigcup_{i\in\mathbb{Z}} M_i) \geq 1 - \sum_{i\in\mathbb{Z}} \mu(M_i) = 1.$$

□

Before we proceed to the next example, we prove a result on random walks in $\mathbb{Z}$.

Proposition 5.6.8 *Let $m \in \mathbb{N}$, and w_i be independent and identically distributed random variables that assume only integers between -1 and m. Define $p_j = P(w_1 = j)$, $j = -1, \ldots, m$. Let $Z_0 = 1$,*

$$Z_n = 1 + \sum_{i=1}^{n} w_i \qquad \text{for } n > 0,$$

and q the probability for the random walk Z_n to hit 0, $q = P(\exists n > 0 : Z_n \leq 0)$. Then, q is the smallest non-negative solution of the equation

$$q = \sum_{j=-1}^{m} p_j q^{j+1}.$$

In particular, $E(w_1) > 0$ implies $q < 1$.

Idea of proof We have $Z_0 = 1$. If we find $w_1 = -1$, then $Z_1 = 0$. Therefore,

$$q = P(w_1 = -1) + \sum_{i=0}^{m} P(w_1 = i)\ P(\exists n > 1 : Z_n \leq 0 | Z_1 = i + 1).$$

What can we say about $P(\exists n > 1 : Z_n \leq 0 | Z_1 = 1)$? The family Z_n forms a homogeneous Markov chain, hence we have again

$$P(\exists n > 1 : Z_n \leq 0 | Z_1 = 1) = P(\exists n > 0 : Z_n \leq 0) = q.$$

Now consider $P(\exists n > 1 : Z_n \leq 0|Z_1 = 2)$. We go at most one step down, i.e., before arriving at 0 we must pass 1. Because of

$$P(\exists n > 1 : Z_n \leq k|Z_1 = k+1) = P(\exists n > 0 : Z_n \leq 0) = q$$

the probability to arrive at 1 is again q. Once we have arrived at 1 we must go down again,

$$P(\exists n > 1 : Z_n \leq 0|Z_1 = 2) = q^2.$$

In a similar manner we find

$$P(\exists n > 1 : Z_n \leq 0|Z_1 = k) = q^k.$$

Now the formula for q assumes the form

$$\begin{aligned} q &= P(w_1 = -1) + \sum_{i=0}^{m} P(w_1 = i)\, P(\exists n > 1 : Z_n \leq 0|Z_1 = i + 1) \\ &= p_{-1} + \sum_{i=0}^{m} p_i\, q^{i+1}. \end{aligned}$$

The probabilities p_i add up to 1. Hence the equation has the solution $q = 1$. The derivatives with respect to q of the right hand side are all non-negative for $q \in [0, 1]$. Hence the right hand side is a convex function and we have at most two solutions. The derivative of the right hand side at $q = 1$ is

$$\frac{d}{dq}\left(p_{-1} + \sum_{i=0}^{m} p_i\, q^{i+1}\right)\Bigg|_{q=1} = \sum_{i=0}^{m} (i+1)p_i = \sum_{i=-1}^{m} ip_i + \sum_{i=-1}^{m} p_i = 1 + E(w_1).$$

Hence we have a solution different from 1 if and only if $E(w_1) > 0$.

Finally we discuss which of the two solutions is the probability $P(\exists n > 0 : Z_n \leq 0)$. This probability depends continuously on the distribution of w_1 (we do not give a proof for this claim). Clearly for $P(w_1 = -1) = 0$ the smaller solution $q = 0$ is the correct one. Hence the smaller solution is the desired probability. □

Example 5.6.9 (Hurley 2 with $\mu(\mathcal{B}(C)) = 0$ for All Chain Components) When Hurley had published his paper in (1990) it was not obvious that this Hurley class is not void. Kůrka found the following example in 1997 [111].

Let $\Gamma = \mathbb{Z}$, $E = \{e = (e_\xi, e_\eta) : e_\xi \in \{0, 1, 2\},\ e_\eta \in \{0, 1\}\}$, $D_0 = \{0, 1\}$, with the local function for $e_0, e_1 \in E$ given by

$$f_0(e_0, e_1) = (f_\xi(e_0), f_\eta(e_1))$$

where

$$f_\xi(e_\xi, e_\eta) = \max\{0, e_\xi - e_\eta\} = (e_\xi - e_\eta)_+,$$

$$f_\eta(e_\xi, e_\eta) = \begin{cases} e_\eta & \text{if } e_\xi = 0 \\ 0 & \text{otherwise} \end{cases} = e_\eta\ \chi_0(e_\xi).$$

As usual, χ denotes the characteristic function. Hence the new state in the ξ-component depends on the cell 0, the new state in the η-component depends on the cell 1.

Information is transmitted only from right to left. The state in the ξ-component is non-increasing; any zero that has occurred at some position in the ξ-component is maintained. We make two further observations:

- States $u \in E^{\mathbb{Z}}$ with vanishing η-component, $u(n)_\eta = 0$, are stationary, i.e., all states $\{0, 1, 2\}^{\mathbb{Z}} \times [\overline{0}]$ are stationary points.
- If the ξ-component vanishes identically, then the automaton acts like a left shift on the η-component.

For the Bernoulli-measure $\mu(\cdot)$ we choose the most simple one and assign the same probability to all state of E.

Claim $Q = [\overline{0}] \times \{0, 1\}^{\mathbb{Z}}$ *is subset of any attractor.*

Proof In the first step, we show that the state u_0 with $u_0(n) = (0, 1)$ for all $n \in \mathbb{Z}$, is contained in every attractor.

The pre-attractor of every attractor contains a cylinder set. The cylinder set contains a state in the set

$$M_n = \{u\ :\ u(i) = (0, 1) \quad \text{for all } i \geq n\}$$

for suitable n. At each step the '1' in the η-component at position $i = n$ is moved to the η-component at position $n-1$ and there the ξ-component is decreased by 1 until, after at most two iterations, it becomes 0. We see that $f^2(M_n) \subset M_{n-1}$, and further, that $\omega(u) = \{u_0\}$ for every u that is contained in one of the M_n. Hence $u_0 \in \mathcal{A}$ for every attractor.

This argument can be easily extended to states with vanishing ξ-component ($(u(i)_\xi = 0$ for all $i \in \mathbb{Z}$), and an eventually periodic η-component,

$$\exists n_0 \in \mathbb{Z},\ n \in \mathbb{N}\ \ \forall i \in \mathbb{N} : u(n_0 + n + i)_\eta = u(n_0 + i)_\eta.$$

As an attractor is closed, and the periodic sequences are dense in $\{0, 1\}^{\mathbb{Z}}$, it follows that Q is a subset of all attractors.

Since Q is a subset of all attractors, there are no disjoint attractors. According to Proposition 5.3.7, we find a minimal quasi-attractor.

Claim *Q is the minimal quasi-attractor.*

Proof Consider the sets

$$U_k = \{u \in E^{\mathbb{Z}} \, : \, u(i)_\xi = 0 \text{ for } |i| \le k\}.$$

These U_k are finite unions of cylinder sets and hence closed and open. As zeros in the ξ-component remain under iteration with f, we have $f(\overline{U_k}) \subset U_k$, i.e., the sets U_k are pre-attractors for $\mathcal{A}_k = \omega(U_k)$. Note that

$$Q = \bigcap_{k=0}^{\infty} U_k.$$

The set Q is contained in each of the pre-attractors U_k and it is invariant. Hence we have $Q \subset \omega(U_k)$ and

$$Q \subset \cap_{k \in \mathbb{N}} \mathcal{A}_k = \cap_{k \in \mathbb{N}} \omega(U_k) \subset \cap_{k \in \mathbb{N}} U_k = Q.$$

The set Q is the intersection of the attractors $\mathcal{A}_k$, and (according to the last claim) Q is subset of any attractor. Therefore Q is the minimal quasi-attractor. □

Below we will show that the basin of attraction of Q has measure zero and hence does not contain open sets. In order to characterize $\mathcal{B}(Q)$, we define the sets

$$M_n^{(m)} = \{u \in E^{\mathbb{Z}} \, : \, \exists l, \, 0 \le l \le m \text{ such that } \sum_{i=0}^{l} \big(u(n+i)_\xi - u(n+i)_\eta\big) \le 0\}.$$

Claim *For all* $n \in \mathbb{N}$, $\mathcal{B}(Q) \subset \cup_{k \in \mathbb{N}_0} f^{-k}(M_n^{(0)})$.

Proof If $u \in \mathcal{B}(Q)$, then $\omega(u) \in Q$. There is an increasing sequence $k_i \in \mathbb{N}$ and $v \in Q$ such that $f^{k_i}(u) \to v$ for $i \to \infty$. Let $n \in \mathbb{N}$ be arbitrary, fixed. Then, in particular, $v(n)_\xi = 0$. For k_i large enough, also $(f^{k_i}(u)(n))_\xi = 0$. Therefore, $f^{k_i}(u) \in M_n^{(0)}$ and $u \in \cup_{k \in \mathbb{N}_0} f^{-k}(M_n^{(0)})$. □

Claim *For* $u \in M_n^{(0)}$ *holds* $(f(u)(n))_\xi = 0$*. In particular,* $f(M_n^{(0)}) \subset M_n^{(0)}$.

Proof If $u \subset M_n^{(0)}$ then $(f(u)(n))_\xi = (u(n)_\xi - u(n)_\eta)_+ = (-u(n)_\eta)_+ = 0$. □

Claim $f^{-1}(M_n^{(m-1)}) = M_n^{(m)}$.

Proof

Step 1. We first show an identity. Let $u \in E^{\mathbb{Z}}$. Then

$$\sum_{i=0}^{l-1} \big((f(u)(n+i))_\xi - (f(u)(n+i))_\eta\big)$$

$$= \sum_{i=0}^{l-1} \big[(u(n+i)_\xi - u(n+i)_\eta)_+ - (f(u)(n+i))_\eta\big]$$

$$= \sum_{i=0}^{l-1} \left[(u(n+i)_\xi - u(n+i)_\eta)_+ - f_\eta(u(n+i+1)) \right]$$

$$= (u(n)_\xi - u(n)_\eta)_+$$

$$+ \sum_{i=1}^{l-1} \left[(u(n+i)_\xi - u(n+i)_\eta)_+ - f_\eta(u(n+i)) \right] - f_\eta(u)(n+l).$$

We compute the expression $(e_\xi - e_\eta)_+ - f_\eta(e)$ for $e = (e_\xi, e_\eta) \in E$ in the following table.

e_ξ	e_η	$e_\xi - e_\eta$	$(e_\xi - e_\eta)_+$	$f_\eta(e)$	$(e_\xi - e_\eta)_+ - f_\eta(e)$
0	0	0	0	0	0
0	1	−1	0	1	−1
1	0	1	1	0	1
1	1	0	0	0	0
2	0	2	2	0	2
2	1	1	1	0	1

We find that $(e_\xi - e_\eta)_+ - f_\eta(e) = e_\xi - e_\eta$ for all $e \in E$, and hence

$$\sum_{i=0}^{l-1} \left((f(u)(n+i))_\xi - (f(u)(n+i))_\eta \right)$$

$$= (u(n)_\xi - u(n)_\eta)_+ + \sum_{i=1}^{l-1} \left[u(n+i)_\xi - u(n+i)_\eta) \right] - f_\eta((u)(n+l)).$$

Step 2. We prove the inclusion $f^{(-1)}(M_n^{(m-1)}) \subset M_n^{(m)}$. Let $u \in f^{-1}(M_n^{(m-1)})$, i.e., $f(u) \in M_n^{(m-1)}$. We show $u \in M_n^{(m)}$ by induction on m.
Basis: $m = 1$. Since $f(u) \in M_n^{(0)}$, $(f(u)(n))_\xi - (f(u)(n))_\eta \leq 0$,

$$(u(n)_\xi - u(n)_\eta)_+ - u(n+1)_\eta \ \chi_0(u(n+1)_\xi) \leq 0.$$

If $u(n+1)_\eta \ \chi_0(u(n+1)_\xi) = 0$, then $u \in M_n^{(0)}$. If $u(n+1)_\eta \ \chi_0(u(n+1)_\xi) = 1$, then $u(n+1) = (0, 1)$ and

$$[u(n)_\xi - u(n)_\eta] + [u(n+1)_\xi - u(n+1)_\eta]$$

$$\leq (u(n)_\xi - u(n)_\eta)_+ - 1 = [(f(u)(n))_\xi - (f(u)(n))_\eta] \leq 0,$$

i.e., $u \in M_n^{(1)}$.

Inductive step: Assume the assertion is true for $m = 1, \ldots, m_0$. Consider $f(u) \in M_n^{(m_0)}$. If already $f(u) \in M_n^{(m)}$ for $m < m_0$, then $u \in M_n^{(m+1)} \subset M_n^{(m_0+1)}$. It remains to consider the case $f(u) \in M_n^{(m_0)} \setminus M_n^{(m_0-1)}$. In particular, $f(u) \notin M_n^{(0)}$ which implies (according to the preceding claim) that $u \notin M_n^{(0)}$. Then,

$$\sum_{i=0}^{m_0-1} \big((f(u)(n+i))_\xi - (f(u)(n+i))_\eta\big) > 0$$

$$\sum_{i=0}^{m_0} \big((f(u)(n+i))_\xi - (f(u)(n+i))_\eta\big) \leq 0$$

$$u(n)_\xi - u(n)_\eta > 0.$$

Since $e_\xi - e_\eta \geq -1$ for all $e = (e_\xi, e_\eta) \in E$, we can further conclude that $(f(u)(n+m_0))_\xi - (f(u)(n+m_0))_\eta = -1$, and hence $f(u)(n+m_0) = (0,1)$. Then we know $u(n+m_0+1) = (0,1)$, because $f_\eta(e) = 1$ only for $e = (0,1)$. Finally we have (using $u(n)_\xi - u(n)_\eta > 0$ and $u(n+m_0+1) = (0,1)$)

$$\begin{aligned}
0 &\geq \sum_{i=0}^{m_0} \big((f(u)(n+i))_\xi - (f(u)(n+i))_\eta\big) \\
&= (u(n)_\xi - u(n)_\eta)_+ + \sum_{i=1}^{m_0} \big[u(n+i)_\xi - u(n+i)_\eta)\big] - f_\eta(u(n+m_0)) \\
&= (u(n)_\xi - u(n)_\eta) \\
&\quad + \sum_{i=1}^{m_0} \big[(u(n+i))_\xi - u(n+i)_\eta)\big] + [u(n+m_0+1)_\xi - u(n+m_0+1)_\eta] \\
&= \sum_{i=0}^{m_0+1} \big[u(n+i)_\xi - u(n+i)_\eta)\big],
\end{aligned}$$

i.e., $u \in M_n^{(m_0+1)}$.

Step 3. Now we prove the converse inclusion $f^{-1}(M_n^{(m-1)}) \supset M_n^{(m)}$. We choose $u \in M_n^{(m)}$ and show $f(u) \in M_n^{(m-1)}$.

We have already shown $f(M_n^{(0)}) \subset M_n^{(0)}$. Let $m > 0$ and, without loss of generality, $u \in M_n^{(m)} \setminus M_n^{(m-1)}$. Then we can conclude—in a similar fashion as before—that $u(n)_\xi - u(n)_\eta > 0$, and $u(n+m) = (0,1)$. Hence we have

$$\begin{aligned}
&\sum_{i=0}^{m-1} \big((f(u)(n+i))_\xi - (f(u)(n+i))_\eta\big) \\
&= (u(n)_\xi - u(n)_\eta)_+ + \sum_{i=1}^{m-1} \big[u(n+i)_\xi - u(n+i)_\eta)\big] - f_\eta(u(n+m))
\end{aligned}$$

$$= \sum_{i=0}^{m-1} \left[u(n+i)_\xi - u(n+i)_\eta)\right] - f_\eta(u(n+m))$$

$$= \sum_{i=1}^{m} \left[u(n+i)_\xi - u(n+i)_\eta)\right] \leq 0$$

and therefore $f(u) \in M_n^{(m-1)}$. □

Claim $\mu(\mathcal{B}(Q)) = 0$.

Proof Fix $n \in \mathbb{N}$. According to the preceding claims, $\mathcal{B}(Q) \subset R$, where

$$R = \bigcup_{m \geq 0} M_n^{(m)} = \{u : \exists m \geq 0 \text{ such that } \sum_{i=1}^{m} \left[u(n+i)_\xi - u(n+i))_\eta\right] \leq 0\}.$$

The Bernoulli measure has been chosen such that $P(e) = 1/6$ for all $e \in E$. In order to estimate $\mu(R)$, we consider random states u: $(y_i)_{i \in \mathbb{Z}}$ is a family of E-valued random variables with $P(y_i = e) = 1/6$ for $e \in E$, and $u(i)$ is given by a realization of the random variable y_i. Then define

$$w_i = u(i)_\xi - u(i)_\eta.$$

The w_i are independently and identically distributed random variables with

$$P(w_i = -1) = P(\{(0,1)\}) = 1/6, \; P(w_i = 0) = P(\{(0,0), (1,1)\}) = 1/3,$$
$$P(w_i = 1) = P(\{(1,0), (2,1)\}) = 1/3, \; P(w_i = 2) = P(\{(2,0)\}) = 1/6,$$

and $E(w_i) = 1/2$. The Bernoulli measure of R is the probability that there is an $m \geq 0$ such that $\sum_{i=1}^{m} [u(n+i)_u - u(n+i)_v)] \leq 0$,

$$\mu(R) \leq P\left(\exists m > 0 : \sum_{i=1}^{m} [u(n+i)_u - u(n+i)_v)] \leq 0\right).$$

Proposition 5.6.8 says that this probability is less than 1, i.e., $\mu(R) < 1$. Since $B(Q) \subset R$, $\mu(B(Q)) < 1$.

The set Q is translation invariant and the translation operator is ergodic, hence $\mu(\mathcal{B}(Q)) \in \{0, 1\}$. Since $\mu(\mathcal{B}(Q)) < 1$ we have $\mu(B(Q)) = 0$. □

Example 5.6.10 (Hurley 3) Consider the Wolfram automaton 128 with local function

$$f_0(x_{-1}, x_0, x_1) = x_{-1}\, x_0\, x_1.$$

The behavior of this automaton can be visualized as if any zero is "eating away" successively all neighboring '1',

time step n	$\cdots 0\ 0\ 0\ 1\ 1\ 1 \cdots$
time step $n+1$	$\cdots 0\ 0\ 0\ 0\ 1\ 1 \cdots$

time step n	$\cdots 1\ 1\ 1\ 0\ 0\ 0 \cdots$
time step $n+1$	$\cdots 1\ 1\ 0\ 0\ 0\ 0 \cdots$

Claim *There are no disjoint attractors.* $[\overline{0}]$ *is contained in every attractor.*

Proof Let $u \in E^{\mathbb{Z}}$ such there is at least one $n \in \mathbb{Z}$ with $u(n) = 0$. Then $\omega(u) = [\overline{0}]$. Hence every nonempty open set contains infinitely many elements u with $\omega(u) = [\overline{0}]$, and $[\overline{0}]$ is contained in every attractor. There are no disjoint attractors. □

Claim *There is an attractor which is a proper subset of the global attractor.*

Proof Let $U = \{u \in E^{\mathbb{Z}} : u(0) = 0\}$. Then U is closed and open and $f(\overline{U}) \subset U$, hence U is a pre-attractor. Therefore there is an attractor $\mathcal{A}$ with $[\overline{1}] \notin \mathcal{A}$. On the other hand $[\overline{1}] \in \omega(E^{\mathbb{Z}})$, i.e., it is an element of the global attractor. □

Claim *The minimal quasi-attractor is already an attractor.*

Proof First look at the global attractor $\omega(E^{\mathbb{Z}})$. We know that every attractor is invariant, in particular backward invariant, every point in the attractor has a preimage under f^n for all $n \in \mathbb{N}$. On the other hand, there may be "Garden of Eden" (GoE) states, i.e., states $u \in E^{\mathbb{Z}}$ with $f(v) \neq u$ for all $v \in E^{\Gamma}$ (for a deeper discussion of GoE states, see Sect. 9.1). The GoE states are not contained in any attractor. Before we continue, we discuss GoE states in the present example.

Suppose there is '0' between two '1'. Then this state has no precursor. Similarly, if there is a pair '00' with a '1' on either side, then there is no precursor. We can go on, every block of finitely many '0' with a '1' on either side tells that the state has no precursor,

$$[\cdots * 1000 \cdots 0001 * \cdots]$$

These states cannot occur in any invariant set.

The remaining states are of one of the following types,

all zero	$[\overline{0}]$
all one	$[\overline{1}]$
first zero, then one	$[\cdots 00001111 \cdots]$
first one, then zero	$[\cdots 11110000 \cdots]$
a block of ones in a sea of zeros	$[\cdots 00011 \cdots 110000 \cdots]$.

For each type of state one finds easily a past history, i.e., $y_n \in E^{\mathbb{Z}}$ with $f^n(y_n)$ is the given state. Hence the global attractor consists of all states with exactly these types.

Now suppose we have *any* attractor $\mathcal{A}$ with a pre-attractor U. Then each element of $\mathcal{A} = \omega(U) \subset \omega(E^{\mathbb{Z}})$ has one of the types above. Choose

$$U = \{u : u(0) = 0\}.$$

Then U is open and closed. Since a '0' always remains, $f(\overline{U}) \subset U$, and U is a pre-attractor. Which elements of $\omega(E^{\mathbb{Z}})$ are contained in $\omega(U)$? We get the following list:

1. $[\bar{0}] \in \omega(U)$.
2. $[\bar{1}]$ is not in U, and hence not in $\omega(U)$.
3. $[\cdots 00001111 \cdots]$ is not in $\omega(U)$ since an infinite past history requires jumps from 0 to 1 way out for cells $n \ll 0$ contradicting the definition of U.
4. For $[\cdots 11110000 \cdots]$ the situation is similar to $[\cdots 00001111 \cdots]$.
5. $[\cdots 00011 \cdots 110000 \cdots]$ is not in $\omega(U)$. Indeed, an infinite past history would require arbitrarily large blocks of '1' which eventually would cover the cell $n = 0$.

Hence we find $\omega(U) = \{[\bar{0}]\}$. The minimal quasi-attractor is already a minimal attractor. □

Example 5.6.11 (Hurley 4) The Wolfram automaton 0 with global function $f(\phi) = [\bar{0}]$ has only one attractor, namely $\{[\bar{0}]\}$. But this attractor is different from $X = E^{\mathbb{Z}}$.

Example 5.6.12 (Hurley 5) Consider the Wolfram automaton with rule 240, the right shift,

$$f_0(x_{-1}, x_0, x_1) = x_{-1}.$$

We construct a special sequence (indexed by $\mathbb{N}$) by chaining together all binary numbers with one, two, ... binary digits.

$$s_0 = [0|1|00|01|10|11|000|001|010|011|100|101|110|111|0000| \cdots].$$

Then define the set M as a set of all $u \in E^{\mathbb{Z}}$ such that a right tail agrees with s_0,

$$M = \{u \in E^{\mathbb{Z}} : \exists n \in \mathbb{Z}\ \forall i \in \mathbb{N} : u(n+i) = s_0(i)\}.$$

For an arbitrary state $u \in E^{\mathbb{Z}}$ and any $v \in M$, and any $n_0 \in \mathbb{N}$ and $\varepsilon > 0$, there is an $n > n_0$, such that $d(u, f^n(v)) < \varepsilon$. To find such n we just shift v so far as to get the correct combination in $\Gamma_{1/\lceil 2\varepsilon \rceil}$. There is such n because this finite sequence occurs again and again in the sequence v.

For every $\varepsilon > 0$ and $u, v \in X = E^{\mathbb{Z}}$ there is an ε-chain from u to v and back:

The behavior of the ε-chain is like jumping onto and from a train. We must "jump onto" at some $w \in M$ and eventually "jump off" (at the right moment). Hence all of $E^{\mathbb{Z}}$ is a single chain component. Since $E^{\mathbb{Z}} \cap \omega(E^{\mathbb{Z}}) \neq \emptyset$, the chain component $E^{\mathbb{Z}}$

has non-empty intersection with the global attractor, and therefore is a subset of this attractor. Hence we find that $\omega(E^{\mathbb{Z}}) = E^{\mathbb{Z}}$ is the global attractor.

Remark 5.6.13 In Corollary 2.4.5 we have introduced the subautomaton on states with finite support. The state space of this subautomaton is a metric space (as a subspace of a metric space) but it is not compact anymore. The lack of compactness leads to mathematical difficulties. Nevertheless one can classify these subautomata in a similar manner, see [94].

Chapter 6
Chaos and Lyapunov Stability

In the preceding section we introduced a classification of cellular automata based on attractors, their number and structure. In the present section we focus on the complexity of the dynamics. The two aspects are not independent, but differ slightly. We start with Devaney's definition of chaos, and relate this definition to the Hurley classification. Thereafter we investigate a class of cellular automata that induce chaotic dynamics: permutive cellular automata. In the last part of this section we focus on one special property of chaotic dynamics: sensitive dependency on initial conditions. It is possible to recognize complex dynamics by inspecting the fate of two neighboring points. If the states stay close under iteration, the long term behavior of the system is rather predictable. If they are driven to different parts of the state space, in the long run, even small perturbations may lead to completely dissimilar states. The mathematical concept of continuity compares the fates of neighboring points under the action of a function. A closer look allows to define different degrees of continuity for the dynamical system. This kind of reasoning leads to the Gilman classification.

6.1 Topological Chaos

There are various non-equivalent definitions of "complex dynamics" and "chaos". The following definition, quoted from Devaney [42], has been used quite often. Devaney requires three properties for a chaotic dynamical system: First of all, the system has to mix well (topological transitivity, see Definition 5.5.4). Any open set is eventually transported to any part of the state space. The state space cannot be separated into regions that are invariant under the dynamics. The second condition requires that periodic points are dense (regularity). Since periodic orbits are predictable and not complex, this condition is rather surprising. A non-periodic point close to a periodic point is—due to continuity—dragged along. As the periodic

K.-P. Hadeler, J. Müller, *Cellular Automata: Analysis and Applications*,
Springer Monographs in Mathematics, DOI 10.1007/978-3-319-53043-7_6

orbits are dense, a non-periodic point is dragged into different directions. Although periodic orbits themselves are predictable, their density forces the rest of the state space to behave in a non-predictable manner. This idea is connected to the third condition: sensitive dependence on initial data. If we take any state as a reference state, we find other states arbitrarily close that will eventually drift far away from the trajectory of the reference state.

Definition 6.1.1 Suppose (X, d, f) is a discrete dynamical system with a compact metric space X and a continuous function f. The system is called chaotic if the following conditions are satisfied.

(1) The system is topologically transitive.
(2) The set of periodic points is dense ("topologically regular").
(3) The system is sensitive with respect to initial conditions, i.e., there is a $\delta > 0$, such that for every point x and every $\varepsilon > 0$ there is y with $d(x, y) < \varepsilon$ and $n \in \mathbb{N}$ with

$$d(f^n(x), f^n(y)) > \delta.$$

Recall Definition 5.5.4, where the property "topologically transitive" has been introduced: For any pair of open sets U, $V \subset X$ there is $n_0 \geq 0$ such that $f^{n_0}(U) \cap V \neq \emptyset$.

Remark 6.1.2 The idea behind properties (1) and (3) is the impossibility of prediction over long times: If we do not know the initial data exactly then any two close approximations will eventually drift far apart [property (3)] and may arrive in any portion of the space [property (1)]. These two properties are not independent: If we start in a small open set and get everywhere after some time steps, then large open sets cannot evolve in a "parallel" fashion, and the converse is also true. Hence the definition of "chaos" as given above is redundant, as has been observed in [6]. These authors showed that topological transitivity and denseness of periodic orbits already imply sensitive dependence on initial data. Hence property (3) need not be required.

As we will see in Proposition 6.1.7, for cellular automata there is an even stronger result. Here property (3) follows from property (1) alone [33].

In order to get some feeling for the definitions, let us consider some classical examples [42].

Example 6.1.3 Let $S^1 = \mathbb{R} \bmod 2\pi$. The mapping $f : S^1 \to S^1$, $\theta \mapsto 2\theta$ defines a chaotic dynamical system. We check the three properties.

1) Topologically transitive: The mapping spreads any open set in S^1 such that eventually an iterated map covers all of S^1.
2) Periodic points are everywhere dense: The point $\theta \in S^1$ has period n, if $2^n \theta = \theta + k 2\pi$ for some $k \in \mathbb{N}$. Hence points $\theta = k 2\pi/(2^n - 1)$, with $n, k \in \mathbb{N}$, are periodic points. These points form a dense set in $[0, 2\pi]$.

3) $|f(\theta_1) - f(\theta_2)| > 1.5\,|\theta_1 - \theta_2|$ if θ_1 and θ_2 are sufficiently close. Hence we have sensitive dependence on initial data.

Example 6.1.4 The discrete logistic equation is defined by the mapping $g : [0, 1] \to [0, 1]$, $g(x) = a x(1 - x)$ with parameter $a \in [0, 4]$. With $y = 2x - 1$ we get the equivalent formulation $g_1 : [-1, 1] \to [-1, 1]$, $y \mapsto \mu - 1 - \mu y^2$ where $\mu \in [0, 2]$. With $z = -y$ we get another formulation $g_2 : [-1, 1] \to [-1, 1]$, $z \mapsto \mu z^2 + 1 - \mu$.

In the classical case of von Neumann and Ulam, with parameter $a = 4$ or $\mu = 2$, i.e., $g_2(z) = 2z^2 - 1$, there is an invariant measure [143]. For this choice of the parameters the dynamical system $([0, 1], g)$ shows chaotic behavior as can be seen by reducing the problem to the preceding one.

Define $h : S^1 \to [-1, 1]$, $\theta \mapsto \cos\theta$. With the function f from the preceding example we have

$$h \circ f(\theta) = \cos 2\theta = 2\cos^2\theta - 1 = g_2 \circ h(\theta).$$

It does not make a difference whether we first iterate with f and then go from S^1 to $[-1, 1]$, or go first from S^1 to $[-1, 1]$ and then iterate with g_2.

The mapping h is not bijective (generally there are two pre-images, one on the "northern hemisphere", the other on the "southern hemisphere"), but we see that the essential properties carry over from (S^1, f) to $([-1, 1], g_2)$ and hence to $([0, 1], g)$.

Example 6.1.5 Let $E = \{0, 1\}$, $E^{\mathbb{Z}}$ be the Cantor space of 0, 1-bisequences, and let $\sigma : E^{\mathbb{Z}} \to E^{\mathbb{Z}}$ be the left shift. We show that $(E^{\mathbb{Z}}, \sigma)$ is chaotic.

1) We have already seen that σ is topologically mixing, and hence also topologically transitive (see Theorem 5.5.8).
2) The periodic points are dense: Let $u \in E^{\mathbb{Z}}$. Restrict this state to the set $[-n, n]$ and extend periodically. Then we get a periodic point and its distance to u is not greater than $1/(n + 1)$.
3) Sensitive dependence on initial data follows from 1) and 2), as mentioned above, but we can show it directly as follows: We use the fact that the metric controls only finite sections of $\mathbb{Z}$. For given u choose a section $[-n, n]$ and define v by $v(i) = u(i)$ in this section and $v(i) = 1 - u(i)$ outside of this section. Then u and v are close but $\sigma^n(u)$ and $\sigma^n(v)$ eventually have the distance 1.

Remark 6.1.6

(1) With the preceding example we have found a first chaotic cellular automaton, the Wolfram automaton with local rule $f(x_{-1}, x_0, x_1) = x_1$. In the next section we will define a class of cellular automata which behave similarly to shift automata and we will show that they are also chaotic.

(2) We consider $\Sigma_2 = \{0, 1\}^{\mathbb{N}_0}$, the Cantor set of all sequences as state space for the left shift. Arguments paralleling those of Example 6.1.5 show that also (Σ_2, σ) is a chaotic dynamical system. It is even the prototype of a chaotic dynamical system.

Dynamical systems that are defined by shifts on strings from an alphabet are called "symbolic dynamics". In many explicit examples one can prove existence of chaos or other properties by proving that the system is conjugated to some appropriate symbolic dynamics.

For a cellular automaton "topologically transitive" implies "sensitive with respect to initial conditions".

Proposition 6.1.7 *Let* (Γ, D_0, E, f_0) *be a cellular automaton with infinite* Γ *and* $|E| \geq 2$. *If the automaton is topologically transitive then it is sensitive with respect to initial data.*

Proof (See [33]) We show that a cellular automaton that is not sensitive with respect to initial data is also not transitive.

Recall that for any $u, v \in E^\Gamma$ the equation $u|_{\Gamma_k} = v|_{\Gamma_k}$ implies $d(u, v) \leq 1/(1+k)$.

Assume that the cellular automaton is not sensitive with respect to initial data. For every $\delta > 0$ there is a state $u \in E^\Gamma$ and an $\varepsilon > 0$ such that for any $v \in E^\Gamma$ with $d(u, v) < \varepsilon$, the distance under iteration with f stays small,

$$d(f^n(u), f^n(v)) \leq \delta \quad \text{for all } n \in \mathbb{N}.$$

In particular, choose $\delta = 1/2$ and find u_0 and ε as stated. Choose $q \in \mathbb{N}$, $q > 1/\varepsilon + 1$. Then $u_0|_{\Gamma_q} = v|_{\Gamma_q}$ implies $d(u_0, v) < \varepsilon$ and this inequality implies $d(f^n(u_0), f^n(v)) < \delta = 1/2$ for all $n \in \mathbb{N}$.

For $v \in B_\varepsilon(u_0)$ holds

$$f^n(v)(0) = f^n(u_0)(0).$$

In the complement of Γ_q the elements $v \in B_\varepsilon(u_0)$ run through all possible values from E. Therefore $f^n(u_0)$ does not depend on the values of u_0 outside of Γ_q. Since the Cayley graph is infinite, there is h in Γ that has a distance of more than $2\,q$ from the origin. Let $e_1, e_2 \in E$, $e_1 \neq e_2$ and define

$$\hat{u}_0(x) = \begin{cases} u_0(g) & \text{for} \quad g \in \Gamma_q \\ u_0(g) & \text{for} \quad g \in \sigma_h \Gamma_q \\ e_1 & \text{otherwise.} \end{cases}$$

Let $u \in E^\Gamma$ be such $d(u, \hat{u}_0) < 1/(4q)$. Then

$$f^n(u)(0) = f^n(u)(h) = f^n(u_0)(0) \quad \text{for all} \quad n \in N.$$

Define

$$U = \{u \,:\, d(u, \hat{u}_0) < 1/(4q)\}, \quad V = \{v \,:\, v(0) = e_1, \;\; v(h) = e_2\}.$$

For $u \in U$ holds $f^n(u)(0) = f^n(u)(h)$ and for $v \in V$ holds $v(0) \neq v(h)$. Hence

$$f^n(U) \cap V = \emptyset \quad \text{for all } n \in \mathbb{N}.$$

□

Recall that the defining property of Hurley's class 5 is "$\omega(X) = X$, and X is the only attractor".

Proposition 6.1.8 *Let the dynamical system (X, d, f) be chaotic. Then it belongs to the Hurley class 5.*

Proof Let $\mathcal{A} \subset X$ be an attractor with pre-attractor U. If $V = X \setminus \overline{U} \neq \emptyset$, then there is (due to the topological transitivity of the dynamical system) $n \in \mathbb{N}$ such that $f^n(U) \cap V \neq \emptyset$. This, however, contradicts the forward invariance of the pre-attractor U. Hence, $\overline{U} = X$. The periodic orbits are dense in X. As $f(X)$ is closed and contains all periodic orbits, $f(X) = X$. From $f(\overline{U}) \subset U$ and $\overline{U} = X$ we obtain $U = X$ and $\mathcal{A} = \omega(U) = X$. The only attractor of the system is X. □

6.2 Permuting Cellular Automata

Here we present some results from a paper by Cattaneo [25] on the relations between the class of permuting cellular automata and chaotic dynamics. We will call a cellular automaton a "one-dimensional cellular automaton" if $\Gamma = \mathbb{Z}$. An interval $[a, b]$ is understood as the integers in this interval, $[a, b] \cap \mathbb{Z}$. We will assume that D_0 is an interval in $\mathbb{Z}$, but do not require that $0 \in D_0$.

Definition 6.2.1 Let $l, r \in \mathbb{Z}$, $l \leq r$ and $D_0 = [l, r]$. The local function f_0 of a one-dimensional cellular automaton is called permuting at $n \in D_0$ if for every local state $u \in E^{D_0}$ holds

$$\{f_0(v) \,:\, v \in E^{D_0},\;\; v|_{D_0'} = u|_{D_0'}\} = E,$$

where $D_0' = D_0 \setminus \{n\}$. It is called right-permuting (left permuting) if it is permuting at $n = r$ ($n = l$).

Remark 6.2.2

(1) A Wolfram automaton is permuting at $n = -1$ if for all $e_1, e_2 \in \{0, 1\}$ it holds that

$$f_0(0, e_1, e_2) \neq f_0(1, e_1, e_2).$$

(2) If the automaton is permuting at n, then information is moved from site n to site 0 (resp. from site $i + n$ to site i, $i \in \mathbb{Z}$) and perhaps permuted. This cellular automaton shares many properties with a shift operator.

We will show that the left- and right-permuting cellular automata are chaotic (in the sense of the definition above). In particular, we show topological transitivity and denseness of periodic points, as we already know that the third property (sensitivity) follows by general arguments. However, in order to prove the denseness of periodic points, some results about surjectivity of one dimensional cellular automata in general and permutive automata in particular are required. We start with these considerations. These results are already shown in the seminal work of Hedlund [85] (Theorems 5.1–5.5).

6.2.1 *Surjective Cellular Automata*

The global functions of right-permuting and of left-permuting cellular automata are surjective. We shall prove this fact later. One-dimensional surjective cellular automata have a very particular property: A state has only finitely many pre-images. We will find that the number of pre-images cannot exceed $|E|^{|D_0|-1}$ if D_0 is an interval (see Proposition 6.2.8). This result is one stepping stone we use below in Sect. 6.2.3 to show that the periodic points of permutive cellular automata are dense (property (1) in the definition of chaos). Further results about surjectivity and injectivity can be found in Sect. 9 and in [27].

In this section, we characterize cellular automata with surjective global functions. It turns out that surjectivity can be checked locally—a global function is surjective if its restrictions on finite grid intervals are surjective. The number of pre-images is independent of the length of the grid interval. From this observation it follows that the number of pre-images of a state is bounded.

First we define the restriction of the function to some finite part of the grid, in a similar way we did before in Sect. 3.4.2.

Definition 6.2.3 Let $(\mathbb{Z}, D_0, E, f_0)$ be any cellular automaton. Define for $\tilde{\Gamma} \subset \Gamma$ the set

$$\tilde{\Gamma}^- = \{x \in \mathbb{Z} |\, \sigma_x(D_0) \subset \tilde{\Gamma}\}.$$

Choose a fixed element $e \in E$. Define an extension operator $F_{\tilde{\Gamma}} : E^{\tilde{\Gamma}} \to E^{\mathbb{Z}}$ by $F_{\tilde{\Gamma}}(u)(x) = u(x)$ for $x \in \tilde{\Gamma}$ and $F_{\tilde{\Gamma}}(u)(x) = e$ otherwise. Finally define the restriction

$$f_{\tilde{\Gamma}} : E^{\tilde{\Gamma}} \to E^{\tilde{\Gamma}^-}, \quad f_{\tilde{\Gamma}} = (f \circ F_{\tilde{\Gamma}})|_{\tilde{\Gamma}^-}.$$

Notation If $\tilde{\Gamma} = \Gamma_n$, we write F_n and f_n instead of F_{Γ_n} and f_{Γ_n}.

Remark 6.2.4

(1) The set $\tilde{\Gamma}^- \subset \mathbb{Z}$ consists of exactly all points x which have a neighborhood $x + D_0$ that is subset of $\tilde{\Gamma}$. Note that $\tilde{\Gamma}^-$ is not necessarily a subset of $\tilde{\Gamma}$, in particular if $0 \notin D_0$.

(2) This construction is independent of the choice of e since in view of the final restriction the image of F_n is used only in Γ_n^- and depends only on the values in Γ_n.

(3) Assume that $D_0 = [l, r] \subset \mathbb{Z}$ is an interval. As $\Gamma = \mathbb{Z}$, for any interval $\tilde{\Gamma} = [a, b] \subset \mathbb{Z}$, we have $|\tilde{\Gamma}^-| = |\tilde{\Gamma}| - |D_0| + 1$. This identity will be frequently used. Note that it is always possible to choose D_0 as an interval.

Proposition 6.2.5 *Let $(\mathbb{Z}, D_0, E, f_0)$ be a cellular automaton. Let f_n be given as in Definition 6.2.3, with $e \in E$ arbitrary. The following statements are equivalent.*

1) $f : E^{\mathbb{Z}} \to E^{\mathbb{Z}}$ *is surjective.*
2) $f_n : E^{\Gamma_n} \to E^{\Gamma_n^-}$ *is surjective for every* $n \in \mathbb{N}$, $\Gamma_n^- \neq \emptyset$.

Proof $1 \Rightarrow 2$: Let $v \in E^{\Gamma_n}$. Extend v and get $\tilde{v} \in E^{\mathbb{Z}}$. By assumption, there is a $u \in E^{\mathbb{Z}}$ with $f(u) = \tilde{v}$. It follows that $f_n(u|_{\Gamma_n}) = v$.

$2 \Rightarrow 1$: Let $v \in E^{\mathbb{Z}}$ be given. We construct a $u \in E^{\mathbb{Z}}$ with $f(u) = v$. For every $v|_{\Gamma_n^-}$ there is a $\tilde{u}_n \in E^{\Gamma_n}$ such that $f(u_n) = v|_{\Gamma_n}$. Extend $\tilde{u}_n$ to $u_n = F_n(\tilde{u}_n) \in E^{\mathbb{Z}}$. Still we have $f(u_n)|_{\Gamma_n^-} = v|_{\Gamma_n^-}$. Since $(E^{\mathbb{Z}}, d)$ is compact, there is a convergent subsequence, $u = \lim_{l\to\infty} v_{n_l}$. Then

$$d(f(u), v) \le d(f(u), f(u_{n_l})) + d(f(u_{n_l}), v).$$

Since u_{n_l} converges to u and f is continuous, the first term converges to zero for $l \to \infty$. Since the sets $\Gamma_{n_l}^-$ exhaust the grid for $l \to \infty$, and $f(u_{n_l})|_{\Gamma_{n_l}^-} = v_{n_l}|_{\Gamma_{n_l}^-}$, also the second term goes to zero, i.e., $d(f(u), v)$ becomes arbitrarily small, hence $f(u) = v$. □

The following two propositions are somewhat technical and prepare for the proof of Proposition 6.2.8.

Proposition 6.2.6 *Let $\Gamma_1 = [l, r]$ is a finite interval in $\mathbb{Z}$, and $\tilde{\Gamma}_1^- \neq \emptyset$. Let furthermore $\tilde{\Gamma}_2 \supset [l-1, r+1]$. Suppose that for a given $u \in E^{\tilde{\Gamma}_1^-}$ the inequality $|f_{\tilde{\Gamma}_1}^{-1}(u)| = k \in \mathbb{N}$ holds. Suppose further that for all $v \in \{w \in E^{\tilde{\Gamma}_2^-} : w|_{\Gamma_1^-} = u\}$ holds $|f_{\Gamma_2}^{-1}(v)| \ge k$. Then $|f_{\Gamma_2}^{-1}(v)| = k$ for all $v \in \{w \in E^{\Gamma_{n+1}^-} : w|_{\Gamma_n^-} = u\}$.*

Proof $u \in E^{\tilde{\Gamma}_1^-}$ is given; we know $|f_{\tilde{\Gamma}_1}^{-1}(u)| = k$. Define the set $U(u)$ of all $w \in E^{\tilde{\Gamma}_2^-}$ that agree on $\tilde{\Gamma}_1^-$ with u, $U(u) = \{w \in E^{\tilde{\Gamma}_2^-} : w|_{\tilde{\Gamma}_1^-} = u\}$.

In order to count the pattern in $\{f_{\tilde{\Gamma}_2}^{-1}(w) : w \in U(u)\}$, we split $\tilde{\Gamma}_2$ into Γ_1 and $\tilde{\Gamma}_2 \setminus \tilde{\Gamma}_1$. As for $v \in \{f_{\tilde{\Gamma}_2}^{-1}(w) : w \in U(u)\}$ we have $v|_{\tilde{\Gamma}_1} \in f_{\tilde{\Gamma}_1}^{-1}(u)$, there are k

possible patterns $v|_{\tilde{\Gamma}_1}$. Furthermore, there are $|E|^{|\tilde{\Gamma}_2\setminus\tilde{\Gamma}_1|}$ ways to occupy the cells in $\tilde{\Gamma}_2 \setminus \tilde{\Gamma}_1$. Hence

$$|\{f_{\tilde{\Gamma}_2}^{-1}(w)\ :\ w \in U(u)\}| = k\,|E|^{|\tilde{\Gamma}_2\setminus\tilde{\Gamma}_1|}.$$

Since $\Gamma = \mathbb{Z}$, we find $|\tilde{\Gamma}_2^- \setminus \tilde{\Gamma}_1^-| = |\tilde{\Gamma}_2 \setminus \tilde{\Gamma}_1|$. Therefore,

$$k\,|E|^{|\tilde{\Gamma}_2\setminus\tilde{\Gamma}_1|} = |\{f_{\tilde{\Gamma}_2}^{-1}(w)\ :\ w \in U(u)\}| = \sum_{v\in U(u)} |f_{\tilde{\Gamma}_2}^{-1}(v)|$$
$$\geq \sum_{v\in U(u)} k = k\,|E|^{|\tilde{\Gamma}_2^-\setminus\tilde{\Gamma}_1^-|} = k\,|E|^{|\tilde{\Gamma}_2\setminus\tilde{\Gamma}_1|}.$$

If there were a $v \in U(u)$ with $|f_{\tilde{\Gamma}_2}^{-1}(v)| > k$, then we would have a contradiction. □

Since we have a one-dimensional grid, $\Gamma = \mathbb{Z}$, we can concatenate patterns in adjacent "intervals" in $\mathbb{Z}$. For example, if $\psi \in E^{\{1,\dots,10\}}$ and $\phi \in E^{\{1,\dots,5\}}$, then we use the notation $[\psi, \phi] \in E^{\{1,\dots,15\}}$. In view of the next proposition it is useful to state the length of some patterns: Let $\Gamma_1 = [1, n+|D_0|-1]$, $u_0 \in E^{\Gamma_1^-}$, $v \in E^{|D_0|-1}$, then $[uvu] \in E^{2n+|D_0|-1}$. If we define $\tilde{\Gamma}_2 = [1, 2n+2|D_0|-2]$, we may also use the shift invariance of the automaton and write $[uvu] \in E^{\tilde{\Gamma}_2^-}$. With this notation, we obtain the following proposition.

Proposition 6.2.7 *Let $|D_0| \geq 2$, $n \in \mathbb{N}$, $\Gamma_1 = [1, n+|D_0|-1]$, $u_0 \in E^{\tilde{\Gamma}_1}$, and $k = |\tilde{f}_{\tilde{\Gamma}_1}^{-1}(u_0)|$. Suppose $k > 0$ and further that for every $v \in E^{|D_0|-1}$, the equality*

$$|\tilde{f}_{\tilde{\Gamma}_2}^{-1}([u_0, v, u_0])| = k$$

holds, where $\tilde{\Gamma}_2 = [1, 2n+2|D_0|-2]$. Then $k = |E|^{|D_0|-1}$.

Proof Let $\tilde{f}_{\tilde{\Gamma}_1}^{-1}(u_0) = \{w_1, \dots, w_k\}$. We show that

$$U := \cup_{v\in E^{|D_0|-1}} \tilde{f}_{\tilde{\Gamma}_2}^{-1}([u_0, v, u_0]) = \{w_1, \dots, w_k\} \times \{w_1, \dots, w_k\} =: W$$

such that W denotes the set of all $[w_i, w_j]$. Let $w = [w_i, w_j] \in W$. Every element in $\tilde{f}_{\tilde{\Gamma}_1}^{-1}(u_0)$ has length $n + |D_0| - 1$, hence w has length $2n + 2|D_0| - 2$ and $\tilde{f}_{[1,2n+2|D_0|-2]}(w)$ has length $2n + |D_0| - 1$. In view of $\tilde{f}_{\tilde{\Gamma}_1}(w_i) = u_0$ the element $\tilde{f}_{[1,2n+2|D_0|-2]}(w)$ has the form

$$\tilde{f}_{[1,2n+2|D_0|-2]}(w) = [u_0, v, u_0],$$

whereby v has length $(2n + |D_0| - 1) - (n + n) = |D_0| - 1$ indicating that $w \in U$.

Conversely, assume $\tilde{w} \in U$ such that $\tilde{f}_{\tilde{\Gamma}_2}(\tilde{w}) = [u_0, v, u_0]$ for some $v \in E^{|D_0|-1}$. Then $\tilde{w}$ has length $2n + 2|D_0| - 2$, and $\tilde{w}$ can be represented as $\tilde{w} = [\tilde{w}_1, \tilde{w}_2]$, where $\tilde{w}_1, \tilde{w}_2$ both have length $n + |D_0| - 1$, and $f_{n+|D_0|-1}(\tilde{w}_i) = u_0$. Hence $\tilde{w} \in W$.

By hypothesis $|\tilde{f}_{\tilde{\Gamma}_2}^{-1}([u_0, v, u_0])| = k$ for all $v \in E^{|D_0|-1}$. Since

$$\tilde{f}_{\tilde{\Gamma}_2}^{-1}([u_0, v, u_0]) \cap \tilde{f}_{\tilde{\Gamma}_2}^{-1}([u_0, v', u_0]) = \emptyset \quad \text{for } v \neq v',$$

we obtain

$$k^2 = |W| = |U| = \sum_{v \in E^{|D_0|-1}} \left|\tilde{f}_{\tilde{\Gamma}_2}^{-1}([u_0, v, u_0])\right| = k\,|E|^{|D_0|-1}$$

and hence $k = |E|^{|D_0|-1}$. □

Proposition 6.2.8 *The global function f of a one-dimensional cellular automaton is surjective if and only if for every $n \geq |D_0|$ and $u \in E^{\Gamma_n^-}$ holds*

$$|f_n^{-1}(u)| = |E|^{|D_0|-1}.$$

Proof $\Leftarrow$: Recall $|E^{\Gamma_n}| = E^{2n+1}$ and $|E^{\Gamma_n^-}| = E^{2n+2-|D_0|}$. If $|f_n^{-1}(u)| = |E|^{|D_0|-1}$ then for every n there is a pre-image of every element in $E^{2n-|D_0|+2}$ under f_n, and hence f_n is surjective. We conclude with Proposition 6.2.5 that also f is surjective.

$\Rightarrow$: We assume that f is surjective. For $m > |D_0|$, let $\tilde{\Gamma} = [1, m]$ and

$$k = \inf\{|f_{\tilde{\Gamma}(m)}^{-1}(u)| \,:\, u \in E^{\tilde{\Gamma}(m)^-},\ m \in \mathbb{N}\}.$$

Since f is surjective, also the $f_{\tilde{\Gamma}(m)}$, $m \in \mathbb{N}$, are surjective, hence $k \geq 1$. Hence there is an m and a pattern $u_0 \in E^{\tilde{\Gamma}(m)^-}$ such that

$$k = |f_{\tilde{\Gamma}(m)}^{-1}(u_0)|.$$

Suppose there is some state $v \in E^{\mathbb{Z}}$ such that the restriction of v to some finite grid domain $\sigma_x(\tilde{\Gamma}(m)^-)$, $x \in \mathbb{Z}$, is a translated version of u_0. In this case we say that v covers u_0. If v covers u_0 then for $\tilde{\Gamma}_2 \supset \sigma_x(\tilde{\Gamma}(m)^-)$ large enough $|f_{\tilde{\Gamma}_2}^{-1}(v|_{\tilde{\Gamma}_2^-})| \geq k$. With Proposition 6.2.6 we find, by induction over the size of the grid domain, that the cardinality of the preimage of any pattern covering u_0 is already k.

In particular, blocks of the form $[u_0, v, u_0]$, with v of length $|D_0| - 1$, have preimages with cardinality k. With Proposition 6.2.7 we conclude $k = |E|^{|D_0|-1}$.

Now we know that the cardinality of preimages of $f_{[1,m]}$ is not less than $|E|^{|D_0|-1}$. Suppose there is $n \geq |D_0|$ and $u_1 \in E^{\Gamma_n^-}$ such that

$$|f_n^{-1}(u_1)| > |E|^{|D_0|-1}.$$

Since $|f_n^{-1}(w)| \geq |E|^{|D_0|-1}$ for all $w \in E^{\Gamma_n^-}$, $|\Gamma_n^-| = |\Gamma_n^-| - |D_0| + 1$, and f_n surjective, we have

$$\begin{aligned} |E^{\Gamma_n}| &= |E^{\Gamma_n^-}|\, |E|^{|D_0|-1} = |\{w \in E^{\Gamma_n^-}\}|\, |E|^{|D_0|-1} \\ &< \sum_{w \in E^{\Gamma_n^-}} |f_n^{-1}(w)| = |f_n^{-1}(\{w \in E^{\Gamma_n^-}\})| \\ &= |\{w \in E^{\Gamma_n}\}| = |E^{\Gamma_n}| \end{aligned}$$

which leads to a contradiction. Hence $|f_n^{-1}(u)| = |E|^{|D_0|-1}$. □

We need not to restrict ourselves to subgraphs $\Gamma_n = [-n, n] \subset \mathbb{Z}$, but the same argument can be used to handle $[1, n] \subset \mathbb{Z}$.

Corollary 6.2.9 *Let* $(\mathbb{Z}, D_0, E, f_0)$ *be a one-dimensional cellular automaton, and* D_0 *is an interval. The global function* f *is surjective if and only if for every* $n \geq |D_0|$ *and* $u \in E^{n+|D_0|-1}$ *holds*

$$|f_{[1,n]}^{-1}(u)| = |E|^{|D_0|-1}.$$

Theorem 6.2.10 *Let* $(\mathbb{Z}, D_0, E, f_0)$ *be a one-dimensional cellular automaton with surjective global function, and* D_0 *is an interval. Then* $v \in E^{\mathbb{Z}}$ *has only finitely many pre-images, their number being bounded by*

$$|f^{-1}(v)| \leq |E|^{|D_0|-1},$$

Proof Suppose the global function f is surjective and further, that there is a $v \in E^{\mathbb{Z}}$ with

$$|f^{-1}(v)| > |E|^{|D_0|-1}.$$

Then there is an $r > |E|^{|D_0|-1}$ and pairwise distinct states $u_1, \ldots, u_r \in E^{\mathbb{Z}}$, with $f(u_i) = v$. Since these states are distinct, there is a Γ_k such that $v_i|_{\Gamma_k} \neq v_j|_{\Gamma_k}$ for $i \neq j$. This finding contradicts the known number of preimages under f_k. □

Proposition 6.2.11 *Let the cellular automaton be left-permuting or right-permuting. Then the global function is surjective.*

Proof We prove that for a left- or right-permuting cellular automaton $|f_{\tilde{\Gamma}}(\tilde{\Gamma})| = |E|^{|\tilde{\Gamma}^-|}$ holds for any interval $\tilde{\Gamma} = [1, n]$.

This identity is trivial for $n = |D_0|$ since in this case $|\tilde{\Gamma}| = |D_0|$. Suppose the identity has been proved for $n-1$. Then it holds for n: From the states $\{1, \ldots, n-1\}$ we get exactly $|E|^{n-1-|D_0|+1}$ combinations. If we fix the state in the cells $\{1, \ldots, n-1\}$ and let the state in the cell n vary then we get $|E|$ combinations. Altogether we get $|f_{[1,n]}(E^{[1,n]})| = |E|^{n-|D_0|+1}$. □

6.2.2 Topological Transitivity

Proposition 6.2.12 *Let* $(\mathbb{Z}, D_0, E, f_0)$ *be a cellular automaton that is left-permuting and* $\min D_0 \neq 0$ *resp. right-permuting and* $\max D_0 \neq 0$*. Then it is topologically transitive.*

Proof (See [57]) Without lack of generality we assume that the automaton is right-permuting. Let $D_0 = [l, r]$, and the automaton be permuting with respect to r. We assume $r > 0$, and discuss at the end of the proof the case $r < 0$.

Let $U, V \subset E^{\mathbb{Z}}$ be nonempty open sets. There are $u \in U$, $v \in V$ and an $\varepsilon > 0$ such that $B_\varepsilon(u) = \{w : d(u, w) < \varepsilon\} \subset U$, $B_\varepsilon(v) = \{w : d(v, w) < \varepsilon\} \subset V$. Let m be the smallest number such that $m > 1/\varepsilon + 1$, and $m + 1$ is a multiple of r. Then define $n = 2(m + 1)/r$. We will construct a state $w_{2m} \in U$ such that $f^n(w_{2m}) \in V$.

All patterns which agree on $\{-m, \dots, m\}$ with u (with v) are in U (in V). Now we construct a finite sequence of states $w_0, \dots, w_{2m+1}$ such that

(1) $w_i \in B_\varepsilon(u)$
(2) $f^n(w_i)|_{\{-m,\dots,-m+i\}} = v|_{\{-m,\dots,-m+i\}}$

Since f is permuting with respect to r, the function f^2 is permuting with respect to $2r$ and f^n is permuting with respect to $n\,r = 2(m + 1)$.

Choice of w_0: Choose an $e \in E$ and define

$$w_0(k) = \begin{cases} u(k) & \text{for } k \neq m+2 \\ e & \text{for } k = m+2. \end{cases}$$

When e runs through the set E, then also $f^n(w_0)(-m)$ runs through the set E. Hence there is an $e \in E$ such that $f^n(w_0)(-m) = v(-m)$.

Choice of w_1: Choose an $e \in E$ and define

$$w_1(k) = \begin{cases} w_0(k) & \text{for } k \neq m+3 \\ e & \text{for } k = m+3. \end{cases}$$

When e runs through the set E then also $f^n(w_1)(-m + 1)$ runs through the set E. Hence we find an $e \in E$ such that $f^n(w_1)(-m + 1) = v(-m + 1)$. Thereby $f^n(w_0)(-m - 1) = f^n(w_1)(-m - 1)$.

Similarly $w_2, \dots, w_{2m+1}$ are constructed recursively, finally we find a w_{2m+1} with $w_{2m+1} \in B_\varepsilon(u)$ and $f^n(w_{2m+1})|_{\{-m,\dots,m\}} = v|_{\{-m,\dots,-m+i\}}$, i.e., $f^n(w_{2m+1}) \in B_\varepsilon(v)$. Now we have shown $f^n(U) \cap V \neq \emptyset$.

The adaptation of the proof for case $r < 0$ concerns the construction of the sequence w_i. We allow for negative m and require that m is the largest number such that $|m| > 1/\varepsilon + 1$ and that $2m - 1$ is a multiple of r. We use $n = (2m - 1)/r$; the remaining part of the proof works as before with some obvious modifications. □

Remark 6.2.13 The proposition excludes the case $r = 0$. In general, a cellular automaton that is right-permuting with $r = 0$ is not topologically transitive as can seen by Wolfram's rule 204, the identity on $\{0, 1\}^{\mathbb{Z}}$.

6.2.3 Denseness of Periodic Points

Proposition 6.2.14 *Let $(\mathbb{Z}, D_0, E, f_0)$ be a cellular automaton with surjective global function. Then every pre-image of a spatially periodic state is spatially periodic (perhaps with a larger period).*

Proof Let $u, v \in E^{\mathbb{Z}}$, and $\sigma_x(u) = u$ with some non-trivial translation $\sigma_x, x \in \mathbb{Z}\backslash\{0\}$. Let $v = f(v)$. In view of

$$f(\sigma_x^i(v)) = \sigma_x^i(f(v)) = \sigma_x^i(u) = u$$

the states $\sigma_x^i(v)$ are pre-images of u. Since there are only finitely many pre-images (Theorem 6.2.10), there is an $i \neq 0$ with $\sigma_x^i(v) = v$. Since $x \neq 0$, the state v has spatial period $i\,x$. □

Definition 6.2.15 The cellular automaton $(\mathbb{Z}, D_0, E, f_0)$ is called right-centered if $D_0 \subset \{x > 0\}$, and left-centered if $D_0 \subset \{x < 0\}$.

Let f be the global function of a cellular automaton. Then the function $\sigma_x^k \circ f$, for $x \neq 0$ and $|k|$ large enough, is the global function of a right- or left-centered cellular automaton. Note that f is not conjugated with $\sigma_x^k \circ f$. Nevertheless, depending on the properties under consideration, it is possible to replace a cellular automaton by a left- or right-centered cellular automaton. An example for such a property is surjectivity: f is surjective if and only if $\sigma_x^k \circ f$ is surjective. Without restriction, we may concentrate on the investigation of surjectivity for right- (left-)centered cellular automata.

For a global function f and the identity I (both are continuous functions) we want to define the difference $I - f$. With this goal in mind we introduce an artificial "addition" on the set E. Let $E = \{\eta_0, \ldots, \eta_{l-1}\}$, and

$$\eta_i \pm \eta_j = \eta_{i \pm j \bmod l}\,.$$

Definition 6.2.16 Let $f, g \in E^{\Gamma}$. Then the functions $f + g \in E^{\Gamma}$ resp. $f - g \in E^{\Gamma}$ are defined as

$$(f \pm g)(x) = f(x) \pm g(x).$$

Proposition 6.2.17 *Let $(\mathbb{Z}, D_0, E, f_0)$ with global function f. For any $n \in \mathbb{N}$ the function $g = I - f^n$ is well-defined and it is also the global function of a cellular automaton.*

If the automaton is left- or right-centered then the function $g = I - f^n$ is surjective.

Proof $I - f^n$ is the global function of a cellular automaton for the following reasons. The difference of two continuous functions is again continuous. The function f^n and the identity are continuous and both commute with the translation operator.

If f is left-centered then $I - f^n$ is right-permuting. The global function of a right-permuting cellular automaton is surjective. A parallel argument shows the claim for right-centered cellular automata. □

Proposition 6.2.18 *Let f be the global function of a left-centered or right-centered cellular automaton $(\mathbb{Z}, D_0, E, f_0)$, not necessarily surjective. Suppose there is a state $u \in E^{\mathbb{Z}}$ and $n \in \mathbb{N}$ such that*

$$f^n(u) = u.$$

Then there is an $x \in \mathbb{Z} \setminus \{0\}$ such that

$$\sigma_x(u) = u.$$

Proof The state u is periodic with period n. Then $u - f^n(u) = [\bar{0}]$. In particular, $(I - f^n)(u)$ is periodic in space, and $I - f^n$ is surjective. Hence u is periodic in space (see Proposition 6.2.14). □

Theorem 6.2.19 *Let $(\mathbb{Z}, D_0, E, f_0)$ be left-permuting or right-permuting. Then the set of (time-) periodic states is dense in $E^{\mathbb{Z}}$.*

Proof Let the automaton be right-permuting. Choose s so large that $g = \sigma_1^s \circ f$ is the global function of a right-centered automaton with neighborhood $\{1, \ldots, k\}$.

Step 1: *Periodic points of g dense $\Rightarrow$ periodic points of f dense.*
Let u be periodic, $g^n(u) = u$. Since g comes from a right-centered cellular automaton, the state u is periodic in space, there is $x > 0$ such that $\sigma_x(u) = u$. For $m = n\,x$ it follows that

$$g^m(u) = g^{nx}(u) = (f^{nx} \circ \sigma_{nxs})(u) = f^m(\sigma_{xns}(u)) = f^m(u).$$

Hence (time-) periodic states of g are (time-) periodic states of f. If the periodic states of g are dense, then also those of f.

Step 2: *Construction of approximative periodic orbits of g.*
Let $[w_{-k}, \ldots, w_k] \in E^{2k+1}$ be a finite sequence of elementary states. We construct a sequence of pairs of states $u_n, v_n \in E^{\mathbb{Z}}$ with the following three properties:

(1) There is an $m \in \mathbb{N}$ such that $g^m(u_n) = v_n$ for all $n \in \mathbb{N}$.
(2) $u_n|_{[-(k+n),(k+n)]} = v_n|_{[-(k+n),k+n]}$
(3) $u_n|_{[-k,k]} = [w_{-k}, \ldots, w_k]$.

We start from the fact that g (being the global function of a right-permuting automaton) is topologically transitive. We define

$$U = V = \{u \in E^{\mathbb{Z}} : u|_{[-k,k]} = [w_{-k}, \ldots, w_k]\}.$$

Then there is an $m > 0$ such that $g^m(U) \cap V \neq \emptyset$, i.e., there is a $u_0 \in U$ such that $v_0 = g^m(u_0) \in V$, and u_0 and v_0 satisfy the conditions (1)–(3) for $n = 0$.
Now we continue recursively. We have already a germ of a periodic state, we extend it on both sides. Suppose we have already constructed u_{n-1} and v_{n-1}. Then we define

$$\tilde{v}_n(x) = \begin{cases} v_{n-1}(x) & \text{for } x \neq k+n \\ u_{n-1}(x) & \text{for } x = k+n. \end{cases}$$

Since g is right-centered and right-permuting, we find $\tilde{u}_n$ with $g^m(\tilde{u}_n) = \tilde{v}_n$ and

$$\tilde{u}_n|_{[-k-n+1,k+n]} = \tilde{v}_n|_{[-k-n+1,k+n]}.$$

We can choose $\tilde{u}_n(x) = u_{n-1}(x)$ for $x \leq k+n$, and then choose step by step $\tilde{u}_n(k+n+1), \tilde{u}_n(k+n+2), \ldots$ in such a way that $g^m(\tilde{u}_n) = \tilde{v}_n$ holds. Then define

$$v_n(x) = \begin{cases} \tilde{v}_n(x) & \text{for } x \neq -(k+n) \\ v_n(x) & \text{for } x = -(k+n). \end{cases}$$

and finally $v_n = g^m(u_n)$. Then (1), (2), (3) hold.

Step 3: Periodic points are dense.
We take the sequences u_n and v_n from step 2 and obtain, by compactness and by taking subsequences, limits

$$u_{n_l} \to \hat{u}, \qquad v_{n_{l_k}} \to \hat{v}.$$

Since on any compact set of $\mathbb{Z}$ the states u_n and v_n agree for large m, n, we have $\hat{u} = \hat{v}$. By continuity of g it follows that $g^m(\hat{u}) = \hat{v}$, and also $\hat{u}|_{[-k,k]} = [w_{-k}, \ldots, w_k]$ holds. For every state there are periodic states that agree with the given state on arbitrarily large domains. Thus, the periodic states are dense. □

Remark 6.2.20

(1) For Wolfram automata the properties "left- or right-permuting" and "chaotic" are equivalent [25].
(2) Also for general grids, there is a connection between the properties "permuting" and "chaotic behavior" [2].

6.3 Lyapunov Stability and Gilman Classification

In 1987 Gilman introduced a classification scheme for cellular automata which is based on the notion of equicontinuity of the global function. In his paper [65] he defined two weaker versions of equicontinuity which are based on measure theory rather than topology. Here, we follow Kůrka [111] who formulated a variant of the Gilman classification in purely topological terms. The Gilman classification scheme is primarily designed for one-dimensional automata. We start with some more definitions.

Definition 6.3.1 Consider a cellular automaton (Γ, D_0, E, f_0) with global function f.

(1) The family $\{f^n\}$ is equicontinuous in $u \in E^\Gamma$ if for every $\varepsilon > 0$ there is $\delta > 0$ such that for any state $v \in B_\delta(u)$, the ball of radius δ centered at u, and any $n \in \mathbb{N}$ holds

$$d(f^n(u), f^n(v)) < \varepsilon.$$

The state u is also called stable in the sense of Lyapunov. Let $Eq(f)$ denote the set of all equicontinuous states.

(2) The function f is called sensitive at $u \in E^\Gamma$ if $u \notin Eq(f)$.

(3) The function f is expanding, if there is a $\delta > 0$ such that for arbitrary $u, v \in E^\Gamma$ with $u \neq v$ there is an $n \in \mathbb{N}$ such that

$$d(f^n(u), f^n(v)) > \delta.$$

Recall Definition 3.4.1, where we defined "expansive". The difference between "expanding" and "expansive" is that expansive functions are bijective and we allow $n \in \mathbb{Z}$ (iteration in forward and backward time). Only bijective functions can be "expansive", while any continuous function can be "expanding".

If we proceed from property (1) through property (2) to property (3) then the dynamics of the cellular automaton considered may get more and more complicated. For a cellular automaton with infinite grid [no isolated points in (E^Γ, d)] the property "expanding" implies "sensitive with respect to initial data at every state in E^Γ", and from "sensitive with respect to initial data at every state in E^Γ" follows $Eq(f) = \emptyset$.

We show that the set of stable points in the sense of Lyapunov is invariant under translations.

Proposition 6.3.2 *Let (Γ, D_0, E, f_0) be a cellular automaton and $g \in \Gamma$. Then,*

$$\sigma_g(Eq(f)) \subset Eq(f).$$

Proof Let $u \in Eq(f)$. We show $\sigma_g(u) \in Eq(f)$ for $g \in \Gamma$. That is, we show that for a given $\varepsilon > 0$ there is a $\delta > 0$ such that $d(\sigma_g(u), \sigma_g(v)) < \delta$ implies

$$d(f^n(\sigma_g(u)), f^n(\sigma_g(v))) < \varepsilon \quad \text{for all } n \in \mathbb{N}.$$

If we want to control the difference of two states in the neighborhood of some $g \in \Gamma$ then we choose a neighborhood of $0 \in \Gamma$ so large that it contains the neighborhood of g and then we try to control the difference in the neighborhood of $0 \in \Gamma$. Precisely we have for $\hat{\varepsilon} = 1/\tilde{\varepsilon} + 1 + d_c(0, g)$ that

$$d(u, v) < \hat{\varepsilon} \quad \Rightarrow \quad d(\sigma_g(u), \sigma_g(v)) < \tilde{\varepsilon},$$

and similarly

$$d(\sigma_g(u), \sigma_g(v)) < \hat{\varepsilon} \quad \Rightarrow \quad d(u, v) < \tilde{\varepsilon}.$$

Therefore we choose $\tilde{\varepsilon} = 1/(\varepsilon + 1 + d_c(0, g))$. Then we find a $\tilde{\delta}$ such that the distance between $f^n(u)$ and $f^n(v)$ becomes small, $d(f^n(u), f^n(v)) < \tilde{\varepsilon}$, for all $n \in \mathbb{N}$ whenever $d(u, v) < \tilde{\delta}$. In particular,

$$d(f^n(\sigma_g(u)), f^n(\sigma_g(v))) < \varepsilon$$

holds for all $n \in \mathbb{N}$. In order to have $d(u, v) < \tilde{\delta}$ we choose $\delta = 1/(\tilde{\delta}+1+d_c(0, g))$. Then we conclude from $d(\sigma_g(u), \sigma_g(v)) < \delta$ that $d(u, v) < \tilde{\delta}$, and furthermore $d(f^n(u), f^n(v)) < \tilde{\varepsilon}$ for all $n \in \mathbb{N}$, and finally $d(f^n(\sigma_g(u)), f^n(\sigma_g(v))) < \varepsilon$ for all $n \in \mathbb{N}$. □

Remark 6.3.3 There are cases where $f(Eq(f))$ is not a subset of $Eq(f)$. Thus, the set of Lyapunov-stable states needs not be invariant. We sketch an example: Let $f : \mathbb{R} \to \mathbb{R}$ be the map depicted in Fig. 6.1. To the left of 0, the function is given by $f(x) = 2x$, such that $\{x < 0\} \cap Eq(f) = \emptyset$. In $[0, 1]$, the function is defined by $f(x) = x$, i.e., $(0, 1) \subset Eq(f)$. Note that $0 \notin Eq(f)$, as any neighborhood of 0 has a non-void intersection with the negative real axis. As third ingredient of our example, we assume a local minimum of f at $x_1 > 1$, and $f(x_1) = 0$. Hence, any small neighborhood of x_1 is mapped into $[0, 1]$ and stays constant under further iteration. We have $x_1 \in Eq(f)$, but $0 = f(x_1) \notin Eq(f)$.

Now we proceed to the Gilman classification (following the Kůrka approach). The proof that the set of all cellular automata decomposes into the given disjoint classes is obvious. After stating the classification theorem, we investigate the properties of these classes in detail and give examples. Class Gilman 1 is investigated for general groups, while in case of the other classes we concentrate on the one-dimensional case, $\Gamma = \mathbb{Z}$.

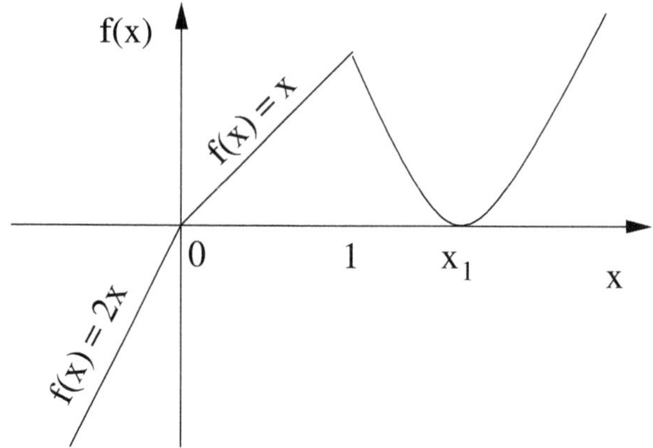

Fig. 6.1 Sketch of the function used in Remark 6.3.3 to indicate that in general $f(Eq(f)) \not\subset Eq(f)$

Theorem 6.3.4 (Gilman Classification) *Any cellular automaton belongs to exactly one of the following disjoint classes.*

Gilman 1: $Eq(f) = E^\Gamma$
Gilman 2: $\emptyset \neq Eq(f) \neq E^\Gamma$
Gilman 3: $Eq(f) = \emptyset$, *but* f *not expanding*
Gilman 4: f *expanding*

6.3.1 Class Gilman 1

Proposition 6.3.5 *Suppose a cellular automaton becomes eventually periodic, in the sense that there are numbers* $m, k \in \mathbb{N}$ *such that for every* $u \in E^\Gamma$ *holds*

$$f^{n+k}(u) = f^n(u) \qquad \text{if} \qquad n > m.$$

Then this automaton is in class Gilman 1.

Proof Without restriction we assume $D_0 \subset \Gamma_{d_0}$. The evolution of the origin $e \in G$, after $m + k$ time steps is known, since the states at $m + 1, \ldots, m + k$ are repeated periodically.

If we want to know $f^n(u)(e)$ for $n \geq m + k$, we need only the values of u on Γ_{d_0+m+k}. Choose l, define $\delta = 1/(l + 1)$, and put $\varepsilon = 1/(d_0 + m + k + l + 1)$. Then $d(u, v) < \varepsilon$ implies

$$f^n(v)(g) = f^n(u)(g) \qquad \text{for } g \in \Gamma_l, \qquad n \geq m + k,$$

and hence $d(f^n(u), f^n(v)) \leq 1/(l + 1) = \delta$ for all large n. □

We have shown that an automaton is in class Gilman 1 if it is eventually uniformly periodic in the sense of Proposition 6.3.5. We show that the converse is also true.

Proposition 6.3.6 *Let an automaton in class* Gilman 1 *be given. There are numbers* $m, k \in \mathbb{N}$ *such that for any* $u \in E^\Gamma$ *holds*

$$f^{n+k}(u) = f^n(u) \qquad \textit{for} \qquad n > m.$$

Proof We know that the family f^n is equicontinuous. Let $\varepsilon = 1$. We can choose $k \in \mathbb{N}$ (independently of n) so large and hence $\delta = 1/(k+1)$ so small that for all $u, v \in E^\Gamma$ with $d(u, v) < \delta$ we have $d(f^n(u), f^n(v)) < \varepsilon$ and hence the values at the origin $e \in G$ satisfy $f^n(u)(e) = f^n(v)(e)$ for all $n \in \mathbb{N}_0$.

Now we fix the state u. The states v with $d(u, v) < \delta$ agree with u on Γ_k and are otherwise arbitrary. The values $f^n(u)(0)$ depend only on the finitely many values in the cells Γ_k. We claim that $f^n(u)(e)$ become eventually periodic with minimal period $\nu(u)$. It is not immediately clear why $f^n(u)|_{\Gamma_k}$ should become periodic, as we only know that site zero is not influenced by states outside of Γ_k, but there may be sites within Γ_k that are influence by sites outside. We construct equivalence classes in E^Γ by the definition $u \sim v$ iff $f^n(u)(e) = f^n(v)(e)$ for all $n \in \mathbb{N}_0$. Due to the fact that $f^n(u)(e)$ depend only on the finitely many values in the cells Γ_k, there are only finitely many equivalence classes. Next, we define $H : E^\Gamma/\sim \to E^\Gamma/\sim$ by $H([u]) = [f(u)]$. This map is well defined, as for $u_1 \sim u_2$ we obtain (by definition of "$\sim$") $f(u_1) \sim f(u_2)$. Since there are only finitely many equivalence classes, $H^n([u])$ becomes eventually periodic with minimal period $\nu(u)$, and hence also $f^n(u)(e)$.

Let the transient phase have maximal length $m(u)$, such that $f^n(u)(e)$ is $\nu(u)$-periodic after $m(u)$ steps but not before. Then,

$$f^{n+\nu(u)}(u)(e) = f^n(u)(e) \qquad \text{for all} \quad n \geq m(u).$$

The numbers $m(u)$ and $\nu(u)$ depend on u, but in fact only on $u|_{\Gamma_k}$. We may as well write $m(u) = m(u|_{\Gamma_k})$ and $\nu(u) = \nu(u|_{\Gamma_k})$. Since E^{Γ_k} is finite, we are able to find the maximal burn in time $\overline{m}$ and a common period $\overline{\nu}$ for all states,

$$\overline{m} = \max\{m(v) \,:\, v \in E^{\Gamma_k}\}, \qquad \overline{\nu} = \prod_{v \in E^{\Gamma_k}} \nu(v).$$

For every $u \in E^\Gamma$ we have

$$f^{n+\overline{\nu}}(u)(e) = f^n(u)(e) \qquad \text{for all} \qquad n > \overline{m}.$$

Since f commutes with the shift operator, this equation holds for all grid points,

$$f^{n+\overline{\nu}}(u) = f^n(u) \qquad \text{for all} \qquad n > \overline{m}.$$

□

Example 6.3.7 (Gilman 1) Consider the Wolfram automaton 204, $f(u) = u$. Then every state is stable in the sense of Lyapunov. This automaton is an example of class Gilman 1.

Note: For the remaining part of this section we assume $\Gamma = \mathbb{Z}$.

6.3.2 Class Gilman 2

This class shares some essential features with the class Gilman 1. We show, that also in this class, the set $Eq(f)$ dominates the dynamics. In order to make this idea precise, we need the following definition. Recall that a nowhere dense set is a set with a closure that has an empty interior.

Definition 6.3.8 A subset $U \subset X$ of a topological space X is called meager, if U is an at most countable union of nowhere dense sets.

A subset $V \subset X$ of a topological space X is called residual, if it is the complement of a meager set. A property that is true for all points of a residual set is called "generic property".

The concept of "generic property" is the topological equivalent of "true for almost all points" in measure theory: a relative small set is allowed to be excluded. A residual set is "almost" the complete topological space.

Proposition 6.3.9 *Let $\Gamma = \mathbb{Z}$. If $Eq(f) \neq \emptyset$, then $Eq(f)$ is a residual set. This is to say that the family f^n is generically equicontinuous.*

Proof We can use the construction in the proof of Proposition 6.1.7 to find $m \in \mathbb{N}$ and a finite pattern $u_0 \in E^{2m+1}$ with the following property: copy this pattern in a state at the sites $[d - m, d + m]$. Then there is a k with $0 < k < m$ such that the dynamics at the sites $d - k, \ldots, d + k$ is completely determined by this pattern. The values of sites outside of the array $[d - m, d + m]$ do not have any influence on the fate of sites $[d - k, d + k]$. We may choose m large enough such that also k becomes large, and $D_0 \subset [-k, k]$. Then no information from the left side of the pattern will move to the right side and vice versa. We call such a pattern that interrupts the information flow, a block. We can use blocks to create new states at which $\{f^n\}$ is equicontinuous.

Let, for $i \in \mathbb{N}$, the set V_i consist of all states $u \in E^{\mathbb{Z}}$ that possess i different copies of the finite block u_0 at the left hand side of $0 \in \mathbb{Z}$, and also i different copies at the right hand side of the origin.

For each $i \in \mathbb{N}$, the set V_i is open and dense. Moreover, we show

$$V = \cap V_i \subset Eq(f).$$

Let $v \in V$. We show that $v \in Eq(f)$. Given $\varepsilon > 0$, we determine $\delta > 0$ such that $d(u, v) < \delta$ results in $d(f^n(u), f^n(v)) < \varepsilon$ for all $n \in \mathbb{N}_0$. As $v \in V$, v has a countable number of blocks to the left and to the right of the origin. Thus, if we want to ensure that $d(f^n(u),f^n(v)) < \varepsilon$ for all n, we only need to find two blocks at the left and the right hand side far away from the origin (minimal distance is $1/\varepsilon$). Let these two blocks be contained in the interval $[-l, l]$. If we require that $d(u, v) < 1/(l+1)$, then u and v agree on an interval that is large enough to cover the two blocks. Then, if we iterate by applying f^n, the images will agree between the two blocks, and thus $d(f^n(u),f^n(v)) < \varepsilon$.

The set $Eq(f)$ contains a residual set, and equicontinuity of $\{f^n\}$ is a generic property of points in $E^{\mathbb{Z}}$. □

Example 6.3.10 (Gilman 2) Consider the Wolfram automaton 232 with local function $f_0(e_{-1}, e_0, e_1) = 1$ for all $e_{-1} + e_0 + e_1 > 1$ and $f_0(e_{-1}, e_0, e_1) = 0$ otherwise. We see that a block of three or more 1 persists. Hence the state $[\bar{1}]$ (identically one) is equicontinuous and $Eq(f) \neq \emptyset$.

The state $[\cdots 0101010 \cdots]$ of alternating "0" and "1" is not stable in the sense of Lyapunov. If we replace a single "1" by "0" then we find three zeros in a row and this block of "0" spreads to eventually everywhere. Hence an arbitrarily small change (corresponding to replacing "1" by "0" far from the tagged origin) produces a distance of at least $1/2$. Thus, $\emptyset \neq Eq(f) \neq E^{\mathbb{Z}}$.

6.3.3 Class Gilman 3

Trivially these automata are sensitive with respect to initial data. This property does not imply that the automata are topologically transitive (although, as we have shown, the converse is true: a topologically transitive automaton is sensitive with respect to initial data, Proposition 6.1.7). We have a closer look at two examples.

Example 6.3.11 (Gilman 3) Let $\Gamma = \mathbb{Z}$, $D_0 = \{0, 1\}$, $E = \{(\zeta, \eta) : \zeta, \eta \in \{0, 1\}\}$. The local function is

$$f_0(e_0, e_1) = ((e_0)_\zeta, (e_1)_\eta).$$

The local (global) function acts on the first component as the identity and on the second as the left shift.

The automaton is not expanding. Consider two states u and v_i, where $u(x) = (0, 0)$ for all $x \in \mathbb{Z}$, and $v_i(x) = (0, 1)$ for $x < i$, and $v_i(x) = (0, 0)$ otherwise. Then $d(u, v_i) \leq 1/(i+1)$, and both u, v_i are fixed points of the automaton. We can choose the number i as large as we want and the fixed points get arbitrarily close. Hence for every $\delta > 0$ there are two states u, v_i such that $d(f^n(u),f^n(v_i)) < \delta$ for all $n \in \mathbb{N}$. The automaton is not expanding.

The automaton is sensitive with respect to initial data. Let $u \in E^{\mathbb{Z}}$. Define $v_i(x) = u(x)$ for $x \neq i$, but $v_i(i)_\zeta = u(i)_\zeta$, $v_i(i)_\eta = 1 - u(i)_\eta$. Then u and v are arbitrarily close, $d(u, v) < 1/(1+i)$, for large i, and $d(f^i(u),f^i(v)) = 1$. Indeed, the automaton is sensitive with respect to initial data.

Example 6.3.12 (Gilman 3) The Wolfram automaton 240, the right shift, is in class Gilman 3. We know already that the shift operator is topologically transitive and sensitive with respect to initial data.

The shift operator is not expanding: For two states which do not agree at $z \ll 0$ (but agree otherwise) and hence have a small distance, the distance gets even smaller when the left shift is applied.

6.3.4 *Class Gilman 4*

Let E be finite, and $\Sigma = E^{\mathbb{N}_0}$, endowed with the product topology. This topology is defined by the Cantor metric as we have seen in case of Σ_2 (see construction in Sect. 3.1.2). We already introduced a shift by a slightly weaker Definition 3.5.1, where we allowed for an infinite alphabet. In this and the following chapters, we will work exclusively with a finite alphabet.

Definition 6.3.13 Let $\Sigma = E^{\mathbb{N}_0}$ for E finite equipped with the Cantor topology, $\sigma : \Sigma \to \Sigma$, $u(\cdot) \mapsto u(\cdot + 1)$. Then the pair (Σ, σ) is called a (unidirectional) shift.

Let $\hat{\Sigma} \subset \Sigma$ be topologically closed and invariant under σ. Then the pair $(\hat{\Sigma}, \sigma)$ is called a subshift.

Recall Definition 3.4.4: Two dynamical systems (X,f) and (Y,g) are called conjugated, if there is a bijective continuous mapping $\Phi : X \to Y$ such that $\Phi \circ f = g \circ \Phi$. The following theorem resembles Theorem 3.5.4; the difference is that we now require that the dynamical system is expanding, and obtain therefore a shift over a finite alphabet. Without this condition, we would have only obtained an infinite alphabet.

Theorem 6.3.14 *Let X be a compact, totally disconnected space and $f : X \to X$ continuous. The following are equivalent*

(1) *(X,f) is expanding*
(2) *(X,f) is conjugated to a subshift with some alphabet E.*

Proof (Following Hedlund [85]) We denote elements in Σ by Roman letters u, v, ..., and states in X by Greek letters ϕ, η,....

(2) $\Rightarrow$(1):
Step 1: A subshift $(\hat{\Sigma}, \sigma)$ is always expanding.
If $\hat{\Sigma}$ is empty or contains just one point then the claim is trivial. Suppose $\hat{\Sigma}$ contains two distinct points u, v. Then there is $x \in \mathbb{N}$ such that $u(x) \neq v(x)$. Then $d(\sigma^x(u), \sigma^x(v)) = 1$, i.e., the subshift is expanding.

Step 2: The property to be expanding is preserved under an isomorphism $\Phi : X \to \tilde{\Sigma}$.

The arguments parallel that of Lemma 3.4.2. Let $\hat{\delta}$ be a constant for the expanding subshifts, i.e., for every pair $u, v \in \hat{\Sigma}$ there is an n, such that

$$d(\sigma^n(u), \sigma^n(v)) > \hat{\delta}.$$

Since X and $\hat{\Sigma}$ are compact and Φ is continuous, also Φ^{-1} is continuous (Proposition A.1.12). Suppose that the property of being expanding does not carry over to (X,f). Then it follows that for every $\delta > 0$ there are two states u and v in $\hat{\Sigma}$ such that

$$d(f^n(u), f^n(v)) < \delta \text{ for all } n \in \mathbb{N}$$

and simultaneously for suitable n

$$d(\sigma^n(\Phi(u)), \sigma^n(\Phi(v))) = d(\Phi(f^n(u)), \Phi(f^n(v))) > \hat{\delta}.$$

We choose $\delta = 1/m$ and find points $u_m = f^n(u)$, $v_m = f^n(v)$, such that $d(u_m, v_m) < 1/m$ and simultaneously $d(\Phi(u_m), \Phi(v_m)) > \hat{\delta}$. As $\hat{Z}$ as well as X are compact, we find a converging subsequence $(u_{m_i}, v_{m_i}) \to (\hat{u}, \hat{v})$. Then we obtain a contradiction: on the one hand $\hat{u} = \hat{v}$ ($d(u_{m_i}, v_{m_i}) \to 0$), and on the other hand $d(\Phi(\hat{u}), \Phi(\hat{v})) > \hat{\delta}$.

(1) $\Rightarrow$ (2):

Step 1: Definition of the isomorphism:

We know that the trajectories of two different initial data move away from each other at least by a distance δ. Since $\hat{\Sigma}$ is compact, the set $\hat{\Sigma}$ can be decomposed into finitely many clopen subsets U_i, $i = 0, \dots, S-1$ that have diameters less than δ (compare the proof of Theorem A.1.14, Step 1).

Let $E = \{0, \dots, S-1\}$ and $\Sigma = E^{\mathbb{N}_0}$. Let

$$\Phi : X \to \Sigma, \quad \theta \mapsto \Phi(\theta)$$

whereby $\Phi(\theta)$ is defined by

$$\Phi(\theta)(n) = e \quad \text{if } f^{(n)}(\theta) \in U_e, \quad e \in E.$$

It immediately follows that $\Phi(f(\theta)) = \sigma(\Phi(\theta))$, i.e., Φ preserves the structure.

Step 2: Φ is continuous.

The sets U_i are open and f is continuous. For every $\eta \in X$ and every $n \in \mathbb{N}_0$ there is an $\varepsilon > 0$ such that the sets $f^l(B_\varepsilon(\eta))$ for $l = 0, \dots, n$ are all in the same U_i. Hence, for v varying in $B_\varepsilon(\eta)$, the value $\Phi(v)$ does not depend on the first n components of the state, and therefore $d(\Phi(v), \Phi(\eta)) < 1/(n+1)$. Thus, Φ is continuous.

Step 3: Φ *is bijective on* $\hat{\Sigma} := \Phi(X)$.

We look at Φ as a mapping from X to $\hat{\Sigma} := \Phi(X)$. Then Φ is trivially surjective. It remains to show that Φ is also injective. Let $\phi, \eta \in X$, and $\phi \neq \eta$. Then there is an n such that $d(f^n(\phi), f^n(\eta)) > \delta$, i.e., $\Phi(\phi)(n) \neq \Phi(\eta)(n)$ and hence the images of the two states are different. □

Remark 6.3.15

(1) If $f : X \to X$ is bijective, then we can use the same method to show that (X,f) is conjugated to the (two-sided) shift on $E^{\mathbb{Z}}$, see Sect. 3.5.1.

(2) Many more results can be shown for expanding cellular automata. Such an automaton is surjective, and the subshift generated by the automaton is topologically transitive (saying that the given automaton has the same property). In particular, there is no pre-attractor other than X, and $\omega(X) = X$. Thus, we have the inclusion

$$\text{Gilman 4} \subset \text{Hurley 5}.$$

Example 6.3.16 (Gilman 4) Consider the Wolfram automaton 150 with the rule

$$f_0(\varphi) = (\varphi_{-1} + \varphi_0 + \varphi_1) \bmod 2.$$

We know that this automaton is linear if we identify E with $\mathbb{F}_2$, the field with two elements (see also Sect. 10), and hence we expect a simple behavior. But linear may be simple if we work over the real or complex numbers but linear is not simple if we work over a finite field.

In the following addition and multiplication are mod 2. Since f_0 is linear, also f is linear, i.e., for any two states u, v we have $f(u+v) = f(u) + f(v)$, and further $f^n(u+v) = f^n(u) + f^n(v)$.

Assume $u \neq v$. Then there is $w \neq 0$ such that $u(x) = v(x) + w(x)$ for all x. We show that there is $x \in Z$, $|x| \leq 1$ and an n such that $f^n(w)(x) = 1$. Assume the contrary, then $f^n(w)(-1) = f^n(w)(0) = f^n(w)(1) = 0$ for all n. This assumption implies $f^n(w)(-2) \neq 1$ and $f^n(2) \neq 1$ for all n since otherwise the "1" would "invade" the interval $[-1,1]$ as can be seen from the following table,

Time	Cell	-3	-2	-1	0	1	2	3	
n	...	*	1	0	0	0	0	0	...
n+1	...	*	*	1	0	0	0	0	...

Hence we have $f^n(w)(x) = 0$ for $|x| \leq 2$ and all $n \in \mathbb{N}_0$. By induction over m we find that $f^n(w)(x) = 0$ for $|x| \leq m$, $m \in \mathbb{N}$, and hence $w = 0$, contrary to the assumption. Let $n \in \mathbb{N}_0$ such that $f^n(w)$ is not zero on all three sites in $\{-1,0,1\}$. Then, $d(f^n(v), f^n(u)) = d(f^n(v), f^n(v) + f^n(w)) \geq 1/3$. Hence the global function is expanding.

Chapter 7
Language Classification of Kůrka

The classifications considered so far focused on the long term dynamics (attractors) respectively the complexity of the dynamics (Lyapunov stability). Culik and Hurd [38] and Kůrka [111] turned their attention to the "automaton" aspect of cellular automata. They developed a classification based on the theory of finite automata resp. formal grammars.

7.1 Grammar

We define a grammar and present some results needed in the sequel. These topics are thoroughly covered in the classical book by Hopcroft and Ullman [90].

The idea of a grammar is to formalize the basic rules of a language. Language is written language. There are finitely many letters (or symbols), and letters are used to form words. Words are finite strings of letters. Forming words is subject to rules. One starts with a simple "nucleus" and then proceeds by using the rules to form more complicated strings.

Example 7.1.1 The letters are $\{A, B, C, D\}$. The rules are:

A can be replaced by C or by AB,
B can be replaced by D.

Starting with the string "A", we can generate the strings

$$A, C, [AB], [AD], [CB], [CD], [ABB], [ABD], [ADB], [ADD], [CBB], [CBD], \ldots$$

In this language, every string of the form $[XY_1Y_2 \cdots Y_n]$ with $X \in \{A, C\}$ and $Y_i \in \{B, D\}$ is an admissible word. Some words are "dead ends" and cannot be modified, e.g. all words of the form $[CDD \cdots D]$. On the other hand, all words containing an A or a B can be further modified.

K.-P. Hadeler, J. Müller, *Cellular Automata: Analysis and Applications*,
Springer Monographs in Mathematics, DOI 10.1007/978-3-319-53043-7_7

The first feature of grammars is a set of letters (or symbols). The example suggests to define two types of letters (with respect to a given grammar): The set V_N contains those letters that can be changed by the application of rules ("non-terminal symbols"), and the set V_T of stop symbols ("terminal symbols"). At the present level the disjoint partition of the alphabet into non-terminal and terminal letters has no meaning. Later on we shall specify different types of grammars; then the distinction between terminal and non-terminal letters becomes relevant. We require $V_N \cap V_T = \emptyset$. The alphabet is $V = V_N \cup V_T$.

If α is a word then $|\alpha|$ is the length of the word, i.e., the number of letters in the word. $\alpha(i)$ for $i \in \{1, \ldots, |\alpha|\}$ denotes the ith symbol of the word. The empty word ϵ does not have any letter and has length 0. Mostly we follow the convention that elements in V_N are denoted by capital letters $A, B, \ldots$ and elements in V_T by lower case letters $w, v, \ldots$.

The set of all words of length n is V^n, and V^0 contains only the empty word. The set of all nonempty words is

$$V^+ = \cup_{n \in \mathbb{N}} V^n$$

and the set of all words (including the empty word ϵ) is

$$V^* = V^0 \cup V^+ = \{\epsilon\} \cup V^+.$$

Definition 7.1.2 Any set $L \subset V^*$ is called a language.

The next feature of a grammar is a set of production rules P (also called syntax). A production rule is an expression of the form

$$\alpha \to \beta$$

with $\alpha \in V^+$ and $\beta \in V^*$. The rule says that a string of the form α can be replaced by β. Here the string β can be ϵ, i.e., the rule may say that α can be deleted.

The last feature of a grammar is the start symbol $S \in V_N$. We begin with a symbol S and apply recursively the rules. Then we get a specific set of words L which is the language generated by this grammar.

Definition 7.1.3 A grammar $\mathcal{G}$ is a tupel (V_N, V_T, P, S) of a set of non-terminal symbols, a set of terminal symbols (these sets are disjoint), a syntax P and a start symbol $S \in V_N$. The language $\mathcal{L}(\mathcal{G}) \subset V_T^*$ is the set of words over V_T that can be generated by $\mathcal{G}$.

Notation If $\alpha, \beta \in V^*$, then $\alpha\beta$ is the word which is produced by concatenation of α and β. Suppose the grammar contains a rule P of the form $\alpha \to \beta$. We apply this rule to a word $[\gamma\alpha\delta]$ and get a word $[\gamma\beta\delta]$. Then we write

$$[\gamma\alpha\delta] \underset{\mathcal{G}}{\Rightarrow} [\gamma\beta\delta].$$

If we need to apply several rules of the grammar $\mathcal{G}$ to get from the string ζ to the string η, then we write

$$\zeta \overset{*}{\underset{\mathcal{G}}{\Rightarrow}} \eta.$$

In the present section, the grammar we use is always non-ambiguous from the context; we skip "$\mathcal{G}$" under "$\Rightarrow$".

We have defined a grammar and a language in a general setting (a so-called type-0 grammar resp. language). We can require additional properties and thus obtain simpler objects that are easier to analyze and more restricted.

Definition 7.1.4 Let $\mathcal{G} = (V_N, V_T, P, S)$ be a grammar. The grammar is called

(1) type-1 or context-sensitive grammar, if for every rule $\alpha \to \beta \in P$ the inequality $|\alpha| \leq |\beta|$ holds.
(2) type-2 or context-free grammar, if for every rule $\alpha \to \beta \in P$ holds, that $|\alpha| = 1$ and $\beta \neq \epsilon$.
(3) type-3 or regular grammar, if for every rule $\alpha \to \beta \in P$ holds:
$|\alpha| = 1, \alpha \in V_N$,
$\beta = [wA]$ with $w \in V_T$ and $A \in \{\epsilon\} \cup V_N$.

Note that in the literature there is an alternative definition for type 3 grammars that also allows for the rule $B \to \epsilon$, i.e., a non-terminal letter may be deleted. Call these grammars type-3'. The difference between type-3 and type-3' is merely that the language of type-3' may include the word ε, which languages generated by type-3 never do. Apart from that, any word of a language defined by type-3' can be also generated by a type-3 grammar and *vice versa*.

Definition 7.1.5 A language $L \subset V^*$ is called regular if there is a regular grammar $\mathcal{G}$ such that $L = \mathcal{L}(\mathcal{G})$.

The language is called right central, if

(1) for every $u \in L$ there is $a \in V$ such that $ua \in L$,
(2) every $u \in L$, and every positive $k < |u|$, the word $u_{[1,k]}$ is in L.

The language is called bounded periodic, if there are $m, n \in \mathbb{N}$ such that

$$\forall u \in L, \ \ i \in \{m, \ldots, |u| - n\} : \ \ u(i) = u(i + n).$$

7.2 Finite Automata

A finite automaton has finitely many states. The state is changed at discrete times according to certain rules (the program). In addition to this internal program, the automaton reads an externally given string—one letter at each time step. In this way, the program of the automaton checks the string. Tagged final states, basically "yes" and "no" answers, allow for an output. The automaton is—at least in some cases—able to determine if a given string has a certain property, in particular: if a string belongs to a certain formal language. Here follows a formal definition.

Definition 7.2.1 A finite automaton is a tupel $(K, \Sigma, \delta, q_0, F)$. Here K is a finite state space, Σ is a finite alphabet, δ is a mapping

$$\delta : K \times \Sigma \to K,$$

q_0 is the initial state and $F \subset K$ is the set of "final" states.

Example 7.2.2 Let $K = \{q_a, q_b\}$, $\Sigma = \{$' ', 'A',..., 'Z', '.'$\}$, and let δ be given by

$\delta(q_*, x) = q_a$ for $x \in \{$' ','A',..., 'Z'$\}$, $q_* \in K$,
$\delta(q_a,$'.'$) = q_b$,
$\delta(q_b,$'.'$) = q_a$.

Furthermore let $q_0 = q_a$ and $F = \{q_b\}$.

How does this automaton work? Whenever the input is a full stop, the automaton jumps to the state q_b and signalizes that a sentence has been completed ($q_b \in F$). Otherwise the automaton is in state q_a.

This automaton can be easily extended. For example, the automaton could check, before jumping to q_b, whether there is any symbol before the full stop.

According to our definition the automaton has a unique state at every time step. Some proofs get simpler if we allow that an automaton can be in one of several states at a given moment. Of course every concretely given automaton is in some state at any given moment but the observer does not know the transition rules which may be, e.g., influenced by an external input. Thus, in mathematical terms, the function δ may be a set-valued function. In this situation we call the automaton "non-deterministic".

Definition 7.2.3 A non-deterministic finite automaton is a tupel

$$M = (K, \Sigma, \delta, q_0, F).$$

Here, K is a finite set of states, $q_0 \in K$, Σ a finite alphabet, δ is a mapping

$$\delta : K \times \Sigma \to \mathfrak{P}(K).$$

The state of the automaton at time $t = 0$ is $\{q_0\}$. The set of final states is $F \subset K$.

An automaton with $|\delta(q, s)| = 1$ for all $q \in K$, $s \in \Sigma$ is called a deterministic finite automaton.

The state of a non-deterministic automaton is an element of the power set of K. If $\tilde{K} \subset K$ is the state at time t, and the input signal is $s \in \Sigma$, then the state at time $t+1$ is defined by the mapping

$$\hat{\delta} : \mathcal{P}(K) \times \Sigma \rightarrow \mathcal{P}(K),$$

$$\hat{\delta}(\tilde{K}, s) = \cup_{q \in \tilde{K}} \delta(q, s).$$

Here we define $\hat{\delta}(\emptyset, s) = \emptyset$.

Notation

(a) Let $w \in \Sigma^*$. Recall that $|w|$ denotes the length of the word, w_i the letter at position i. Let $w_{i \ldots j}$, $1 \leq i \leq j \leq |w|$, be the subword consisting of the letters $w_i, \ldots, w_j$.

(b) We extend the domain of δ from $K \times \Sigma$ to $K \times \Sigma^*$. Let $q \in K, w \in \Sigma^*$. For $w = \epsilon$, we define $\delta(q, \epsilon) = q$. If $w \neq \epsilon$, we let $\delta(q, w) = \delta(\delta(q, w|_{1,\ldots,|w|-1}), w|_{|w|})$. Note that the extended function coincides with the original function δ on $K \times \Sigma$.

(c) As for grammars above, Σ^* denotes the set of all finite words, including the empty word, with letters from the alphabet Σ. By $T(M) \subset \Sigma^*$ we denote the set of all words in Σ^* with the property that the state of the automaton is in the set F after the word has been read,

$$T(M) = \{w \in \Sigma^* \, : \, \delta(q_0, w) \in F\};$$

and similarly for non-deterministic automata

$$T(M) = \{w \in \Sigma^* \, : \, \delta(q_0, w) \cap F \neq \emptyset\}.$$

The following proposition shows that there is no essential difference between non-deterministic and deterministic finite automata.

Proposition 7.2.4 *Let M be a non-deterministic finite automaton. Then there is a deterministic finite automaton M' with $T(M) = T(M')$.*

Proof Let $M = (K, \Sigma, \delta, q_0, F)$ be a non-deterministic finite automaton. We construct a deterministic finite automaton $M' = (K', \Sigma, \delta', q_0', F')$ such that $T(M) = T(M')$. We extend the state space K in such a way that the relevant sets in $\mathfrak{P}(K)$ appear as points in K'.

Let $K' = \mathfrak{P}(K)$ (i.e., points in K' are *subsets* of K), $q_0' = q_0$, $F' = \{X \subset K : X \cap F \neq \emptyset\}$.

Now we define the function δ'. For $K_1 \subset K$, $K_2 \subset K$, and $s \in \Sigma$ let

$$\delta'(K_1, s) = K_2$$

if and only if

$$\hat{\delta}(K_1, s) = K_2.$$

Then the equality $T(M) = T(M')$ is evident. □

7.3 Finite Automata and Regular Languages

In this section we show that regular languages and finite automata are essentially the same thing. The essential property is that in a regular language words can be extended only "at the end". Hence one does not need a memory if one wants to check whether a word belongs to the language.

Theorem 7.3.1 *Let $\mathcal{G} = (V_N, V_T, P, S)$ be a regular (type-3)-grammar. There exists a finite automaton $M = (K, V_T, \delta, S, F)$ with*

$$T(M) = \mathcal{L}(\mathcal{G}).$$

Proof We construct a non-deterministic automaton. Let κ be a further symbol that is not contained in $V_N \cup V_T$. Choose $K = V_N \cup \{\kappa\}$. The initial state is $q_0 = S$, and $F = \{\kappa\}$. Recall that in a regular syntax a rule $\alpha \to \beta \in P$ has the following structure: $\alpha = B \in V_N$, and either $\beta = [wC] \in V_T \times V_N$, or $\beta = w \in V_T$. We define the function δ as follows

$$\delta(B, w) = \begin{cases} \{C \in V_N : B \to [wC] \in P\} & \text{if } B \to w \notin P, \quad w \in V_T \\ \{\kappa\} \cup \{C \in V_N : B \to [wC] \in P\} & \text{if } B \to w \in P, \quad w \in V_T. \end{cases}$$

$\mathcal{L}(\mathcal{G}) \subset T(M)$: Let $[a_1 a_2 \ldots a_n] \in \mathcal{L}(\mathcal{G})$. Then there are rules in P that allow for

$$S \Rightarrow [a_1 A_1] \Rightarrow [a_1 a_2 A_3] \Rightarrow \cdots \Rightarrow [a_1, \ldots, a_{n-1} A_n] \Rightarrow [a_1, \ldots, a_n].$$

Since $S \to [a_1 A_1] \in P$, we have $A_1 \in \delta(S, a_1)$. A_1 is an admissible state of the automaton after one step. As $A_1 \to [a_2 A_2] \in P$, $A_2 \in \delta(A_1, a_2)$, and A_2 is in the state of the automaton after step two. In the last (nth) step, we find A_n in the state of the automaton. Hence, $\kappa \in \delta(A_n, a_n)$, and the automaton accepts the string, $[a_1 a_2 \ldots a_n] \in T(M)$.

$\mathcal{L}(\mathcal{G}) \supset T(M)$: On the other hand, if $[a_1, \ldots, a_n] \in T(M)$, then there is a sequence of states, S, A_1, A_2,…,A_n such that $A_1 \in \delta(S, a_1)$, $A_2 \in \delta(A_1, a_2)$, $A_2 \in \delta(A_1, a_2)$,…, $\kappa \in \delta(A_{n-1}, a_n)$. Therefore P contains the rules $S \to a_1 A_1$, $A_1 \to a_2 A_2$,…,$A_n \to a_n$. Hence $[a_1, \ldots, a_n] \in \mathcal{L}(\mathcal{G})$. □

Theorem 7.3.2 *Let $M = (K, \Sigma, \delta, q_0, F)$ be a finite automaton. Then there is a regular grammar $\mathcal{G}$ with*

$$T(M) = \mathcal{L}(\mathcal{G}).$$

Proof We can assume, without lack of generality, that the automaton is a deterministic finite automaton (Proposition 7.2.4). We define a regular grammar $\mathcal{G} = (V_N, V_T, P, S)$ as follows

- $V_N = K$
- $V_T = \Sigma$
- $S = q_0$
- P is defined by

  (1) $B \to [aC] \in P$ if $\delta(B, a) = C$ and $C \notin F$
  (2) $B \to a \in P$ if $\delta(B, a) = C$ and $C \in F$

It is easy to see (as in the proof of the preceding theorem) that $\mathcal{L}(\mathcal{G}) = T(M)$. □

Languages that are not too complex are regular. Examples are right centered, bounded periodic languages.

Proposition 7.3.3 *Let L be a right centered, bounded periodic language. Then, L is regular.*

Proof We define a regular grammar generating the given language L. Recall that bounded periodic implies that there are $m, n \in \mathbb{N}$ such that every word $u \in L$ satisfies

$$\forall i \geq m,\ i + n \leq |u| :\ u(i + n) = u(i).$$

As the language is right centered, every subword of a word is an element of L, and any word can be extended by at least one letter of the alphabet. The words of a right centered bounded periodic language can be build by cutting off sequences of the form

$$a_1 a_2 \cdots a_{m-1} a_m b_1 \cdots b_n b_1 \cdots b_n b_1 \cdots b_n \cdots$$

after position $i \in \mathbb{N}$. We note first that there are only finitely many words with length $l = m + n$. Enumerate the words in L which have length l; let these words be $w_1, \ldots, w_k$. Then define the set

$$V_l = \cup_{i=1}^{k} \{a : \exists j \in \{1, \ldots, l\} \text{ such that } a = w_l(j)\}.$$

This set is finite because the number of symbols in finitely many words of finite length is finite. Further define

$$V_N = \{B_{i,j} : i \in \{1, \ldots, k\}, j \in \{1, \ldots, l\}\} \cup \{S\}$$

(whereby we can assume that the symbols in V_N are different from those in V_T). Let S be the start symbol.

Now we define the production rules P

$$\begin{array}{llll} S & \to & [w_i(1)B_{i,1}] & \text{for} \quad i = 1, \ldots, k \\ S & \to & w_i(1) & \text{for} \quad i = 1, \ldots, k \\ B_{i,j} & \to & [w_i(j+1)B_{i,j+1}] & \text{for} \quad i = 1, \ldots, k, j = 1, \ldots, l-1 \\ B_{i,j} & \to & w_i(j+1) & \text{for} \quad i = 1, \ldots, k, j = 1, \ldots, l-1. \\ B_{i,l} & \to & [w_i(m+1)B_{i,m+1}] & \text{for} \quad i = 1, \ldots, k \\ B_{i,l} & \to & w_i(m+1) & \text{for} \quad i = 1, \ldots, k \end{array}$$

It is evident that $L = \mathcal{L}(\mathcal{G})$ and that $\mathcal{G} = (V_N, V_T, P, S)$ is a regular grammar. □

7.4 Cellular Automata and Language: Kůrka Classification

Any dynamical system (X, d, f) in a metric Cantor space X can be associated with a cellular automaton (Theorem 3.5.12). We have mentioned in passing shifts and subshifts as another model for dynamical systems (Definition 6.3.13). A subshift defines a language. In particular all finite initial sequences of the subshift are included in the language. For cellular automata these languages have different levels of complexity and these lead to another classification of cellular automata.

Recall Definitions 3.5.1, 6.3.13: A shift is $\Sigma = E^{\mathbb{N}_0}$, the set of all sequences over a finite alphabet E together with the shift operator σ, and a subshift a subset of Σ that is topologically closed and forward invariant under the shift operator. In accordance with Definition 6.3.13, we use in the present section always the Cantor topology.

Note the difference in numbering: The first symbol of an element of a shift has number 0 while the first symbol of a word has number 1.

Again, let E^+ be the set of all finite words with symbols from E and E^* this set with the empty word included. A subshift may agree nowhere with any word or pattern from a finite set $F \subset E^+$ or it may agree everywhere locally with words or patterns from F. In the first case we say that the subshift belongs to the black list with respect to F, in the second case it belongs to the white list. Any black list defines a subshift; a white list has to satisfy certain consistency conditions, and is better defined as the complement of a black list.

Definition 7.4.1 Let E be a finite set, $F \subset E^+$ finite, and Σ_F given by

$$\Sigma_F = \{u \in E^{\mathbb{N}_0} : \forall i \in \mathbb{N}_0 \ \forall v \in F : u|_{i,\cdots,i+|v|-1} \neq v\}.$$

Σ_F is called a shift of finite type.

The next proposition indicates that a black list can be converted into a white lists of words with a fixed length.

Proposition 7.4.2 *For any finite $F \subset E^+$ the set Σ_F is a subshift. The subshift is forward invariant w.r.t. the shift operator. Given F, we find $k \in \mathbb{N}$ and $\hat{F} \subset E^k$ such that*

$$\Sigma_F = \{u \in E^{\mathbb{N}_0} \ :\ \forall n \in \mathbb{N}_0,\ \ u|_{[n,n+k-1]} \in \hat{F}\}.$$

Proof Σ_F is topologically closed: Consider a converging sequence $u_n \in \Sigma_F$, $u_n \to u$. If there is $i_0 \in \mathbb{N}_0$ and $v \in F$ such that $u|_{i_0,\cdots,i_0+|v|-1} \neq v$, then there is a large n such that $u_n(i) = v_n(i)$ for $i \leq i_0 + |v| - 1$, in contradiction to the assumption.

Σ_F is forward invariant under σ: Clearly, if $u \in \Sigma_F$ nowhere agrees with the pattern given in F, then also $\sigma(u)$.

The set F consist of finite words which may have different lengths. Let k be the maximal length of any word in F. Extend the words in F (using all possibly extensions) to length k,

$$\tilde{F} = \{w \in E^k \ :\ \exists i \in \{1,\ldots,k\} \text{ such that } w|_{[1,i]} \in F\}.$$

Then, $\Sigma_F = \Sigma_{\tilde{F}}$. Now define $\hat{F} = E^k \setminus \tilde{F}$, and

$$\Lambda = \{u \in E^{\mathbb{N}_0} \ :\ \forall n \in \mathbb{N}_0 \text{ holds true that } u|_{[n,n+k-1]} \in \hat{F}\}.$$

$\Sigma_F = \Lambda$ as

$$u \in \Sigma_F \Leftrightarrow \forall n \in \mathbb{N}_0 : u|_{[n,n+k-1]} \notin \tilde{F} \Leftrightarrow \forall n \in \mathbb{N}_0 : u|_{[n,n+k-1]} \in \hat{F} \Leftrightarrow u \in \Lambda.$$

□

We say that a subshift of finite type has order k if we find a white list in E^k that describes the subshift.

Given a subshift $\{\hat{\Sigma}, \sigma\}$, we now define a language $\mathcal{L}(\hat{\Sigma})$ associated with this subshift.

Construction 7.4.3 Let E be a finite alphabet, and $\hat{\Sigma}$ a subshift over this alphabet. The language $\mathcal{L}(\hat{\Sigma}) \subset E^*$ associated with the subshift consists of all finite words that agree somewhere with some state $u \in \hat{\Sigma}$,

$$\mathcal{L}(\hat{\Sigma}) = \{v \in E^+ \ :\ \exists u \in \hat{\Sigma} \text{ such that } u|_{[0,|v|-1]} = v\}.$$

Note that, due to the forward invariance of finite subshifts $\hat{\Sigma}_F$, we find

$$\mathcal{L}(\hat{\Sigma}_F) \supset \{v \in E^+ \ :\ \exists u \in \hat{\Sigma}_F\ \exists i_0 \in \mathbb{N}_0 : u|_{i_0,\cdots,i_0+|v|-1} = v\}.$$

Moreover, the language associated with a subshift always is right-centered.

Definition 7.4.4 A subshift $\{\hat{\Sigma}, \sigma\}$ is called regular (bounded periodic), if $\mathcal{L}(\hat{\Sigma})$ is regular (bounded periodic).

Proposition 7.4.5 *A subshift Σ_F of finite type is regular.*

Proof Suppose the subshift has order k, and let $\hat{F}$ be the corresponding white list (as constructed in the proof of Proposition 7.4.2). This white list defines a subshift Λ that coincides with Σ_F (again, see proof of Proposition 7.4.2). We remove all elements in $\hat{F}$ that never appear in elements of Σ_F,

$$\hat{F}' = \{w \in \hat{F} \,:\, \exists u \in \Sigma_F,\, i_0 \in \mathbb{N}_0 : u|_{i_0,\ldots i_0+k-1} = w\}.$$

The white list $\hat{F}'$ obviously also generates Λ. Assume that $\hat{F}'$ has cardinality l,

$$\hat{F}' = \{w_1, \cdots w_l\}.$$

Let S and $B_{i,j}$, $i = 1, .., l$, $j = 1, .., k$, be further symbols that do not occur in the strings of $\hat{F}'$. Then define

$$V_N = \{B_{i,j} \,:\, i = 1, \ldots, l, j = 1, \ldots, k\} \cup \{S\}.$$

Let V_T be the set of all symbols that occur in the words of $\hat{F}$.

We define the production rules P in two steps. The rules

$$\begin{array}{llll}
S & \to & [w_i(1)B_{i,2}] & \text{for} \quad i = 1, \ldots, l \\
S & \to & w_i(1) & \text{for} \quad i = 1, \ldots, l \\
B_{i,j} & \to & [w_i(j)B_{i,j+1}] & \text{if} \quad j < k \\
B_{i,k} & \to & w_i(k) & \text{for} \quad i' \in \nu(i)
\end{array}$$

produce all words and subwords (the initial part of a word) contained in $\hat{F}$. What is missing are rules that allow to prolong an admissible word beyond the length k (see Fig. 7.1). Suppose we reach the end of a word $w_i \in \hat{F}$, which patterns $w_{i'} \in \hat{F}$

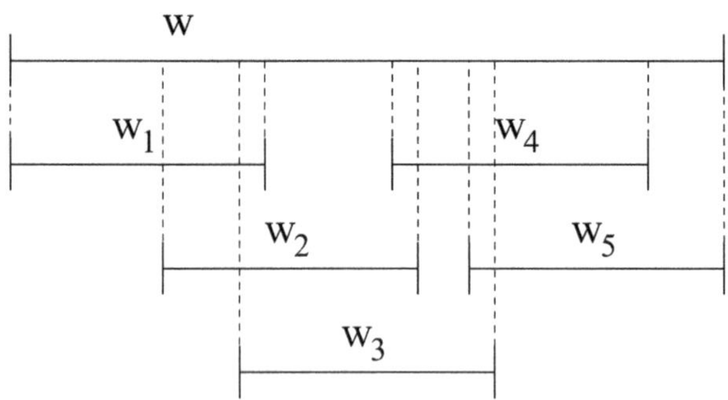

Fig. 7.1 Every subword of the word $w \in \Sigma_F$ is covered by words $w_i \in \hat{F}$. The words w_i overlap

can be used to extend the word? The last part of w_i and the first part of $w_{i'}$ must coincide. We define the set

$$\text{extend}(i) = \{(i',j) \ : \ w_i|_{[|w_i|-j,|w_i|]} = w_{i'}|_{[1,j]}, \text{ and } 1 \le j < |w_{i'}|\}$$

and add the production rule

$$B_{i,k} \quad \to \quad [w_i(k)B_{i',j+1}] \qquad \text{for} \qquad (i',j) \in \text{extend}(i).$$

Note that any pattern $w \in \hat{F}$ can be extended by at least one other pattern $w' \in F$ as we reduced the set of admissible strings to those which actually appear in elements of Σ_F.

This grammar is regular and it generates the language which is induced by the subshift. □

Note that a subshift of finite type is regular, but the converse is not necessarily true. There are regular subshifts that are not of finite type. The regular subshifts coincide with the so-called sofic subshifts [38].

By now, we know subshifts, languages and their relation. Now we build a bridge between dynamical systems on Cantor spaces and subshifts. In particular we construct a subshift from a cellular automaton.

Construction 7.4.6 Let (X,d,f) be a dynamical system where X is a metric Cantor space, and let B be a finite set endowed with the discrete topology. Let $\beta \in C(X,B)$. Then $\{\beta^{-1}(b) \ : \ b \in B\}$ defines a disjoint decomposition of X into clopen sets. Define a function $H_\beta : X \to B^{\mathbb{N}_0}$ by

$$H_\beta(u)(i) = \beta(f^i(u)).$$

Let σ be the shift operator acting on $B^{\mathbb{N}_0}$ as introduced above. Then

$$H_\beta \circ f = \sigma \circ H_\beta.$$

$\Sigma_\beta(f) = H_\beta(X)$ is a subshift (induced by f and β).

Definition 7.4.7 Let (X,d,f) be a dynamical system on a metric Cantor space X, B a finite set equipped with the discrete topology, $\beta \in C(X,B)$ and $\Sigma_\beta(f)$ the subshift induced by β and f. The language $\mathcal{L}_\beta(f) \subset B^*$ is the language defined by the subshift $\Sigma_\beta(f)$,

$$\mathcal{L}_\beta(f) = \mathcal{L}(\Sigma_\beta(f)) = \{v \in B^+ \ : \exists u \in \Sigma_\beta(f) \ \text{ such that } v = u|_{[0,|v|-1]}\}.$$

A dynamical system (X,d,f) is called regular (bounded periodic), if for every finite and topologically discrete set and for every $\beta \in C(X,B)$ the language $\mathcal{L}_\beta(f)$ is regular (bounded periodic).

Proposition 7.4.8 *A cellular automaton with global function f is regular if and only if every projection $\beta_n : E^\Gamma \to E^{\Gamma_n}$, $u \mapsto u|_{\Gamma_n}$ yields a regular subshift $\Sigma_{\beta_n}(f)$.*

Proof If the cellular automaton is regular then $\mathcal{L}_{\Sigma_{\beta_n}}(f)$ is clearly regular. On the other hand, assume that β_n yields a regular language $\mathcal{L}_{\Sigma_{\beta_n}}(f)$ for all $n \in \mathbb{N}$. We show that any continuous mapping $\eta : E^\Gamma \to B$ to a bounded set B (endowed with the discrete topology) yields a regular language, that is, a regular subshift $\tilde{\Sigma} = \Sigma_\eta(f)$. The sets

$$V_b = \{u \in E^\Gamma \,:\, \eta(u) = b\}, \qquad b \in B$$

form a partition of E^Γ into disjoint clopen sets. Therefore the V_b have a positive minimal distance δ. Let $k \in \mathbb{N}$ large enough such that $\delta > 1/(1+k)$. If $u|_{\Gamma_k} = v|_{\Gamma_k}$ then u, v are in the same V_b. The corresponding projection $\beta_k : E^\Gamma \to E^{\Gamma_k}$ defines the subshift $\hat{\Sigma} = \Sigma_{\beta_k}(f)$. Then we know that $\mathcal{L}_{\beta_k}(f)$ is regular (and the subshift $\hat{\Sigma}$) is regular).

In order to show that $\mathcal{L}_\eta(f)$ is regular we define $h : E^{\Gamma_k} \to B$ as follows. Choose $v \in E^{\Gamma_k}$, and select any $u \in E^\Gamma$ with $u|_{\Gamma_n} = v$. Then put $h(v) = \eta(u)$. This definition is independent of the choice of u. Then define

$$H : \hat{\Sigma} \to \hat{\Sigma}, u \mapsto v$$

by $v(i) = h(u(i))$. By construction, the mapping H is surjective (but in general not injective). Since $\hat{\Sigma}$ is regular, there is a regular (type-3) grammar $\hat{\mathcal{G}} = (\hat{V}_N, \hat{V}_T, \hat{P}, \hat{S})$ with $\mathcal{L}_{\hat{\beta}}(f) = \mathcal{L}(\mathcal{G})$. We define $\tilde{V}_T = h(V_T)$, and production rules $\tilde{P}$ from $\hat{P}$ by replacing every symbol $x \in \hat{V}_T$ by $h(x)$. Then we obtain a regular grammar $\tilde{\mathcal{G}} = (\hat{V}_N, \tilde{V}_T, \tilde{P}, \hat{S})$ for which $\mathcal{L}_\beta(f) = L(\tilde{\mathcal{G}})$ holds. □

Theorem 7.4.9 (Kůrka Classification) *A cellular automaton* (Γ, D_0, E, f_0) *with global function* f *is an element of exactly one of the (disjoint) classes*

(Kůrka 1) (Γ, d, f) *is bounded periodic.*
(Kůrka 2) (Γ, d, f) *is regular but not bounded periodic.*
(Kůrka 3) (Γ, d, f) *is not regular.*

Kůrka compares in his paper [111] the three classifications (Hurley, Gilman and Kůrka), and investigates which classes have empty and non-empty intersections; many interesting examples can be found there.

We focus on the investigation of the classes Kůrka 1 for general groups and Kůrka 2 for $\Gamma = \mathbb{Z}$ in more detail, and give examples for all three classes.

7.4.1 Class Kůrka 1

Proposition 7.4.10 *The class Kůrka 1 is exactly the class Gilman 1.*

Proof We show that a cellular automaton with global function $f : E^\Gamma \to E^\Gamma$ is bounded periodic if and only if $Eq(f) = E^\Gamma$.

Kůrka 1 $\Rightarrow$ $Eq(f) = E^\Gamma$.

Every finite B and every $\beta \in C(E^\Gamma, B)$ induce a bounded periodic language $\mathcal{L}_\beta(f)$. This is true, in particular, for $B = E^{\Gamma_k}$ and the projection $\beta_k : E^\Gamma \to E^{\Gamma_k}$, $u \mapsto u|_{\Gamma_k}$. There are $m, n \in \mathbb{N}_0$ such that $\beta_k(f^n(u))$ becomes n-periodic after m time steps.

Now assume $Eq(f) \neq E^\Gamma$. Then there is a $\delta > 0$ and there are sequences $u_i \in E^\Gamma$, $v_i \in E^\Gamma$ with $d(u_i, v_i) < 1/i$ and $l_i \in \mathbb{N}$ such that

$$d(f^{l_i}(u_i), f^{l_i}(v_i)) > \delta.$$

Choose $k > 1/\delta + 1$. The language $\mathcal{L}_{\beta_k}(f)$ is bounded periodic. We find $m, n \in \mathbb{N}$ such that all words become n-periodic after m symbols. Let $j > k + (m+n)\,2d_0$, where $D_0 \subset \Gamma_{d_0}$. Then

$$f^{\tilde{n}}(u_j)|_{\Gamma_k} = f^{\tilde{n}}(v_j)|_{\Gamma_k} \qquad \text{for } \tilde{n} = 0, \ldots, m+n$$

and therefore—since the language is then periodic—for all times. Hence we have $d(f^{\tilde{n}}(u_j), f^{\tilde{n}}(v_j)) < 1/(k+1) < \delta$ for all times $\tilde{n} \in \mathbb{N}_0$, contrary to the assumption that there is an l_i such that the distance is not less than δ.

$Eq(f) = E^\Gamma \Rightarrow$ Kůrka 1.

Suppose $Eq(f) = E^\Gamma$. According to Proposition 6.3.6, there are $m, n \in \mathbb{N}$, such that $f^i(u)$ becomes n-periodic after at most m steps (m and n are independent of $u \in E^\Gamma$). Then, the same holds for $(\beta \circ f^i(u))_{i \in \mathbb{N}}$ where β may be any continuous function $\beta : E^\Gamma \to B$, and B a finite set equipped with the discrete topology. Hence every language $\mathcal{L}_\beta(f)$ is bounded periodic, and the automaton is bounded periodic. □

Example An example is the identity function, see also Example 6.3.7.

Note: For the remaining part of this section we assume $\Gamma = \mathbb{Z}$.

7.4.2 Class Kůrka 2

Class Kůrka 2 contains some important subclasses of cellular automata which can be characterized in terms of dynamical systems. For the exact description we need one more definition.

Definition 7.4.11 (Shadowing Property) A dynamical system (X, d, f) has the shadowing property if for every $\varepsilon > 0$ there is a $\delta > 0$, such that for every δ-chain $\{x_i\}_{i=0,\ldots,n}$ ($n \in \mathbb{N}$ arbitrary) there is a $y \in X$ with

$$d(x_i, f^i(y)) < \varepsilon, \qquad \text{for } i = 0, \ldots, n.$$

The definition says that for given ε a (given) δ-chain with sufficiently small δ "shadows" a (section of) a true trajectory. The shadowing property indicates that small perturbations are not really relevant: Every δ-chain is essentially a trajectory.

The next propositions clarify the relation between the shadowing property, regularity and the property "all points are Lyapunov stable" for one-dimensional cellular automata. We will show that Lyapunov stability of all points implies the shadowing property which in turn implies regularity. For the next proposition, note again that we work in this section always with the Cantor topology.

Proposition 7.4.12 *Every one-dimensional cellular automaton with $Eq(f) = E^{\mathbb{Z}}$ has the shadowing property.*

Proof We have $Eq(f) = E^{\mathbb{Z}}$. For every $\epsilon = 1/(n+1)$ there is a $\delta = 1/(k+1)$ such that $u|_{\Gamma_k} = v|_{\Gamma_k}$ implies $f^j(u)|_{\Gamma_n} = f^j(v)|_{\Gamma_n}$. Let $u_1, \ldots, u_m$ be a δ-chain. Then $f(u_i)|_{\Gamma_k} = u_{i+1}|_{\Gamma_k}$ and therefore

$$f^j(u_i)|_{\Gamma_n} = u_{i+j}|_{\Gamma_n}.$$

Then $(u_i)_{i=1,\ldots,m}$ is shadowed by $(f^i(u_1))_{i=0,\ldots,m-1}$. □

Recall Definition 3.4.4: A topological dynamical system (X,f) can be embedded into the topological dynamical system (Y,g), if there is a continuous, injective map $\tau : X \to Y$ such that $g \circ \tau = \tau \circ f$. We already showed that a one-dimensional cellular automaton with an arbitrarily large neighborhood D_0 can be embedded into an automaton with neighborhood $\hat{D}_0 = \{-1, 0, 1\}$ at the expense of a larger state space (Proposition 3.4.10). This proposition will be used to show that for one-dimensional cellular automata the shadowing property implies regularity.

Proposition 7.4.13 *If a cellular automaton has the shadowing property then it is regular.*

Proof Step 1: Construction of a subshift of finite type.

We can assume that $D_0 = \{-1, 0, 1\}$ (Proposition 3.4.10). We choose $B = E^{\Gamma_k}$, and define the associated projection $\beta_k : E^{\mathbb{Z}} \to B$, where $\beta_k(u) = u|_{\Gamma_k}$. Then we show that the language $\mathcal{L}_{\beta_k}(f)$ is regular.

Let $n > k$; we will choose n later. In order to define the subshift we use the definition of $f_{n+1} : E^{\Gamma_{n+1}} \to E^{\Gamma_n}$, introduced in Definition 6.2.3:

$$\tilde{\Sigma}_n = \{(u_i)_{i\in\mathbb{N}} \in (E^{\Gamma_{n+1}})^{\mathbb{N}} \; : f_{n+1}(u_i) = u_{i+1}|_{\Gamma_n}\}.$$

$\hat{\Sigma}_n$ is a subshift of finite type (with order 2): Define a finite set of words of length two as

$$\hat{F} = \{[u\, v] \; : \; u, v \in E^{\Gamma_{n+1}}, \quad v|_{\Gamma_n} = f_{n+1}(u)\}$$

That is, v is an "allowed" successor of u. Evidently $\Sigma_F \subset \hat{\Sigma}_n$ with $F = E^{\Gamma_{n+1}} \setminus \hat{F}$. On the other hand, suppose $u \in \mathcal{L}(\hat{\Sigma}_n)$. Then, $u = [u_1 u_2 \cdots u_l]$, where $[u_i u_{i+1}] \in \hat{F}$. Hence, $\Sigma_F \supset \hat{\Sigma}_n$, and therefore $\hat{\Sigma}_n$ is a shift of finite type with order 2. Let $H : \hat{\Sigma}_n \to \Sigma = (E^{\Gamma_k})^{\mathbb{N}}$ be given by $H(u)(i) = u(i)|_{\Gamma_k}$.

Step 2: $\Sigma_{\beta_k}(f) = H(\hat{\Sigma}_n)$ for n large enough.

First we show $\Sigma_{\beta_k}(f) \subset H(\hat{\Sigma}_n)$: Let $u = (u_i)_{i\in\mathbb{N}_0} \in \Sigma_{\beta_k}(f)$, i.e., $u_i = \beta_k(f^i(v))$ for some $v \in E^{\mathbb{Z}}$. Define $w = (w_i)_{i\in\mathbb{N}_0}$ by $w_i = \beta_{n+1}(f^i(v)) \in E^{\Gamma_{n+1}}$. Then,

$$f_{n+1}(w_i) = w_{i+1}|_{\Gamma_n}.$$

i.e., $w \in \hat{\Sigma}_n$. As $u_i = w_i|_{\Gamma_k} = H(w)(i)$, we obtain $u \in H(\hat{\Sigma}_n)$.

The non-trivial part $H(\hat{\Sigma}_n) \subset \Sigma_{\beta_k}(f)$ remains to be shown. At this point, the shadowing property comes in. For $\varepsilon = 1/(1+k)$ exists a $\delta > 0$ such that a (finite) δ-chain is ε-close to a trajectory. We choose n large enough such that $\delta > 1/(1+n)$.

Let $u = (u_i)_{i\in\mathbb{N}_0} \in \hat{\Sigma}_n$, in particular $u_i \in E^{\Gamma_{n+1}}$. Then we know that $f_{n+1}(u_i) = u_{i+1}|_{\Gamma_n}$. Hence every chain $v_i \in E^{\mathbb{Z}}$ with $v_i|_{\Gamma_{n+1}} = u_i$ is an infinite δ-chain.

Consider the finite initial section $(v_i)_{i=1,\ldots,m}$. This section is a δ-chain of length m. Due to the shadowing property, we find $w_m \in E^{\mathbb{Z}}$, such that

$$d(f^i(w_m), v_i) < \epsilon \qquad \text{for } i = 1, \ldots, m \quad \Rightarrow \quad f^i(w_m)|_{\Gamma_k} = v_i|_{\Gamma_k}.$$

Since $E^{\mathbb{Z}}$ is compact, there is a convergent subsequence of $(w_m)_{n\in\mathbb{N}}$ that converges to some element w. Let $i \in \mathbb{N}_0$ be fixed. $f^i(w_m)|_{\Gamma_k} = v_i$ for $m > i$ implies $f^i(w)|_{\Gamma_k} = v_i|_{\Gamma_k}$. As i has been arbitrary, $H(u)(i) = \beta_k(f^i(w))$ for all $i \in \mathbb{N}_0$.

Step 3: Regularity of (E^Γ, f).

Since $\hat{\Sigma}_n$ is a subshift of finite type, $H(\hat{\Sigma}_n) = \Sigma_\beta(f)$ has the same property. But every subshift of finite type is regular (Proposition 7.4.5). □

Remark 7.4.14 The mapping H preserves regularity, but not necessarily the property "finite type".

Corollary 7.4.15 *Let* $(\mathbb{Z}, D_0, E, f_0)$ *be a cellular automaton with the shadowing property. If* $Eq(f) = E^\Gamma$ *does not hold, then the automaton is in class Kůrka 2.*

Example 7.4.16 (Kůrka 2) We consider the automaton we already used as an example for Gilman 2 (Example 6.3.10), the Wolfram rule 232: $f_0(e_{-1}, e_0, e_1) = 1$ if $e_{-1} + e_0 + e_1 > 1$, and $f_0(e_{-1}, e_0, e_1) = 0$ otherwise. Since this automaton is in class Gilman 2, it is not in class Kůrka 1. We show that this automaton is regular.

First of all, this cellular automaton will not change a block of at least two 0 or at least two 1. These blocks act as barriers that no information can pass. Between these blocks are alternating 01-pattern (anything that is not an alternating 01-pattern necessarily shows a 00 or 11-block). The typical pattern resp. time evolution thus looks like

```
··· 1 1 | 0 1 0 1 0 1 0 1 0 1 | 0 0 ···
··· 1 1 | 1 0 1 0 1 0 1 0 1 0 | 0 0 ···
··· 1 1 | 1 1 0 1 0 1 0 1 0 0 | 0 0 ···
··· 1 1 | 1 1 1 0 1 0 1 0 0 0 | 0 0 ···
··· 1 1 | 1 1 1 1 0 1 0 0 0 0 | 0 0 ···
··· 1 1 | 1 1 1 1 1 0 0 0 0 0 | 0 0 ···
··· 1 1 | 1 1 1 1 1 0 0 0 0 0 | 0 0 ···
```

The initial 00 and 11 blocks are indicated by vertical lines. We are only interested in the evolution of the pattern in between these lines. The alternating region oscillates; at the same time the 00-block and the 11-block expand until eventually the alternating part vanishes and a stationary pattern appears. Only the completely alternating states form an orbit with period two. All other states converge (in the Cantor sense) to a stationary state. For the language $\mathcal{L}(\Sigma_{\beta_k})$ this indicates that there is a clearly predictable pattern. Of course, $V_N = E^{\Gamma_k}$ as any initial state is allowed. In case a state ("letter" in the sense of the language) has a two-block at the boundaries, e.g. $[00 * 00]$, $[00 * 11]$, etc., the fate of the symbol does not depend on the state outside of Γ_k, the word is—up to its length—completely determined. This part can be easily modeled by a regular grammar. If the state is alternating between 0 and 1 at the boundary, e.g. $[010 * 101]$ or $[101 * 101]$ etc., either in the next step it is still alternating (where 0 is replaced by 1 and *vice versa*), or a block appears:

$$[0101*] \to [1010*] \text{ or } [0010*].$$

Also this event can be easily modeled by a regular language.

All in all, $\mathcal{L}(\Sigma_{\beta_k})$ is regular, but the CA is not in Gilman 1 i.e., Kuřka 1, and hence it is in class Kůrka 2.

We see that every word eventually becomes constant or stays periodic with period 2. It almost looks like uniform bounded periodic. However, it may happen that an arbitrary long time is required until a state becomes constant: consider the alternating state. If we toggle one bit at position i, then it takes $i+k$ time steps until the state is constant on Γ_k. There is no uniform bound in time until a word becomes constant.

7.4.3 Kůrka 3

For this class, we only present an example.

Example 7.4.17 Let $\Gamma = \mathbb{Z}$, $D_0 = \{0, 1, 2\}$, $E = \{0, 1\}$, and $f_0(e_0, e_1, e_2) = e_1\, e_2$. Note that this is not a Wolfram-automaton, as $D_0 \neq \{-1, 0, 1\}$. All states tend to zero except those having on the right an uninterrupted sequence of "1"; any "0" is spreading to the left. Consider the subshift generated by the projection to $\Gamma_0 = \{0\}$, i.e. by $\beta_0 : E^\Gamma \to E$, $u \mapsto u(0)$. We ask: is it possible that in this subshift a word appears where n "1" are followed by m "0", and then again by a "1",

$$\underbrace{11\cdots11}_{n}\underbrace{00\cdots00}_{m}1\cdots.$$

Such a sequence appears at site zero, if we start with a state that has a block of 1 around state zero, followed by some 0's, and eventually we have again a block one 1. We have typically the following time evolution

$$
\begin{array}{c}
\cdots 1\,1\,1\,1\,1\,1\,1\,1\,1\,0\,1\,1\,1\,1\,1 \cdots \\
\cdots 1\,1\,1\,1\,1\,1\,1\,0\,0\,1\,1\,1\,1\,1\,1 \cdots \\
\cdots 1\,1\,1\,1\,1\,0\,0\,0\,1\,1\,1\,1\,1\,1\,1 \cdots \\
\cdots 1\,1\,1\,0\,0\,0\,0\,1\,1\,1\,1\,1\,1\,1\,1 \cdots \\
\cdots 1\,0\,0\,0\,0\,0\,1\,1\,1\,1\,1\,1\,1\,1\,1 \cdots \\
\cdots 0\,0\,0\,0\,0\,1\,1\,1\,1\,1\,1\,1\,1\,1\,1 \cdots
\end{array}
$$

In all columns where a sequence appears that moves from 1 to 0 and back to 1, we find $m \geq n$. The reason is, that—per time step—the left block of 1 moves two sites to the right, while the right block of 1 only moves one site to the right. This observation can be used to show that always $m \geq n$, where n can be (in dependence on the initial state) arbitrarily large.

An automaton that produces this language needs to memorize an arbitrarily large natural number n. This is not possible for a finite automaton, and hence the language is not regular.

Chapter 8
Turing Machines, Tiles, and Computability

Cellular automata can be seen as dynamical systems or as algebraic or combinatorial objects, but one can also take the word *automaton* literally and use methods from logic and languages.

In the present section we derive the necessary theoretical tools: first we consider Turing machines and decidability. Turing machines, which are automata in their own right, are too different from cellular automata to be of immediate use in the theory of cellular automata. There is one exception: for certain classes of cellular automata one can show that they are computation-universal, i.e., they can simulate Turing machines.

There are many interesting results at the interface of cellular automata and Turing machines. They show that certain properties of cellular automata are in general not decidable. Mostly these theorems do not use Turing machines but use tilings instead. Tilings have a geometry much closer to that of cellular automata and are better suited than Turing machines to prove results on cellular automata.

We will use the results obtained in the present section to prove theorems about cellular automata in the next section.

8.1 Turing Machines

A Turing machine is a device that lives on a discrete tape, i.e., a linearly ordered storage medium (Fig. 8.1). Each site on the tape carries an element of a given finite alphabet. The machine itself has an internal state (from a finite set of possible internal states) and is located at a certain site on the tape. Using some mechanism, the machine can read the entry at this site, and change the entry. It either stays at the location or it moves to a neighboring site left or right of its present position.

The machine has the following processing cycle: it first reads the entry at a given position of the tape. Depending on its own internal state and the entry on the tape,

K.-P. Hadeler, J. Müller, *Cellular Automata: Analysis and Applications*,
Springer Monographs in Mathematics, DOI 10.1007/978-3-319-53043-7_8

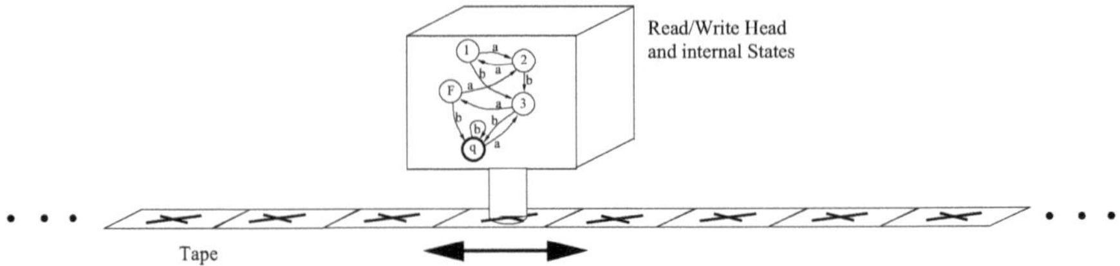

Fig. 8.1 Structure of a Turing machine

the machine changes the entry on the tape and either stays or goes one step to the left or to the right. At the same time it may also change its own state.

There is one special internal state called the acceptance state "F": if the machine arrives at this state then it indicates that the processing of the data is done. The user may read off the result.

This machine is an abstract version of a computer. It is used to express algorithms formally. Since we believe in the omnipotence of our computers, we believe that anything that can be done can be done by such a device. To be more precise: a problem that can be decided using an algorithm which can be handled by a Turing machine in a finite number of steps is *defined* to be decidable. If there is no such Turing machine then we call the problem undecidable.[1]

Let us now define these machines. Since our focus is on cellular automata, we keep the following rather informal. A more precise presentation of these topics can be found in the classical textbook [90] or in [86, 87].

First we define a tape, then we define the Turing machine, and then we describe how it works.

Definition 8.1.1 A tape is a map $C : \mathbb{Z} \to \mathfrak{A}$, where $\mathfrak{A}$ denotes a finite alphabet.

Definition 8.1.2 A (one-tape) Turing machine consists of six ingredients. Let Q be a finite set of the internal states of the machine with $q^0, F \in Q$ being the initial state and the accepting state, respectively. $\mathfrak{A}$ is a finite alphabet, and $a_0 \in \mathfrak{A}$ is a tagged symbol. Finally, a function $\delta : Q \times \mathfrak{A} \to Q \times \mathfrak{A} \times \{L, S, R\}$ describes the dynamics of the Turing machine. The complete Turing machine is the six-tupel $(Q, \mathfrak{A}, q^0, F, a_0, \delta)$.

We now explain how the Turing machine $(Q, \mathfrak{A}, q^0, F, a_0, \delta)$ acts on its tape. Assume the machine is located at the site $k \in \mathbb{Z}$. The machine is in state $q \in Q$ and reads $C(k)$. The value $\delta(q, C(k)) = (q', a', x)$ tells the machine what to do: the machine changes $C(k)$ to a', such that the new state of the tape at location k is given by a'. The state of the machine itself is changed to q'. The machine will either stay at the site k if $x = S$, go to the left if $x = L$ (the new location on the tape is $k - 1$) or go the right if $x = R$ (the new location on the tape is $k + 1$). We obtain a sequence

[1] At this stage of reasoning one will ask whether the problem of decidability is decidable. See the comments on universal Turing machines later.

of states $q^n \in \mathfrak{A}$, a sequence of locations $k^n \in \mathbb{Z}$ and a sequence of states of the tape, $C^n \in \mathfrak{A}^{\mathbb{Z}}$, where $n \in \mathbb{N}$ denotes time. If $q^n = F$ then the machine stops.

The definition of δ does not imply that the Turing machine literally stops once the state "F" is hit. The interpretation of the sentence "the Turing machine stops" is, that the machine assumes the state "F", indicating that the result is there. What the machine does afterwards is of no further interest—in our setting, the machine is allowed to go on and on, even destroying the result again.

Since $Q \times \mathfrak{A}$ only consists of $|Q|\,|\mathfrak{A}|$ elements, it is convenient to represent the function δ as a table: the first column denotes the value from Q, the second the symbol from the alphabet that the machine reads from the tape, the third column denotes the state that the machine assumes at the next time step, the fourth the symbol which the machine prints on the tape and the last column represents the command for the movement of the machine on the tape.

The definition does not contain a readout, i.e., the Turing machine does not know which information on the tape, at the moment of stopping, is the result that may be of interest to a potential user. See the following example.

Example We want to define a Turing machine that is able to decide whether a natural number is even or odd. The set of the natural numbers can be coded, e.g., by a decimal representation. The property reads: "x is even", $x \in \mathbb{N}$. The alphabet is $\mathfrak{A} = \{0, 1, 2, 3, 4, 5, 6, 7, 8, 9, b, F, E, O\}$, i.e., the ciphers, "$b$" for "blank", "$F$" for "failure", "$E$" for "even" and "$O$" for "odd". The machine may assume five states $Q = \{q_0, q_{even}, q_{odd}, q_f, F\}$. We define the function δ (in the form of a table, see Table 8.1). We define $\mathfrak{A}_1 = \{1, 3, 5, 7, 9\}$, $\mathfrak{A}_2 = \{0, 2, 4, 6, 8\}$ and $\mathfrak{A}_3 = \{F, E, O\}$.

In Fig. 8.2 it is shown what happens, if we start with a tape on which "132bbb" is written. The machine searches for the last digit of a natural number and writes 'E' on the tape if the number is even or 'O' if the number is odd. If there is a sign on the tape that is not acceptable for a decimal number, e.g., an 'F', then the machine prints

Table 8.1 The definition of the function δ for our example

Q	$\mathfrak{A}$	q'	a'	$\{L, R, S\}$
q_0	$a \in \mathfrak{A}_1$	q_{odd}	a	R
q_0	$a \in \mathfrak{A}_2$	q_{even}	a	R
q_0	$a \in \mathfrak{A}_3$	q_f	a	R
q_0	$a = b$	q_f	a	R
q_{even}	$a \in \mathfrak{A}_1$	q_{odd}	a	R
q_{even}	$a \in \mathfrak{A}_2$	q_{even}	a	R
q_{even}	$a \in \mathfrak{A}_3$	q_f	a	R
q_{even}	$a = b$	F	E	S
q_{odd}	$a \in \mathfrak{A}_1$	q_{odd}	a	R
q_{odd}	$a \in \mathfrak{A}_2$	q_{even}	a	R
q_{odd}	$a \in \mathfrak{A}_3$	q_f	a	R
q_{odd}	$a = b$	F	O	S
q_f	$a \in \mathfrak{A}$	F	F	S

Fig. 8.2 A Turing machine at work: it decides whether a natural number is even or odd

time step	C(-1)	C(0)	C(1)	C(2)	C(3)	C(4)
0		q_0 ↓				
	b	1	3	2	b	b
1			q_{odd} ↓			
	b	1	3	2	b	b
2				q_{odd} ↓		
	b	1	3	2	b	b
3					q_{even} ↓	
	b	1	3	2	b	b
4					F ↓	
	b	1	3	2	E	b

F for failure and stops. This simple machine decides whether a natural number can be divided by two or not. In this example the readout is the letter at the site at which the machine has assumed the internal state 'F'.

Now we can turn to define "decidable" and "undecidable". We are given a countable set of objects and a property of these objects (which, of course, needs to be well-defined). We may code these objects as finite words from an alphabet and we can put any of these words on a tape. We may find a Turing machine that reads the tape and can tell if the object has the property under consideration. For example, we can code all natural numbers by words of finite length using the alphabet $\{0, 1, 2, \ldots, 9\}$. We may ask, if a number, represented by a certain string, is even or odd. In a pragmatic point of view, the property is decidable if there is a Turing machine that will stop after a finite number of steps and give the desired answer. We formulate this idea in a definition.

Definition 8.1.3 Consider a set Ω and a property $\mathfrak{P}$. This property is decidable, if we can code the elements of Ω by finite (but arbitrarily long) words in a finite alphabet $\mathfrak{A}$ on a tape and if we find a Turing machine $(Q, \mathfrak{A}, q^0, F, a_0, \delta)$ that stops after finitely many steps and indicates when stopped if the object written on the tape (a finite word in the alphabet $\mathfrak{A}$) possesses the property or not. In this case the characteristic function $\chi : \Omega \to \{0, 1\}$

$$\chi(x) = \begin{cases} 1 & x \text{ has the property } \mathfrak{P} \\ 0 & \text{otherwise} \end{cases}$$

is called computable. If there is no such Turing machine, then the property is called undecidable and its characteristic function uncomputable.

The example suggests how to implement algorithms by Turing machines. Turing machines that represent a certain algorithm are called elementary Turing machines. Notice that the definition aims at objects that can be coded by countable sets (finite words over a finite alphabet, like the natural numbers). We mention in passing that if functions defined on the real numbers are considered, the definition requires an extension. Neither input (real value) nor output (real number) can be coded exactly by finite words over a finite alphabet [173].

There are various extensions of Turing machines as defined here, e.g. Turing machines with several tapes, Turing machines with a stack, etc. Most of these variants are equivalent in the sense that one of them may simulate the others.

8.2 Universal Turing Machine

By now we know elementary Turing machines that can decide about properties. If a property is given and we have a Turing machine at hand that can decide about this property then we are fine. But otherwise, how do we know whether the property is decidable or not? In other words, we have the difficulty that we cannot prove that there is no Turing machine that can decide about the given property. Here the idea of a universal Turing machine comes in, i.e., of a Turing machine that can simulate all other Turing machines. If a universal Turing machine cannot decide about the given property, then no Turing machine can and the property is undecidable.

Since the exact description of a universal Turing machine requires a lot more technical details that are of minor interest in other respects, we only give an informal sketch similarly to describing an algorithm by pseudo code. We start with an arbitrary Turing machine TM_1, and construct a universal Turing machine UTM that is able to simulate the machine TM_1. Concerning the UTM, only the initial state of the tape will depend on TM_1; the internal states and the program are independent of TM_1. That is, if we have another Turing machine $\overline{\mathrm{TM}}_1$, we can use the same construction, and end up with the same UTM with only another initial state on its tape. In this sense, the UTM is indeed "universal" and is able to simulate any Touring machine.

The construction is done in several steps: first we reformulate the Turing machine TM_1 as a more convenient Turing machine TM_2 (steps 1 and 2). Then we code TM_2 in an appropriate way (steps 3 and 4). Finally we describe how the UTM works (step 5).

In the following construction we must overcome a difficulty. Whatever the UTM is, it has a finite number of internal states. But the number of internal states of a given TM, though finite, may be arbitrarily large. Hence some information about the internal states of the TM must be stored on the tape of the UTM. Let $\mathfrak{A}_1$, Q_1, and δ_1 denote the alphabet, set of internal states and transition function of the Turing machine TM_1.

Step 1: *Choose a new alphabet.*
Since everything that can be coded can be coded by a binary sequence, we may use $\mathfrak{A}_1 = \{0, 1, b\}$ as the alphabet: 0 and 1 code the information, "b" is blank.

Step 2: *Work with a one-sided tape.*
Our UTM will have distinct regions on its tape: one for coding the transition function δ_1 of TM_1, and one for coding the content of its tape. In order to separate these two regions properly, we show that there is a Turing machine TM_2 that uses a one-sided tape and is equivalent with TM_1. The idea is to fold the tape such that each entry of the one-sided type carries two entries of the original tape. An additional marker o tells, where the origin of the one-sided type is located. The alphabet is $\mathfrak{A}_2 = \mathfrak{A}_1 \times \mathfrak{A}_1 \times \{o, \emptyset\}$. Furthermore, we augment the state space Q_1 by the information whether TM_1 is located on the tape at the right or at the left side of zero (see Fig. 8.3), $Q_2 = Q_1 \times \{l, r\}$. In this way, we obtain a Turing machine $\text{TM}_2 = (Q_2, \mathfrak{A}_2, q_2^0, F_2, a_0^{(2)}, \delta)$ that only uses a one-sided tape. Notice that the alphabet $\mathfrak{A}_2$ consists of $2\,|\mathfrak{A}|^2 = 18$ letters.

Step 3: *Encoding the Turing machine TM_2.*
We first code the alphabet, the states and the actions of the Turing machine. Then we code the function δ_2. We number the letters in $\mathfrak{A}_2$,

$$\mathfrak{A}_2 = \{a_1, \ldots, a_{18}\}.$$

The letter a_i will be represented by a string "111...1" of length i. Similarly, the internal states: we do not know how many internal states TM_2 has, but we know that there are finitely many. We number the states

$$Q_2 = \{q_i \,:\, i = 1, \ldots, m\},$$

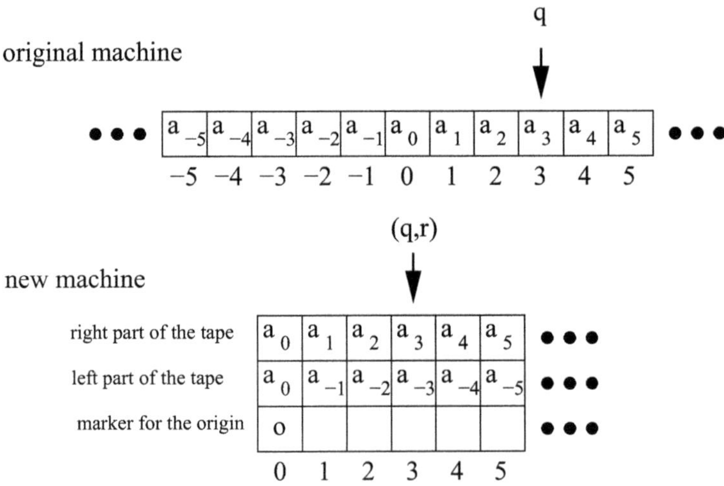

Fig. 8.3 A Turing machine that uses a one-sided tape to simulate a two-sided tape

with the initial state q_1, and represent q_i by a string "111…1" of length i. Then we code the actions: "L" is represented by "1", "R" by "11" and "S" by "111".
For coding the function (the table) δ_2, we need a format that allows a unique identification of the coded elements. A string "11" may denote a_2, q_2 or "R". The context will tell us, which interpretation is appropriate. Therefore, we need a further letter, say "σ".
For convenience, let us denote "L"$= \alpha_1$, "R"$= \alpha_2$ and "S"$= \alpha_3$. We now code one line of the table defining δ. Let us assume

$$\delta_2(q_i, x_j) = (q_k, x_l, \alpha_m).$$

We start with two σ to indicate that a line is beginning. Then, we write the numbers i, j, k, l and m, coded by strings of '1' of the appropriate length and separated by σ. In this way, we code one line after the other. This procedure yields a string that carries all the information to reconstruct δ_2. To indicate, where the defining string starts respectively stops, we begin with three σ in a row and end with three σ. The initial state q_1 is the first state. Without restriction, we chose $q_2 = F$ to be the final/accepting state.
Consider a very simple example: Let $\mathfrak{A} = \{a, b\}$, $Q = \{q_1, F\}$ and δ given by

q_1	a	q_1	b	R
q_1	b	F	b	R
F	a	F	b	S
F	b	F	b	S

We code the letter 'a' by '1', the letter 'b' by '11', the state 'q_1' by '1', the state 'F' by '11', the action 'R' by '11', and the action 'S' by '111'. Thus, the first line of the table, indicating

$$\delta(q_1, a) = (q_1, b, R)$$

is coded by

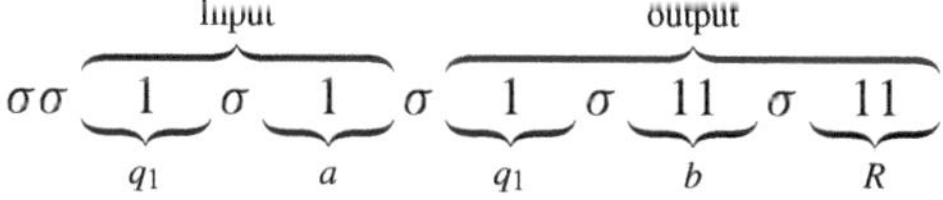

and the code for the complete function δ reads

$$\sigma\sigma\sigma 1\sigma 1\sigma 1\sigma 11\sigma 11\sigma\sigma 1\sigma 11\sigma 11\sigma 11\sigma 11\sigma\sigma 11\sigma 1\sigma 11\sigma 11\sigma 111\sigma\sigma 11\sigma 11\sigma 11\sigma 11\sigma 111\sigma\sigma\sigma$$

Step 4: Add markers indicating state and location of TM_2.
As far as the construction of δ_2 is concerned, the internal states of TM_2 are coded as strings as in step 3. But now we must code the actual internal state of TM_2 in UTM. A simple extension of the alphabet $\{1, \sigma\} \cup \mathfrak{A}_2$ will not do, we need

markers to distinguish the machine from the states. We place a marker m_2 at the location where the corresponding state appears for the first time as an argument of δ. This marker indicates the state of TM_2. And, of course, we need a marker m_3 that indicates the location of TM_2 on the tape. We will need one additional marker m_1 that helps counting the states or symbols. We use the alphabet

$$\mathfrak{A}_3 = (\{1, \sigma\} \times \{m_1, m_2, `\ '\}) \cup (\mathfrak{A}_2 \times \{m_3, `\ '\}).$$

Step 5: How does our UTM operate?

We describe one cycle of the UTM, i.e., the simulation of one step of TM_2. This process is shown in Fig. 8.4. The marker m_2 tells, in which state the TM_2 is, and the marker m_3 tells the location of TM_2 on the tape. In the example, TM_2 is in state q_2, and the letter on the tape that TM_2 will read off is a_1. At the beginning of the cycle, the UTM is located at the site marked by m_2. Now the process starts:

1. The UTM first goes to the marker m_2 and memorizes the entry. It is one of the 18 letters, such that the UTM is able to memorize this entry in its own finite internal states. In the example, it stores the letter a_1.
2. The UTM moves back to marker m_2. It then moves to the right, until it finds the memorized entry in the coding area, see Fig. 8.4b. It places the marker m_1 there, and reads off the entry to write on the tape (again, one out of 18) and the action to do (one out of three). Both letters can be stored in the internal state of the UTM. The state to which TM_2 will go next cannot be stored, because TM_2 may have arbitrarily many states. In our example, the UTM recognizes

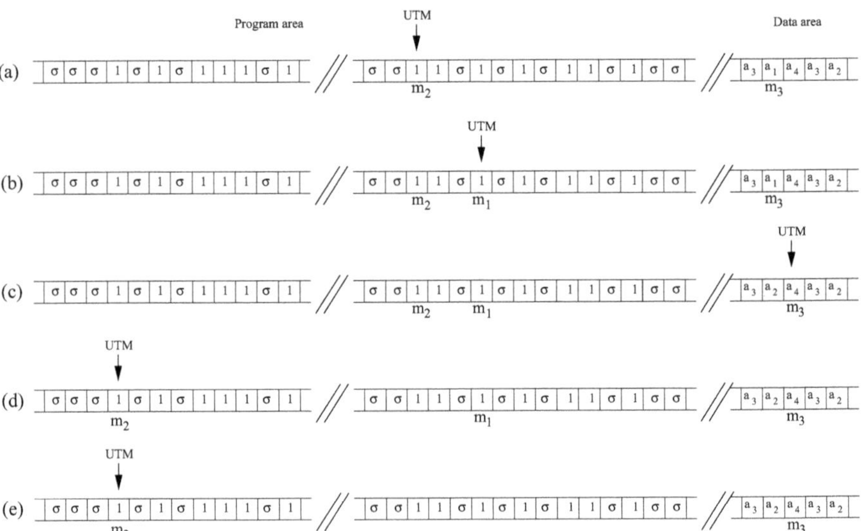

Fig. 8.4 Working cycle of the universal Turing machine. (a)–(e) visualize the steps (1)–(5), see text

that the new symbol will be a_2 and the action will be "move one step to the left".

3. Now the UTM changes the entry at the site marked by m_3 accordingly and moves the marker m_3, see Fig. 8.4c.
4. The UTM returns to the coding area, removes the marker m_2 and places this marker at the beginning of the first line that encodes δ_2, at the very first state (at the first '1' after the start of the coding area, indicated by the three subsequent σ, see Fig. 8.4d).
5. What comes now is the most subtle part: we must select the next state of TM_2. Our internal memory (the internal states of the UTM) is not sufficient to do this directly. The UTM solves the problem in the following way: it goes to m_1 and moves m_1 by one step to the right. If m_1 is still located over a '1', it goes to the marker m_2 and moves it to the beginning of the first line that codes the reaction of TM_2 on the next state (notice that this is not the beginning of the next line in the table, i.e., the next $\sigma\sigma$, we must to skip 18 lines, corresponding to the 18 symbols). This move is repeated until m_1 is located over a σ. Then m_2 is at the correct position, i.e., it indicates the state of TM_2. The UTM deletes the marker m_1 and moves to the marker m_2. It is prepared for the next cycle, see Fig. 8.4e.

Of course, our construction can be made more efficient. There are universal Turing machines with an alphabet consisting of five symbols and two states [183]. However, our point is to show that universal Turing machines exist. Obviously, a universal Turing machine is able to compute everything that can be computed. This leads to the following (somewhat informal) definition.

Definition 8.2.1 A structure that can simulate any Turing machine is called computation-universal.

We know from the above that there are computation-universal Turing machines, namely the universal Turing machines. It is quite easy to check the "program" of the universal Turing machine as sketched above, i.e., the coding sequence for the function δ: if we get a string consisting of σ and '1', we can check if it has the appropriate format for coding a transition function δ. We will use this fact later.

8.3 Computational Universality of Cellular Automata

In this section we investigate cellular automata with respect to computational universality.

Theorem 8.3.1 *Let $(Q, \mathfrak{A}, q^0, F, a_0, \delta)$ be a Turing machine. There is a one-dimensional cellular automaton $(\mathbb{Z}, D_0, E, f_0)$ that simulates the Turing machine. Let C^n and q^n denote the state of the tape and the internal state, respectively, of the machine after time step n. Then, there is an initial state u_0 with trajectory $\{u_n\}_{n\in\mathbb{N}}$ for the cellular automaton and there are functions $\varphi_1 : E^\Gamma \to \mathfrak{A}^\mathbb{Z}$ and $\varphi_2 : E^\Gamma \to Q$ such* *that*

$$\varphi_1(u_n) = C^n, \qquad \varphi_2(u_n) = q^n.$$

Proof Typically, a proof of such a theorem provides an explicit construction of the cellular automaton. Since the grid as well as the tape have cells enumerated by $\mathbb{Z}$, we can identify the location of a tape site and a corresponding cell.

We choose the state of a cell to consist of three components. The first component represents the state of the corresponding cell on the tape of the Turing machine. The second component stands for the internal state $q \in Q$ of the Turing machine, if the machine is located over the corresponding location on the tape. The third component is a marker, telling whether the Turing machine is located over the corresponding location. Thus,

$$E = \mathfrak{A} \times Q \times \{h, \emptyset\}.$$

The neighborhood of the cellular automaton consists of the cell itself and the two adjacent cells, $D_0 = \{-1, 0, 1\}$. Now we define the local function f_0. Assume that the state of the cell x is given by $e_0 = (a_0, q_0, m_0)$, the state of its left neighbor $x-1$ by $e_{-1} = (a_{-1}, q_{-1}, m_{-1})$ and that of the right neighbor $x+1$ by $e_1 = (a_1, q_1, m_1)$. The markers $m_i \in \{h, \emptyset\}$ indicate whether the Turing machine is located over the corresponding cell ($m_i = h$) or not ($m_i = \emptyset$). Although our Turing machine is located over one well defined tape location, the rules of the cellular automaton do not prevent more than one cell with a marker $(e)_3 = h$. However, we will choose an initial state where only one cell has $(e)_3 = h$. Hence we can ignore the other cases (just ensure that the local function is well defined).

If the marker h is placed over the cell $(e_0)_3 = h$, the cellular automaton will change the first component of the state of this cell to $(\delta((e_0)_1, (e_0)_2))_1$. If $(\delta((e_0)_1, (e_0)_2))_3 = S$, also the component that simulates the internal state q of the Turing machine is changed to $(\delta((e_0)_1, (e_0)_2))_2$. Accordingly, if the location of the Turing machine is at the left (right) neighbor, and the machine steps to the right (left), the state of the cell has to be chosen appropriately. In all other cases, the state is not changed at all. Therefore we define

$$f(e_{-1}, e_0, e_1)$$

$$= \begin{cases} (\delta((e_0)_1, (e_0)_2)_1, \delta((e_0)_1, (e_0)_2)_2, h) & \text{if } (e_0)_3 = h, \delta((e_0)_1, (e_0)_2)_3 = S \\ (\delta((e_0)_1, (e_0)_2)_1, (e_0)_2, \emptyset) & \text{if } (e_0)_3 = h, \delta((e_0)_1, (e_0)_2)_3 \neq S \\ ((e_0)_1, (e_0)_2, \delta((e_{-1})_1, (e_{-1})_2)_2, h) & \text{if } (e_0)_3 = \emptyset, (e_1)_3 = \emptyset, (e_{-1})_3 = h, \\ & \quad \delta((e_{-1})_1, (e_{-1})_2)_3 = R \\ ((e_0)_1, (e_0)_2, \delta((e_1)_1, (e_1)_2)_2, h) & \text{if } (e_0)_3 = \emptyset, (e_{-1})_3 = \emptyset, (e_1)_3 = h, \\ & \quad \delta((e_1)_1, (e_1)_2)_3 = L \\ e_0 & \text{otherwise} \end{cases}$$

The functions δ_1 and δ_2 are given by $\delta_1(u) = \{(u(i))_1\}_{i\in\mathbb{Z}}$, $\delta_2(u) = u(i_0)_2$ where i_0 is the integer with smallest absolute value such that $(u(i))_3 = h$. If there are two integers i_0 with $(u(\pm i_0))_3 = h$ or if there is no such integer, δ_2 returns the value q_0 (a dummy value, since these cases never occur in the setting interesting for us). Induction shows, that this cellular automaton simulates the Turing machine in the stated sense. □

This automaton is a straightforward implementation of a Turing machine as a cellular automaton. The converse question is more difficult: Given a cellular automaton, is it computationally universal? An example is Conway's "Game of Life" for which the answer is positive. The proof is based on the construction of a "machine" (similar to a self-reproducing automaton) with memory cells, clocks to synchronize processes, data wires, etc. build in the state space of the cellular automaton. These components are connected such at the end a huge machine is created that simulates the Turing machine. The proof (or, better, the construction) can be found in [1, 12]. Similar constructions are known for other cellular automata. Since among Turing machines there are also universal Turing machines, we have the following corollary.

Corollary 8.3.2 *There are cellular automata that are computation-universal.*

The proof of the theorem suggests to distinguish two classes of cellular automata, dumb ones and clever ones. The dynamics of the dumb ones, i.e., those that cannot simulate a universal Turing machine, can perhaps be completely understood. The dynamics of the other ones is too complex to be understood completely. Very soon one runs into trouble with the notions of computability and decidability. Therefore we discuss these notions in the next section.

8.4 Undecidable Problems

We introduce undecidable problems by a sequence of paradoxes and examples, from the very informal approach known since ancient times to the halting problem for Turing machines. The starting point is the paradox of Epimenides. We will see that also in today's reference problem, the halting problem, the structure of Epimenides paradox has been maintained. Its structure is only hidden in an abstract formulation.

8.4.1 *The Paradox of Epimenides*

The Cretan philosopher Epimenides stated "Cretans always lie". This paradox is at the center of all undecidable problems or uncomputable functions. We should understand this paradox in depth (Fig. 8.5). There are two ingredients: One is the speaker himself. He is a philosopher, Epimenides of Knossos, who lived about 600

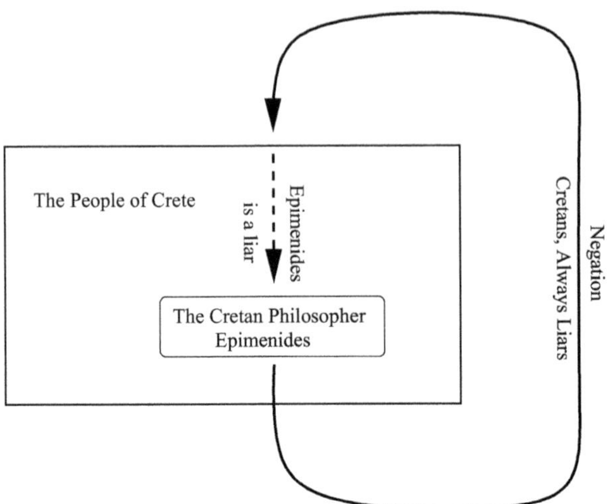

Fig. 8.5 Epidemides' paradox illustrates the basic structure of all undecidable problems: there is a set of objects about which a statement is made. The statement may apply to the system that makes the statement. If the statement includes a negation, we are likely to discover an undecidable problem

BC. Being a Cretan, he states a property of all Cretans. This is the first important observation: we have a system (the Cretan philosopher) that investigates a class of objects (the people of Crete). These two levels are mixed: The statement about Cretan people (the objects) is at the same time a statement about the stating system (the Cretan philosopher). Or, in other words, Epimenides appears in two different roles: first, as someone who states the sentence. Second, as an object to which the statement refers.

The second important fact is that the statement includes something like a negation: if we call Cretans "liars" we imply that the negations of their statements are true. These two ingredients lead to an oscillatory behavior: if the statement is true then Epimenides is a liar, hence the statement is false, hence Epimenides is not a liar, hence the statement is true, ... Evidently, the statement "the sentence of Epimenides is true" leads to a contradiction as well as the statement "the sentence of Epimenides is false". This problem is undecidable, both options (true and false) lead to contradictions. This is exactly the kind of structure that underlies undecidable problems.

8.4.2 Russel's Paradox

In the well known paradox of Russel, we consider sets. There are two classes of sets: the simple sets that only contain elements or other sets, but do not contain themselves, and the strange sets which have themselves as an element. Let $\mathcal{S}$ denote the set of all simple sets. Is $\mathcal{S} \in \mathcal{S}$? If $\mathcal{S}$ is not contained in itself, it belongs to the class-one objects, hence it is element of $\mathcal{S}$, and hence it is a strange set which does not belong to $\mathcal{S}$, and as a consequence $\mathcal{S}$ is a simple set. Neither $\mathcal{S} \in \mathcal{S}$ nor $\mathcal{S} \notin \mathcal{S}$ can be true. Also here, we find a negation—the definition of the simple sets says " ...states that are *not* an element of themselves...". The two levels (the stating system and the object) are mixed, since we state a definition about sets using the structure of sets.

The theory of sets was the first mathematical theory where paradoxes of this kind have been discovered (around 1900). There are many more paradoxes from this theory, see e.g. [116, pp. 185].

8.4.3 Richard's Paradox

This paradox reaches a new level of abstraction in introducing a new aspect into the game of Epimenides: the idea of coding sequences by natural numbers, and, in this way, mixing statements about natural numbers with the numbers themselves.

The starting point is the observation, that we can make statements about natural numbers using a natural language (English or Latin or whichever) and a finite alphabet. For instance, we can define

A number is even, if it can be divided by two.
A number is prime, if it can be only divided by one and by itself.

And so on, and so on. Next, we take a set of definitions about natural numbers and order them in a lexicographic way: first by length, and all statements with the same length by the given order of the alphabet. Then, we number these ordered definitions. We find perhaps:

the number of " *A number is even, if it can be divided by two* " is 17536;
the number of "*A number is prime, if it can be only divided by one and by itself*" is 77785.

The number of a definition may or may not have the property that is stated in the corresponding definition. E.g., the number 17536 is even, i.e., it has the property of its corresponding definition, while 77785 can be divided by five, i.e., it is not prime and hence does not have property of its definition. We state a new definition:

A number is Richard's, if it does not possess the property of the definition that is coded by this number.

Thus, 17356 is not Richard's while 77785 is. Let $x \in \mathbb{N}$ be the code of the definition of Richard's numbers. Is x Richard's or not? ...... You see. We meet our old pal Epimenides again. This paradox has been developed by Jules Richard in 1905. Gödel in his famous incompleteness theorems (published 1931) did nothing else but use the formal setting of number theory to transform this argument from its rather informal setting into a strict proof [139]. However, we will not go into this highly technical business but directly jump to Turing machines.

8.4.4 The Word Problem

The arguments of Richard can be translated into the formalism of Turing machines. We construct a language $\hat{L}$ of higher complexity as Turing machines. That is, no Turing machine is able to decide for all possible words if a given word is an element of $\hat{L}$. We focus on the word problem:

Problem 8.4.1 (Word Problem) *Given an alphabet $\mathfrak{A}$, a language $L \subset \mathfrak{A}^*$ and word $w \in \mathfrak{A}^*$, is there a Turing machine that is able to decide the property $w \in L$?*

In order to construct the language $\hat{L}$, we associate the language of accepted words with a Turing machine.

Definition 8.4.2 Consider a Turing machine $(Q, \mathfrak{A}, q^0, F, a_0, \delta)$. Let w be a finite word over the alphabet $\mathfrak{A}$. Consider the Turing machine on a blank tape on which the word w is written somewhere. If we start the Turing machine at the first letter of the word, and the Turing machine finally stops (i.e., it eventually hits the acceptance state "F"), we say that the word is accepted. If the Turing machine never stops, the word is not accepted. The set of all accepted words is called the language $\mathcal{L}$(TM) of accepted words of the Turing machine.

In order to construct a language that is too complex for a Turing machine, we choose an appropriate alphabet. Since we can code anything by a binary code, we use the alphabet $\mathfrak{A} = \{0, 1, b\}$. The finite words over this alphabet can be lexicographically ordered. We then can number these words,

$$\begin{aligned}
w_1 &= \text{"0"} \\
w_2 &= \text{"1"} \\
w_3 &= \text{"b"} \\
w_4 &= \text{"00"} \\
w_5 &= \text{"01"} \\
w_6 &= \text{"0b"} \\
w_7 &= \text{"10"} \\
w_8 &= \text{"11"} \\
w_9 &= \text{"1b"} \\
w_{10} &= \text{"b0"} \\
w_{11} &= \text{"b1"} \\
w_{12} &= \text{"bb"} \\
w_{13} &= \text{"000"} \\
&\dots
\end{aligned}$$

Due to our universal Turing machine, see Sect. 8.2, the function δ of any Turing machine can be coded as a finite string of the symbols 1 and σ. All finite words of these two signs can be lexicographically ordered and numbered like the words w_i above. Let us denote these strings with $y_1, y_2, y_3, \ldots$ Not all of these strings correspond to valid codes of a Turing machine. However, it is easy to check if a string obeys the necessary format that can be interpreted as a Turing machine. We define: words y_j that do not code for a Turing machine correspond to the Turing machine that stops immediately, $\delta(a, q_0) = (a, F, S)$.

Now we find a connection between the objects w_i and the systems y_i processing these objects: the index i. This connection may appear weak, but in fact it is not weaker than Richard's definition. The identification by the index i mixes two levels. Now we introduce a negation by defining the set

$$\hat{L} = \{w_i \;:\; w_i \text{ is not accepted by the Turing machine coded by } y_i,\;\; i \in \mathbb{N}\}.$$

This language (set of finite words) is not empty, since there is a Turing machine that only stays in its starting state, $\delta(a, q_0) = (a, q_0, S)$; no word is accepted by this machine. Now we ask, if there is a Turing machine that accepts exactly the words in the language $\hat{L}$. Assume that there is such a Turing machine. This machine is coded by some string y_{i_0}, $i_0 \in \mathbb{N}$. Is $w_{i_0} \in \hat{L}$ or not? If $w_{i_0} \in \hat{L}$, then w_{i_0} is not accepted by the Turing machine y_{i_0}, i.e., our Turing machine does not accept all words in $\hat{L}$, a contradiction to the assumption. If $w_{i_0} \notin \hat{L}$, then w_{i_0} is accepted by the Turing machine though it is not a word of the language $\hat{L}$, also in contradiction to the assumption. Hence, both possibilities $w_{i_0} \in \hat{L}$ and $w_{i_0} \notin \hat{L}$ lead to contradictions. There is no Turing machine that accepts exactly the language $\hat{L}$.

From that, we can conclude that there is also no Turing machine that can decide if a word w is in the language $\hat{L}$ or not: if we would have such a machine, then, in the first step we let this machine decide if a given word is in $\hat{L}$ or not. If the word is in $\hat{L}$, our machine would stop. If not, it would jump to an endless dummy loop. Hence, this machine would accept exactly $\hat{L}$, but such a machine is not possible. Thus, the characteristic function of $\hat{L}$ is not computable.

Theorem 8.4.3 *The word problem for Turing machines is not decidable.*

8.4.5 *The Halting Problem*

The word problem is still too close to Epimenides to be of great value for proving the undecidability of problems. We move forward to the halting problem.

Problem 8.4.4 *The general halting problem poses the following question. Does a Turing machine, started on a given state of the tape, eventually stop?*

Obviously, this problem is a close relative of the word problem. And it is also undecidable. Because, if we could solve the halting problem we could solve the word problem.

Theorem 8.4.5 *The general halting problem is undecidable.*

Proof We sketch the idea of the proof. This proof is based on a construction of a Turing machine that can solve the word problem, using as the central decision mechanism an algorithm (Turing machine) that solves the halting problem. The construction is quite straightforward:

Step 1 Given a word w, compute its number in the lexicographic order, i.e., determine $i \in \mathbb{N}$ such that $w = w_i$.
Step 2 Construct the code of the corresponding Turing machine y_i.
Step 3 Ask the algorithm that solves the halting problem, if the Turing machine y_i, started on a tape which is blank but one word, w_i, will stop.
Step 4 If the Turing machine y_i will stop, then $w_i \notin \hat{L}$, otherwise $w_i \in \hat{L}$.

□

We may define the halting problem in a more restricted way.

Problem 8.4.6 *The restricted halting problem poses the question whether a given Turing machine, started on an empty tape, will eventually stop (or not).*

Theorem 8.4.7 *The restricted halting problem is undecidable.*

Proof If the restricted halting problem were decidable, we could program a Turing machine that starts on an empty tape, writes in a first step a given word on the tape and transfers the control to the Turing machine that is the interesting one. Thus, in case we can predict if a Turing machine, started on a blank tape, eventually stops, we can also predict if this Turing machine stops if started on a tape on which a finite word is written (and that is blank elsewhere). □

Often, if one wants to prove that a given problem is undecidable, one reduces it to the halting problem. We already used this idea in the preceding proofs: If a given problem is decidable, construct an algorithm that solves the halting problem, using the solution of the given problem as a centerpiece. If this is possible, the problem is not decidable since the halting problem is not decidable. We used this technique to show that the halting problem is undecidable: if it is decidable the word problem is also decidable, and this is not true.

The strength of the halting problem is its relation to a dynamic process (as opposed to the static problem of Epimenides). It seems that many problems in mathematics and theoretical computer sciences are closer to dynamics than to linguistics.

8.4.6 The Immortality Problem

Another useful variation of this theme is the immortality problem: In this problem the aim is to develop an algorithm that is able to decide if—for a given Turing machine—there is at least one state in $\mathfrak{A}^{\mathbb{Z}}$ for which the machine never stops. Note that we not only allow for finite words, but for any state. The proof below indicates that this problem is undecidable, independently of the initial state of the tape [92].

Theorem 8.4.8 *The immortality problem is undecidable.*

Proof Assume the immortality problem is decidable. Then the halting problem is decidable: We embed a given Turing machine (the "smaller machine") into a larger one. Essentially, this larger machine first erases all information from the tape, and then lets the smaller machine run. Thus, the larger machine, started with any state on the tape, will stop if and only if the smaller machine, started on a blank tape, eventually stops. Or, conversely, the larger machine possesses an immortal state (better: all initial states are immortal), if and only if the smaller machine, started on a blank tape, never stops. If we can decide the immortality problem, we can decide the halting problem.

We indicate how to delete the initial information on the tape in an appropriate way. This is not completely straightforward as the tape is infinite. First erase the information at the origin of the tape (the site where the larger Turing machine starts), and from both neighbors. At these neighboring sites, the larger machine places two markers. These markers indicate the rightmost and the leftmost site the (larger) machine did visit. Then it returns to the origin and starts with the program of the smaller machine. Each time the Turing machine hits a site with these markers indicating the outmost visited places, the marker is shifted outwards, and the information on this place is erased. In this way, the smaller Turing machine will behave as on an initially empty tape. □

8.4.7 Non-computability of ω-Limit Sets for Cellular Automata

We can draw first conclusions about cellular automata. We utilize the embedding of a Turing machine into a cellular automaton constructed in Theorem 8.3.1. We define the local function in such a way that the trajectory becomes constant if and only if the Turing machine stops. We equip the state space of the cellular automaton with the Cantor topology. If we can compute the ω-limit sets for all initial data then we can decide the halting problem.

Corollary 8.4.9 *If we consider the cellular automaton as a topological dynamical system, equipped with the Cantor topology, then the ω-limit set of a point is, in general, not computable.*

Notice that this argument does not carry over to the Besicovitch and Weyl topologies as we cannot follow the states of single cells. It is not clear to the authors if the ω-limit sets with respect to these topologies are computable. Kari [102] proved a more general theorem about the undecidability of ω-limit sets (a so-called Rice theorem): given any property of ω-limit sets that is nontrivial (in the sense that there is at least one automaton for that all ω-limit sets do have this property, and one for which the ω-limit sets do not have this property), then there is at least one cellular automaton where it is undecidable whether the ω-limit sets have the property.

8.5 Tiles

Turing machines are not very close to cellular automata. Tessellations or tilings are closer—in the beginning of their history, cellular automata have been also called tessellation automata [129]. There is a vast literature on patterns and tessellations, see e.g. the book of Grünbaum and Shephard [79]. Here we are interested in one simple tiling structure—the Wang tiling—and its relation to undecidable problems.

8.5.1 *Definitions and Examples*

Wang [172] used a simple structure to develop algorithms that can tessellate the plane. A tessellation or tiling consists of tiles that carry a color at each face. Similar to dominoes, two tiles can be placed next to each other if adjacent faces carry the same color. So, if a Cayley graph $\Gamma(G)$ with generators $\sigma_1, \ldots, \sigma_m$ is tessellated then at each vertex there is an element t (a tile) from a tile set T, this tile has $2m$ faces indexed by $\sigma_i^{\pm 1}$, $i = 1, \ldots, m$, and each face carries a color. Two tiles at neighboring vertices fit together if they carry the same color. If we want a more formal definition, we can consider a tile as a map from the set of faces to the set of colors [3]. Then the set T is just numbering the tiles that are used for the particular tessellation or tiling.

Definition 8.5.1 Consider a Cayley graph $\Gamma(G)$ generated by $\sigma_1, \ldots, \sigma_m$, and a finite set of colors C. A tile is a map

$$t : \{\sigma_i^{\pm 1} \; : i = 1, \ldots, m\} \to C.$$

A tile set T is a finite set of tiles described by the function H,

$$H : T \times \{\sigma_i^{\pm 1} \; : i = 1, \ldots, m\} \to C, \quad (t, \tau) \mapsto H(t, \tau)$$

that indicates the color of a tile $t \in T$ at face $\tau \in \{\sigma_i^{\pm 1} \; : i = 1, \ldots, m\}$.

A tiling or tessellation of the Cayley graph Γ by the tile set T is a map

$$\phi : \Gamma \to T.$$

The tiling is called coherent if adjacent faces carry the same color,

$$H(\phi(g), \tau) = H(\phi(\tau g), \tau^{-1}) \quad \text{for } g \in \Gamma, \tau \in \{\sigma_i^{\pm 1} : i = 1, \ldots, m\}.$$

A tiling of a subgraph $\tilde{\Gamma}$ of Γ is a map $\tilde{\phi} : \tilde{\Gamma} \to T$, such that

$$H(\tilde{\phi}(g), \tau) = H(\tilde{\phi}(\tau g), \tau^{-1}) \quad \text{for } g \in \tilde{\Gamma}, \tau \in \{\sigma_i^{\pm 1} : i = 1, \ldots, m\}, \tau g \in \tilde{\Gamma}.$$

A tessellation is called periodic, if the subgroup of G

$$\Lambda(\phi) = \{h \in G : \phi(h g) = \phi(g) \ \forall g \in G\}$$

has a finite index $[G : \Lambda(\phi)]$.

The definition of "periodic tiling" is analogous to the definition of "periodic state of a cellular automaton", see Definition 2.3.3. If $\Gamma = \mathbb{Z}^2$, this definition is equivalent to $\phi(x + a, y) = \phi(x, y)$ and $\phi(x, y + b) = \phi(x, y)$ for some $a, b \in \mathbb{N}$ and all $(x, y) \in \mathbb{Z}^2$.

A (coherent) tiling need not respect symmetries of the Cayley graph. For instance, if $\Gamma = \mathbb{Z}^2$ then T may contain some tile, but not the tile rotated by 90°. In Fig. 8.6 we could produce other tilings by applying elements of the symmetry group

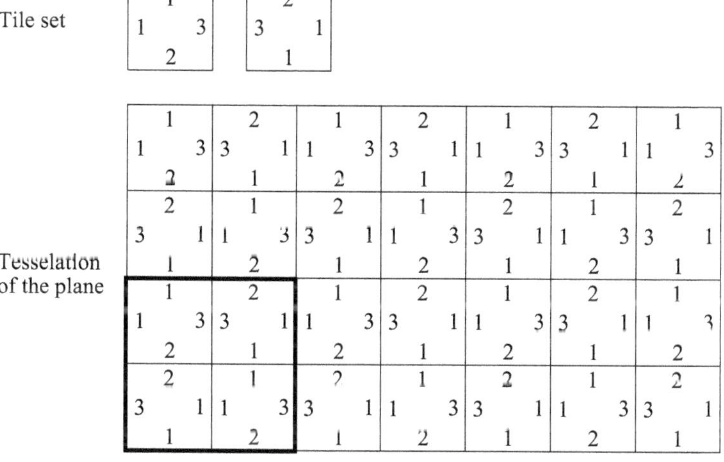

Fig. 8.6 Periodic tiling of the plane by tiles from a set with two elements. The structure that repeats periodically is indicated by the *fat square*

of $\mathbb{Z}_2$ to the given tiling. In general, the tiling has a smaller symmetry group than the Cayley graph.

Similar to the notions of a cellular automaton with resting state and the subautomaton of states with finite support, we consider tilings with a "blank" tile. We call a tiling finite if only finitely many tiles are not blank.

Definition 8.5.2 Consider a Cayley graph Γ and a tile set T. Let $b \in T$ denote a tagged tile (called the "blank tile" or "white tile"). A tiling is called finite if there are only finitely many sites that carry tiles other than b,

$$|\{g \in \Gamma \, : \, \phi(g) \neq b\}| < \infty.$$

A tiling is called non-trivial, if there is at least one $g \in \Gamma$ with

$$\phi(g) \neq b.$$

It is a pure convention which tile to consider as "blank". It can be very well the case that a "blank" tile carries colors. Note that this definition implicitly requires that the blank tile allows for a consistent tessellation of the plane. Given this setting, some basic questions can be raised immediately.

Problem 8.5.3 (Domino Problem) *Given a Cayley graph Γ (or a subgraph) and a tile set T, is there a coherent tiling by T?*

Problem 8.5.4 (Finite Domino Problem) *Given a Cayley graph Γ and a tile set T, where $b \in T$ is a tagged tile ("blank"). Is there a nontrivial finite coherent tiling of Γ?*

Problem 8.5.5 (Completion Problem) *Given a (subgraph of a) Cayley graph Γ and a tile set T and a starting configuration on a subset of sites. Is it possible to extend this starting configuration to a coherent tiling?*

Wang conjectured that any tile set T that allows a coherent tiling of the plane $\mathbb{Z}^2$ also allows a (coherent) periodic tiling. If this conjecture were true, the problem of whether a tile set allows a coherent tiling of the plane could be decided by the following algorithm [172].

Algorithm 8.5.6 *For $n \in \mathbb{N}$, let $Q_n = \{(i,j) \, : \, |i|, |j| < n\} \subset \mathbb{Z}^2$, and $\mathfrak{T}_n$ be the set of all coherent tilings on Q_n with tiles from T.*

Start with $n = 1$.

Step 1: Determine $\mathfrak{T}_n$.

Step 2: If $\mathfrak{T}_n = \emptyset$, then there is no coherent tiling. Stop.

Step 3: Determine if one of the elements of $\mathfrak{T}_n$ allows a periodic tiling of the plane. If this is the case, then: Stop.

Step 4: Increase n by one, proceed with Step 1.

Theorem 8.5.7

(a) *For n finite, each step of Algorithm 8.5.6 requires to consider only a finite number of finite configurations.*

(b) *Assume that a tile set T either allows a coherent periodic tiling or no coherent tiling at all. Then, Algorithm 8.5.6 stops after a finite number of steps.*

Proof Part (a) follows from $|T|, |Q_n| < \infty$. If T allows a periodic tiling, then the algorithm will stop after a finite number of steps. Assume that T does not allow a periodic tiling but does not stop in finite time. Then there is a sequence $\tilde{\tau}_n \in \mathfrak{T}_n$. We extend $\tilde{\tau}_n$ to a tiling on $\mathbb{Z}^2$ (which, in general, is not coherent). We simply define

$$\tau(i,j) = \begin{cases} \tilde{\tau}_n(i,j) & \text{if} \quad (i,j) \in Q_n \\ t_0 & \text{otherwise} \end{cases}$$

for an arbitrary, fixed element $t_0 \in T$. We note again that only the restriction of τ_m to Q_n for $m > n$ is a coherent tiling but not τ_m considered on $\mathbb{Z}^2$.

Thus, $\{\tau_i\}_{i\in\mathbb{N}}$ is a sequence in the compact Cantor space $T^{\mathbb{Z}^2}$. There is a converging subsequence of $\{\tau_i\}_{i\in\mathbb{N}}$ that becomes eventually constant on any finite region $Q_n \subset \mathbb{Z}^2$. The limiting element $\tau \in T^{\mathbb{Z}^2}$ is a coherent tiling if we restrict τ to Q_n for any $n \in \mathbb{N}$. Hence, τ is a coherent tiling on $\mathbb{Z}^2$.

Since we assumed that the algorithm does not stop, τ is not periodic, and there is no periodic tiling of the plane. This is a contradiction to the assumption that the tile set allows a periodic tiling if it allows a coherent tiling of the plane. □

Remark 8.5.8 We will show later, that the assumption in this proposition is wrong. There are tile sets on $\mathbb{Z}^2$ that only allow for aperiodic tessellations.

8.5.2 Tessellations of Free Groups

Since in $\mathbb{Z}$ Wang's idea is true—a tile set that allows for a tessellation of $\mathbb{Z}$ also allows for a periodic tessellation—it is evident how to solve the domino problem on $\mathbb{Z}$. There is one property of $\mathbb{Z}$ that carries over to the canonical Cayley graph of a freely generated group: a path that does not visit any point twice is determined by its starting point and its end point (see e.g. Fig. 2.5). In this aspect free groups are essentially one-dimensional.

We focus on free groups over two symbols $G = < a, b >$, as the generalization to n symbols is straightforward. Given a tile set T. We describe an algorithm to decide whether a coherent tiling is possible with this tile set. We construct a directed graph G_0 with the tiles as vertices. The edges are labeled by a and b. A directed edge labeled with a goes from t_1 to t_2 if these two tiles can be neighbors in the Cayley-graph $\Gamma(G, \{a, b\})$ at two sites g_1, g_2 such that a carries g_1 into g_2, i.e., if

$$H(t_1, a) = H(t_2, a^{-1}).$$

Similarly a b-edge goes from t_1 to t_2, if $H(t_1, b) = H(t_2, b^{-1})$.

Definition 8.5.9 Consider the graph defined above. Let $\tau \in \{a, b\}$ denote a generator.

- A τ-path from $t \in T$ to $\hat{t} \in T$ is a finite sequence of $t_i \in T$ such that $t_0 = t$, $t_n = \hat{t}$, $n > 0$, and τ-edges connect t_i and t_{i+1}, $i = 0, \ldots, n-1$.
- A bi-infinite τ-path is a sequence $\{t_n\}_{n \in \mathbb{Z}}$ such that there is a τ-edge from t_n to t_{n+1}.
- $Q_\tau^+(t) = \{t' \in T \,:\, \text{there is a } \tau\text{-path from } t \text{ to } t'\}$
- $Q_\tau^-(t) = \{t' \in T \,:\, \text{there is a } \tau\text{-path from } t' \text{ to } t\}$
- $P_\tau = \{t \in T \,:\, t \in Q_\tau^+(t)\}$.

The sets $Q_\tau^+(t)$, $Q_\tau^-(t)$, P_τ, $\tau \in \{a, b\}$, are computable as the graph is finite. Also the next proposition is an immediate consequence of the fact that the graph is finite.

Proposition 8.5.10 *Let $\tau \in < a, b >$. An element $t \in T$ is a member of a bi-infinite τ-path, if and only if $Q_\tau^+(t) \cap P_\tau \neq \emptyset$ and $Q_\tau^-(t) \cap P_\tau \neq \emptyset$.*

This observation is a basic ingredient for the following algorithm (see Fig. 8.7).

Algorithm 8.5.11 *Starting from G_0 introduced above, we construct recursively nested subgraphs. Let G_1 denote the subgraph of G_0 consisting of all those vertices that are elements of bi-infinite a- and, at the same time, bi-infinite b-paths. In the same way, define G_i as the subgraph of G_{i-1} consisting of the vertices that are elements of bi-infinite a- and b-paths, where the paths are completely contained in G_{i-1}. Proceed recursively, and stop if either $G_i = G_{i-1}$, or if $G_i = \emptyset$.*

Theorem 8.5.12 *Consider the construction given in Algorithm 8.5.11. The algorithm stops after a finite number of steps i_0. Furthermore, a coherent tiling exists if and only if the resulting set of vertices in G_{i_0} is not empty.*

Proof If $G_i \neq G_{i+1}$, then we decrease the size of the subgraph by at least one. As the original graph is finite, this procedure stops after at most $|G_0|$ steps. Let $G_{i_0} = G_{i_0+1}$. We show that $G_{i_0} \neq \emptyset$ if and only if there is a coherent tiling.

"$\Leftarrow$": Assume that there is a coherent tiling $\phi : \Gamma \to T$. Let $D = \{\phi(g) \,:\, g \in \Gamma\} \subset T$ denote the set of all tiles used. This set is a subset of the set of vertices of the graph G_0. Each of these tiles is a member of bi-infinite a- and b-paths, where the path does not leave the set D. Let $t \in D$. There is $g \in \Gamma$ such that $t = \phi(g)$. Then, $t_i = \phi(ga^i)$, $i \in \mathbb{Z}$ is the required bi-infinite a-path. Similarly, the bi-infinite b-path is constructed. Obviously, these paths do not leave $D \subset T$. Thus, D is a subset of the set of vertices for all G_i, $i \in \mathbb{N}_0$, and the set of vertices never becomes empty.

"$\Rightarrow$": In the present case, a coherent tiling can be constructed as follows. For each vertex t in G_{i_0} choose a bi-infinite a-path $p_t^a(i) \in T$ and a bi-infinite b-path $p_t^b(i) \in T$, both bi-infinite paths are contained in the graph G_{i_0}. Next, select any vertex t of G_{i_0} as a starting point. We construct the tiling by defining $\phi(e) = t$, and recursively

$$\phi(a^{\pm 1} w) = p^a_{\phi(w)}(\pm 1), \qquad \phi(b^{\pm 1} w) = p^b_{\phi(w)}(\pm 1)$$

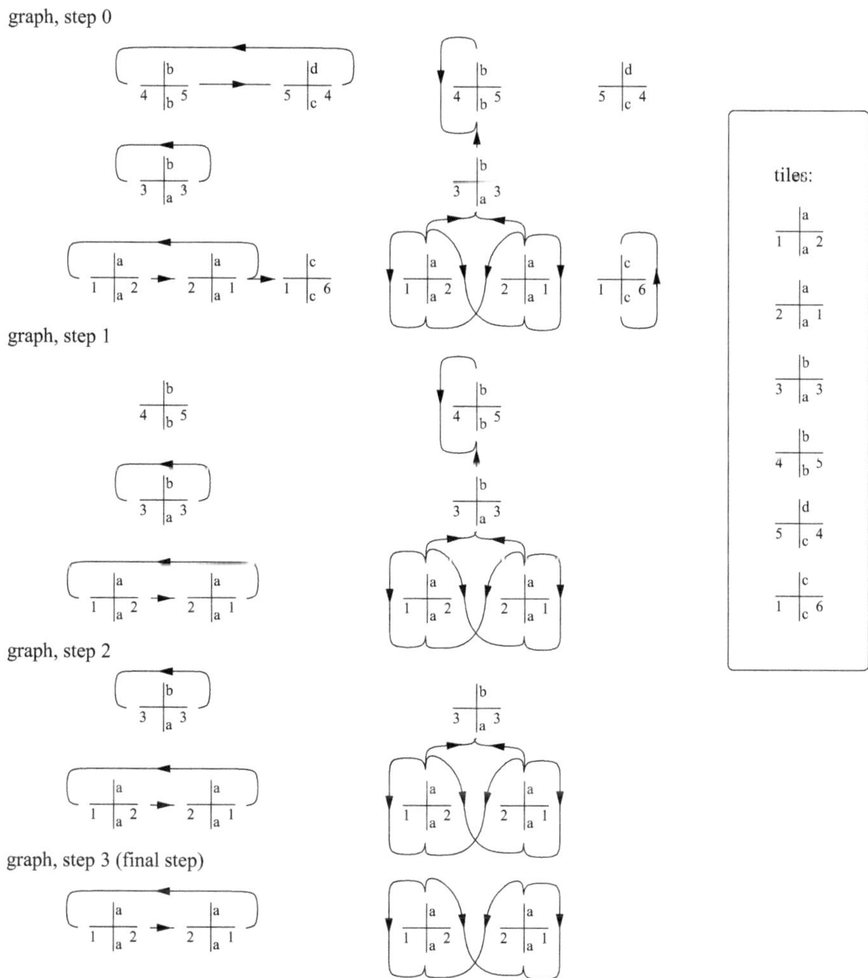

Fig. 8.7 Algorithm 8.5.11 in action on a free group over two symbols $G = < a, b >$. We do not show a and b-edges in the same graph, but show two graphs instead. *Left*: a-paths, *right*: b-paths

where $a^{\pm 1}w$ and $b^{\pm 1}w$ are irreducible words in the free group—words that cannot be shortened, i.e., words where never a and a^{-1} (resp. b and b^{-1}) are next to each other. The tiling obtained is coherent by construction. □

This construction can be easily extended to free groups with an arbitrary number of generators.

Corollary 8.5.13 *The domino problem is decidable on Cayley graphs over finitely generated free groups* $< \sigma_1, \ldots, \sigma_n >$.

Remark 8.5.14 The methods developed here can be used to prove a slightly more general result: Let G_1 and G_2 are groups for which the domino problem is decidable. Then, the domino problem is decidable for the free product $G_1 * G_2$.

Since the domino problem is decidable for $\mathbb{Z}$, it is decidable for any finitely generated free group (this is the theorem we just did prove). However, it is also decidable for any free product of finite groups and $\mathbb{Z}$. It is still possible to generalize this result without large effort. Using the arguments of Proposition 3.4.12, we find that the domino problem is decidable for a group that has a subgroup of finite index for which the domino problem is decidable. It is an open problem, if the set of finitely generated groups obtained in this way (all groups that have a subgroup of finite index that is in turn a free product of finite groups and free groups) are exactly all groups for which the domino problem is decidable, or if there are more [3]. Results are only for few groups [3, 11, 120], and there seem to be arguments for polynomial growing groups [5].

8.5.3 *Aperiodic Tessellations on Z^2*

Although Wang's conjecture (a tile set that allows a tiling of the graph also admits a periodic tiling) is true in one dimension, it is wrong in general. Berger [11] constructed a tile set of over 50,000 tiles that did not allow a periodic tiling of $\mathbb{Z}^2$ but did allow a coherent and necessarily aperiodic tiling. Using this construction he proved undecidability of the domino problem in $\mathbb{Z}^2$. Later, Robinson [148] found a simpler example with 32 tiles. Kari showed that one only needs 14 tiles [103]. The central ingredients of his construction are Beatty sequences and balanced words.

Definition 8.5.15 The Beatty sequence associated with $r \in \mathbb{R}_+ \setminus \{0\}$ is defined as

$$(A(r))_n = \lfloor n\,r \rfloor, \quad n = 1, 2, 3, \ldots.$$

For convenience, the set

$$A(r) = \{(A(r))_n \; : \; n \in \mathbb{N}\}$$

is also called "the Beatty sequence".

In order to construct such a sequence we take a real number, multiply the number by n and take the integer part. This procedure yields a map from the natural numbers into the natural numbers. Let us look at two examples: we choose $r = \sqrt{2} + 1 = 2.4142\ldots$ and $s = 1 + 1/\sqrt{2} = 1.7071\ldots$. For these numbers we obtain

n	1	2	3	4	5
$n\,r$	2.4142	4.82843	7.24264	9.65685	12.07107
$\lfloor n\,r \rfloor$	2	4	7	9	12
$n\,s$	1.70711	3.41421	5.12132	6.82843	8.53553
$\lfloor n\,s \rfloor$	1	3	5	6	8

Why is this procedure remarkable? If we inspect the sequences $\lfloor n\,r \rfloor$ and $\lfloor n\,s \rfloor$, we find that in the table above all integers up to 9 appear exactly once. If r is irrational, i.e., $r \in \mathbb{R}_+ \setminus \mathbb{Q}$, $r > 1$, and $s = r/(r-1)$, then the Beatty sequences $A(r)$ and $A(s)$ decompose the set $\mathbb{N}$ in two complementary subsets. We state this fact as a proposition.

Proposition 8.5.16 *Let $r \in \mathbb{R}_+ \setminus \mathbb{Q}$ and $s \in \mathbb{R}_+$ such that*

$$\frac{1}{r} + \frac{1}{s} = 1.$$

The Beatty sequences corresponding to r and s are complementary, i.e.,

$$A(r) \cap A(s) = \emptyset, \qquad A(r) \cup A(s) = \mathbb{N}.$$

Proof (see also [89, pp. 94]) Let $\tilde{A}_n = \#\{a \,:\, a \in A(r),\ \ a < n\}$, i.e., the number of multiples of r less than n. Thus, $\tilde{A}_n = \lfloor n/r \rfloor$. Similarly, $\tilde{B}_n = \#\{b \,:\, b \in A(s),\ b < n\} = \lfloor n/s \rfloor$. We prove an inequality for $\tilde{A}_n + \tilde{B}_n$. Let N be an integer. Then, N/r resp. N/s are not integers and

$$\frac{N}{r} - 1 < \tilde{A}_n = \left\lfloor \frac{N}{r} \right\rfloor < \frac{N}{r}, \qquad \frac{N}{s} - 1 < \tilde{B}_n = \left\lfloor \frac{N}{s} \right\rfloor < \frac{N}{s}.$$

If we add these two inequalities, we find

$$N\left(\frac{1}{r} + \frac{1}{s}\right) - 2 < \tilde{A}_n + \tilde{B}_n < N\left(\frac{1}{r} + \frac{1}{s}\right).$$

Since $1/r + 1/s = 1$, we find $N - 2 < \tilde{A}_n + \tilde{B}_n < N$ and hence

$$\tilde{A}_n + \tilde{B}_n = N - 1.$$

Hence, in the interval $(N-2, N)$, there is exactly one of the values $n\,r$ or $n\,s$. This proves the result. □

Already Beatty [9, 10, 146] considered also bi-sequences, in the following way.

Definition 8.5.17 Let $r \in \mathbb{R}_+ \setminus \{0\}$. The Beatty bi-sequence $A(r) \subset \mathbb{Z}$ associated with r is defined by

$$(A(r))_z = \lfloor zr \rfloor, \quad z \in \mathbb{Z}$$

$$A(r) = \{\lfloor zr \rfloor \,:\, z \in \mathbb{Z}\}.$$

There is a similar result for bi-sequences.

Theorem 8.5.18 *Let $r \in \mathbb{R}_+ \setminus \mathbb{Q}$ and $s \in \mathbb{R}$ such that*

$$\frac{1}{r} + \frac{1}{s} = 1.$$

The Beatty bi-sequences corresponding to r and s are complementary in the following sense:

$$A(r) \cap A(s) = \{0\}, \qquad A(r) \cup A(s) = \mathbb{Z} \setminus \{-1\}.$$

Proof For $z \notin \mathbb{N}$ the numbers $(A(r))_z$, $(A(s))_z$ are not in $\mathbb{N}$. Hence the previous proposition shows $A(r) \cap A(s) \cap \mathbb{N} = \emptyset$, $A(r) \cup A(s) \supset \mathbb{N}$, and we only need to consider $z \leq 0$. Here we can use the same argument as before. For $z < -1$, let $\hat{A}_z = \#\{a : a \in A(r),\ a \in [z, -1]\} = \lfloor(-z)/s\rfloor$ and $\hat{B}_z = \#\{b : b \in A(s),\ b \in [z, -1]\} = \lfloor(-z)/s\rfloor$. With the same arguments as before we find

$$\hat{A}_z + \hat{B}_z = |z| - 1 \qquad \text{for } z \in \mathbb{Z}, \quad z < -1.$$

This inequality tells us, that all integers smaller than -1 appear only once in the sequences $(A(r))$ and $(A(s))$. The cases $z = 0$ and $z = -1$ are special. Since for $z = 0$ we find $a_0 = b_0 = 0$, the number 0 appears twice (once in $(A(r))$ and once in $(A(s))$). The number -1, however, does not appear at all, since $-s < -1$ as well as $-r < -1$ and therefore $(A(r))_{-1}$, $(A(s))_{-1} < -1$. □

Theorem 8.5.18 is Beatty's original result. We will use some sequences $B(r)$ derived from the Beatty sequences, the so.called balanced words. An overview about balanced words and their properties can be found in [169]. For a geometric interpretation of these sequences, see Fig. 8.8.

Definition 8.5.19 For $r \in \mathbb{R}_+ \setminus \{0\}$, the "balanced representation of r" is defined by the bi-sequence

$$(B(r))_z = (A(r))_z - (A(r))_{z-1}, \qquad z \in \mathbb{Z},$$

$$B(r) = \{(B(r))_z : z \in \mathbb{Z}\}.$$

Proposition 8.5.20 *For $r \in \mathbb{R}_+$ holds* $\lim_{k\to\infty} \frac{1}{2k+1} \sum_{z=-k}^{k} B(r)_z = r$ *and*

$$B(r) \subset \{\lfloor r \rfloor, \lfloor r \rfloor + 1\}. \qquad (*)$$

Furthermore, if $r \in \mathbb{R}_+ \setminus \mathbb{Q}$, then the sequence $((B(r))_i)$ is not periodic.

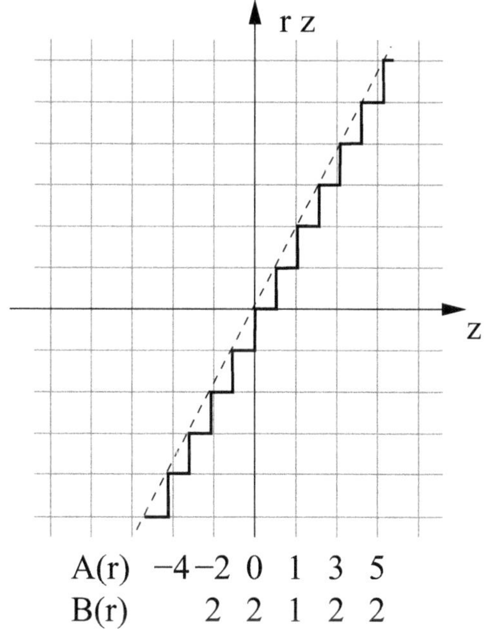

Fig. 8.8 The balanced word representation of r ($r = 1.8783$ in our example) is based on a *line* with slope r (*dashed line*). This *line* is discretized ($\lfloor zr \rfloor$, *solid line*) and yields the Beatty sequence $A(r)$. The balanced word representation $B(r)$ consists of the increments of the discretized line at subsequent integers. The *gray vertical and horizontal lines* indicate $\mathbb{Z}^2$

Proof Note that

$$\sum_{z=-k}^{k} B(r)_z = A(r)_k - A(r)_{-k-1} = \lfloor r\,k \rfloor - \lfloor -r\,(k+1) \rfloor \in (2k+1)\,r + [-2,2].$$

Therefore, $\lim_{k\to\infty} \frac{1}{2k+1} \sum_{z=-k}^{k} B(r)_z = r$.
We next show the inclusion $(*)$. For all $z \in \mathbb{Z}$ we have

$$r-1 = ((z+1)r-1) - zr < \lfloor (z+1)r \rfloor - zr < \lfloor (z+1)r \rfloor - \lfloor zr \rfloor = (B(r))_{z+1}$$
$$(B(r))_{z+1} = \lfloor (z+1)r \rfloor - \lfloor zr \rfloor < (z+1)r - \lfloor zr \rfloor < (z+1)r - (zr-1) = r+1,$$

and, since $(B(r))_z$ is always an integer, we find the inclusion $(*)$. Now we show that, if r is not rational, then the sequence $(B(r))_z$ is not periodic, not even, when we restrict to $z \in \mathbb{N}$. Assume that this is not true, i.e., that there is $r \in \mathbb{R}_+ \setminus \mathbb{Q}$ and $i_0 \in \mathbb{N}$, such that $(B(r))_{i+i_0} = (B(r))_i$ for all $i \in \mathbb{N}$. Then

$$\begin{aligned}(A(r))_{2i_0} - (A(r))_{i_0} &= \sum_{j=1}^{i_0} (B(r))_{j+i_0} \\ &= \sum_{j=1}^{i_0} (B(r))_j = (A(r))_{i_0} - (A(r))_0 = (A(r))_{i_0}.\end{aligned}$$

Hence, $(A(r))_{2i_0} = 2(A(r))_{i_0}$. By induction, we find $\lfloor l i_0 r \rfloor = (A(r))_{li_0} = l(A(r))_{i_0} = l\lfloor i_0 r\rfloor$ for $l \in \mathbb{N}$. This is only possible if $i_0 r \in \mathbb{Z}$, i.e., $r \in \mathbb{Q}$, in contradiction to the assumption. □

We proceed to actually constructing an aperiodic tiling. In a first step, we allow infinite sets of colors and tiles (Construction 8.5.21), then we check coherence (in the sense of Definition 8.5.1). Later we check whether the tile set is actually finite (Example 8.5.24) and only allows aperiodic tilings of the plane (Proposition 8.5.25). However, before we start, we give an informal description of the central ideas developed by Kari [103, 104]. These very ideas have been used to construct the aperiodic tiling (this section) as well as to show that the domino problem is in general not decidable for $\mathbb{Z}^2$ (next section).

Idea of the Construction Consider a linear function $f : \mathbb{R} \to \mathbb{R}$. We intend to describe a tessellation $\phi : \mathbb{Z}^2 \to T$ that simulates an orbit $\{f^i(x)\}_{i\in\mathbb{N}}$. To simplify notation, we identify the faces with the four compass points (N, S, W, E) as indicated in Fig. 8.9; a point in $\mathbb{Z}^2$ is given by (z_x, z_y). Via the balanced word representation, the south-colors of the line z_y code $f^{z_y}(x) =: r_{z_y}$ and the north-colors $f^{z_y+1}(x) = f(r_{z_y})$:

$$\phi(z_x, z_y)_S = B(r_{z_y})_{z_x}, \qquad \phi(z_x, z_y)_N = B(f(r_{z_y}))_{z_x}.$$

Hence, for $n \to \infty$,

$$\frac{1}{2n+1}\sum_{z_x=-n}^{n} \phi(z_x, z_y)_S \to r_{z_y}, \qquad \frac{1}{2n+1}\sum_{z_x=-n}^{n} \phi(z_x, z_y)_N \to f(r_{z_y}).$$

These limits are *global* properties: We need to know a complete row of a tessellation to compute the limits. In order to define one tile *local* relations are required. As f is

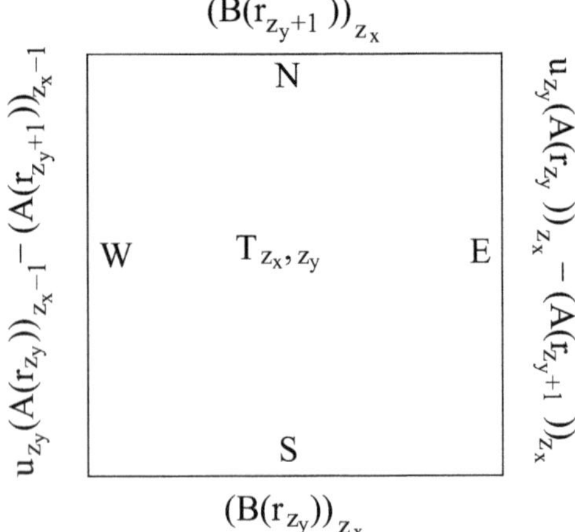

Fig. 8.9 The *colors* of the tile attached to the point $(z_x, z_y) \in \mathbb{Z}^2$

linear, we obtain

$$f\left(\frac{1}{2n+1}\sum_{z_x=-n}^{n}\phi(z_x,z_y)_S\right)=\frac{1}{2n+1}\sum_{z_x=-n}^{n}f(\phi(z_x,z_y)_S)\to f(r_{z_y}).$$

At this point, we have a local connection between south and north colors: in average, $\phi(z_x,z_y)_N \approx f(\phi(z_x,z_y)_S)$. The east and west face transport the information from the south face to the north face. Here, the second idea comes in: we require

$$\phi(z_x,z_y)_N = f(\phi(z_x,z_y)_S)+\phi(z_x,z_y)_W-\phi(z_x,z_y)_E.$$

If we do so, averaging over the north colors of a row can be replaced by averaging $f($ south colors), as the west- and east colors form a telescope sum and hence play no role. Seemingly we have one equation and two degrees of freedom (W and E color); however, as the east color of one tile is the west color of the neighboring tile in a coherent tessellation, there is only one effective degree of freedom, and the E/W color are more or less determined. We obtain

$$\begin{aligned}\phi(z_x,z_y)_W-\phi(z_x,z_y)_E &= \phi(z_x,z_y)_N - f(\phi(z_x,z_y)_S)\\ = B_{z_x}(f(r_{z_y})) - f(B_{z_x}(r_{z_y})) &= \left[f(A_{z_x-1}(r_{z_y}))-A_{z_x-1}(f(r_{z_y}))\right]\\ &\quad -\left[f(A_{z_x}(r_{z_y}))-A_{z_x}(f(r_{z_y}))\right].\end{aligned}$$

Symmetry suggests to choose $\phi(z_x,z_y)_W = f(A_{z_x-1}(r_{z_y})) - A_{z_x-1}(f(r_{z_y}))$ and $\phi(z_x,z_y)_E = f(A_{z_x}(r_{z_y})) - A_{z_x}(f(r_{z_y}))$.

Of course, in the following we just define a tile set according to these lines of reasoning. It is then necessary to prove that we only obtain a finite set of tiles, and that any tessellation possible possesses the required properties.

Construction 8.5.21 Let $r\in\mathbb{R}_+\setminus\mathbb{Q}$, and $(u_z)_{z\in\mathbb{Z}}$ with $u_z\in\mathbb{Q}_+\setminus\{0\}$ be a bi-sequence. Define the bi-sequence $(r_z)_{z\in\mathbb{Z}}$ by $r_0=r$ and $r_z=u_z\,r_{z-1}$.

Starting point for the construction of the tiling are the sequences $B(r_z)$. These numbers form a two-dimensional array of natural numbers $B(r_{z_1})_{z_2}$. We use these numbers as the colors in the north and the south of our tiles. We need to add the colors in the west and the east. Why we chose these colors the way we do becomes clear only below (see Theorem 8.5.25). For $(z_x,z_y)\in\mathbb{Z}^2$ we define the tile T_{z_x,z_y} by (with $N\equiv\sigma_x$, $S\equiv\sigma_x^{-1}$, $E\equiv\sigma_y$, $W\equiv\sigma_y^{-1}$)

$$\begin{aligned}(T_{z_x,z_y})_N &= (B(r_{z_y+1}))_{z_x}\\ (T_{z_x,z_y})_S &= (B(r_{z_y}))_{z_x}\\ (T_{z_x,z_y})_W &= u_{z_y}(A(r_{z_y}))_{z_x-1}-(A(r_{z_y+1}))_{z_x-1}\\ (T_{z_x,z_y})_E &= u_{z_y}(A(r_{z_y}))_{z_x}-(A(r_{z_y+1}))_{z_x}\end{aligned}$$

(see Fig. 8.9). We find at once that T_{z_x,z_y} yields a coherent tiling of the plane $\mathbb{Z}^2$. Below we will use the relation

$$(T_{z_x,z_y})_N = (T_{z_x,z_y})_W - (T_{z_x,z_y})_E + u_{z_y}(T_{z_x,z_y})_S \tag{†}$$

which follows from

$$\begin{aligned}
&(T_{z_x,z_y})_W - (T_{z_x,z_y})_E + u_{z_y}(T_{z_x,z_y})_S \\
&= [u_{z_y}(A(r_{z_y}))_{z_x-1} - (A(r_{z_y+1}))_{z_x-1}] \\
&\quad -[u_{z_y}(A(r_{z_y}))_{z_x} - (A(r_{z_y+1}))_{z_x}] + u_{z_y}[(B(r_{z_y}))_{z_x}] \\
&= -u_{z_y}(B(r_{z_y}))_{z_x} + (B(r_{z_y+1}))_{z_x} + u_{z_y}(B(r_{z_y}))_{z_x} \\
&= (T_{z_x,z_y})_N.
\end{aligned}$$

Since $B(r)$ is not periodic for r not rational, the tiling (with possibly an infinite tile set) constructed above is not periodic. At last we must answer two questions: Is the tile set in fact finite? Is there, in addition to the aperiodic tiling, also a periodic tiling?

Remark 8.5.22 Which conditions lead to a finite tile set? First of all, we have seen that $B(r) \subset \{\lfloor r \rfloor, \lfloor r \rfloor + 1\}$. Hence, if the bi-sequence $(r_z)_{z\in\mathbb{Z}}$ is bounded, also the number of colors in the north and the south of tile elements are bounded. Which values may the west and the east of our tile set assume?

Proposition 8.5.23 *Let $n, m \in \mathbb{N}$ and $u = n/m$ and*

$$\mathcal{S}(u) = \{u\lfloor r \rfloor - \lfloor u\, r \rfloor \,:\, r \in \mathbb{R}_+\}$$

Then

$$\mathcal{S}(u) \subset \left\{-\frac{n-1}{m}, -\frac{n-2}{m}, \dots, \frac{m-1}{m}\right\}.$$

Proof The inequality

$$u\lfloor r \rfloor - 1 \le u\,r - 1 < \lfloor u\,r \rfloor \le u\,r < u(\lfloor r \rfloor + 1)$$

implies the inequality

$$-u < u\lfloor r \rfloor - \lfloor u\,r \rfloor < 1.$$

Because the numbers $u\lfloor r \rfloor - \lfloor u\,r \rfloor$ are multiples of $1/m$, they are elements of $\mathcal{S}(u)$ whatever the value of $r \in \mathbb{R}_+$ is. □

This proposition and the definition of the tile set above indicate, that the colors in the west and the east of each tile are elements of a finite set, if $(u_z)_{z\in\mathbb{Z}}$ only assume a finite number of values in $\mathbb{Q}\setminus\{0\}$. All in all we find that the following two conditions ensure that the tile set is finite:

1. The bi-sequence $(r_z)_{z\in\mathbb{Z}}$ is bounded.
2. The bi-sequence $(u_z)_{z\in\mathbb{Z}}$ only assumes a finite number of values.

Example 8.5.24 Let $r_0 = \sqrt{2} \in [2/3, 2]$. Then define $r_{z+1} = f(r_z)$ for $z \geq 0$ where the function f is defined as

$$f : [2/3, 2] \to [2/3, 2], \qquad x \mapsto \begin{cases} 2x & \text{for } x \in [2/3, 1] \\ 2x/3 & \text{for } x \in (1, 2]. \end{cases}$$

For z negative, the idea is to iterate with f^{-1}. As f is not invertible, g takes over the role of f^{-1} by selecting a preimage; define $r_{z-1} = g(r_z)$, $z \leq 0$, where

$$g : [2/3, 2] \to [2/3, 2], \qquad x \mapsto \begin{cases} 3x/2 & \text{for } x \in [2/3, 4/3] \\ x/2 & \text{for } x \in (4/3, 2]. \end{cases}$$

Define a bi-sequence (u_z) by $u_z = r_{z+1}/r_z$. Due to the definition of f and g we have for $z \geq 0$ that $r_{z+1}/r_z = f(r_z)/r_z \in \{2/3, 2\}$, and for $z < 0$, $r_{z+1}/r_z = r_{z+1}/g(r_{z+1}) \in \{2/3, 2\}$. In any case, u_z only assumes values in $\{2/3, 2\}$.

We study the example in detail. The functions f, g that generate the sequences r_z and u_z are shown in Fig. 8.10. We have $r_0 = r = \sqrt{2}$. Then $u_0 = 2/3$ and

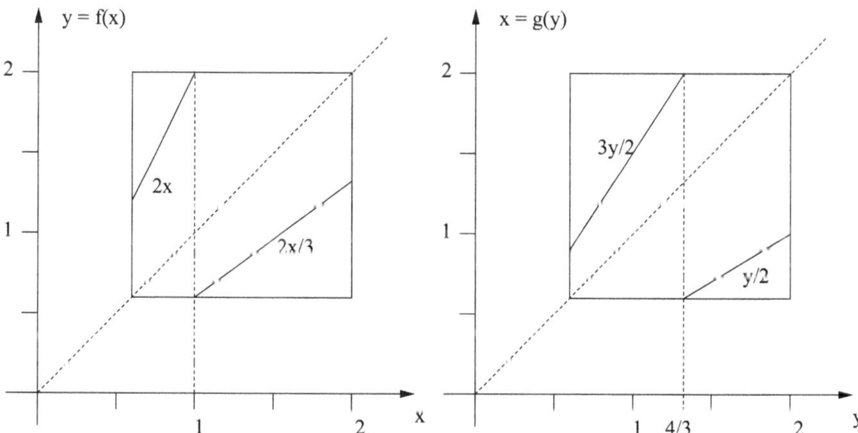

Fig. 8.10 *Left hand side*: the function f that generates the bi-sequence $\{r_z\}_{z\in\mathbb{N}_0}$. *Right hand side*: the function g that is used for the backward iteration, i.e., to generate $\{r_z\}_{z\in\mathbb{Z}}$, $z < 0$. In both figures, the invariant region $[1/2, 2]$ is indicated by a *box*, and the *line* $\{x = y\}$ is shown. The *dashed vertical line* indicates the location of the discontinuity of f, g, respectively

$r_1 = (2^1/3^1)\,\sqrt{2}$; further $u_1 = 2$ and $r_2 = (2^2/3^1)\,\sqrt{2}$ etc. Our construction yields two bi-sequences,

$$(r_z)_{z\in\mathbb{Z}}, \qquad (u_z)_{z\in\mathbb{Z}}$$

with the properties

$$r_z \notin \mathbb{Q}, \quad u_z \in \{2, 2/3\}, \quad r_{z+1} = u_z r_z, \;\; z \in \mathbb{Z}.$$

How does the set of tiles of the example look like? Since f, g leave the interval $[2/3, 2]$ invariant and $r_0 \in [2/3, 2]$, we have $r_z \in [2/3, 2]$ for all $z \in \mathbb{Z}$. Since $r_z \notin \mathbb{Q}$, we never obtain $r_z = 2$. Proposition 8.5.20 implies that $B(r) \subset \{\lfloor r\rfloor, \lfloor r\rfloor + 1\}$. According to Construction 8.5.21, the integers in the sequences $B(r_z)$ are used as colors in the south and the north, and hence these colors assume values in $\{0, 1, 2\}$.

We note that there will be no tile with color "0" in the south and the north: If the colors in south and north are 0, then $r_z, r_{z+1} \in (2/3, 1)$. This is not possible as the images of $(2/3, 1)$ under f and g are subset of $[1, 2]$; moreover, $B(r) \subset \{1, 2\}$ for $r \in (1, 2)$.

In order to obtain the colors in the west and in the east, we recall that these colors are given by

$$u_{z_y}(A(r_{z_y}))_{z_x} - (A(r_{z_y+1}))_{z_x} = u_{z_y}(A(r_{z_y}))_{z_x} - (A(u_{z_y}\, r_{z_y}))_{z_x} \in \mathcal{S}(u_z).$$

The set $\mathcal{S}(u)$ has been determined in Proposition 8.5.23. In the present case, $u_z \in \{2/3, 2\}$. Proposition 8.5.23 implies

$$\mathcal{S}(2) = \{-1, 0\} \quad \text{and} \quad \mathcal{S}(2/3) = \{-1/3, \hat{0}, 1/3, 2/3\},$$

where we use 0 and $\hat{0}$ to distinguish between $0 \in \mathcal{S}(2)$ and $\hat{0} \in \mathcal{S}(2/3)$.

By now we know all possible colors in north/south, resp. east/west. On valid tiles, only some combinations of colors appear. In order to determine all tiles, we draw two graphs. In the construction we have shown the formula (†). We use $u = 2$ for the first graph, and $u = 2/3$ for the second graph. Let—for a valid tile—the colors in north, south, west and east be a, b, q_1 and q_2, respectively. We draw a directed edge from q_1 to q_2 if

$$q_2 = q_1 + u\,b - a. \tag{+}$$

In this case we label the edge by a/b, where $a, b \in \{0, 1, 2\}$ and either $q_1, q_2 \in \mathcal{S}(2)$ (first part of the tile set constructed in the first graph) or $q_1, q_2 \in \mathcal{S}(2/3)$ (second part of the tile set constructed in the second graph), see Fig. 8.11. However, not all combinations of $a, b \in \{0, 1, 2\}$ are possible. We already have indicated that $a = b = 0$ never appears.

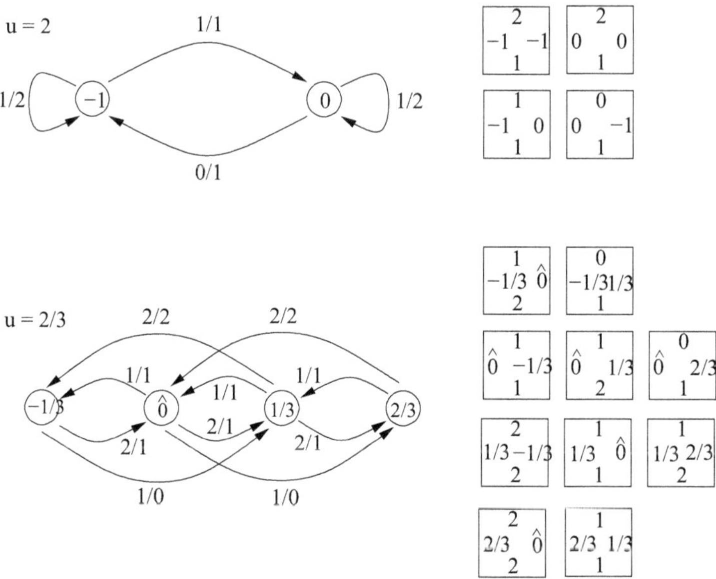

Fig. 8.11 The transition graph for our example. At the right hand side the corresponding tile sets are listed. Notice that the tile with color "0" at each side appears in the tile set of both graphs

For $z \geq 0$, $r_{z+1} \in (1,2)$, and $B(r_{z+1}) \subset \{1,2\}$. We draw an edge from the color at the west of a tile to the color at the east of the tile, and attach to this edge the possible colors at the north and the south. Every edge in any of the two graphs corresponds to exactly one tile from the tile set. Since the graph has 14 edges, the tile set we constructed consists of 14 tiles.

We know the tile set. Now we find out how the tiles are placed on the Cayley graph $\mathbb{Z}^2$. We construct the tiling of the plane in several steps. Each tile has the form T_{z_x,z_y}. We start with the sequences $(r_z)_{z\in\mathbb{Z}}$ and $(u_z)_{z\in\mathbb{Z}}$, as indicated in the first table.

z	-2	-1	0	1	2
r_z	$(3^1/2^2)\sqrt{2}$ $= 1.06067$	$(3^0/2^1)\sqrt{2}$ $= 0.70711$	$r = (2^0/3^0)\sqrt{2})$ $= 1.41421$	$(2^1/3^1)\sqrt{2})$ $= 0.94281$	$(2^2/3^1)\sqrt{2}$ $= 1.88562$
u_z	$u_{-2} = 2/3$	$u_{-1} = 2$	$u_0 = 2/3$	$u_1 = 2$	$u_2 = 2/3$

This table is one-dimensional. In order to obtain a two-dimensional object, we expand each entry r_z using the corresponding Beatty sequence and the balanced words. In the next table the row index is z as before (which we now call z_y (north-south) because of its role in the tile construction) and the column index is z_x (west-east). For each choice z_y, z_x the table shows the value for A and for B.

z_y	z_x	−5	−4	−3	−2	−1	0	1	2	3	4
−2	$A((3^1/2^2)\sqrt{2})$	−6	−5	−4	−3	−2	0	1	2	3	4
	$B((3^1/2^2)\sqrt{2})$		1	1	1	1	2	1	1	1	1
−1	$A((3^0/2^1)\sqrt{2})$	−4	−3	−3	−2	−1	0	0	1	2	2
	$B((3^0/2^1)\sqrt{2})$		1	0	1	1	1	0	1	1	0
0	$A((3^0/2^0)\sqrt{2})$	−8	−6	−5	−3	−2	0	1	2	4	5
	$B((3^0/2^0)\sqrt{2})$		2	1	2	1	2	1	1	2	1
1	$A((2^1/3^1)\sqrt{2})$	−5	−4	−3	−2	−1	0	0	1	2	3
	$B((2^1(3^1)\sqrt{2})$		1	1	1	1	1	0	1	1	1
2	$A((2^2/3^1)\sqrt{2})$	−10	−8	−6	−4	−2	0	1	3	5	7
	$B((2^2/3^1)\sqrt{2})$		2	2	2	2	2	1	2	2	2

The next table is an expanded version of the previous table. The values for A show up as before, the values for B are shifted to the right, and the numbers in each third row show the colors at the west and east according to the formulas in Construction 8.5.21, given by $u_{z_y}(A(r_{z_y}))_{z_x} - (A(u_{z_y}\, r_{z_y}))_{z_x}$. Hence the first complete tile shown in the table has a 1 at the north, a 0 at the south, and $-1/3$ at the west and $1/3$ at the east (boldface entries).

z_y	z_x	−4	−3		−2		−1		0		1		2		3		4	
−2	$A((3^1/2^2)\sqrt{2})$	−5		−4		−3		−2		0		1		2		3		4
	$B((3^1/2^2)\sqrt{2})$		**1**		1		1		2		1		1		1		1	
$u_{z_y} = \frac{2}{3}$		$-\frac{1}{3}$		$\frac{1}{3}$		0		$-\frac{1}{3}$		0		$\frac{2}{3}$		$\frac{1}{3}$		0		$\frac{2}{3}$
−1	$A((3^0/2^1)\sqrt{2})$	−3		−3		−2		−1		0		0		1		2		2
	$B((3^0/2^1)\sqrt{2})$		**0**		1		1		1		0		1		1		0	
$u_{z_y} = 2$		0		−1		−1		0		0		−1		0		0		−1
0	$A((3^0/2^0)\sqrt{2})$	−6		−5		−3		−2		0		1		2		4		5
	$B((3^0/2^0)\sqrt{2})$		1		2		1		2		1		1		2		1	
$u_{z_y} = \frac{2}{3}$		0		$-\frac{1}{3}$		0		$-\frac{1}{3}$		0		$\frac{2}{3}$		$\frac{1}{3}$		$\frac{2}{3}$		$\frac{1}{3}$
1	$A((2^1/3^1)\sqrt{2})$	−4		−3		−2		−1		0		0		1		2		3
	$B((2^1(3^1)\sqrt{2})$		1		1		1		1		0		1		1		1	
$u_{z_y} = 2$		0		0		0		0		0		−1		−1		−1		−1
2	$A((2^2/3^1)\sqrt{2})$	−8		−6		−4		−2		0		1		3		5		7
	$B((2^2/3^1)\sqrt{2})$		2		2		2		2		1		2		2		2	

Can we find a condition that ensures that there is no periodic tiling of the plane with a given tile set? In the present case we use the special structure of our tile set to exclude a periodic tiling: we know that a tessellation simulates an orbit $\{f^i(x)\}_{i\in\mathbb{N}}$. If this tessellation is periodic, then there is $n \in \mathbb{N}$ such that $f^n(x) = x$. We only need to exclude that any iterative of f possesses a fixed point to exclude a periodic tessellation.

Proposition 8.5.25 *The tile set constructed in Example 8.5.24 does not allow a periodic tiling.*

Proof We recall formula (†). Assume that there is a periodic tiling, i.e., there are $a, b \in \mathbb{N}$ such that $T_{z_x+a,z_y} = T_{z_x,z_y}$ and $T_{z_x,z_y+b} = T_{z_x,z_y}$ for all $(z_x, z_y) \in \mathbb{Z}^2$. Let $n_{z_y} = \sum_{i=1}^{a}(T_{z_x+i,z_y})_N$. Since the tiling is periodic, the colors at $(T_{z_x,z_y})_W$ and $(T_{z_x+a,z_y})_E$ are the same. Thus,

$$\begin{aligned} n_{z_y+1} &= \sum_{i=1}^{a}(T_{z_x+i,z_y+1})_N \\ &= \sum_{i=1}^{a}\big((T_{z_x+i,z_y+1})_W - (T_{z_x+i,z_y+1})_E + u_{z_y+1}(T_{z_x+i,z_y+1})_S\big) \\ &= u_{z_y+1}\sum_{i=1}^{a}(T_{z_x+i,z_y+1})_S \\ &= u_{z_y+1}\sum_{i=1}^{a}(T_{z_x+i,z_y})_N = u_{z_y+1}n_{z_y}. \end{aligned}$$

We iterate this formula b times and find, due to the assumed periodicity,

$$n_{z_y} = n_{z_y+b} = \Pi_{j=1}^{b}u_{z_y+j}\; n_{z_y}.$$

We exclude $n_{z_y} = 0$. The colors in south and north only assume values $\{0, 1, 2\}$. The condition $n_{z_y} = 0$ forces these colors to be always 0 (also in case $a = 1$ since in this case south and north colors have to be identical). As no tile has 0 in the south as well as in the north this is impossible. Hence

$$\Pi_{j=1}^{b}u_{z_y+j} = 1.$$

Since $u_i \in \{2, 2/3\}$, the expression $\Pi_{j=1}^{b}u_{z_y+j}$ assumes the form $2^\alpha\, 3^{-\beta}$, with $\alpha \in \mathbb{N}$, $\beta \in \mathbb{N}_0$, $\alpha \geq \beta$, and hence is not equal to 1. □

Thus, we have the following theorem.

Theorem 8.5.26 *There is a finite tile sets that allows for an aperiodic but not for a periodic coherent tiling of the plane.*

In general, the tiling constructed above will be different for different choices of the parameter r. It is possible to produce uncountably many aperiodic tilings of the plane that are essentially different.

8.5.4 Undecidability of the Domino Problem in $\mathbb{Z}^2$

As Wang's hypothesis proved to be wrong in $\mathbb{Z}^2$, the way is open to derive undecidability results. We first consider the domino problem, and show that it is undecidable. The first proof, due to Berger [11], is quite lengthy; it has been considerable shortened by Robinson [148]. However, recently a new proof, much shorter and simpler, has been published by Kari [104]. He uses similar ideas as in his proof for the existence of tile sets that allow to tessellate the plane exclusively in an aperiodic way, which we reviewed in the last pages. Before we start with the proof for the domino problem, we reformulate the immortality problem for Turing machines in terms of affine maps. In a second step, we connect the immortality problem for affine maps with the domino problem.

Definition 8.5.27 Consider a set of $m \in \mathbb{N}$ pairwise disjoint unit squares $U_i \subset \mathbb{R}^2$ and, associated with each U_i, an affine map $f_i : U_i \to \mathbb{R}^2$. The joint function is defined by

$$f : \cup_{i=1}^m U_i \to \mathbb{R}^2, \quad x \mapsto f_i(x) \text{ iff } x \in U_i.$$

We call a point $x \in \cup_{i=1}^m U_i$ mortal, if there is $k \in \mathbb{N}$, such that the kth iterate of x is not a member of the set $\cup_{i=1}^m U_i$,

$$f^k(x) \notin \cup_{i=1}^m U_i,$$

and immortal otherwise.

The immortality question asks if an affine map of certain type possesses an immortal point.

Problem 8.5.28 (The Immortality Problem for Affine Maps) *Consider a set of $m \in \mathbb{N}$ pairwise disjoint closed squares $U_i = z_i + [0,1]^2 \subset \mathbb{R}^2$, where $z_i \in \mathbb{Z}^2$. Associated with each U_i, an affine map $f_i : U_i \to \mathbb{R}^2$, $x \mapsto M_i x + b_i$, where $M_i \in \mathbb{Q}^{2\times 2}$ and $b_i \in \mathbb{Q}^2$. Decide, if the joint function allows for an immortal point.*

Note that the assumptions in this problem are strong enough to only allow for a countable number of joint functions, such that the decision problem is well defined.

In order to link the immortality problem for affine maps with that for Turing machines, we represent Turing machines as affine maps. This technique has been developed e.g. in [104, 108]. We briefly sketch the construction. Without restriction, we use integers $\mathfrak{A} = \{0, \ldots, M\}$ for the alphabet, and $Q = \{0, \ldots, K\}$ for the internal states. The number $0 \in Q$ corresponds to q^0, and $K \in Q$ to F. Furthermore, we do not assume that the head of the machine moves over the tape, but that the tape moves under the head, such that the head is always located at the origin (site zero).

The first step in the construction consists of coding the state of the tape and the internal state jointly in a two-dimensional vector with entries in $\mathbb{R}$. Let $C^{(n)} : \mathbb{Z} \to \mathfrak{A}$ be the state of the tape, and $q^{(n)}$ the internal state of the machine in time step n. Let

B an integer larger than $2\max\{M,K\}+1$. We define

$$u^{(n)} = 2\sum_{i=0}^{\infty} B^{-i+1}\, C^{(n)}(i), \qquad v^{(n)} = 2\,B\,q^{(n)} + 2\sum_{i=-\infty}^{-1} B^{i+1}\, C^{(n)}(i).$$

Obviously, $x^{(n)} = (u^{(n)}, v^{(n)})^T \in \mathbb{R}^2$ codes the state of the tape and the internal state at time n. We note three facts: first,

$$\lfloor x^{(n)} \rfloor := \begin{pmatrix} \lfloor u^{(n)} \rfloor \\ \lfloor v^{(n)} \rfloor \end{pmatrix} = \begin{pmatrix} 2\,B\,C^{(n)}(0) & +2\,C^{(n)}(1) \\ 2\,B\,q^{(n)} & +2\,C^{(n)}(-1) \end{pmatrix}$$

contains all information about the local state of the machine and the sign on the tape under the head, i.e., all information needed to evaluate the next move $\delta(q^{(n)}, C^{(n)}(0))$ of the machine. Second, the vector $\lfloor x^{(n)} \rfloor$ only assumes a finite number of values (less or equal B^2). There are a finite number of disjoint unit squares $U_1,\ldots,U_m$, that cover the coded state, $x^{(n)} \in U_1 \cup \cdots \cup U_m$. Third, not all points in U_i correspond to states of the Turing machine, due to ambiguities of the kind $0.999\cdots = 1.000\cdots$. We have chosen $B \geq 2\max\{M,K\}+1$, and multiplied $q^{(n)}$ as well as the entries of $C^{(n)}$ with a factor 2, and, in this way, exclude all such points: Consider the v-component of our construction above. A valid state of the Turing machine is mapped by $\sum_{i=0}^{\infty} B^{-i+1}\, C(i)$ into an interval $k + [0, 0.5]$ for some $k \in \mathbb{N}$. Another state is mapped to $\sum_{i=0}^{\infty} B^{-i+1}\, C'(i) \in k' + [0, 0.5]$. Hence, $2\sum_{i=0}^{\infty} B^{-i+1}\, C(i) \in 2k + [0,1]$ and $2\sum_{i=0}^{\infty} B^{-i+1}\, C'(i) \in 2k' + [0,1]$, such that these two unit intervals never overlap, if the states C and C' of the Turing machine are different.

Assume $x^{(n)} \in U_i$. We define a map $f_i : U_i \to \mathbb{R}^2$, such that $x^{(n+1)} = f_i(x^{(n)})$. The most important fact that allows to construct f_i is, that we can read off the state of the tape at the location of the head (i.e., location zero) $a_0^{(n)} \in \mathfrak{A}$, and the local state $q^{(n)}$ from the index i. Thus, we are able to compute the action of the Turing machine $\delta(a_0^{(n)}, q^{(n)}) = (\hat{q}, \hat{a}, D) \in Q \times \mathfrak{A} \times \{L, S, R\}$. This action is the same for all states of the Turing machine corresponding to points in U_i. Now we establish the vector $x^{(n+1)}$ in several steps. First, we remove the local state from $x^{(n)}$, and change $C^{(n)}(0)$ to $\hat{a}$,

$$\begin{aligned} x^{(n)} \mapsto y_1 &= x^{(n)} + \begin{pmatrix} 2\,B(\hat{a} - C^{(n)}(0)) \\ -2\,Bq^{(n)} \end{pmatrix} \\ &= 2\begin{pmatrix} B\hat{a} & +C^{(n)}(1) & +C^{(n)}(2)/B & +\cdots \\ B0 & +C^{(n)}(-1) & +C^{(n)}(-2)/B & +\cdots \end{pmatrix}. \end{aligned}$$

Note that $C^{(n)}(0)$ is fixed by the choice of U_{i_0}. The vector $\left(B(\hat{a} - C^{(n)}(0)),\ -Bq^{(n)}\right)^T$ looks as it were a function of $x^{(n)}$, since $C^{(n)}(0)$ and $q^{(n)}$ appear in its entries. However, this information is already contained in the choice of U_{i_0}. We can write down this vector once we know U_{i_0} before we know the exact value of x. If i_0 is given, this vector is constant in x. We will use this fact below.

Next, we shift the tape accordingly. This is the most subtle part in the construction. If the head stays ("S"), nothing is to do,

$$\begin{aligned} y_1 \mapsto y_2 &= y_1 + 0 \\ &= 2 \begin{pmatrix} B\hat{a} & +C^{(n)}(1) & +C^{(n)}(2)/B & +\cdots \\ B0 & +C^{(n)}(-1) & +C^{(n)}(-2)/B & +\cdots \end{pmatrix}. \end{aligned}$$

If the tape moves to the left (recall that we move the tape instead of the head), we remove $\hat{a}$ from the first component, place it in the second component, and shift both vectors,

$$\begin{aligned} y_1 \mapsto y_2 &= \begin{pmatrix} B & 0 \\ 0 & 1/B \end{pmatrix} (y_1 + 2\, B\hat{a} \begin{pmatrix} -1 \\ 1 \end{pmatrix}) \\ &= 2 \begin{pmatrix} BC^{(n)}(1) & +C^{(n)}(2) & +C^{(n)}(3)/B & +\cdots \\ B0 & +\hat{a} & +C^{(n)}(-1)/B & +\cdots \end{pmatrix}. \end{aligned}$$

Similarly, in case of a right shift, we define

$$\begin{aligned} y_1 \mapsto y_2 &= \begin{pmatrix} 1/B & 0 \\ 0 & B \end{pmatrix} y_1 + 2\, BC^{(n)}(-1) \begin{pmatrix} -1 \\ 1 \end{pmatrix} \\ &= 2 \begin{pmatrix} BC^{(n)}(-1) & +\hat{a} & +C^{(n)}(1)/B & +\cdots \\ B0 & +C^{(n)}(-1) & +C^{(n)}(-2)/B & +\cdots \end{pmatrix}. \end{aligned}$$

Last, we set the new local state,

$$x^{(n+1)} = y_2 + 2 \begin{pmatrix} 0 \\ B\hat{q} \end{pmatrix}.$$

All information we required for these transformations are coded in $\lfloor x^{(n)} \rfloor$. I.e., if we know that $x^{(n)} \in U_i = z_i + [0, 1]^2$, which allows us to find a matrix M_i and a vector $r_i \in \mathbb{R}^2$, such that

$$x^{(n+1)} = f_i(x^{(n)}) := M_i x^{(n)} + r_i.$$

We will use the properties of M_i and r_i below: (1) M_i as well as r_i have only rational entries, and (2) M_i are non-negative diagonal matrices.

All in all, we have constructed a disjoint, finite set of unit squares $U_1, \ldots, U_m$ in $\mathbb{R}^2$, and associated affine maps $f_1, \ldots, f_m$, such that the dynamics of the given Turing machine can be determined by the dynamics of the joint function

$$f : \cup_{i=1}^{m} U_i \to \mathbb{R}^2, \quad x \mapsto f_i(x) \text{ iff } x \in U_i.$$

Based on this construction, the immortality problem of Turing machines and that for affine maps can be linked.

Proposition 8.5.29 *Consider—for a given Turing machine—the constructed system of affine maps f_i on disjoint unit squares U_i, $i = 1, \ldots, m$. Remove all squares U_i that correspond to the internal state F.*

The Turing machine possesses an immortal state if and only if the system of affine maps possesses an immortal point.

Proof We already know that an immortal point in the Turing machine implies an immortal point in the system of affine maps. Now we look at the reverse direction: Let $x_0 \in U_{i_0}$ be the initial point of an immortal trajectory. We show that this immortal trajectory implies an immortal state of the Turing machine. The problem in this is, that this point x_0 does not necessarily represent a state of the Turing machine; if we reconstruct the state of the tape using expansion with respect to the basis B (if this expansion is not unique, take any), there may appear numbers larger M, i.e., signs that are not in the alphabet of the Turing machine. The reconstructed state of the tape is in general not valid. However, we know that the trajectory always stays in the union of U_i. As the signs the Turing machine ever reads from the tape are expressed within the integral part of the trajectory, they are valid. The local state of the machine is valid as well. Thus, we are allowed to replace all signs not in the alphabet by zero, say, and obtain a valid, immortal state of the Turing machine. □

Corollary 8.5.30 *The immortality problem for affine maps is in general undecidable.*

Now we connect the immortality problem for affine maps and tilings of $\mathbb{Z}^2$ similarly to the construction of an aperiodic tiling: the tiling simulates a linear function (the law for r_z, see Example 8.5.24).

Construction 8.5.31 Given a finite set of pairwise disjoint unit squares $U_1, \ldots, U_m$ in the plane, and corresponding affine functions $f_1, \ldots, f_m$, we construct a tile set that allows a tiling of $\mathbb{Z}^2$ if and only if the affine system of functions possesses an immortal orbit. The functions $f_i : U_i \to \mathbb{R}^2, x \mapsto M_i x + b_i$ have only rational entries in M_i and b_i, and M_i are non-negative diagonal matrices.

The definition of the colors is again based on the balanced representation, this time for vectors $x \in \mathbb{R}^2$. Using the floor function component wise, we define

$$A_k(x) = \lfloor kx \rfloor \in \mathbb{Z}^2, \qquad B_k(x) = A_k(x) - A_{k-1}(x) \in \mathbb{Z}^2, \qquad k \in \mathbb{Z}.$$

Associated with one single map $f_{i_0} : U_{i_0} \to \mathbb{R}^2$, $x \mapsto M_{i_0}x + b_{i_0}$, a vector $x \in U_{i_0}$, and an integer $k \in \mathbb{Z}$, we define tiles by

$$\begin{aligned}
(T_{x,k})_S &= B_k(x) \in \mathbb{Z}^2\\
(T_{x,k})_N &= B_k(f_{i_0}(x)) \in \mathbb{Z}^2\\
(T_{x,k})_W &= (i_0, f_{i_0}(A_{k-1}(x)) - A_{k-1}(f_{i_0}(x)) + (k-1)b_{i_0} \in \mathbb{Q}^3\\
(T_{x,k})_E &= (i_0, f_{i_0}(A_k(x)) - A_k(f_{i_0}(x)) + kb_{i_0} \in \mathbb{Q}^3.
\end{aligned}$$

The first component of the W and E color ensures that only tiles connected with U_i can be aligned horizontally. The deeper reason for this choice is the following formula, which parallels the formula (†) in Construction 8.5.21. In the following computation, we neglect the first component in the west- and east colors.

$$\begin{aligned}
& f_{i_0}((T_{x,k})_S) + (T_{x,k})_W\\
&= f_{i_0}(B_k(x)) + f_{i_0}(A_{k-1}(x)) - A_{k-1}(f_{i_0}(x)) + (k-1)b_{i_0}\\
&= f_{i_0}(A_k(x) - A_{k-1}(x)) + f_{i_0}(A_{k-1}(x)) - A_{k-1}(f_{i_0}(x)) + (k-1)b_{i_0}\\
&= f_{i_0}(A_k(x)) - f_{i_0}(A_{k-1}(x)) + f_{i_0}(A_{k-1}(x)) - A_{k-1}(f_{i_0}(x)) + kb_{i_0}\\
&= f_{i_0}(A_k(x)) - A_k(f_{i_0}(x)) + A_k(f_{i_0}(x)) - A_{k-1}(f_{i_0}(x)) + kb_{i_0}\\
&= f_{i_0}(A_k(x)) - A_k(f_{i_0}(x)) + B_k(f_{i_0}(x)) + kb_{i_0}\\
&= (T_{x,k})_E + (T_{x,k})_N
\end{aligned}$$

We next check for conditions such that this definition only yields a finite number of tiles. As $x \in U_{i_0}$ and $k \in \mathbb{Z}$ are arbitrary, this is by no means clear. According to Proposition 8.5.20, $B_k(x)$ and $B_k(f_{i_0}(x))$ only assume a finite number of values, as x as well as $f_{i_0}(x)$ are members of a compact set. Thus, in north and south we only find a finite number of colors. Concerning the colors in west and east, the first component for the color is the given and fixed number i_0. The second component of the color at the west side reads

$$\begin{aligned}
c_k &= f_{i_0}(A_k(x)) - A_k(f_{i_0}(x)) + kb_{i_0}\\
&= M_{i_0}A_k(x) + b_{i_0} - \lfloor M_{i_0}kx + kb_{i_0} \rfloor + kb_{i_0}\\
&= M_{i_0}\lfloor kx \rfloor - \lfloor M_{i_0}kx + kb_{i_0} \rfloor + (k+1)b_{i_0}
\end{aligned}$$

As the matrices M_{i_0} are diagonal matrices, we can decouple both coordinates. The entries of the matrices as well as the entries of the vectors b_i are rational. All in all, we obtain entries of the form

$$\alpha\lfloor kx \rfloor - \lfloor \alpha kx + k\beta \rfloor + (k+1)\beta$$

where $\alpha = n_1/m_1$, $\beta = n_2/m_2$ are rational numbers depending only on i_0. This expression can be rewritten as

$$\frac{n_1m_2\lfloor kx\rfloor - m_1m_2\lfloor n_1kx/m_1 + kn_2/m_2\rfloor + (k+1)n_2m_1}{m_1m_2}.$$

Now

$$\begin{aligned}
&-n_1m_2 - m_1m_2 + n_2m_2\\
&= n_1m_2(kx-1) - m_1m_2(n_1kx/m_1 + kn_2/m_2 + 1) + (k+1)n_2m_1\\
&\leq n_1m_2\lfloor kx\rfloor - m_1m_2\lfloor n_1kx/m_1 + kn_2/m_2\rfloor + (k+1)n_2m_1\\
&\leq n_1m_2(kx+1) - m_1m_2(n_1kx/m_1 + kn_2/m_2 - 1) + (k+1)n_2m_1\\
&= n_1m_2 + m_1m_2 + n_2m_2.
\end{aligned}$$

Thus, in case of diagonal matrices and rational coefficients of the functions f_i, the tile set defined above only consists of a finite number of tiles. Using this construction, we link the domino problem to the immortality problem for affine maps.

Theorem 8.5.32 *The tile set constructed above allows for a tiling of the plane if and only if the corresponding affine map possesses an immortal state.*

Proof

Step 1: Immortal orbit for the system of affine maps $\Rightarrow$ Coherent tiling of $\mathbb{Z}^2$. Assume the system of affine maps exhibits an immortal orbit starting at $x \in U_{i_0}$. Let the corresponding trajectory be $\{x_l\}$, with $x_l \in U_{i_l}$, and $x_{l+1} = f_{i_l}(x_l)$. Using the tile set defined above, we establish a coherent tiling of $\mathbb{Z}^2$. To start, we define for the upper half-plane a tessellation based on the given sequence, for the lower half-plane we just use a dummy tile,

$$\phi : \mathbb{Z}^2 \to T, \qquad (k,l) \mapsto \begin{cases} T_{x_l,k} & \text{if} \quad l \geq 0 \\ T_{x_0,0} & \text{otherwise.} \end{cases}$$

Due to our construction, the tiling of the upper half-plane is coherent, while that of the lower half-plane is not. However, if we consider the sequence of tilings created by repeatedly shifting the whole pattern downwards by one step, a larger and larger region around the origin of $\mathbb{Z}^2$ becomes a coherent tiling. If we equip the state space for tilings $T^{\mathbb{Z}^2}$ by the Cantor topology, we know that there is a converging subsequence; the limit of this subsequence is a coherent tiling of the plane.

Step 2: Coherent tiling of $\mathbb{Z}^2$ $\rightarrow$ Immortal orbit for the system of affine maps. We start with a coherent tiling $\phi : \mathbb{Z}^2 \to T$. From this tiling, we construct an immortal orbit. Let us consider one row of the tiling $\phi(k,l)$, l fixed and $l \in \mathbb{Z}$. We average over a finite number of tiles in this row, from site $\underline{k}$ to site $\overline{k}$, and use the relation $f_{i_l}((\phi(l,k)_S) + (\phi(l,k))_W = (\phi(l,k))_E + (\phi(l,k))_N$ derived above (again dismissing the first component in west- and east colors),

$$
\begin{aligned}
& f_{i_l}\left(\frac{1}{\overline{k}-\underline{\mathrm{k}}}\sum_{k=\underline{k}}^{\overline{k}}(\phi(l,k))_S\right)+\frac{1}{\overline{k}-\underline{\mathrm{k}}}(\phi(l,\underline{k}))_W \\
&= M_{i_l}\frac{1}{\overline{k}-\underline{\mathrm{k}}}\sum_{k=\underline{k}}^{\overline{k}}(\phi(l,k))_S + b_{i_l} + \frac{1}{\overline{k}-\underline{\mathrm{k}}}(\phi(l,\underline{k}))_W \\
&= \frac{1}{\overline{k}-\underline{\mathrm{k}}}\sum_{k=\underline{k}}^{\overline{k}}\left(M_{i_l}(\phi(l,k))_S + b_{i_l}\right) + \frac{1}{\overline{k}-\underline{\mathrm{k}}}(\phi(l,\underline{k}))_W \\
&= \frac{1}{\overline{k}-\underline{\mathrm{k}}}\sum_{k=\underline{k}}^{\overline{k}} f_{i_l}((\phi(l,k))_S) + \frac{1}{\overline{k}-\underline{\mathrm{k}}}(\phi(l,\underline{k}))_W \\
&= \frac{1}{\overline{k}-\underline{\mathrm{k}}}\sum_{k=\underline{k}}^{\overline{k}}\Big(-(\phi(l,k))_W + (\phi(l,k))_E + (\phi(l,k))_N\Big) + \frac{1}{\overline{k}-\underline{\mathrm{k}}}(\phi(l,\underline{k}))_W \\
&= \frac{1}{\overline{k}-\underline{\mathrm{k}}}\sum_{k=\underline{k}}^{\overline{k}}(\phi(l,k))_N + \frac{1}{\overline{k}-\underline{\mathrm{k}}}(\phi(l,\overline{k}))_E + \frac{1}{\overline{k}-\underline{\mathrm{k}}}\left(\sum_{k=\underline{k}}^{\overline{k}-1}(\phi(l,k))_E - (\phi(l,k+1))_W\right) \\
&= \frac{1}{\overline{k}-\underline{\mathrm{k}}}\sum_{k=\underline{k}}^{\overline{k}}(\phi(l,k))_N + \frac{1}{\overline{k}-\underline{\mathrm{k}}}(\phi(l,\overline{k}))_E.
\end{aligned}
$$

As $(\phi(l,k))_S = B_{k'}(x)$ for some $k' \in \mathbb{Z}$ and some $x \in U_{i_l} = z_{i_l} + [0,1]^2$ (where i_l is determined by the first component of the E/W colors), we have $(\phi(l,k))_S \in \{z_{i_l} + (a,b)\ a,b \in \{0,1\}\}$. We find a sequence k_m such that $k_m \to \infty$ monotonously, and

$$
\lim_{m\to\infty}\frac{1}{2k_m+1}\sum_{k=-k_m}^{k_m}(\phi(l,k))_S = x_0.
$$

Then, $x_0 \in U_{i_l} = z_{i_l} + [0,1]^2$. We proceed with our considerations, and apply f_{i_l} to x_0,

$$
\begin{aligned}
f_{i_l}(x_0) &= \lim_{m\to\infty} f_{i_l}\left(\frac{1}{2k_m+1}\sum_{k=-k_m}^{k_m}(\phi(l,k))_N\right) \\
&= \lim_{m\to\infty}\frac{1}{2k_m+1}\sum_{k=-k_m}^{k_m}(\phi(l,k))_N =: x_1.
\end{aligned}
$$

The colors in the south of row l are identical with the colors in the north in row $l+1$. Thus, using the very same averaging process, we obtain

$$\begin{aligned} f_{i_{l+1}}(x_1) &= f_{i_{l+1}}\left(\lim_{m\to\infty}\frac{1}{2k_m+1}\sum_{k=-k_m}^{k_m}(\phi(l,k))_S\right) \\ &= \lim_{m\to\infty} f_{i_{l+1}}\left(\frac{1}{2k_m+1}\sum_{k=-k_m}^{k_m}(\phi(l+1,k))_N\right) \\ &= \lim_{m\to\infty}\frac{1}{2k_m+1}\sum_{k=-k_m}^{k_m}(\phi(l+1,k))_S =: x_2. \end{aligned}$$

Proceeding recursively, we constructed the desired immortal orbit. □

Our tile set is only finite for affine linear maps with rational coefficients. However, as these maps are sufficient to decide the immortality problem for any Turing machine, we have the following corollary.

Corollary 8.5.33 *The domino problem is in general not decidable for a two-dimensional grid.*

8.5.5 Undecidability of the Finite Domino Problem in $\mathbb{Z}^2$

The completion and the domino problem are in general undecidable. With respect to cellular automata, we need information on the finite domino problem. We will prove that this problem is undecidable in general by considering the case $\Gamma = \mathbb{Z}^2$ with von Neumann neighborhood. The proof requires three steps. In the first step, the dynamics of a Turing machine is reformulated as a tiling. The Turing machine produces a time series of states on the tape. This is already a two-dimensional structure. In the construction of the Turing machine we define a set of tiles such that a completion problem on $\mathbb{Z}\times\mathbb{N}$ ($\mathbb{Z}$ for the tape and $\mathbb{N}$ for the time) corresponds to the actual simulation of the Turing machine.

If the Turing machine stops after a finite number of steps, only a finite number of sites on the tape have been changed until this time point. If we reset the states on the tape after stopping the machine, and remove the machine itself, then we obtain a finite set of non-blank sites on the tape in space and time if and only if the machine stops. We need a signal that transmits the message that the Turing machine stopped. This signal is transmitted horizontally (using specialized tiles), in the given time step at which the Turing machine did stop. It is transmitted to all sites that need this information, i.e., to all sites affected by the Turing machine. It is not clear how far the signal has to run. It must not run to infinity, as we aim at a finite tessellation. At this point, the second ingredient of construction comes in. The trace of the Turing machine can be framed by a finite rectangle if and only if the machine eventually stops. The second ingredient is a tile set that only allows for finite rectangles as the only finite coherent tessellations.

In the third and last step we combine the Turing-machine construction and the rectangle-construction to derive a set of tiles that allows for a finite nontrivial tessellation of $\mathbb{Z}^2$ if and only if the Turing machine eventually stops. The Turing machine sends a horizontal signal after it stops; this signal only reaches from one boundary of a rectangle to the other boundary. In this way we show that the halting problem can be reformulated as the finite domino problem. Therefore, the finite domino problem is not decidable.

We follow Durand [48] to prove that the finite domino problem is undecidable. Originally this theorem has been proved by Kari [101] by similar arguments.

(a) Coding the dynamics of a Turing machine by a tessellation

Consider a Turing machine $(Q, \mathfrak{A}, q, F, b, \delta)$. We define a completion problem on $\mathbb{Z}^2$ that codes the dynamics of this Turing machine. The horizontal direction corresponds to the tape, the vertical direction to time. We define the state at time zero as a tessellation of the horizontal line through the origin $\{(z, 0) \mid z \in \mathbb{Z}\}$. We are interested in an appropriate tessellation of the grid $\mathbb{Z} \times \mathbb{N} \subset \mathbb{Z}^2$ with von Neumann neighborhood. The colors of the tiles are tuples (a, q, X) with $a \in \mathfrak{A} \cup \{\emptyset\}$, $q \in Q \cup \{\emptyset\}$, $X \in \{L, R, S, \emptyset\}$.

In order to better understand the idea of this construction, let us consider a tile of a valid tessellation at position $(i, n) \in \mathbb{Z} \times \mathbb{N}$. The color in the south (bottom) indicates the state of the machine at time $n - 1$. It is used as a kind of input. The tile "computes" the new state in time n, and indicates this state in the "output' color at its north; the west and east colors are necessary to communicate between neighboring tiles in case the head of the Turing machine moves one step (see also Fig. 8.12). The position of the head is indicated by the second component of the "input" and "output" color: this component is only unequal $\emptyset$ if the head is located at the given site; if so, it assumes a value $q \in Q$ which is the local state of the machine. The first entry of the south (north) colors represents the sign on the tape at the location of the tile and the given time $n - 1$ (resp. n).

Construction 8.5.34 (Construction of a Turing Machine) Consider the following completion problem on $\mathbb{Z} \times \mathbb{N} \subset \mathbb{Z}^2$ with von Neumann neighborhood. The colors have three components (a, q, X) with $a \in \mathfrak{A}$, $q \in \{\emptyset\} \cup Q$ and $X \in \{L, R, S, \emptyset\}$.

(a) The first component of the color of a tile in position $(z, n) \in \mathbb{Z} \times \mathbb{N}$ indicates the state of the tape in location z at time step n (south) and time step $n + 1$ (north). The state and location of the Turing machine is indicated in the second and third components of the color (see Fig. 8.12). The blank tile carries $(b, \emptyset, \emptyset)$ at each face.

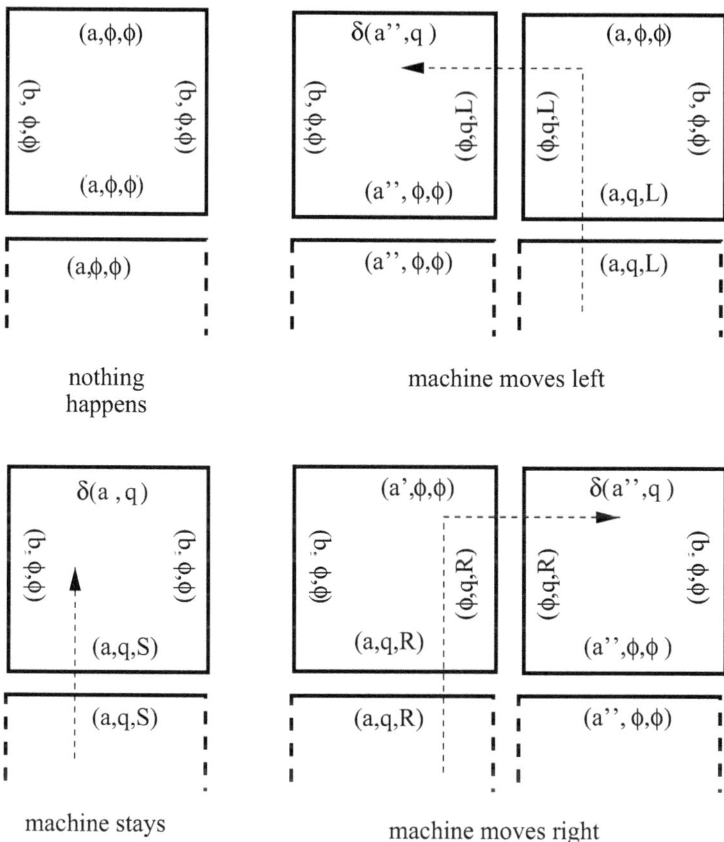

Fig. 8.12 Tiles for Construction 8.5.34

(b) If the Turing machine is not located at site z and not at the sites $z \pm 1$, then the state of the tape does not change in the next step and the machine will not visit this site. We define tiles

$$(t)_N = (t)_S = (a, \emptyset, \emptyset) \quad \text{with } a \in \mathfrak{A}, \qquad (t)_W = (t)_E = (b, \emptyset, \emptyset)$$

These tiles read the sign of the tape in the south and copy this sign to the north. In this way, the state of the type (away from the head of the Turing machine) is just copied from one row to the next row, that is, from one time step to the next.

(c) Assume that the Turing machine is located at the site $z \in \mathbb{Z}$ at time step n, has state q and reads the sign a, i.e., the south color of the tile is (a, q, S). We then transport the result of the Turing machine with these data $\delta(a, q)$ to the next row, while the neighboring sites are not changed, i.e.,

$$(t)_S = (a, q, S), \quad (t)_N = \delta(a, q), \quad (t)_W = (t)_E = (b, \emptyset, \emptyset).$$

(d) If the color in the south is (a, q, X) with $X \in \{L, R\}$, then the machine moves to the *right* or to the *left*. The actual tile is designed to pass the information to the tile at the west (or east) side. The appropriate neighbor (*left or right*) is informed that the Turing machine moves to the corresponding site. If the Turing machine shall move to the left we need the tile

$$(t)_S = (a, q, L), \quad (t)_W = (\emptyset, q, L), \quad (t)_E = (b, \emptyset, \emptyset), \quad (t)_N = (a, \emptyset, \emptyset),$$

and if it shall move to the right we use

$$(t)_S = (a, q, R), \quad (t)_W = (b, \emptyset, \emptyset), \quad (t)_E = (\emptyset, q, R), \quad (t)_N = (a, \emptyset, \emptyset).$$

(e) Finally, we require the counterparts of tiles constructed in (d) that receive the information. If the machine comes from the right (i.e., if it moves to the left), then the appropriate tile is

$$(t)_S = (a, \emptyset, \emptyset), \quad (t)_W = (b, \emptyset, \emptyset), \quad (t)_E = (\emptyset, q, L), \quad (t)_N = \delta(a, q),$$

and similarly, if the machine moves to the right,

$$(t)_S = (a, \emptyset, \emptyset), \quad (t)_W = (\emptyset, q, R), \quad (t)_E = (b, \emptyset, \emptyset), \quad (t)_N = \delta(a, q).$$

(f) In order to stop the machine after a finite number of steps, we introduce a special set of tiles that transport the information "machine did stop" over a complete line, and produces blank tiling from this step onward. We remove all tiles from the step (c)–(e) in this construction with colors at the south face that have an "F" in the state-component, i.e., all tiles which carry colors at the south face of the form $(*, F, *)$. Instead, we add the tiles

$$(t)_S = (a, F, X) \quad \text{with } a \in \mathfrak{A},\ X \in \{L, R, S\},$$

$$(t)_W = (\emptyset, F, \emptyset), \quad (t)_E = (\emptyset, F, \emptyset), \quad (t)_N = (b, \emptyset, \emptyset)$$

and

$$(t)_S = (a, \emptyset, \emptyset), \quad (t)_W = (\emptyset, F, \emptyset), \quad (t)_E = (\emptyset, F, \emptyset), \quad (t)_N = (b, \emptyset, \emptyset),$$

where $a \in \mathfrak{A}$. As a consequence, if the machine hits state F, the next line only consists of these tiles only, and after the next line all tiles are blank tiles forever.

Remark 8.5.35

(1) In order to simulate a Turing machine starting from a blank tape by tessellation of $\mathbb{Z}\times\mathbb{N}$ using the tile set constructed above, we place at row 1 only blank tiles

$$(t)_S=(b,\emptyset,\emptyset),\quad (t)_W=(b,\emptyset,\emptyset),\quad (t)_E=(b,\emptyset,\emptyset),\quad (t)_N=(b,\emptyset,\emptyset),$$

except for the tile at location $(0,1)\in\mathbb{Z}\times\mathbb{N}$, where we place the tile that represents the initial state of the Turing machine,

$$(t)_S=(\emptyset,\emptyset,\emptyset),\quad (t)_W=(\emptyset,\emptyset,\emptyset),\quad (t)_E=(\emptyset,\emptyset,\emptyset),\quad (t)_N=\delta(a_0,q_0).$$

The completion of this initial row is equivalent with the simulation of the Turing machine.

(2) If we remove in step (f) all tiles with a color $(*,F,*)$ in the south, then there is no valid tiling if and only if the machine eventually hits the acceptance state. Thus, the completion problem has a solution if and only if the Turing machine stops. Hence the completion problem is in general undecidable.

(b) Rectangle-Construction

The second ingredient in the proof of the undecidability of the finite domino problem is a construction of a tile set that forces nontrivial tilings to be rectangular. As an additional feature of this construction, one single site within a finite rectangle is tagged.

Construction 8.5.36 (Rectangle Construction) Consider $\mathbb{Z}^2$ with the von Neumann neighborhood. A tile set is defined as follows: There are three types of colors. The first type is the symbol $\{\emptyset\}$ indicating "blank". Next, there are eight "border colors"

$$\{N+,N-,E+,E-,S+,S-,W+,W-\}$$

which indicate that the tile is adjacent to a blank tile in an appropriate relative position. Finally there are colors X, Y indicating directions east-west and north-south. The set of colors is

$$C^{til}=\{\emptyset\}\cup\{N+,N-,E+,E-,S+,S-,W+,W-\}\cup\{X,Y\}.$$

There are 20 tiles as shown in Fig. 8.13. The blank tile is the tile with the color $\emptyset$ at every face. A tile that carries at least one border color is called a border tile. In particular, there are four corner tiles (carrying colors $N+$, $W+$ a.s.o.). The distinguished tile is the tile that carries X at the east and west face and Y at the north and south face.

Proposition 8.5.37 *A nontrivial, finite tiling of $\mathbb{Z}^2$ by the tile set given in Construction 8.5.36 consists of a disjoint union of rectangles bounded by border colors. In each of the rectangles, there is unique distinguished tile.*

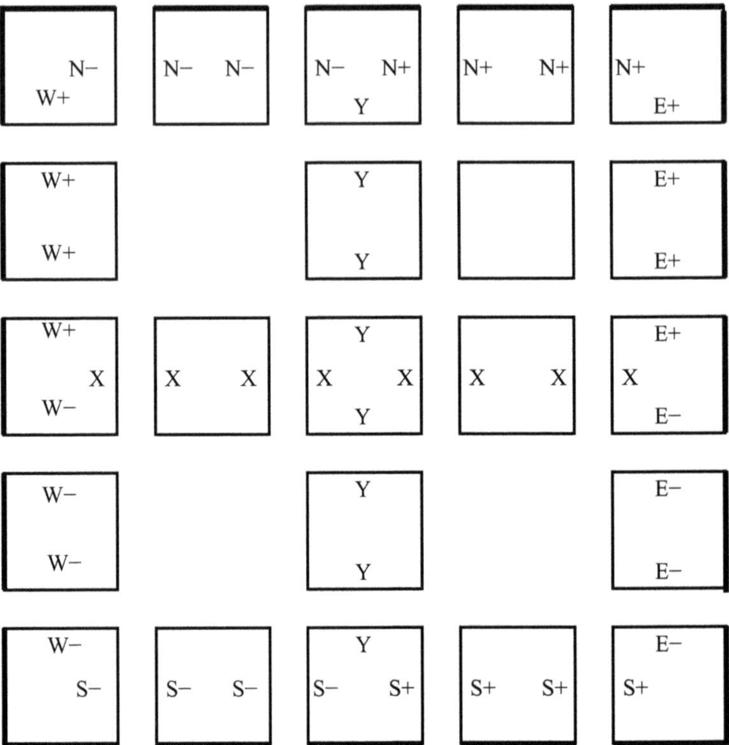

Fig. 8.13 Tiles for the Construction 8.5.36. The *fat lines* indicate the borders of the *bounding box*, they do not indicate *colors*. An edge at which no color is indicated carries the color Ø

Proof Assume we have a nontrivial finite tiling. At least one tile is not blank. Assume that it does not carry a border color. Then, at least one face carries X or Y. Suppose it is X. Then the tile carries X on the east and on the west face and its two neighbors carry also X. Hence the tile sits in a row of $X-X$ tiles which on both sides eventually arrives at blank tiles. But before that there are border tiles (tiles that carry a border color) on each side. Hence there are border tiles.

Any border tile leads to a row or column of border tiles which ends with a corner tile on either side. Thus, starting from one border tile, we get the boundary of a rectangle made up of border tiles. Since each of the four edges of a rectangle contains a single transition from $-$ to $+$, the length of the edge is at least three. The transition from N^- to N^+ selects a column, and similarly a row is selected. Where this distinguished row and column intersect, there (and only there within the rectangle) is the distinguished tile. Of course there may be several separate rectangles. □

A Cayley graph is completely symmetric, no cell is tagged in a natural way. We have one cell that corresponds to the neutral element of the group, but—using the graph homomorphism induced by the action of the group on the graph—we can move this element to any place we want.

The above construction produces a finite tiling that breaks this symmetry. This symmetry breaking allows to select one specific cell.

(c) Undecidability of the finite domino problem

Now we combine the last two constructions of tile sets in order to show that the finite domino problem is not decidable in general.

Construction 8.5.38 Consider the Turing machine $(Q, \mathfrak{A}, q^0, F, b, \delta)$. We construct a tiling that corresponds to this Turing machine. The colors have four components, the color set C is a subset of the product

$$C = C^* \times \mathfrak{A} \times (\{\emptyset\} \cup Q) \times \{\emptyset, L, R, S\}.$$

The first entry of the color is the "rectangle-component", the component that mimics the rectangle construction. The set C^* contains all tiles and all faces of the Rectangle-Construction. The entries two, three and four are the "TM-component" that are used by the simulation of the Turing machine. The blank tile has the color $(\emptyset, b, \emptyset, \emptyset)$ at all faces.

We always require that restriction to the first component of the colors (at the four faces) yields a valid tile of the Rectangle-Construction that carries a symbol from $\{X, Y, \emptyset\}$ at each face. Restriction to the last three components yields a valid tile of the Turing machine construction. Valid tiles are described as follows in (a), (b) and (c):

(a) The tiles that correspond to the initial state of the Turing machine,

$$(t)_S = (\emptyset, b, \emptyset, \emptyset),\ (t)_W = (X, b, \emptyset, \emptyset),\ (t)_E = (\emptyset, b, \emptyset, \emptyset),\ (t)_N = (X, b, \emptyset, \emptyset),$$

$$(t)_S = (Y, b, \emptyset, \emptyset), (t)_W = (t)_E = (X, b, \emptyset, \emptyset), (t)_N = (Y, \delta(a_0, q_0)).$$

(b) tiles that allow the line "F" to touch the west and east boundary of a rectangle (and to stop the Turing machine there):

$$(t)_S = (W+, b, \emptyset, \emptyset),\ (t)_W = (\emptyset, b, F, \emptyset),\ (t)_E = (\emptyset, b, \emptyset, \emptyset),\ (t)_N = (W+, b, \emptyset, \emptyset),$$

$$(t)_S = (E+, b, \emptyset, \emptyset),\ (t)_E = (\emptyset, b, F, \emptyset),\ (t)_W = (\emptyset, b, \emptyset, \emptyset),\ (t)_N = (E+, b, \emptyset, \emptyset),$$

$$(t)_S = (c, a, F, c'),\ (t)_E = (\emptyset, b, F, \emptyset),\ (t)_W = (\emptyset, b, F, \emptyset),\ (t)_N = (\emptyset, b, \emptyset, \emptyset),$$

where $a \in \mathfrak{A}$ and $c' \in \{L, R, S, \emptyset\}$.

(c) All tiles that are trivial in the TM-component, and any color in the rectangle-component, apart from the unique tile that carries the rectangle colors X and Y at the same time (the tagged time at which we intend to start a Turing machine).

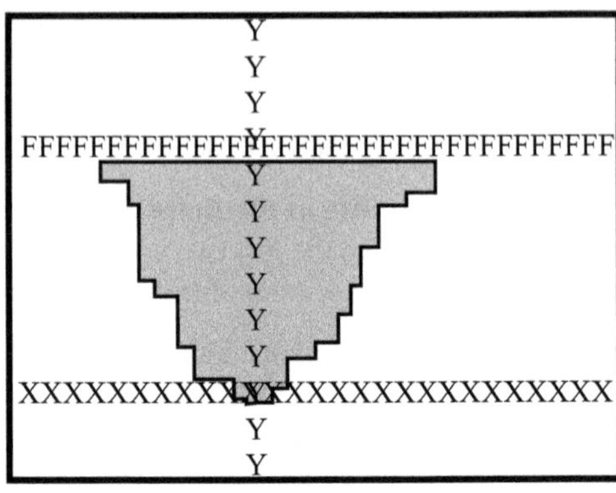

Fig. 8.14 Sketch of the Construction 8.5.38. The Turing machine starts to work at the intersection of the Y- and the X-line. The states that are changed by the Turing machine are indicated by *grey shading*. If the Turing machine stops, a line of F's resets the *colors to blank*. Only this line is allowed to touch the *left and right* border of the *outer rectangle*

Remark 8.5.39

(1) Assume that we have a nontrivial, finite tiling. Then the Construction 8.5.36 together with (b) and (c) shows that there is at least one rectangle in the first component of the colors.
(2) The parts (a) and (b) guarantee that a Turing machine starts at the unique interior point that the Rectangle-Construction selects by the intersection of the X and Y-colored lines. This Turing machine will change the states in the following rows.
(3) If the Turing machine never stops, tiles with nontrivial (non-white) colors extend to the north of the plane. If the Turing machine does not stop, there is no nontrivial finite tiling.
(4) If the Turing machine stops, the construction will lead to a row of tiles from the east to the west border of the rectangle with F in the state-component of the tiles in east and west direction (Fig. 8.14). The north direction of the Turing machine part of the color is blank. Thus, in this case there is a finite tiling of the plane. We have shown the following corollary.

Corollary 8.5.40 *The construction above allows for a nontrivial finite tiling if and only if the Turing machine eventually stops. Since the halting problem is not decidable, the finite domino problem is also not decidable.*

8.5.6 Group or Graph

By now we know that for some Cayley graphs the domino problem is decidable, for others not. It is a natural question to ask if this decidability is a property of the underlying group or of the special Cayley graph at hand.

Definition 8.5.41 The domino problem is called solvable for a Cayley graph $\Gamma(G, \sigma)$, if—for any given tile set—the domino problem is decidable.

Proposition 8.5.42 *Let $\sigma = \{\sigma_1, \ldots, \sigma_n\}$, $n \geq 2$, and $\tilde{\sigma} = \sigma \setminus \{\sigma_n\}$ be both generator sets for the same group G. If the domino problem is solvable for $\Gamma(G, \sigma)$, it is also solvable for $\Gamma(G, \tilde{\sigma})$.*

Proof Let $\tilde{T}$ be a tile set for the Cayley graph $\Gamma(G, \tilde{\sigma})$, i.e., $\tilde{t} \in \tilde{T}$ is a map

$$\tilde{t} : \{\sigma_1^{\pm 1}, \ldots, \sigma_{n-1}^{\pm 1}\} \to C$$

where C is the set of admissible colors. Let $c \in C$, arbitrary but fixed. We construct a tile set T for $\Gamma(G, \sigma)$ by the union of all tiles t

$$t : \{\sigma_1^{\pm 1}, \ldots, \sigma_n^{\pm 1}\} \to C, \quad t(\tau) = \begin{cases} \tilde{t}(\tau) & \text{if} \qquad \tau \in \{\sigma_1^{\pm 1}, \ldots, \sigma_{n-1}^{\pm 1}\} \\ c & \text{otherwise} \end{cases}.$$

The tile set T allows a tessellation of $\Gamma(G, \tilde{\sigma})$ if and only if $\tilde{T}$ allows a tessellation of $\Gamma(G, \tilde{\sigma})$. As the domino problem on $\Gamma(G, \sigma)$ is solvable, also that on $\Gamma(G, \tilde{\sigma})$ is solvable. □

Proposition 8.5.43 *Let $\sigma = \{\sigma_1, \ldots, \sigma_n\}$, $n \geq 2$, and $\tilde{\sigma} = \sigma \cup \{\theta\}$ with $\theta = \sigma_1\sigma_2$ be generator sets for a group G. If the domino problem is solvable for $\Gamma(G, \sigma)$, then the domino problem is also solvable for $\Gamma(G, \tilde{\sigma})$.*

Proof Let $\tilde{T}$ be a tile set for the Cayley graph $\Gamma(G, \tilde{\sigma})$, i.e., $\tilde{t} \in \tilde{T}$ is a map

$$\tilde{t} : \{\sigma_1^{\pm 1}, \ldots, \sigma_n^{\pm 1}, \theta^{\pm 1}\} \to C$$

where C is the set of admissible colors. We construct a tile set T that allows a tessellation of $\Gamma(G, \sigma)$ if any only if $\tilde{T}$ allows a tessellation of $\Gamma(G, \tilde{\sigma})$. We use for T the colors $C^* = C \times (C \cup \{\emptyset\})$. The idea is to replace the edge $\theta^{\pm 1}$ by a path of length 2 in the second component of the color (see Fig. 8.15). Accordingly, one tile in $\tilde{T}$ corresponds to a set of tiles in T, parametrized by $c \in C$, defined by

$$\begin{aligned} (t(\tau))_1 &= \tilde{t}(\tau) \quad \text{for } \tau \in \{\sigma_1^{\pm 1}, \ldots, \sigma_n^{\pm 1}\} \\ (t(\tau))_2 &= \emptyset \quad \text{for } \tau \in \{\sigma_3^{\pm 1}, \ldots, \sigma_n^{\pm 1}\} \\ (t(\sigma_1^{-1}))_2 &= (t(\sigma_2))_2 = c \\ (t(\sigma_1))_2 &= \tilde{t}(\theta) \\ (t(\sigma_2^{-1}))_2 &= \tilde{t}(\theta^{-1}) \end{aligned}$$

As in the last proposition, we are able to decide if $\tilde{T}$ allows a tessellation of $\Gamma(G, \tilde{\sigma})$ by testing whether T can produce a tessellation on $\Gamma(G, \sigma)$. □

Theorem 8.5.44 *Given a finitely generated group G, the domino problem is either solvable for all Cayley graphs over G or for none.*

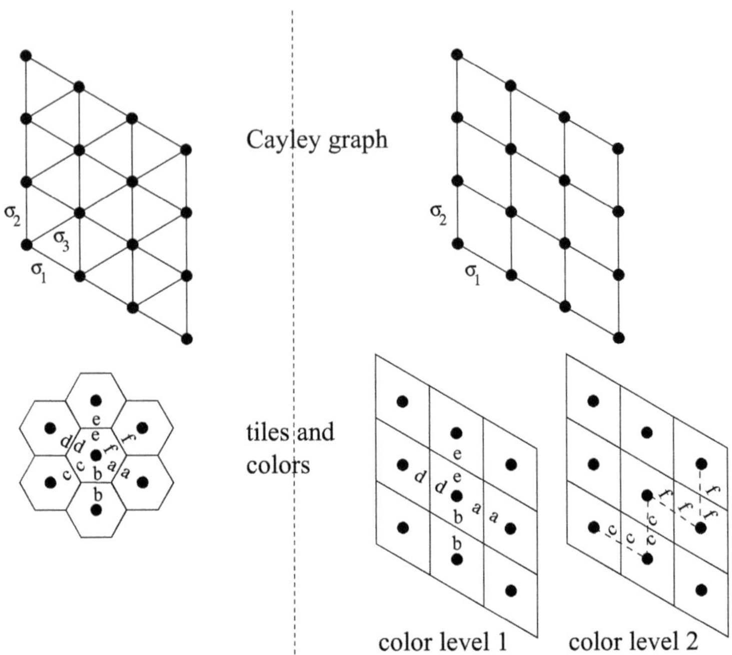

Fig. 8.15 Idea of the proof of Proposition 8.5.43. *Left*: Cayley graph $\Gamma(G, \tilde{\sigma})$. *Right*: Cayley graph $\Gamma(G, \sigma)$. The missing edge $\sigma_3 = \sigma_1 + \sigma_2$ on the *right hand side* is replaced by an additional level in the *color*, which transports the information between the tiles that are not direct neighbors any more

Proof Let $\sigma = \{\sigma_1, \ldots, \sigma_n\}$ be a generator set of G, and $\tilde{\sigma} = \sigma \cup \{\theta\}$ for some $\theta \in G$. This θ is a word $\theta = \tau_1\tau_2 \cdots \tau_m$, where each factor is one of the generators σ or its inverse. By means of Propositions 8.5.42 and 8.5.43 we find that solvability of the domino problem on $\Gamma(G, \sigma)$ is equivalent with solvability on $\Gamma(G, \tilde{\sigma})$:

"$\tilde{\sigma} \Rightarrow \sigma$": If $\Gamma(G, \tilde{\sigma})$ is solvable, then also $\Gamma(G, \sigma)$ (Proposition 8.5.42).
"$\sigma \Rightarrow \tilde{\sigma}$": If $\Gamma(G, \sigma)$, then also $\Gamma(G, \lambda_1)$ with $\lambda_1 = \sigma \cup \{\tau_1\tau_2\}$ (Proposition 8.5.43).

Therefore also the domino problem for $\lambda_2 = \lambda_1 \cup \{\tau_1\tau_2\tau_3\}$ (Proposition 8.5.43). Proceeding in the same way, we find that also the generator set

$$\{\sigma_1, \ldots, \sigma_n, \ \tau_1\tau_2, \ \tau_1\tau_2\tau_3, \ \tau_1\tau_2\tau_3\tau_4, \ \ldots, \ \tau_1\tau_2 \cdots \tau_m\}$$

corresponds to a Cayley graph for which the domino problem is solvable. Using Proposition 8.5.42 we are allowed to remove the superfluous elements in this set of generators one after the other, until we obtain $\tilde{\sigma}$. Hence, solvability of the domino problem for $\Gamma(G, \sigma)$ and $\Gamma(G, \tilde{\sigma})$ are equivalent.

In a similar way, we find that solvability for two arbitrary sets of generators σ and $\hat{\sigma}$ are equivalent. First add to σ one element after the other of $\hat{\sigma}$, then remove from $\sigma \cup \hat{\sigma}$ one element after the other of σ. In each step we only add or remove one element, and thus find that solvability of the domino problem is equivalent for all Cayley graphs that occur. □

This theorem says that solvability of the domino problem is a property of a group and not of a given Cayley graph over this group. The problem to characterize all groups for that the domino problem is solvable is still open.

8.5.7 Domino Problem and Monadic Second Order Logic

At the end of the present chapter, we briefly discuss an intriguing connection between the domino problem and monadic second order logic on graphs. Therefore, we first sketch informally the idea of second order logic, and reformulate the domino problem in this context. We then cite some relevant results and indicate their relation to the domino problem. We will not prove anything in this section. For a thorough introduction to logic we refer to the book by Enderton [52] or the book by Courcelle and Engelfriet [36]; a nice presentation of this topic in the present context can be found in the paper [28].

Zero Order Logic Zero order logic (or propositional logic) deals with elementary (atomic) statements about objects (propositions) and their relations. Examples of atomic statements may be "Bob fails the exam" or "Bob passes the exam". That is, we are able to assign "true" or "false" to an atomic statement. To proceed to a more formal level, we replace these atomic statements by variables P, Q etc., that assume the values "true" or "false". Next, we may draw elementary conclusions from our statements:

Bob fails the exam or Bob passes the exam
Bob does not fail the exam
Therefore: Bob passes the exam.

Using our variables (P replaces "Bob fails the exam" and Q replaces "Bob passes the exam") we may write

$$((P \vee Q) \wedge \neg P) \rightarrow Q.$$

That is, we may construct sentences (that may be true or false) using our variables, logical connectives (and $\wedge$, or $\vee$, not $\neg$, implies $\rightarrow$) and some technical signs as brackets or commas. As sentences have finite length, for given values (true/false) of the variables we can decide if the sentence is true/false. No problem is here.

First Order Logic At the next level, the first order logic, relations/predicates (membership in a set, $\in$) and quantifiers ("exists" $\exists$ and "for all" $\forall$) are added to the zero order logic. "Sokrates is a philosopher" or "philosophers are humans" or "Sokrates is green" specify properties of Sokrates or Philosophers (that may be true or false). More formal, $x \in P$ (often written as $P(x)$) indicates that object x has property P. If $x \in Q$ means that object x is green, a is the constant/object "Sokrates", then $a \in Q$ does mean "Sokrates is green" (which is false). If we define $x \in R$ as "x is philosopher", $x \in S$ as "x is human", we may rewrite "Sokrates is a philosopher" as $a \in R$ and "philosophers are humans" as

$$\forall x \in R : x \in S.$$

Note that in $\forall x \in R$ the quantifier only specifies an object variable, and no predicate. This is the characteristics of first order logic: We have predicates, and we may state formulas with quantifiers, but we are not allow to quantify about predicates. A typical statement about real numbers

$$\forall x, y \in \mathbb{R} : ((x < y) \rightarrow (\exists z \in \mathbb{R} : x < z < y))$$

is a first order statement that uses relations and quantifiers.

First Order Theory Almost always we are interested to investigate an object. This object is described by some elementary statements that are true for this object (by definition), the axioms. E.g., an undirected graph is simple to characterize: it is only a set Γ with a relation describing edges, $\mathrm{EDG} \subset \Gamma \times \Gamma$

$$\begin{aligned} &\forall x \in \Gamma : \neg((x, x) \in \mathrm{EDG}) \\ &\forall x, y \in \Gamma : ((x, y) \in \mathrm{EDG} \rightarrow (y, x) \in \mathrm{EDG}) \end{aligned}$$

Another example is a strict order $<$ over a set Ω. We may write $P = \{(x, y) \in \Omega^2 : x < y\}$, and find

$$\begin{aligned} &\forall x \in \Omega : \neg(x, x) \in P \\ &\forall x, y, z \in \Omega : ((x, y) \in P \wedge (y, z) \in P) \rightarrow (x, z) \in P \\ &\forall x, y \in \Omega : ((x, y) \in P \vee x = y \vee (y, x) \in P) \end{aligned}$$

Now we forget the definition of P, and take these three statements as axioms. We are able to investigate the consequences that superimpose the properties of an order relation. E.g., we may also ask if there is a minimal element:

$$\exists a \in \Omega : \forall x \in \Omega : (x, a) \notin P.$$

If we can decide this statement, we are able to decide if a given total order of a given set possesses a minimal element. A first-order theory, for that all sentences can be decided is called decidable.

Second Order Logic The second order logic, now, allows also to quantify about predicates: Let Ω be a set. A statement as

$$\forall P \subset \Omega \, \forall x \in \Omega : (x \in P \vee \neg x \in P)$$

is not possible in first order but in second order logic. If we restrict ourselves to quantify about sets, as in this example, we talk about monadic second order logic. The full second-order logic also allows to quantify about functions; however, we are only interested in the monadic version. E.g., we may formulate the Peano axioms for natural numbers. We have some set X, the en element $0 \in X$, and a successor-relation $s(x)$:

$$\forall x \in X : \neg(s(x) = 0)$$
$$\forall x \in X \exists y \in X : (x \neq 0) \rightarrow (x = s(y))$$
$$\forall x, y \in X : (s(x) = s(y)) \rightarrow (x = y)$$
$$\forall Y \subset X : [(0 \in Y) \wedge (\forall x \in X : (x \in Y \rightarrow s(x) \in Y))] \rightarrow (\forall y \in X : y \in Y)$$

Note that the last axiom, the induction postulate, can only be formulated in second order logic, as we quantify over subsets of X. We may state properties of the successor-function in form of monadic second order sentences about X and $s(x)$. Büchli [19] did show, that all these statements are decidable.

Domino Problem We now return to the domino problem. This problem can be expressed in monadic second order logic over the Cayley graph: Let $\Gamma = \Gamma(G)$ denote a Cayley graph with generators $\sigma_1, \ldots, \sigma_m \in G$. We first aim to formulate that two subsets $X, Y \subset \Gamma$ form a partition:

$$\forall x \in \Gamma : (x \in X \vee x \in Y) \wedge (\neg(x \in X \wedge x \in Y)).$$

It is straightforward to extend this formula to describe a partition into k different subsets $X_1, \ldots X_k$ if k is fixed. Let us abbreviate the corresponding statement (that is true if and only if we have a valid partition) by $\text{PART}(X_1, \ldots, X_k)$. In the domino problem, we are given a finite, non-empty tile set $T = \{t_1, \ldots t_k\}$. A consistent tiling assigns each node to a tile. That is, if we define $X_i \subset \Gamma$ as the set of nodes assigned to tile $t_i \in T$, $i = 1, \ldots, k$, then $(X_i)_{i=1,\ldots,k}$ forms a partition of Γ. Now we turn to the consistency of the tessellation. Let $\sigma \in \{\sigma_1^{\pm 1}, \ldots, \sigma_m^{\pm 1}\}$. The consistency of the tessellation implies that if $x \in X_i$, σx cannot belong (in general) to an arbitrary X_j, but only to those X_j that have the appropriate color at this face. We formulate this fact as a relation. Let

$$\text{Tile} = \{(i, \sigma, \ell) : i, \ell = 1, \ldots, k,\ \sigma \in \{\sigma_1^{\pm 1}, \ldots, \sigma_m^{\pm 1}\},\ t_i(\sigma) = t_\ell(\sigma^{-1})\}.$$

That is, we code the property of the tile set in this relation/set. We may check if a tiling (given by the partition $(X_i)_{i=1,\ldots,k}$) is consistent at node $x \in \Gamma$ by the statement

$$\bigwedge_{i=1}^{k} \bigwedge_{\sigma \in \{\sigma_1^{\pm 1}, \ldots, \sigma_m^{\pm 1}\}} [x \in X_i \rightarrow (\ \vee_{\ell=1}^{k}((i, \sigma, \ell) \in \text{Tile} \wedge \sigma x \in X_\ell)\ )].$$

Let us abbreviate this statement by $\text{CONSIST}(x, X_1, \ldots X_k)$. Then there is a consistent tiling of the Cayley graph if and only if the statement

$$\exists X_1, \ldots, X_k \subset \Gamma\ \forall x \in \Gamma : \text{Part}(X_1, \ldots X_k) \wedge \text{CONSIST}(x, X_1, \ldots X_k)$$

is true. The domino problem can be formulated in terms of the monadic second order theory over Cayley graphs.

Monadic Second Order Theory Over Cayley Graphs A Cayley graph can be related to a language: the set of all finite words in the generators of the underlying finitely generated group that can be reduced by the group action to the identity of the group is a language *WP* (*WP* stands for "Word Problem"). The complexity of the language *WP* is, of course, related to the algebraic structure of the group, and the complexity of the monadic second order theory over the Cayley graph of this group. With this relation it is possible to classify Cayley-graphs according to the complexity of the language *WP*. In a seminal paper, Muller and Shupp [130] did prove that the monadic second-order theory of a group with a context-free language *WP* (context-free group in short) is decidable. The converse is also true: if the monadic theory of a Cayley graph is decidable, then the group is context-free [113]. In particular, we find that the domino problem is decidable for context-free groups. By the way, Muller and Shupp also apply their theory to cellular automata and conclude that injectivity resp. surjectivity of a cellular automaton is decidable for context-free groups [130]. Later, Muller and Shupp [131] showed that a finitely generated group is context-free if and only if it is virtually free (has a free subgroup of finite index). That is, the result we derived above in constructing an explicit algorithm to decide if a tile set allows a coherent tessellation can be recovered by general statements about the monadic second order logic on Cayley graphs.

At the present time, the converse is not clear: If, in general, the monadic second order theory is not decidable for a given Cayley graph, it nevertheless could be the case that the domino-problem is special enough to allow for a decision [3]. There is no example contradicting the idea that it is necessary for a group to be context-free to allow for the decidability of the domino-problem, and there are only few results that show for some non-virtually free groups that the domino problem is indeed undecidable [3, 5, 11, 120].

Chapter 9
Surjectivity and Injectivity of Global Maps

If we encounter a function then we may ask whether it is continuous and next whether it is bijective, surjective or injective. With respect to cellular automata we have already dealt with continuity. This section is concerned with the second question. Also physicists may want to know whether the dynamical system defined by a cellular automaton is reversible. Physicists use cellular automata as simple models for microscopic processes. Hence there should be some interest to identify those cellular automata that can be reversed in time, i.e., which have a bijective global function [163].

We will discuss two classes of results: first of all, a fundamental theorem—the Garden of Eden theorem—tells, that for a certain class of grids injectivity and surjectivity are not independent. This finding is non-trivial since in general for functions on infinite sets it is not possible to conclude anything about surjectivity from injectivity or *vice versa*. Since the global function is defined by coupling many copies of the local function which has a finite domain and range, cellular automata are close enough to functions on finite sets to allow for a connection between surjectivity and injectivity. However, such connection is known only for Cayley graphs that are not too complex.

The second class of results deals with the question whether one can *decide* if a given cellular automaton is reversible and, if so, whether one can compute its inverse function. For this task one must distinguish between one-dimensional (Abelian) grids and grids in higher dimensions: these problems can be solved for $\Gamma = \mathbb{Z}$, but not for $\Gamma = \mathbb{Z}^d$, $d > 1$.

In contrast to the topological theory there is a direct influence of the grid. Topologically, all state spaces are metric Cantor spaces and homeomorphic to each other and therefore the special structure of the grid does not matter. But if we consider algebraic and combinatorial properties, then the choice of the grid does matter.

K.-P. Hadeler, J. Müller, *Cellular Automata: Analysis and Applications*,
Springer Monographs in Mathematics, DOI 10.1007/978-3-319-53043-7_9

9.1 The Garden of Eden

Let G be a finitely generated group, with neutral element $e \in G$. In this section, we will always assume $e \in D_0$ and $|G| = \infty$.

Definition 9.1.1 Let (Γ, E, D_0, f_0) be a cellular automaton, and let $\tilde{\Gamma} \subset \Gamma$ be a finite subgraph.

(a) Let $u \in E^{\tilde{\Gamma}}$ have the following property. If $v \in E^{\Gamma}$ contains u in the sense that there is a shift operator σ_g with

$$\sigma_g v|_{\tilde{\Gamma}} = u,$$

then v has no preimage. In this case u is called a Garden of Eden (GOE) pattern.

(b) Two patterns $u_1, u_2 \in E^{\tilde{\Gamma}}$ are called mutually erasable, if for all $\tilde{u}_1, \tilde{u}_2 \in E^{\Gamma}$ the implication

$$\tilde{u}_1|_{\tilde{\Gamma}} = u_1, \quad \tilde{u}_2|_{\tilde{\Gamma}} = u_2, \quad \tilde{u}_1|_{\Gamma \setminus \tilde{\Gamma}} = \tilde{u}_2|_{\Gamma \setminus \tilde{\Gamma}} \quad \Rightarrow \quad f(\tilde{u}_1) = f(\tilde{u}_2)$$

holds. We write for two mutually erasable patterns in short $u_1 \sim_{\tilde{\Gamma}} u_2$.

The relation $\sim_{\tilde{\Gamma}}$ is an equivalence relation, given a finite subgraph $\tilde{\Gamma}$. If $\tilde{\Gamma}$ carries two different patterns that are mutually erasable then there is an equivalence class that contains more than one pattern. In the following it is tacitly assumed (unless stated otherwise) that "two mutually erasable patterns" are "two different mutually erasable patterns".

Remark 9.1.2 Surjectivity does not allow GOE patterns, and injectivity does not allow mutually erasable patterns. The two concepts of GOE patterns and of mutually erasable patterns are local versions of non-surjectivity and non-injectivity.

For functions on a finite set, surjectivity and injectivity are not independent. One could conjecture that the functions of cellular automata have similar properties because the local functions map finite sets to finite sets. Moore and Myhill pursued this idea in their famous papers [128, 138]. We present results for cellular automata on Cayley graphs due to Machi and Mignosi [118]. Gottschalk [71] defined in 1973 "surjunctive groups" as those groups on which injectivity implies surjectivity. A complete, elementary characterization of surjunctive groups is not available by now, but several classes of groups are proven to be surjunctive groups: Abelian groups, residually finite groups or sophic groups (definitions and proves can be found in the book by Ceccherini-Silberstein and Coornaert [27]).

We need some further definitions (compare with Definition 6.2.3).

Definition 9.1.3

(a) Let (Γ, E, D_0, f_0) be a cellular automaton, and let $\tilde{\Gamma} \subset \Gamma$ be a subgraph. Define sets $\tilde{\Gamma}^- \subset \tilde{\Gamma} \subset \tilde{\Gamma}^+$ as[1]

$$\tilde{\Gamma}^+ = \bigcup_{g \in \tilde{\Gamma}} \sigma_g(D_0)$$

and

$$\tilde{\Gamma}^- = \bigcup_{g \in \tilde{\Gamma},\, \sigma_g(D_0) \subset \tilde{\Gamma}} \{g\}.$$

The difference

$$\partial\tilde{\Gamma} = \tilde{\Gamma} \setminus \tilde{\Gamma}^-$$

is called the boundary of $\tilde{\Gamma}$.

(b) Let $\Gamma_1, \Gamma_2 \subset \Gamma$. A mapping $\eta : \Gamma \to \Gamma$ with $\eta(\Gamma_1) \subset \Gamma_2$ is said to embed Γ_1 in Γ_2, if the following two conditions hold:

(1) the map η is bijective.
(2) the map η respects the neighborhood in the sense that $\eta(\sigma_g(D_0)) = \sigma_{\eta(g)}(D_0)$ holds for all $g \in \Gamma_1$.

(c) If $\eta_1, \ldots, \eta_m$ embed Γ_1 in Γ_2, and $\eta_i(\Gamma_1) \cap \eta_j(\Gamma_1) = \emptyset$ for $i \neq j$, we say that Γ_2 contains m copies of Γ_1.

(d) Let $u \in E^{\Gamma_1}$. The pattern $v \in E^{\Gamma_2}$ contains m copies of u, if there are m copies of Γ_1 in Γ_2 with embeddings $\eta_1, \ldots, \eta_m$ and

$$u(g) = v(\eta_i(g)) \quad \text{for} \quad g \in \Gamma_1, \quad \text{for} \quad i = 1, \ldots, m.$$

The following proposition follows immediately from the definition of Γ^-.

Proposition 9.1.4 *Let $u \in E^\Gamma$ and let $\tilde{\Gamma} \subset \Gamma$. Then $f(u)|_{\tilde{\Gamma}^-}$ depends only on $u|_{\tilde{\Gamma}}$ (and is independent of $u|_{\Gamma \setminus \tilde{\Gamma}}$).*

Proposition 9.1.5 *Let $\Gamma \subset \Gamma$. Suppose that E^{Γ^+} does not contain two mutually erasable patterns. Let $v_1, v_2 \in E^\Gamma$ with*

$$v_1|_{\tilde{\Gamma}^+ \setminus \tilde{\Gamma}^-} = v_2|_{\tilde{\Gamma}^+ \setminus \tilde{\Gamma}^-}, \quad v_1|_{\tilde{\Gamma}^-} \neq v_2|_{\tilde{\Gamma}^-}.$$

Then

$$f(v_1)|_{\tilde{\Gamma}} \neq f(v_2)|_{\tilde{\Gamma}}.$$

If $\tilde{\Gamma}$ is finite, then $|f(E^\Gamma)|_{\tilde{\Gamma}^-}| = |E|^{|\tilde{\Gamma}^-|}$.

[1] The sets $\tilde{\Gamma}^+$ and $\tilde{\Gamma}^-$ can be interpreted as the closure and the interior of $\tilde{\Gamma}$.

Proof Consider $w_1, w_2 \in E^\Gamma$, with

$$w_i|_{\tilde{\Gamma}^+} = v_i|_{\tilde{\Gamma}^+}, \quad i = 1, 2; \quad w_1|_{\Gamma\setminus\tilde{\Gamma}^+} = w_2|_{\Gamma\setminus\tilde{\Gamma}^+}.$$

Suppose $f(v_1)|_{\tilde{\Gamma}} = f(v_2)|_{\tilde{\Gamma}}$. Then $f(w_1) = f(w_2)$, and $v_i|_{\tilde{\Gamma}^+}$, $i = 1, 2$, are mutually erasable patterns, in contradiction to the assumption. □

The following lemma is at the center of the Garden of Eden theorems. By counting patterns it is possible to relate local injectivity and local surjectivity. If a cellular automaton would map E^{Γ_n} to E^{Γ_n} this equivalence would be immediately clear. However, a pattern on Γ_n determines the image only on Γ_n^-. The main difficulty will be the control of the boundary of a region $\partial\Gamma_n$ in the counting argument. We will find that this boundary should be small in comparison with Γ_n in order to guarantee the equivalence of local injectivity and local surjectivity. This, in turn, is the case for groups of non-exponential growth.

The following lemma expresses this line of reasoning in a certain inequality, and the remaining part of Sect. 9.1 is devoted to the question of which Cayley graphs satisfy this inequality.

Lemma 9.1.6 *Consider two finite subgraphs $\Gamma_1, \Gamma_2 \subset \Gamma$. Suppose that Γ_2 contains m copies of Γ_1 with embeddings $\eta_1, \ldots, \eta_m$. Let furthermore $a = |E|$, $l = |\Gamma_1|$, $b = |\Gamma_2|$, $c = |\Gamma_2 \setminus \Gamma_2^-| = |\partial\Gamma_2|$. Suppose the inequality*

$$a^{b-c} > (a^l - 1)^m \, a^{b-ml} \tag{*}$$

is satisfied. Then the following statements hold.

(1) *If E^{Γ_1} contains two mutually erasable patterns, then $E^{\Gamma_2^-}$ contains a GOE pattern.*
(2) *If E^{Γ_1} contains a GOE pattern, then $E^{\Gamma_2^+}$ contains two mutually erasable patterns.*

Proof The proof is based on counting patterns.

1) We assume that Γ_1 carries two mutually erasable patterns.

Step 1: Equivalence relation on E^{Γ_1}.
From the definition of $\sim_{\Gamma_1}$ it follows that E^{Γ_1} contains at most $a^l - 1$ equivalence classes.

Step 2: Equivalence relation on E^{Γ_2}.
Using $\sim_{\Gamma_1}$, we define an equivalence relation $\backsim_{\Gamma_2}$ on E^{Γ_2}: Given $u, v \in E^{\Gamma_2}$, define $\tilde{u}_i, \tilde{v}_i \in E^{\Gamma_1}$ by

$$\tilde{u}_i(g) = u(\eta_i(g)), \quad \tilde{v}_i(g) = v(\eta_i(g)), \quad i = 1, \ldots, m.$$

Then define $u \backsim_{\Gamma_2} v$ if

(i) $\tilde{u}_i \sim_{\Gamma_1} \tilde{v}_i$ for $i = 1, \ldots, m$.
(ii) $u(g) = v(g)$ for $g \notin \cup_{i=1}^{m} \eta_i(\Gamma_1)$.

The number of equivalence classes of $\backsim_{\Gamma_2}$ is at most

$$\underbrace{(a^l - 1)^m}_{\text{equivalence classes in } \eta_i(\Gamma_1)} \quad \times \quad \underbrace{a^{b-ml}}_{\text{possible combinations in } \Gamma_2 \backslash (\cup_{i=1}^{m} \eta_i(\Gamma_1))} .$$

If two patterns are in the same equivalence class with respect to $\tilde{\Gamma}_2$ then their images under f agree on Γ_2^-. Hence there are at most $(a^l - 1)^m\, a^{b-ml}$ different patterns in $f(E^\Gamma)|_{\Gamma_2^-}$.

Step 3: Existence of a GOE pattern.

The total number of patterns in $E^{\Gamma_2^-}$ is a^{b-c}. Since we assume $a^{b-c} > (a^l - 1)^m\, a^{b-ml}$, there is at least one pattern in $E^{\Gamma_2^-}$ that is not contained in $f(E^\Gamma)|_{\Gamma_2^-}$, i.e., is a GOE pattern.

2) Assume the contrary of (2), i.e., Γ_1 is the support of a GOE pattern but $E^{\Gamma_2^+}$ does not contain two mutually erasable patterns.

Since $E^{\Gamma_2^+}$ does not contain mutually erasable patterns, the image of $E^{\Gamma_2^+}$ contains at least

$$|E|^{|\Gamma_2^-|} = a^{b-c}$$

different patterns (see Proposition 9.1.5). Let h be the number of non-GOE patterns in E^{Γ_2} (which is simply the number of images in E^{Γ_2}). Then,

$$h = \left| f(E^\Gamma)|_{\Gamma_2} \right| \geq a^{b-c}.$$

On the other hand, the number of patterns in $E^{\Gamma_2^-}$ that do not contain any copy of the GOE pattern in E^{Γ_1} is at most

$$\underbrace{(a^l - 1)^m}_{\text{non-GOE patterns in } E^{\cup_{i=1}^{m} \eta_i(\Gamma_1)}} \quad \times \quad \underbrace{a^{b-ml}}_{\text{combinations in } \Gamma_2 \backslash (\cup_{i=1}^{m} \eta_i(\Gamma_1))} .$$

Thus,

$$h \leq (a^l - 1)^m\, a^{b-ml}.$$

Combining the two inequalities for h, we find $a^{b-c} \leq h \leq (a^l - 1)^m\, a^{b-ml}$. This inequality contradicts the assumption. Therefore $E^{\Gamma_2^+}$ contains at least two mutually erasable patterns.

□

Remark 9.1.7 The inequality (*) in Lemma 9.1.6 can be written in a convenient way taking the logarithm with basis a,

$$\begin{aligned} 1 &> \frac{b}{b-c} - \frac{ml}{b-c} + \frac{m}{b-c}\log_a(a^l-1) \\ &= \frac{b}{b-c}\left[1 - \frac{ml}{b}\left(1 - \frac{\log_a(a^l-1)}{\log_a(a^l)}\right)\right] \\ &= \frac{b}{b-c}\left[1 - \frac{ml}{b}\,\frac{\log_a(a^l) - \log_a(a^l-1)}{\log_a(a^l)}\right] \\ &= \frac{b}{b-c}\left[1 + \frac{ml}{b}\left(\frac{\log_a(1-a^{-l})}{\log_a(a^l)}\right)\right] \\ &= \frac{b}{b-c}\left[1 - \frac{m}{b}\;|\log_a(1-a^{-l})|\right]. \end{aligned}$$

The right hand side of this inequality can be split into two factors,

$$q = 1 - \frac{m}{b}\;|\log_a(1-a^{-l})|, \qquad r = \frac{b}{b-c}.$$

The aim is to find conditions such that q is strictly smaller than 1 and to choose subgraphs such that r becomes arbitrarily close to 1. If this is possible, then the inequality is satisfied, and the existence of GOE patterns is equivalent to the existence of mutually erasable patterns.

The variables q and r address slightly different things: $r = |\Gamma_2|/(|\Gamma_2| - |\partial\Gamma_2|)$ measures the relative size of the boundary of a subgraph in comparison to the size of the subgraph itself. This interpretation suggests that most likely balls $B_n(g) = \{h : d_c(g,h) \leq n\}$ minimize r. The number q addresses the question of how many copies of Γ_1 can be placed in Γ_2. The graph Γ_1 should be large (l large), and at the same time also the number of copies placed in Γ_2 should be large (m large). For Cayley graphs, we will see that it is possible to satisfy the condition $q < 1$. However, the condition that r should be small imposes restrictions on the grid.

We now return to the fundamental inequality of Lemma 9.1.6. The growth function is the appropriate tool to identify large classes of groups for which the inequality holds.

Proposition 9.1.8 *For given $k \in \mathbb{N}$, it is possible to cover a Cayley graph $\Gamma(G)$ by a countable number of balls with radius $2k$, such that the balls with the same centers and radius k do not intersect. Equivalently, there are $g_i \in G$, $i \in \mathbb{N}$, such that $d_c(g_i, g_j) > 2k$ for $i \neq j$, and for every $g \in G$ there is $i \in \mathbb{N} : d_c(g, g_i) \leq 2k$.*

Proof We show by induction over m that each Γ_m can be covered by balls with the desired property. Since the Γ_m exhaust Γ, the claim follows. For $m = 1$ the set Γ_m can be covered by a ball of radius $2k$ with center $g = e$. Suppose that the set Γ_m can be covered by balls with the desired property (with centers $g_1, \ldots, g_M$, where M

depends on m). Now proceed to $m+1$. Enumerate the finitely many points in Γ_{m+1} that are not yet covered. Choose the first of these as g_{M+1} and then the next. If it is already covered, neglect it. Otherwise choose it as the next g_i, a.s. □

Proposition 9.1.9 *Let Γ be a Cayley graph of an infinite, finitely generated group. For $k \leq n$ let $m = m(k, n)$ be the number of copies of Γ_k contained in Γ_n. Then*

$$\frac{m(k,n)}{|\Gamma_n|} \geq \frac{1}{\gamma(3k)\gamma(2k)} \quad \text{for } n > 3k.$$

Proof Assume that we have constructed the points g_i as in the proof of Proposition 9.1.8. The number of non-overlapping balls of radius k in Γ_n is not less than the number of points g_i contained in Γ_{n-k},

$$m \geq |\{g_i \,:\, g_i \in \Gamma_{n-k}\}|.$$

Furthermore, if $h \in \Gamma_{n-3k}$, then there is at least one point g_{i_0} at a distance not greater than $2k$. Therefore, this g_{i_0} is member of Γ_{n-k}. Hence,

$$\gamma(n-3k) = |\Gamma_{n-3k}| \leq |\{g_i \,:\, g_i \in \Gamma_{n-k}\}| \; \gamma(2\,k) \leq m(k,n)\; \gamma(2k).$$

Recall $\gamma(n_1+n_2) \leq \gamma(n_1)\,\gamma(n_2)$ (proof of Proposition 2.5.2). Since for $n > 3k$ we obtain $\gamma(n) = \gamma(n-3k+3k) \leq \gamma(n-3k)\gamma(3k)$, we find

$$m(k,n) \geq \frac{\gamma(n-3k)}{\gamma(2k)} \geq \frac{\gamma(n)}{\gamma(3k)\gamma(2k)}.$$

□

Now we show the main GOE theorem for Cayley graphs.

Theorem 9.1.10 *Let Γ be a Cayley graph of an infinite, finitely generated group that is not of exponential growth. Consider any cellular automaton on this Cayley graph. Then there exist GOE patterns if and only if there are mutually erasable patterns.*

Proof We start from the situation in Lemma 9.1.6 and Remark 9.1.7. For the subgraphs Γ_1 (small) and Γ_2 (large) we choose, as in Proposition 9.1.9, Γ_k and Γ_n with $3k < n$. Then, $a = |E|$, $b = \gamma(n)$, $l = \gamma(k)$ are known, and for m/b there is a bound from Proposition 9.1.9. We choose $\varepsilon > 0$ such that

$$\varepsilon < \frac{|\log_a(1-a^{-\gamma(k)})|}{\gamma(3k)\gamma(2k)}$$

and find

$$q = 1 - \frac{m(k,n)}{|\Gamma_n|} |\log_a(1 - a^{-\gamma(k)})| \leq 1 - \frac{1}{\gamma(3k)\gamma(2k)} |\log_a(1 - a^{-\gamma(k)})| < 1 - \varepsilon.$$

There is a (minimal) ν such that $D_0 \subset \Gamma_\nu$. For $n > \max\{3k, \nu\}$ we have $\Gamma_{n-\nu} \subset \Gamma_n^-$ and hence $\gamma(n) - |\partial\Gamma_n| \geq \gamma(n-\nu)$. Next we have

$$r < \frac{\gamma(n)}{\gamma(n) - |\partial\Gamma_n|} \leq \frac{\gamma(n)}{\gamma(n-\nu)}.$$

Now assume that the lim inf of the r.h.s. is larger than one. In this case, the group is of exponential growth. The lim inf is not less than 1. Thus, there is a sequence $n_\ell \to \infty$ such that

$$\lim_{\ell\to\infty} \frac{\gamma(n_\ell)}{\gamma(n_\ell) - |\partial\Gamma_{n_\ell}|} = 1.$$

Hence for large n_ℓ the factor q is close to 1 and qr is less than 1. As we only need a sequence of Γ_{n_ℓ} that exhaust Γ (the GOE-pattern as well as the mutually erasable pattern are local), this observation yields the theorem. □

Remark 9.1.11

(1) The essential point in the preceding proof are estimates for the ratio of $\partial\Gamma_n$ and Γ_n. The control of this ratio is loosely related to Følner sequences. Given a finitely generated group G, a Følner sequence is an exhausting sequence of finite subsets $F_i \subset G$ that are rather stable under the group action,

$$\forall g \in G: \quad \lim_{i\to\infty} \frac{|(g\,F_i)\Delta F_i|}{|F_i|} = 0,$$

where Δ denotes, as usual, the symmetric difference. A finitely generated group is amenable if and only if such a sequence exists. Amenability says that some kind of mean value on functions can be defined that is invariant under the group action. Along the lines of this observation, it is possible to show that amenable groups are surjunctive [27, Theorem 5.9.1]. Amenability, in turn, is connected with the Besicovitch topology as introduced before—we may say, that topology constructs the pseudo-distance as a kind of mean value for the difference of two global states. Results about the connection between surjunctivity and Besicovitch/Weyl topology can be e.g. found in the article of Capobianco [22].

(2) There are examples that show that the claim of the GOE theorem does not hold for exponentially growing groups (see [118]). It is known that a cellular automaton on an exponentially growing group may have mutually erasable patterns without showing GOE patterns. For the first time we find a property of cellular automata that depends essentially on the group. Until now, we mainly considered topological properties. These properties seem mainly to be independent on the dimension and complexity of the grid, as long as it is finitely generated.

The GOE theorem only targets finite patterns. How are infinite configurations (finite support) or states, related to finite patterns?

Proposition 9.1.12 *A cellular automaton is not surjective if and only if there are finite GOE patterns.*

Proof Surjectivity means that every state in E^Γ has a preimage. If there are finite GOE-patterns, then it is clear that they imply the existence of infinite GOE-pattern. Also the converse is true. Assume that $u \in E^\Gamma$ has no preimage, but each (finite) configuration on Γ_n has a (finite) preimage on Γ_n^+. Then, let $v_n \in E^\Gamma$ be such that the image under the global function f agrees with u on Γ_n,

$$f(v_n)|_{\Gamma_n} = u|_{\Gamma_n}.$$

There is a converging subsequence, and the limit v satisfies $f(v) = u$. Thus, u has a preimage in contradiction to the assumption. □

The following observation is a consequence of the definition of mutually erasable patterns.

Proposition 9.1.13 *A cellular automaton that exhibits mutually erasable patterns is not injective.*

Remark 9.1.14

(1) A cellular automaton that is not injective does not necessarily possess (finite) mutually erasable patterns. A simple example is an automaton with $\mathbb{Z}$ as a grid, $E = \{0, 1\}$, $D_0 = \{0, 1\}$ and $f_0(b_0, b_1) = (1-b_0)(1-b_1)+b_0b_1$. Then the state "all 1" and the state "all 0" are both mapped to the state "all 1". The automaton is not injective. Now assume that there are mutually erasable patterns and show that this assumption leads to a contradiction. Let $u_1, u_2 \in E^{\mathbb{Z}}$ be two states that disagree only at a finite number of cells, and $f(u_1) = f(u_2)$. Let $z \in \mathbb{Z}$ be the rightmost coordinate with $u_1(z) \neq u_2(z)$. Assume $u_1(z) = 0$, $u_2(z) = 1$ (the other case works similarly). If $u_1(z+1) = u_2(z+1) = 1$, then

$$f(u_1)(z) = f_0(0, 1) = 0 \neq f(u_2)(z) = f_0(1, 1) = 1.$$

This is not possible. The assumption $u_1(z+1) = u_2(z+1) = 0$ leads in a similar way to a contradiction. Hence this automaton has no mutually erasable patterns and it is not injective.

(2) Consider a cellular automaton with resting state b, and the subautomaton restricted to states with finite support. If the latter is not injective, then there are mutually erasable configurations: If there are two states $u_1, u_2 \in E^\Gamma$ with $|\{g \in \Gamma\ :\ u_1(g) \neq b \text{ or } u_2(g) \neq b\}| < \infty$, then there is a finite subgraph $\hat{\Gamma}$

such that outside of $\hat{\Gamma}$ the states u_1, u_2 are constant b,

$$u_1|_{\Gamma\setminus\hat{\Gamma}} = u_1|_{\Gamma\setminus\hat{\Gamma}} = b.$$

Thus, u_1 and u_2 are mutually erasable with $\tilde{\Gamma} = \hat{\Gamma}^+$.

(3) A cellular automaton that is not surjective has GOE patterns; on a Cayley graph that is not exponentially growing this implies that there are mutually erasable configurations, and hence the cellular automaton is not injective. Thus, we have found the following corollary.

Corollary 9.1.15 *Consider cellular automata on a graph that does not grow exponentially.*

(1) *If the cellular automaton is injective then it is also surjective and hence bijective.*

(2) *The restriction of a cellular automaton with resting state to the subautomaton of states with finite support is injective if and only if it is surjective. In this case injectivity, surjectivity and bijectivity are all equivalent.*

Remark 9.1.16 Consider a bijective cellular automaton. As the state space E^Γ is compact (with the Cantor metric) and the global function is bijective and continuous, the function is a homeomorphism, f^{-1} is continuous. Of course, f^{-1} is also shift invariant. The Curtis-Hedlund-Lyndon Theorem ensures that f^{-1} is also a global function of a cellular automaton.

9.2 Algorithms for One-Dimensional Cellular Automata

Algorithms to determine surjectivity or injectivity of cellular automata are constructed in an almost identical way. The centerpiece is a finite graph. The structure of this graph determines, whether the cellular automaton is surjective respectively injective or not. We will first discuss these ideas for a rather simple but relevant problem: we will determine if a cellular automaton has stationary states, and—if there are some—we will construct them. Afterward we discuss the algorithms to determine injectivity and surjectivity.

9.2.1 Stationary Points

Before we consider one-dimensional cellular automata, we first stay with cellular automata on general Cayley graphs. We show below that the domino problem and the existence problem for stationary points are equivalent. Therefore we cannot expect that stationary points can be determined on general Cayley graphs; for some cellular automata on $\mathbb{Z}^2$ it is not possible to decide whether there are stationary states. Then we turn to one dimensional cellular automata $(\mathbb{Z}, D_0, E, f_0)$. Based on the ideas developed to prove the equivalence of stationary points and tessellations, we present an explicit algorithm that allows to decide if there are stationary points and, in case, to determine them.

We connect the domino problem and the computation of stationary points of a cellular automaton by two constructions that show that both tasks are essentially identical. First we start with a domino problem and construct a cellular automaton that has the coherent tilings, coded as states of the cellular automaton, as stationary points. Afterward we reverse the direction, and show that we can code a cellular automaton by tiles such that coherent tilings correspond to stationary states.

Construction 9.2.1 Let $\Gamma = \Gamma(G, \sigma)$ be a Cayley graph, and T a tile set with colors C. We construct a cellular automaton: choose $\Gamma = \Gamma(G, \sigma)$, $D_0 = \{e\} \cup \{\tau \in G : \tau \in \sigma \text{ or } \tau^{-1} \in \sigma\}$. We define $E = T \times \{\pm 1\}$. The local states $e \in E$ and the local function have two components. The first component of the local function indicates a tile and is left unchanged, $(f_0(u))_1 = (u(e))_1$ for all $u \in E^{D_0}$. The second component indicates if the tessellation given by the first component is (locally) coherent: The local function checks if tiles at the given site $(u(e))_1$ fit with the neighboring tiles $(u(\tau))_1$, $\tau \in D_0 \setminus \{e\}$. If this is the case, i.e., if

$$(u(e))_1(\tau) = (u(\tau))_1(\tau^{-1}) \text{ for all } \tau \in D_0 \setminus \{e\}$$

then the second component is set to 1: $(f(u))_2 = 1$. Otherwise, the second component is toggled, $(f(u))_2 = 1 - (u)_2$.

For the given Cayley graph and tile set, we have constructed a cellular automaton such that tilings correspond to states and coherent tilings correspond to stationary states.

Construction 9.2.2 We are given a cellular automaton $(\Gamma(G, \sigma), D_0, E, f_0)$, and construct a tile set, such that a coherent tiling of (another) Cayley graph $\tilde{\Gamma}(G)$ is equivalent with a stationary state. We assume without loss of generality that $D_0 = \Gamma_{d_0}$ for some $d_0 \in \mathbb{N}$, in particular $\sigma \subset D_0$.

In order to construct the domino problem, we first construct a Cayley graph: Consider site $e \in G$. A neighborhood for the tiling problem is given by all points $\tau \in G$ with a neighborhood overlapping with D_0,

$$\Sigma = \{\tau \in G : D_0 \cap \sigma_\tau(D_0) \neq \emptyset\}.$$

The tiles are based on the elements of stationary states $\Lambda = \{u \in E^{D_0} : f_0(u)(e) = u(e)\}$. The aim is to tessellate $\Gamma(G)$ by elements in Λ. Therefore, the configurations in $u|_{\sigma_g(D_0)}$, $g \in G$, have to agree at the overlapping sites of their respective neighborhood. Accordingly, we define the color set

$$C = \cup_{\tau \in \Sigma \setminus \{e\}} E^{D_0 \cap \sigma_\tau(D_0)}.$$

Each element $u \in \Lambda$ is assigned to the tile t defined by

$$t : \Sigma \setminus \{e\} \to C, \quad t(\tau) = u|_{D_0 \cap \sigma_\tau(D_0)}.$$

If the tiles constructed above allow a coherent tiling of the graph, we have a stationary point: Assume we have a coherent tiling $g \mapsto t_g$. We may define a state $u \in E^\Gamma$ by $u(g) = t_g(e)$ (note that $t \in E^{D_0}$). We claim that $u|_{\sigma_g D_0} = t_g$ for all $g \in G$. Let $g' = \tau g \in \sigma_g D_0$. Then, $\tau \in \Sigma$ and $t_g(\tau) = t_{g'}(\tau^{-1})$. Now,

$$t_g(\tau) = t_g|_{D_0 \cap \sigma_\tau D_0}, \quad t_{g'}(\tau^{-1}) = t_{g'}|_{D_0 \cap \sigma_{\tau^{-1}} D_0}.$$

Hence,

$$u(g') = t_{g'}(e) = t_{\tau g}|_{D_0 \cap \sigma_{\tau^{-1}} D_0}(e) = t_g|_{D_0 \cap \sigma_\tau D_0}(\tau) = t_g(g').$$

Thus $u|_{\sigma_g D_0} \in \Lambda$ for all $g \in G$, and $f(u) = u$.

If we start with a stationary state $u \in E^\Gamma$, it is clear that $t_g = u|_{\sigma_g D_0}$ yields a coherent tessellation.

As a coherent tessellation of $\tilde{\Gamma}(G, \sigma)$ is in one-to-one correspondence with the stationary points of $(\Gamma(G), D_0, E, f_0)$, the two constructions yield the following theorem.

Theorem 9.2.3 *Given a Cayley graph $\Gamma(G)$. It is decidable if an arbitrary cellular automaton over this graph possesses stationary points if and only if the domino problem for G is solvable.*

Remark 9.2.4 It is decidable if a cellular automaton over a free group possesses stationary points or not.

The construction of the tile set in Construction 9.2.2 allows to use Proposition 8.5.10 to determine stationary points. The construction becomes particularly simple for $\Gamma = \mathbb{Z}$. If we review the idea behind Proposition 8.5.10 we find a finite graph that allows to determine the stationary states: Consider the cellular automaton $(\mathbb{Z}, D_0, E, f_0)$. Let $D_0 = [-d_0, d_0]$. The vertices of the graph are all possible local patterns in E^{D_0} that are restrictions of a stationary state, i.e., all $u \in E^{D_0}$ with

$$f_0(u) = u(0).$$

We draw a directed edge from the vertex (pattern) u_1 to the vertex (pattern) u_2, if u_2 can be the right neighbor of u_1 in the sense, that $u_1(i+1) = u_2(i)$ for $i = -d_0, \ldots, d_0 - 1$.

This design is also the basis for the algorithms to decide surjectivity and injectivity. The centerpiece is a finite graph. The vertices are (sets of) states of the neighborhood that fulfill certain properties locally. The edges of the graph allow to string together the local structures to a global state (see Fig. 9.1). At this point, the fact that we consider a one-dimensional automaton becomes essential.

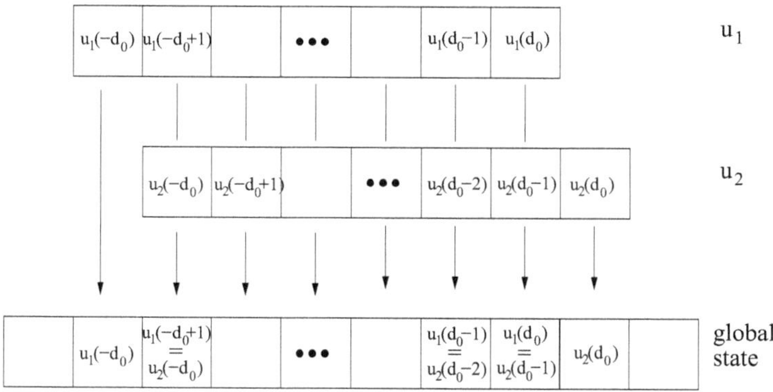

Fig. 9.1 Assembling neighboring vertices of the graph $(\mathcal{N}, \mathcal{E})$ to a part of a global state

Definition 9.2.5 Let $(\mathbb{Z}, D_0, E, f_0)$ be given with $D_0 = \Gamma_{d_0}$. Define a directed graph $(\mathcal{N}, \mathcal{E})$ as follows. The vertices $u \in \mathcal{N}$ are those elements $u \in E^{D_0}$ for which $f_0(u) = u(0)$. An edge runs from $u_1 \subset \mathcal{N}$ to $u_2 \subset \mathcal{N}$ if

$$u_1(i+1) = u_2(i), \quad i = -d_0, \ldots, d_0 - 1.$$

Let $\mathcal{P}$ be the set of all bi-infinite paths in the graph $(\mathcal{N}, \mathcal{E})$,

$$\mathcal{P} = \{p : \mathbb{Z} \to \mathcal{N} : \forall i \in \mathbb{Z} \text{ there is an edge from } p(i) \in \mathcal{N} \text{ to } p(i+1) \in \mathcal{N}\}.$$

Let $\mathcal{S}$ be the set of the stationary states of the cellular automaton,

$$\mathcal{S} = \{u \in E^{\mathbb{Z}} : f(u) = u\}.$$

Example 9.2.6 Let us illustrate the definition of $(\mathcal{N}, \mathcal{E})$ by the three Wolfram automata 51, 178, and 217. The local functions are given in Table 9.1.

For the rule $f_0^{(a)}$, we find $\mathcal{N} = \emptyset$: this automaton only toggles all colors. Therefore there is no pattern $(u_{-1}, u_0, u_1) \in E^3$ with $f_0^{(a)}(u_{-1}, u_0, u_1) = u_0$. The other two rules lead to a non-empty graph $(\mathcal{N}, \mathcal{E})$. The vertices of $\mathcal{N}$ can be read from Table 9.1: those patterns for which the value in the x-column is identical with the value of the local function. The graphs are shown in Fig. 9.2.

Any bi-infinite path $p \in \mathcal{P}$ corresponds to a stationary state (Fig. 9.1): defining $u(i) = p(i)(0)$, we obtain a state $u \in E^{\mathbb{Z}}$. Since every restriction of this state to a finite neighborhood is a member of $\mathcal{N}$, the state u is stationary. Also the other direction works. If we have a stationary state $u \subset E^{\mathbb{Z}}$, then $p : \mathbb{Z} \in E^{D_0}$, with $p(i) = u|_{i+D_0}$ is a bi-infinite path in $(\mathcal{N}, \mathcal{E})$, i.e., $p(i) \in \mathcal{N}$ and $p(i) \to p(i+1)$. We have the following corollary.

Table 9.1 Three local functions (Wolfram rules 51, 178, and 217) used in Example 9.2.6

x	y	z	$f_0^{(a)}(x,y,z)$	$f_0^{(b)}(x,y,z)$	$f_0^{(c)}(x,y,z)$
0	0	0	1	0	1
0	0	1	1	1	0
0	1	0	0	0	0
0	1	1	0	0	1
1	0	0	1	1	1
1	0	1	1	1	0
1	1	0	0	0	1
1	1	1	0	1	1

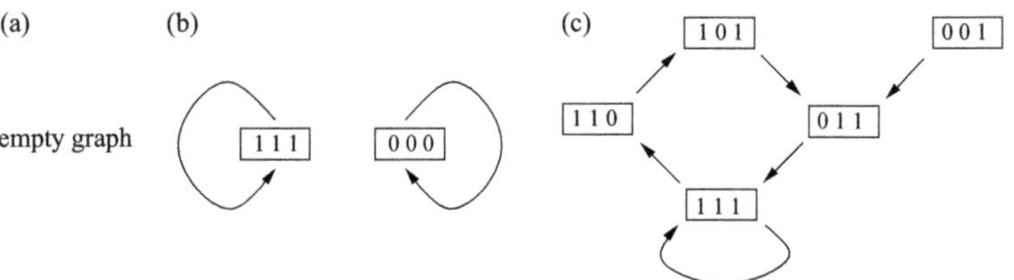

Fig. 9.2 The graph $(\mathcal{N}, \mathcal{E})$ for the Examples 9.2.6 (a), (b) and (c)

Corollary 9.2.7 *The map* $\mathcal{S} \to \mathcal{P}$, $\quad u \mapsto p$ *with*

$$p(i) = u|_{i-d_0,\cdots,i+d_0}, \qquad i \in \mathbb{Z}$$

is bijective.

Before we determine all stationary states in theses examples, we further discuss bi-infinite paths in a finite graph. These are closely related to periodic paths: As the graph is finite, the front- as well as the rear part of a bi-infinite path will loop through periodic orbits.

Definition 9.2.8 Let $p : \mathbb{Z} \to N$ be a periodic path with minimal period $k \in \mathbb{N}$, i.e., k is the smallest positive number with $p(i+k) = p(i)$ for all $i \in \mathbb{Z}$. This periodic path is called double point free, if $p(i) = p(i+j)$ for some j implies that k divides j, i.e., no vertex is visited twice during one period. Let PO the set of all double point free periodic paths in the graph $(\mathcal{N}, \mathcal{E})$.

Remark 9.2.9

(1) Since the paths in PO are double point free and the graph $(\mathcal{N}, \mathcal{E})$ is finite, PO is finite (or empty).

(2) $\mathcal{P} \neq \emptyset$ is equivalent with $PO \neq \emptyset$.

(3) In order to determine the stationary states of the cellular automaton, we need not only to know whether PO is empty or not, but also check whether different paths in PO have vertices in common.

Definition 9.2.10 Consider the directed graph $(\mathcal{N}, \mathcal{E})$. A periodic path $p \in PO$ is called isolated, if no other element in PO ever visits a vertex that is visited by p. Two different paths p_1, p_2 are connected by a one-way connection, if there is a path in $(\mathcal{N}, \mathcal{E})$ from a vertex in p_1 to a vertex in p_2, but no path from p_2 to p_1. We say that two different paths $p_1, p_2 \in PO$ are communicating, if we find a path from a vertex in p_1 to p_2 and back.

The proof of the following theorem is obvious.

Proposition 9.2.11 *Given a cellular automaton* $(\mathbb{Z}, D_0, E, f_0)$, *the set* $\mathcal{S}$ *of its stationary states satisfies:*

1. $\mathcal{S} = \emptyset$, *iff* $PO = \emptyset$.
2. $\mathcal{S} \neq \emptyset$ *and finite, iff* $PO \neq \emptyset$ *and all elements in* PO *are isolated.*
3. $\mathcal{S} \neq \emptyset$ *and countably infinite, iff* $PO \neq \emptyset$, *there is a one-way connection from one element in* PO *to another element, and no elements in* PO *are communicating.*
4. $\mathcal{S} \neq \emptyset$ *and uncountably infinite, iff* $PO \neq \emptyset$ *and there are communicating elements in* PO.

Example 9.2.6 (Continuation.) We can see from the graph $(\mathcal{N}, \mathcal{E})$ for $f_0^{(h)}$ that PO consists of two isolated periodic paths (case 2 of Proposition 9.2.11). This automaton has exactly two stationary states, both constant in space. In the third example, $f_0^{(c)}$, the set PO consists of two periodic paths

$$(111) \to (110) \to (101) \to (011) \to (111) \qquad \text{and} \qquad (111) \to (111)$$

and these have the vertex (111) in common, i.e., they are communicating (case 4 of Proposition 9.2.11). The automaton has uncountably many stationary states. A stationary state is characterized by the property: any two '0' are separated by a block of at least three '1'.

9.2.2 Surjectivity

We already know a necessary and sufficient criterion for a cellular automaton to be surjective: any finite pattern has exactly $|E|^{|D_0|-1}$ preimages (Proposition 6.2.8 resp. [85]). It is this property that is checked by the algorithm presented in this section. Following [47, 158], the algorithm is based on a graph, similar to the algorithm that allows to construct stationary points.

De Bruijn graphs are used to describe overlaps in finite sequences. Suppose we have m symbols $s_1, \dots, s_m$ and we consider sequences of length n. The vertices of the graph are the sequences of length n. A directed edge goes from $a = (a_1, \dots, a_n)$ to b if b can be represented as $b = (a_2, \dots, a_n, s)$ where s is any of the s_i. A de Bruijn graph is characterized by the parameters m (number of symbols) and n (length of words). We adapt this tool to cellular automata.

Definition 9.2.12 Consider the automaton $(\mathbb{Z}, D_0, E, f_0)$ where, without restriction of generality, $D_0 = \{-d_0, \ldots, d_0\}$. Let $D_1 = \{-d_0, \ldots, d_0 - 1\}$. The vertices of the de Bruijn graph $(\mathcal{N}, \mathcal{E})$ are the elements of $\mathcal{N} = E^{D_1}$. A directed edge goes from $u_1 \in \mathcal{N}$ to $u_2 \in \mathcal{N}$ if

$$u_1(i+1) = u_2(i), \qquad i = -d_0, \ldots, d_0 - 2.$$

Let $u_1 \to u_2$ be an edge. Define a local state

$$u = (u_1(-d_0), u_1(-d_0+1), \ldots, u_1(d_0-1), u_2(d_0-1)) \in E^{D_0}.$$

The edge $u_1 \to u_2$ is labeled by $f_0(u) \in E$. We call this labeled, directed graph the de Bruijn graph of the cellular automaton.

Consider a finite path p given by $v_1, \ldots, v_k$ in the de Bruin graph. Each edge $v_i \to v_{i+1}$ carries a label $e_i \in E$. We define the image of the path as the sequence $w = (e_1, \ldots, e_{k-1}) \in E^{k-1}$. This definition makes sense. At each edge, only one symbol is added. Hence the path is described by the array of length $2d_0 + k - 1$

$$(v_1(-d_0), \ldots, v_1(d_0-1), v_2(d_0-1), \ldots, v_k(d_0-1)).$$

If we apply the local rule to successive local states obtained by restriction to arrays of length $2d_0 + k + 1$ then we get the symbols in the word $w \in E^{k-1}$.

If we prescribe a sequence of labels w, and find a path that indeed has this sequence as labels, we construct the preimage of the word w. This observation allows to study surjectivity of a cellular automaton by studying the properties of the de Bruin graph.

Example 9.2.13 We show the de Bruin graph of Wolfram rule 90 in Fig. 9.3. The local function is given by

E^{D_0}	000	001	010	011	100	101	110	111
f_0	0	1	0	1	1	0	1	0

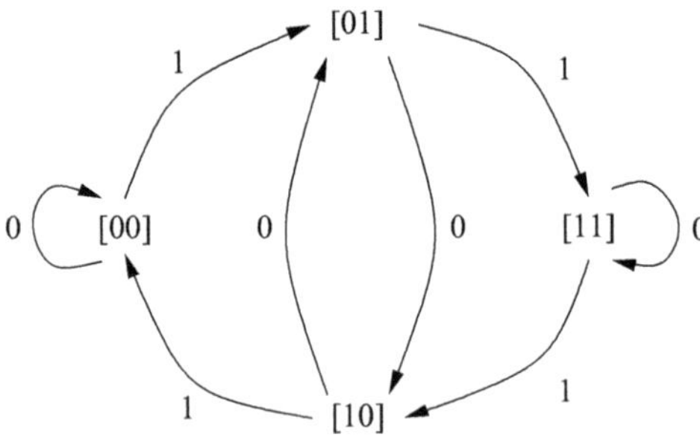

Fig. 9.3 The de Bruin graph for Wolfram rule 90

The vertices of the de Bruin graph for $E = \{0, 1\}$ and $D_0 = \{-1, 0, 1\}$ are members of E^2. There are four vertices, each vertex has two successors and two predecessors. There are eight edges corresponding to the eight elements of E^{D_0}. The edges carry the values of the local function. The de Bruijn graph is a complete characterization of the local function.

Definition 9.2.14

(a) Let w be a finite word with symbols from E. The number $\mu(w)$ is the number of paths in the de Bruijn graph $(\mathcal{N}, \mathcal{E})$ with image w.
(b) The de Bruijn graph is called balanced, if $\mu(w_1) = \mu(w_2)$ for all finite words (independently of their lengths).
(c) The de Bruijn graph is called unambiguous, if for any given start vertex, end vertex, and image, there is at most one path. Otherwise the graph is called ambiguous.

The present approach could also be formulated in the language of the theory of automata [158]. The terms "balanced" and "ambiguous" remind of that theory.

Theorem 9.2.15 *A cellular automaton $(\mathbb{Z}, D_0, E, f_0)$ with $D_0 = \Gamma_{d_0}$ is surjective if and only if the corresponding de Bruijn graph is balanced.*

Proof "surjective ⇒ balanced".
Let $n \in \mathbb{N}$ arbitrarily but fixed, and $w \in E^n$ be given. The number $\mu(w)$ counts the preimages, i.e., the number of paths with w as an image. This is also the number of patterns $u \in E^{n+2d_0}$ that are mapped by $f_{[1,n+2d_0]}$ (recall Definition 6.2.3) to the word w. In Corollary 6.2.9 we showed that for a surjective cellular automaton this number is always $|E|^{|D_0|-1}$. Hence the corresponding de Bruijn graph is balanced.

"balanced ⇒ surjective".

If μ is the same for all w, it cannot be zero. With this information, we can construct a preimage u for any state $v \in E^{\mathbb{Z}}$: Let $w_n = v|_{[-n,n]}$. We find (finite) preimages $u_n \in E^{[-n-d_0,n+d_0]}$ of the (finite) words w_n. Extend u_n to states $\tilde{u}_n \in E^{\mathbb{Z}}$ that coincide with u_n on $[-n - d_0, n + d_0]$ and are defined arbitrarily outside this region. In the Cantor metric, there is a converging subsequence $\tilde{u}_{n_i} \to u$. As on any finite interval $[-a, b] \subset \mathbb{Z}$ the state u equals $\tilde{u}_{n_i}$ for i sufficiently large, the image of u coincides with v on $[-a + d_0, b - d_0]$. Since a, b are arbitrary, $f(u) = v$. The automaton is surjective. □

Remark 9.2.16 The proof of this theorem tells in particular that $\mu(w) = |E|^{|D_0|-1}$ for a balanced de Bruin graph.

Theorem 9.2.17 *A cellular automaton $(\mathbb{Z}, D_0, E, f_0)$ with $D_0 = \Gamma_{d_0}$ is surjective if and only if the corresponding de Bruijn graph is unambiguous.*

Before we prove this theorem, we prove a technical lemma.

Lemma 9.2.18 *Let $v \in E^m$ be a pattern, and let $\eta(l)$, for $l > m$, be the number of pattern $u \in E^l$ that contain v somewhere, i.e., there is $i \in \mathbb{N}$ with $u(i+j) = v(j)$, $j = 1, \dots, m$. Then,*

$$\eta(l) \geq |E|^{l-m} + \sum_{i=1}^{m} (|E|-1)\eta(l-i),$$

and the asymptotic behavior of $\eta(l)$ is given by

$$\lim_{l\to\infty} \eta(l)/|E|^l = 1.$$

Proof If $|E| = 1$ then we have $\eta(l) = 1$, and the result is trivial. Let $k := |E| > 1$.

Step 1: Proof of the inequality.
Let M_l be the set of all patterns $u = (u_1, \dots, u_l) \in E^l$ that contain v. We split M_l into several disjoint sets. Define $M_{l,1} = \{u \in M_l : u_1 \neq v_1\}$, and, for $i = 2, \dots, m$

$$M_{l,i} = \{u \in M_l \,:\, u_1 = v_1, \dots, u_{i-1} = v_{i-1}, u_i \neq v_i\}$$

and for $i = m+1$,

$$M_{l,m+1} = \{u \in M_l \,:\, u_1 = v_1, \dots, u_{m-1} = v_{m-1}, u_m = v_m\}.$$

Then, $M_{l,i} \cap M_{l,i'} = \emptyset$ for $i \neq i'$. The set M_l is the union of the disjoint sets $M_{l,1}, \dots, M_{l,m+1}$.
First we estimate the number of patterns in $M_{l,i}$ for $i \leq m$. The first $i-1$ symbols agree with v, and the symbol at location i disagrees with v. The patterns in $M_{l,i}$ have the form

$$(v_1, \dots, v_{i-1}, e, w) \in M_{l,i},$$

where $e \in E \setminus \{v_i\}$. Thus,

$$|M_{l,i}| \geq (k-1)\eta(l-i).$$

Next we find $|M_{l,m+1}| = k^{l-m}$, as in this set v appears at position 1, and from position $m+1$ any pattern is allowed. All in all we find

$$\eta(l) = \sum_{i=1}^{m+1} |M_{l,i}| \geq k^{l-m} + \sum_{i=1}^{m} (k-1)\eta(l-i).$$

Step 2: Convergence of an auxiliary sequence $\eta(i)$.
We divide both sides of the inequality by k^l,

$$\eta(l)/k^l \geq k^{-m} + \sum_{i=1}^{m} \frac{(k-1)}{k^i} \, (\eta(l-i)/k^{l-i}),$$

define another sequence $\zeta(l)$ recursively,

$$\zeta(l) = k^{-m} + \sum_{i=1}^{m} \frac{(k-1)}{k^i} \, \zeta(l-i), \qquad \zeta(i) = \eta(i)/k^i \text{ for } i = 1, \ldots, m$$

and see that $\eta(l)$ is a minorant for $\eta(l)/k^l$,

$$\zeta(l) \leq \eta(l)/k^l \leq 1.$$

The remaining effort is to show that this sequence converges. We write the recursion as a first order vector recursion for the variable $Z_l = (\zeta(l), \ldots, \zeta(l-m))^T$,

$$Z_l = AZ_{l-1} + b$$

where the matrix A is a Frobenius companion matrix

$$A = \begin{pmatrix} (k-1)/k & (k-1)/k^2 & (k-1)/k^3 & \cdots & (k-1)/k^{m-1} & (k-1)/k^m \\ 1 & 0 & 0 & \cdots & 0 & 0 \\ 0 & 1 & 0 & \cdots & 0 & 0 \\ \vdots & \vdots & \vdots & & & \\ 0 & 0 & 0 & \cdots & 1 & 0 \end{pmatrix}$$

and the vector b is

$$b = \begin{pmatrix} k^{-m} \\ 0 \\ \vdots \\ 0 \end{pmatrix}.$$

The row sums of the matrix A are 1 and (finite geometric sequence)

$$(k-1) \sum_{i=1}^{m} \frac{1}{k^i} = \frac{(k-1)(1-(1/k)^m)}{k(1-1/k)} < 1.$$

Since the matrix is non-negative and irreducible, it follows that $\rho(A) < 1$, and hence

$$Z_l = (I - A)^{-1}(I - A^l)b + A^l Z_0.$$

Again, using $\rho(A) < 1$, we find

$$Z_l \to (I - A)^{-1}b.$$

Step 3: $\eta(l)/k^l \to 1$.
The limit of the recursion $\theta = \lim_{l\to\infty} \zeta(l)$ is a fixed point of the equation

$$\theta = k^{-m} + \sum_{i=1}^{m} \frac{(k-1)}{k^i}\theta = k^{-m} + \theta(1 - k^{-m}).$$

The only solution is $\theta = 1$. As $\zeta(l) \leq \eta(l)/k^l \leq 1$, the result follows. □

Proof (of Theorem 9.2.17) "⇒" We show that a surjective cellular automaton leads to an unambiguous de Bruijn graph. Assume that the graph is ambiguous, i.e., there is a word $w \in E^n$ with length $n \in \mathbb{N}$ that is the image of two different paths with identical start and end vertex,

$$p_0 \stackrel{w_1}{\to} p_1 \stackrel{w_2}{\to} p_2 \stackrel{w_3}{\to} \cdots p_{n-1} \stackrel{w_n}{\to} p_n$$

and

$$q_0 = p_0 \stackrel{w_1}{\to} q_1 \stackrel{w_2}{\to} q_2 \stackrel{w_3}{\to} \cdots q_{n-1} \stackrel{w_n}{\to} q_n = p_n.$$

Consider the first path. We extend this path such that it ends where it started from, namely with vertex p_0. In order to construct this extension, first note that—according to Definition 9.2.12—the path corresponds to a pattern $u \in E^{n+|D_0|-1}$, with $f_{[1,n+|D_0|-1]}(u) = w$. Now, append the first $|D_0| - 1$ entries of u at the end,

$$\tilde{u} = (u_1, \ldots, u_n, u_1, \ldots, u_{|D_0|-1}).$$

This pattern corresponds to a longer path $\tilde{p}_0, \ldots, \tilde{p}_m$, where $\tilde{p}_i = p_i$ for $i = 1, \ldots, n$, and $\tilde{p}_m = \tilde{p}_0$. Similarly, we extend the other path, obtaining $\tilde{q}_1, \ldots, \tilde{q}_m$ where $\tilde{q}_i = q_i$ for $i = 1, \ldots, n$, for $i = 1, \ldots, n$, and $\tilde{q}_i = \tilde{p}_i$ for $i = n+1, \ldots, n+|D_0|-1$. In particular, $\tilde{q}_1 = \tilde{q}_m = p_0$. Note that the images of the two extended paths are the same: the images of the first n nodes are w, and the remaining part of the two paths are identical and have thus identical images.

All in all, we obtain a configuration $\tilde{w}$, and two paths that both start and end with the very same vertex p_0, that are not identical, and that have $\tilde{w}$ as image.

Consider the configuration $\hat{w}$ that is defined by r times repeating $\tilde{w}$,

$$\hat{w} = \underbrace{[\tilde{w}, \tilde{w}, \cdots, \tilde{w}]}_{r \text{ times}}.$$

We can construct 2^r different paths with $\hat{u}$ as an image (i.e., different preimages of $\hat{w}$), as we may select in each repetition of $\tilde{w}$ one of the two constructed paths independently. If r is chosen large enough, this number becomes larger than $|E|^{|D_0|-1}$. According to Remark 9.2.16, this is impossible.

"$\Leftarrow$" Assume that the de Bruijn graph is unambiguous, but the cellular automaton is not surjective. Then there is a finite pattern w without preimage. Any state that contains w has no preimage.

Choose $l \in \mathbb{N}$ large enough, such that patterns of length l are mapped to configurations longer than w. The set E^l is split into two disjoint subsets: the set P_l of patterns with preimage and the set N_l of patterns without preimage. First we estimate $|P_l|$: If we pick one vertex p_0 in the de Bruijn graph, we find $|E|^{l-|D_0|+1}$ patterns in $E^{l+|D_0|-1}$ that start and end with p_0. We obtain $E^{l+|D_0|-1}$ distinct paths. As the automaton is unambiguous, all these paths have different images. Thus, $|P_l| \geq |E|^{l-|D_0|-1} = c|E|^l$ with $c > 0$,

$$\lim_{l\to\infty} |P_l|/|E|^l \geq c > 0.$$

Now we turn to $|N_l|$: we obtain a lower bound for the size of this set, if we count all configurations that contain u. Due to Lemma 9.2.16,

$$\lim_{l\to\infty} |N_l|/|E|^l \to 1,$$

and hence

$$1 = (|P_l| + |N_l|)/|E|^l = \lim_{l\to\infty} (|N_l|/|E|^l + |P_l|/|E|^l) \geq c + 1.$$

We have obtained a contradiction, the automaton is surjective. $\square$

The property of unambiguity of the de Bruijn graph can be checked in an algorithmic way. In order to do so, we construct a subgraph of the Cartesian product of the de Bruin graph with itself.

Definition 9.2.19 Consider a de Bruin graph $(\mathcal{N}, \mathcal{E})$. Define the de Bruin product graph in the following way: the vertices are given by $\mathcal{N}' = \mathcal{N} \times \mathcal{N}$. A directed edge, labeled by $e \in E$, goes from (p_1, q_1) to (p_2, q_2) if and only if

$$p_1 \stackrel{e}{\to} p_2, \qquad q_1 \stackrel{e}{\to} q_2.$$

The diagonal of the product graph is the set

$$\Delta = \{(p, p) \,:\, p \text{ is vertex in the de Bruijn graph}\}.$$

Theorem 9.2.20 *The cellular automaton is surjective, if and only if there is no path leaving and returning to the diagonal* Δ *of the product graph defined in Definition 9.2.19.*

Proof Assume that the de Bruijn graph is ambiguous. Thus, there are two different paths $\{p_i\}$ and $\{q_i\}$ in the de Bruijn graph with the same image (and thus the same length), the same start point and end point. If we consider (p_i, q_i), this finite sequence forms a path in the de Bruijn product graph, starting at and returning to, but not staying in Δ. Hence, if and only if there is no path leaving the diagonal and returning to the diagonal then the de Bruijn graph is not ambiguous. □

Example (Continuation of Example 9.2.13) We show the de Bruin product graph for Wolfram rule 90 in Fig. 9.4. How can we produce this graph? First, group all edges of the de Bruin graph according to its label. We have one group labeled by 0,

$$[00] \stackrel{0}{\to} [00], \quad [11] \stackrel{0}{\to} [11], \quad [01] \stackrel{0}{\to} [10], \quad [10] \stackrel{0}{\to} [01],$$

and one group labeled by 1

$$[00] \stackrel{1}{\to} [01], \quad [01] \stackrel{1}{\to} [11], \quad [10] \stackrel{1}{\to} [00], \quad [11] \stackrel{1}{\to} [10].$$

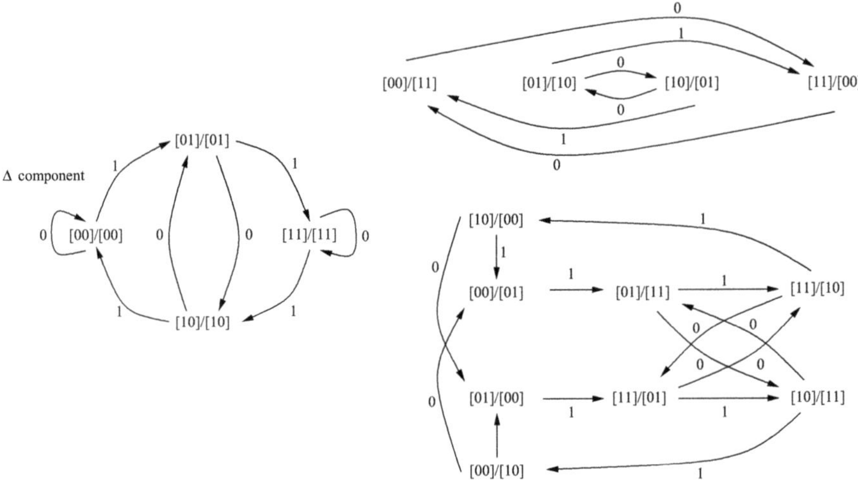

Fig. 9.4 The de Bruin product graph for Wolfram rule 90

Any pair of edges within these two groups yields a valid edge within the product de Bruin graph. In this way, we find e.g.

$$[00]/[00] \stackrel{0}{\rightarrow} [00]/[00], \quad [00]/[11] \stackrel{0}{\rightarrow} [00]/[11], \quad [00]/[01] \stackrel{0}{\rightarrow} [00]/[10], \quad \cdots .$$

The resulting set of edges are used to draw the product de Bruin graph. The product de Bruin graph of rule 90 consists of three connected components, among them is the diagonal Δ. There is no path leaving and returning to Δ. Hence, the cellular automaton with rule 90 is surjective.

9.2.3 *Injectivity and Bijectivity*

In order to check injectivity of a cellular automaton, we again use the de Bruijn product graph as constructed in Definition 9.2.19. If the cellular automaton is not injective, we find a state with two different preimages. Then we have a bi-infinite path in the de Bruijn product graph, not completely contained in the diagonal Δ. In order to express the consequences of this fact in terms of graph theory, we define connected components.

Definition 9.2.21 Given a directed graph with vertices V and edges E, we call two vertices $v_1, v_2 \in V$ communicating, if there is a path from v_1 to v_2 as well as from v_2 to v_1. A connected component $C \subset V$ is given by a non-empty subset of communicating vertices that is maximal: if $v_0 \in C$, then

$$C = \{v \in V \,:\, v \text{ communicates with } v_0\}.$$

As the de Bruijn product graph is finite, any bi-infinite path not completely contained in the diagonal Δ is equivalent with the existence of a connected component unequal Δ - either these is a second connected component or Δ is part of a larger connected component.

Corollary 9.2.22 *A cellular automaton is injective if and only if the diagonal Δ of its de Bruin product graph is the only connected component in this graph.*

As the connected components of a finite graph can be determined in finitely many steps, it is decidable whether the corresponding cellular automaton is injective.

Example (Continuation of Example 9.2.13) The de Bruin product graph for Wolfram rule 90 in Fig. 9.4 has three connected components. Hence, the cellular automaton is not injective.

Example 9.2.23 In Fig. 9.5 the de Bruin product graph for Wolfram rule 85 is depicted. This automaton shifts the state to the left, and performs a bit-wise complement. It is obviously bijective. If we consider the de Bruin product graph,

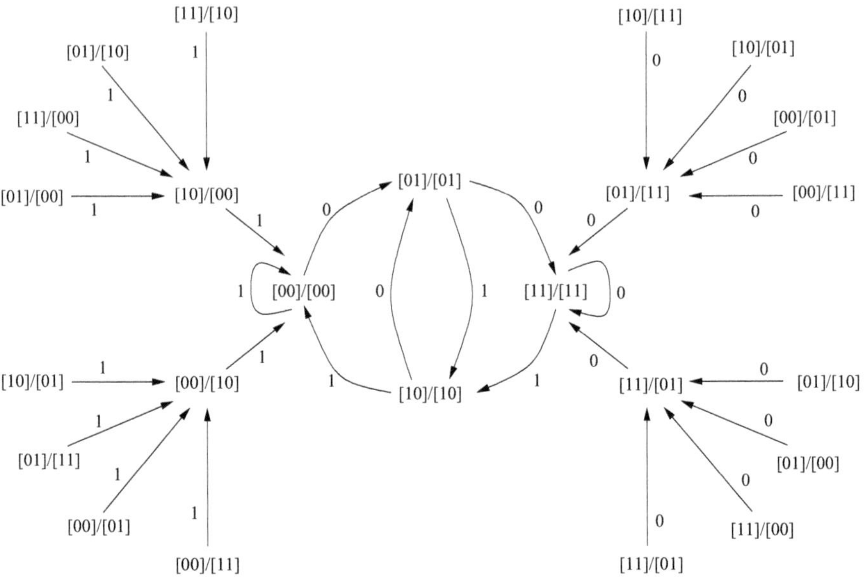

Fig. 9.5 The de Bruin product graph for Wolfram rule 85

we indeed find that there is only one connected component which coincides with the diagonal Δ. Any path starting outside of Δ enters Δ after at most two steps.

Note that the criterion for non-injectivity is not equivalent with the criterion for non-surjectivity: Non-injectivity is also given if we find a connected component separated from the diagonal, while non-surjectivity requires a path leaving and returning to the diagonal. Hence, non-surjectivity implies non-injectivity, resp. injectivity implies surjectivity. This conclusion is a special case of Corollary 9.1.15.

An injective cellular automaton is already bijective. The next logical step is the characterization of the inverse of its global function. The next theorem is not restricted to $\Gamma = \mathbb{Z}$, but valid for a general spatial structure.

Theorem 9.2.24 *Let* (Γ, D_0, E, f_0) *be a cellular automaton with a bijective global function* $f : E^\Gamma \to E^\Gamma$. *Then the inverse function* $f^{-1} : E^\Gamma \to E^\Gamma$ *is again the global function of a cellular automaton on* Γ.

Proof Since E^Γ is compact, and f continuous, also f^{-1} is continuous (Theorem A.1.12). Since f commutes with the shift operator, so does f^{-1}. Hence, the Curtis-Lyndon-Hedlund Theorem implies that f^{-1} is the global function of a cellular automaton explicitly. □

Let us return to the one-dimensional case, $\Gamma = \mathbb{Z}$. There are numerous interesting properties of reversible cellular automata, see e.g. the overview article of Toffoli and Margolus [163]. In the following, we determine the inverse automaton of Wolframs rule 85 in an rather algorithmic way.

Example (Continuation of Example 9.2.23) In order to obtain the inverse cellular automaton for rule 85, we check the images of E^{Γ_n}. We use the notation introduced in Definition 6.2.3. The idea for the construction of the inverse automaton has already been described in the proof of the Curtis-Lyndon-Hedlund Theorem: determine the sets U_e, $e \in E = \{0, 1\}$, of states with

$$U_e = \{u \in E^{\mathbb{Z}} \, : f(u)(0) = e\}.$$

Then, determine Γ_n such that

$$\{u|_{\Gamma_n} \, : \, u \in U_0\} \cap \{u|_{\Gamma_n} \, : \, u \in U_1\} = \emptyset.$$

Then, $D_0' = \Gamma_n$ is suited as a neighborhood for the inverse automaton. We may as well determine $f_n : E^{\Gamma_n} \to E^{\Gamma_{n-1}}$, and group the images into two classes:

$$V_{e,n} = \{v \, : \, v = f(u), \;\; u \in E^{\Gamma_n}, \;\; u(0) = e\} \qquad \text{for } e \in \{0, 1\}.$$

If n is sufficiently large, then $V_{0,n} \cap V_{1,n} = \emptyset$. In this case, $D_0' = \Gamma_{n-1}$ is suited as neighborhood of for the inverse automaton. The local function of the inverse automaton maps all patterns in $V_{0,n}$ to 0, and all patterns in $V_{1,n}$ to 1.

In our particular case, we find for $n = 1$

$u \in E^{\Gamma_1}$	$f_1(u)$
000	1
001	0
010	1
011	0
100	1
101	0
110	1
111	0

Hence, for $n = 1$ we find $V_{1,n} = \{0, 1\}$ and $V_{0,n} = \{0, 1\}$. These two sets contain common elements, and hence $n = 1$ is too small. Next we test $n = 2$:

$u \in E^{\Gamma_1}$	$f_1(u)$	$u \in E^{\Gamma_1}$	$f_1(u)$	$u \in E^{\Gamma_1}$	$f_1(u)$	$u \in E^{\Gamma_1}$	$f_1(u)$
00000	111	01000	111	10000	111	11000	111
00001	110	01001	110	10001	110	11001	110
00010	101	01010	101	10010	101	11010	101
00011	100	01011	100	10011	100	11011	100
00100	011	01100	011	10100	011	11100	011
00101	010	01101	010	10101	010	11101	010
00110	001	01110	001	10110	001	11110	001
00111	000	01111	000	10111	000	11111	000

Therefore,

$$V_{0,n} = \{111, 110, 101, 100\}, \quad V_{1,n} = \{011, 010, 001, 000\}.$$

Since $V_{0,n} \cap V_{1,n} = \emptyset$ for $n = 1$, $D'_0 = \Gamma_1$ is a neighborhood suited for the inverse cellular automaton. We find, that the inverse for Wolframs rule 85 is again a Wolfram automaton, with number 15.

Remark 9.2.25 We may interpret the entries of the table above in terms of the de Bruin graph: e.g., $f_2(01010) = 101$ and $f_2(10010) = 101$ correspond to

$$01 \xrightarrow{1} 10 \xrightarrow{0} 01 \xrightarrow{1} 10, \text{ and } 10 \xrightarrow{1} 00 \xrightarrow{0} 01 \xrightarrow{1} 10.$$

Since $f(01010) = f(11010)$, we may combine these two de Bruin-paths to one path in the de Bruin product graph,

$$01/01 \xrightarrow{1} 10/00 \xrightarrow{0} 01/01 \xrightarrow{1} 10/10.$$

Now we know for the de Bruin product graph displayed in Fig. 9.5, that after at most two steps any path enters the diagonal component Δ. For $101 \in E^{\Gamma_1}$, two preimages $u, \tilde{u} \in E^{\Gamma_2} = E^{\{-2,\ldots,2\}}$ can only differ in the two leftmost entries; at the latest the third entry is the same, $u(0) = \tilde{u}(0)$. More general, if $v \in E^{\Gamma_1}$ is given, and $u \in E^{\Gamma_2}$ is any preimage of v, we already know $u(0)$. In consequence, $D'_0 = E^{\Gamma_1}$ is suited as the neighborhood of the inverse cellular automaton. The structure of the de Bruin product graph already allowed us to determine the size of D'_0.

We generalize this observation: Consider the de Bruin graph of a reversible, one-dimensional cellular automaton $(\mathbb{Z}, D_0, E, f_0)$ with $D_0 = \Gamma_{d_0}$. Let m denote the maximal length of a path that does not enter the diagonal Δ, such that at least the last vertex of any path of length $m + 1$ is located in Δ, and let m' the smallest number such that $2m' + 1 > m$. Then, all preimages $u \in E^{\Gamma_{m'+d_0}}$ of a pattern $v \in E^{\Gamma_{m'}}$ have a fixed tail $u|_{m'+1-d_0,\ldots,m'+d_0}$. In particular $u(m' + 1 - d_0)$ does not depend on the special preimage u of v. If $w \in E^{\mathbb{Z}}$, and we know $f(w)|_{\Gamma_{m'}}$, then we know $w(m' + 1 - d_0)$. Hence, if we know $f(w)|_{D'_0}$, with

$$D'_0 = \{-2m' - 1 + d_0 \ldots, d_0 - 1\},$$

we already know $w(0)$. The set $D'_0 \subset \mathbb{Z}$ is a possible neighborhood for the inverse automaton. In particular,

$$|D'_0| = 2m' + 1.$$

We can estimate m as follows: The product de Bruin graph has altogether at most $|E|^{2(|D_0|-1)}$ vertices, and the diagonal has $|\Delta| = |E|^{|D_0|-1}$ vertices. The longest possible path that does not enter the diagonal has length $|E|^{2(|D_0|-1)} - |E|^{|D_0|-1}$. Hence we have an upper bound for the size of the inverse cellular automaton.

Corollary 9.2.26 *Let $(\mathbb{Z}, D_0, E, f_0)$, $D_0 = \Gamma_{d_0}$ a reversible cellular automaton, and let $(\mathbb{Z}, D_0', E, \tilde{f}_0)$ the inverse cellular automaton. It is possible to choose the neighborhood D_0' of the inverse cellular automaton as $D_0' = \{-2m'-1+d_0, \ldots, d_0-1\}$, where m' is a positive number with $2m' + 1 > m$ and*

$$m = |E|^{2(|D_0|-1)} - |E|^{|D_0|-1}.$$

Hence,

$$|D_0'| \leq 2(|E|^{2(|D_0|-1)} - |E|^{|D_0|-1}) + 3.$$

Remark 9.2.27 Czeizler and Kari [39, 40] derived sharper bounds for one-dimensional cellular automata: they show that, if $|D_0| = m$, then the size $\tilde{m}$ of the neighborhood $\tilde{D}_0$ for the inverse cellular automaton can be bounded by

$$\tilde{m} \leq |E|^{m-1} - |E| + 1.$$

They also show that this estimate is sharp in the case $|E| = 2$.

For reversible Wolfram automata, this estimate indicates that the inverse automaton is again a Wolfram rule. Indeed, if we check for reversible Wolfram automata, we only find six rules,

$$15, \ 51, \ 85, \ 170, \ 204, \ 240.$$

These automata are identity, left- and right shift, and their combination with a bitwise complement. It is immediately clear that these automata are reversible, and that their inverses are again among these six automata.

9.3 Undecidability Higher Dimensional Cellular Automata

Given a cellular automaton, we want to know to which of the Hurley classes it belongs, whether it is surjective etc. Surprisingly, even such seemingly simple properties like surjectivity are undecidable if we go to higher dimensions.

9.3.1 Stationary Points

Due to Theorem 9.2.3, we know that the stationary points of a cellular automaton are computable if and only if the domino problem for the underlying graph is solvable. For $\mathbb{Z}^2$ this is not the case (Corollary 8.5.33).

Corollary 9.3.1 *In general, the stationary points of a cellular automaton defined on $\mathbb{Z}^2$ cannot be computed.*

9.3.2 Surjectivity

In order to check whether a cellular automaton is surjective, one could consider a sequence of finite subgraphs Γ_n that become larger and larger, and check whether the restriction of the global function to E^{Γ_n} has the complete set $E^{\Gamma_n^-}$ as its image. If this is not the case for some $n \in \mathbb{N}$, then the procedure stops, and the cellular automaton is recognized as not surjective. But if it is surjective? Can we determine *a priori* the necessary size of a finite subgraph such that the restriction of all cellular automata of a certain "size" (given by the number of local states, say) to this finite subgraph is surjective if and only if the global function is surjective? We show that there is no function that gives the necessary size of the subgraph for a given "size" of the automaton.

We will prove that for a cellular automaton on $\mathbb{Z}^2$, with von Neumann neighborhood and resting state, it is in general undecidable whether the global function is surjective or not. The proof is based on the rectangle Construction 8.5.36. Here we add a second color component which we call the "path color". In addition to the components that indicate the border of a rectangle and the X, Y component the second component tells whether a rectangle can be filled by a continuous path.

Using these extended colors, we take an arbitrary tile set with a blank tile, and define a cellular automaton that is not surjective if and only if the tile set allows for a finite, non-trivial tiling. Since the finite domino problem is not decidable, also the injectivity problem of this automaton is not decidable.

Extension of the rectangle-construction.
What we will use is the "plane-filling-property" of a tile set we now aim to construct. The colors will have three components. The first two components are those we already considered in the Rectangle Construction 8.5.36. In consequence, finite tessellations consist of unions of disjoint rectangles. We add a third component. This component will be chosen in such a way that the interior of any finite rectangle is filled by a path, as indicated in Fig. 9.7.

Construction 9.3.2 We use the tile set of the Rectangle-Construction 8.5.36 and add a further component to the colors, the "path-color". The path color can be either "head" or "tail" of an arrow, "A", or $\emptyset$. A tiling of the plane is valid, if the restriction to the color components of the rectangle-construction yields a valid tiling, and in

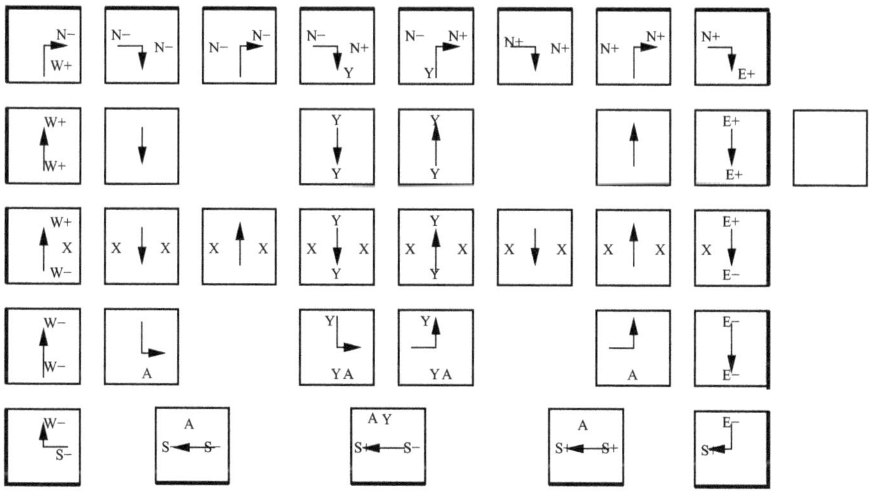

Fig. 9.6 Tile set for the Construction 9.3.2

addition, if for the new color components neighboring faces both carry Ø, both carry "A", or if "head" on the one side points to "tail" on the other side.

The set of valid tiles are shown in Fig. 9.6. We denote this tile set by ρ. The tile that only carries blank colors (rectangle-colors as well as path-color) is the blank tile.

Proposition 9.3.2 *For the tile set shown in Fig. 9.6, there is a finite, nontrivial tessellation of the plane. Any finite, nontrivial tiling of the plane consists of the union of disjoint rectangles. The arrows of the path-color form a closed loop that runs through each point of the rectangle once.*

Proof First of all, Fig. 9.7 presents an example of a finite tessellation of the plane by this tile set, hence the statement of this proposition is not about an empty set.

Given a finite, nontrivial tessellation, then especially the rectangular part of the colors form a valid, finite and nontrivial tiling. Thus, the tessellation consists of the union of a finite number of disjoint rectangles. Consider one of these rectangles. Since the arrows of west-, east- and southern boundaries are given, we already know the path there: starting on the upper, right corner, the path runs along the right, lower and left boundary to the left, upper corner.

What happens in the remaining part of the rectangle? The path can only move up, down and to the right. Since all tiles that (1) carry arrows, (2) carry no border colors and (3) allow to change the direction (arrow angled) carry on the lower rim the color A, or at the northern boundary, the path goes up and down from the upper edge of the rectangle to the lowest non-border row of the rectangle. In this way the path fills necessarily the complete rectangle in the sense that each site is visited exactly once by the path, and the path closes to a loop. □

Similarly, we find the following proposition.

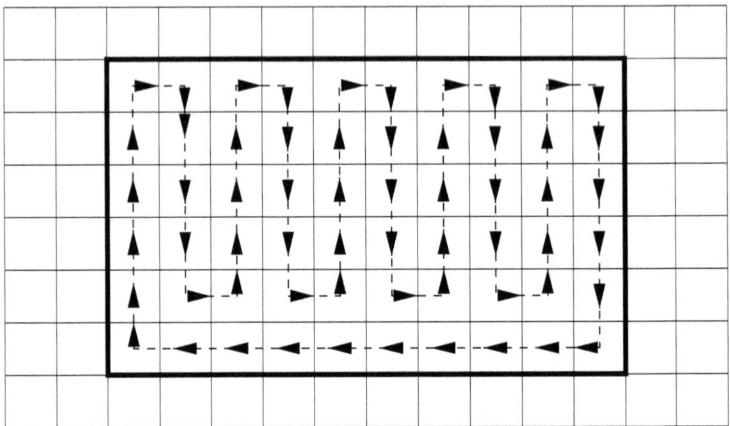

Fig. 9.7 Plane filling property for rectangle construction

Proposition 9.3.3 *Consider a finite, possibly non-valid tessellation by the tile set defined above. If this tessellation exhibits a closed path such that the tiling is valid along the complete path, then this path forms a valid rectangle.*

Proof If we have a closed paths of arrows, this path visits only a finite number of sites. Thus, there is a site with minimal and with maximal y coordinate. The tile at the place of the maximal coordinate carries necessarily the color N+ or N-, i.e., is part of an upper border of the rectangle. Since the tessellation is valid along the path, this maximum necessarily forms the upper edge of a U-turn of a path. Since the tiling along the path is valid, the path runs parallel in vertical direction. After a finite number of sites, the two ends of the path we consider in the moment will arrive at the lower edge of the path. Then, the two lines need to turn again.

Thus, the path will go up and down. Since the tiling is valid along the path, the upper edges will form the upper border of a rectangle. The lower border carries the color A, that forces a lower bound of a rectangle at the lower edge. Since it cannot extend in x direction to infinity, a left and right border appear at the left- and the right side. In this way, we arrive at the valid rectangle, filled by a closed path. □

Remark 9.3.5

(1) Since the path goes up and down and up and down, the number of sites in horizontal direction is even.

(2) The tagged site (the tile that carries colors X and Y) is not neighbored by the lower boundary, since there is no tile with path-color A on the lower edge. However, for our proof it is not necessary to select a site in the second row from below.

Given any tile set $\mathcal{T}$ with a blank tile, we construct a cellular automaton that possesses mutually erasable patterns if and only if the tile set allows for a finite, non-trivial tiling of the plane. The tile set ρ is defined in Construction 9.3.2. A local state $e \in E$ of the cellular automaton we construct now, is a three-tupel, $e = (t, r, b)$. The first component is a tile from $\mathcal{T}$, the second a tile from ρ, and the third a bit component, $b \in \{0, 1\}$. Not all combinations are allowed. The local function will check if the tiling, given in the $\mathcal{T}$-component, is locally, at a given site, coherent, and will act accordingly on the bit component (where the direction given in the ρ-component is taken into account).

Construction 9.3.6 We define a cellular automaton (Γ, D_0, E, f_0) associated with the tile set $\mathcal{T}$ that possesses a blank tile b.

Let $\Gamma = \mathbb{Z}^2$ and D_0 the von Neumann neighborhood. Let $E \subset \mathcal{T} \times \rho \times \{0, 1\}$. A member $a = (t, r, b) \in \mathcal{T} \times \rho \times \{0, 1\}$ is an element of E if one of the following two conditions are satisfied:

(i) If r is blank then $t \in \mathcal{T}$ is the blank tile b.
(ii) If r is one of the two tiles that carries both colors X and Y, then $t \in \mathcal{T}$ is not blank.

Finally we define the local function f_0. This function changes only the third component of a local state $e = (t, r, b)$, the bit component b, according to the following three rules:

(i) If in the first or second component there is a tiling error, then the bit component is not changed.
(ii) If the second component contains no arrow, then the bit component is not changed.
(iii) Assume that there is no tiling error in the first and second components, and the second component contains an arrow. Let $a_1 = (t_1, r_1, b_1)$ denote the state of the actual cell and $a_2 = (t_2, r_2, b_2)$ the state of the cell at which the arrow points. Then, the function performs an exclusive-or of the bit components of these two cells, i.e.,

$$f_0(\ldots) = (t_1, r_1, b_1\,(1 - b_2) + (1 - b_1)b_2)\,).$$

Proposition 9.3.6 *The cellular automaton constructed above carries two mutually erasable patterns if and only if $\mathcal{T}$ allows for a nontrivial and finite tessellation of the plane.*

Proof

(1) Two mutually erasable pattern force the existence of a finite, nontrivial tessellation.

Assume that there is a finite subgraph $\tilde{\Gamma} \subset \Gamma$ that carries a mutually erasable pattern. Then there are $u_1, u_2 \in E^{\Gamma}$ such that $u_1(g) = u_2(g) = (b, b, 0)$ for $g \not\in \tilde{\Gamma}$ and $u_1 \neq u_2$ such that $f(u_1) = f(u_2)$.

Since f only changes the bit component (the third component) of the state, the first and second components of u_1 and u_2 agree,

$$(u_1(g))_1 = (u_2(g))_1, \quad (u_1(g))_2 = (u_2(g))_2, \quad \forall g \in \Gamma.$$

However, there is at least one element $g \in \tilde{\Gamma}$ such that $u_1(g) \neq u_2(g)$. Thus, without restriction, $(u_1(g))_3 = 0$ and $u_2(g))_3 = 1$. Since the image of the two states at g agree, the bit component of at least one of the two sites has changed. Thus, $(u_1(g))_2$ contains an arrow. Let $b^{(1)}$ ($b^{(2)}$) be the bit component of the state of the cell to which this arrow points, if we consider u_1 (u_2). Since we know that $f(u_1)(g) = f(u_2)(g)$, and hence

$$b^{(1)} = (1-0)b^{(1)} + 0\,(1-b^{(1)}) = (1-b^{(2)})1 + (1-1)b^{(2)} = 1 - b^{(2)},$$

We know that the bit components (according to u_1 and u_2) of the cell to which the arrow points, are different. By means of finite induction, we find that all bit components along the path described by the arrows are different. Since the path fills the complete rectangle, the tiling along the path is valid for the rectangle-colors as well as for the $\mathcal{T}$-colors. The validity of the rectangle-colors along a closed path forces this path to fill a rectangle. Since there is exactly one tile which carries the colors X and Y, the $\mathcal{T}$-tiles cannot be all trivial. Furthermore, the $\mathcal{T}$-tiling is valid and blank at the border sites of the rectangle. Hence it can be extended by blank tiles to the complete grid, i.e., we find a non-trivial, finite tiling of the plane by $\mathcal{T}$.

(2) *A finite, nontrivial tessellation forces the existence of mutually erasable patterns.*
We construct two local patterns that are mutually erasable. Take a rectangular tiling of a finite (rectangular) subgraph $\tilde{\Gamma}$ such that $\tilde{\Gamma}^-$ covers the finite, non-trivial tiling of the plane by $\mathcal{T}$. Outside of $\tilde{\Gamma}$ choose the state $(b, b, 0)$. Define u_1 inside of the rectangle by the $\mathcal{T}$-tiling (first components of the states), by the rectangular-tiling (second component of the states) and zero (bit component). For u_2, we keep u_1 on all sites, where the rectangular-component does not carry an arrow. The bit components of sites that carry an arrow, are turned to one. Then,

$$f(u_1) = f(u_2) = u_1.$$

Since u_1 and u_2 disagree only in a finite number of sites, there are two mutually erasable patterns. □

Since the finite domino problem is not decidable, we cannot decide if a cellular automaton in two dimensions has mutually erasable pattern. We can draw the following conclusion.

Theorem 9.3.8 *(1) In general, it is not decidable whether a cellular automaton is surjective. (2) If the cellular automaton has a resting state, it is not decidable whether the global function of the subautomaton with finite support is surjective or injective.*

Remark 9.3.9

(1) Kari [101] showed that also injectivity is in general undecidable for $\Gamma = \mathbb{Z}^2$.
(2) The undecidability of injectivity on $\mathbb{Z}^2$ is equivalent with the undecidability of bijectivity. Let us consider a bijective cellular automaton $(\mathbb{Z}^2, D_0, E, f_0)$. Then, there is an inverse cellular automaton $(\mathbb{Z}^2, \tilde{D}_0, E, \tilde{f}_0)$. To find an *a priori* bound for the size of $\tilde{D}_0$, similar to Proposition 9.2.26, is impossible: otherwise one could test all cellular automata on this maximal neighborhood (these are finitely many) if they are indeed an inverse for the given cellular automaton, and in this way decide about bijectivity (and hence also about injectivity).
(3) For many cellular automata, however, it is possible to decide if they are injective or surjective, see e.g. [60].

Chapter 10
Linear Cellular Automata

A cellular automaton (Γ, D_0, E, f_0) is called "linear" if the set E of elementary states is a ring, and the global function f is a morphism on the product space of this structure.[1]

Many results for linear cellular automata require only weak assumptions on the algebraic structure of E. The basic ideas and central theorems can be best explained if we restrict ourselves to a simple case: E is a finite field of the form $\mathbb{F}_p$ for p prime. This assumption is not necessary for many results obtained in this chapter, but it simplifies the presentation. If not stated otherwise, this assumption holds throughout the section.

10.1 Representation of Linear Cellular Automata

When dealing with algebraic structures, suitable notation is important as it allows intuitive formulations.

Notation E is assumed to carry a ring structure. It is convenient to represent states $u \in E^\Gamma$ as formal sums

$$\sum_{g \in \Gamma} u(g)\, g$$

in which $g \in \Gamma$ are multiplied by coefficients $u(g) \in E$. To identify the local state of site g, one inspects the coefficient of g in this formal sum.

[1]The term "linear cellular automaton" is also used for cellular automata with grid $\Gamma = \mathbb{Z}$ as the points are linearly ordered. We do not use this definition.

K.-P. Hadeler, J. Müller, *Cellular Automata: Analysis and Applications*,
Springer Monographs in Mathematics, DOI 10.1007/978-3-319-53043-7_10

The set $E[\Gamma]$ is defined as

$$E[\Gamma] = \left\{ \sum_{g\in\Gamma} u(g)\, g \; : \; u \in E^\Gamma \right\}.$$

The identification of E^Γ and $E[\Gamma]$ is made explicit by the definition of the bijective function Λ,

$$\Lambda : E^\Gamma \to E[\Gamma], \quad u \mapsto \sum_{g\in\Gamma} u(g)\, g.$$

To shorten the notation, we are allowed to skip all terms $0\,g$ in this sum. E.g.,

$$\Lambda(u) = \sum_{i=1}^{n} \lambda_i\, g_i \qquad \text{for } \lambda_i \in E, \quad g_i \in \Gamma$$

implies that $u(g) = 0$ for all $g \notin \{g_i \,:\, i = 1,\dots,n\}$.

We assume that the neutral element $0 \in E$ is a resting state. The set $E[\Gamma]_c$ denotes the subset in $E[\Gamma]$ of all states with finite support, this is, $E[\Gamma]_c = \Lambda((E^\Gamma)_c)$.

Remark 10.1.1 $E[\Gamma]_c \subset E[\Gamma]$ becomes a ring if we define

$$\left(\sum_{g\in\Gamma} u(g)\, g\right) + \left(\sum_{g'\in\Gamma} v(g')\, g'\right) := \sum_{g\in\Gamma} (u(g) + v(g))\, g,$$

$$\left(\sum_{g\in\Gamma} u(g)\, g\right) \left(\sum_{g'\in\Gamma} v(g')\, g'\right)$$

$$:= \sum_{g,g'\in\Gamma} (u(g)\, v(g'))\, (g\, g') = \sum_{h\in\Gamma} \left(\sum_{g\in\Gamma} u(g) v(g^{-1}h)\right) h.$$

This definition of a multiplication cannot be extended from $E[\Gamma]_c \times E[\Gamma]_c$ to $E[\Gamma] \times E[\Gamma]$ as the inner sum on the right hand side should be finite. Later we will use that the multiplication can be extended to $E[\Gamma]_c \times E[\Gamma]$ and to $E[\Gamma] \times E[\Gamma]_c$.

If $E[\Gamma]$ should be considered as a ring, then we can use the point wise multiplication

$$\left(\sum_{g\in\Gamma} u(g)\, g\right) \odot \left(\sum_{g'\in\Gamma} v(g')\, g'\right) := \sum_{g\in\Gamma} (u(g)\, v(g))\, g.$$

Here $E[\Gamma]$ is considered as a product ring, and the set $E[\Gamma]_c$ becomes a subring. As we will see below, the first definition of a ring structure on $E[\Gamma]_c$ is more useful.

We have a suitable notation for the state space. Now we proceed to the definition and representation of linear cellular automata. A local function is a map from E^{D_0} to E. That is, a finite tupel $(u_g)_{g\in D_0}$ of elements in E are mapped to $f_0((u_g)_{g\in D_0}) \in E$. It can happen that this function can be represented as a weighted sum, $f_0((u_g)_{g\in D_0}) = \sum_{g\in D_0} \lambda_g u_g$ with coefficients $\lambda_g \in E$. Cellular automata with this property are of particular interest and are called "linear".

In our definition of a linear cellular automaton we require $E = \mathbb{F}_p$. Some results carry over to more general rings E. We do not state these general results but just add appropriate comments.

Definition 10.1.2 A cellular automaton (Γ, D_0, E, f_0) is called linear, if $E = \mathbb{F}_p$, p prime, and f_0 has a representation

$$f_0(u) = \sum_{g\in D_0} \lambda_g u(g) \qquad \text{for} \quad u \in E^{D_0}.$$

The next lemma justifies the name "linear".

Lemma 10.1.3 *Let (Γ, D_0, E, f_0) be a cellular automaton with $E = \mathbb{F}_p$, p prime. The cellular automaton is linear if and only if for all $u, v \in E^\Gamma$ and $a, b \in E$, it is true that*

$$f(au + bv) = af(u) + bf(v).$$

Proof If the cellular automaton is linear, then the equation is clearly true. Now, assume that the equation is true. Define for $g \in \Gamma$ the state $\delta_g \in E^\Gamma$ by $\delta_g(g') = 1$ if $g = g'$, and $\delta_g(g') = 0$ otherwise, and define furthermore $\lambda_g := f(\delta_g)(0)$. Now, let $u \in E^{D_0}$. The state $\tilde{u} = \sum_{g\in D_0} u(g)\delta_g \in E^\Gamma$ extends u to Γ. Then,

$$f_0(u) = f(\tilde{u})(0) = \sum_{g\in D_0} u(g)f(\delta_g)(0) = \sum_{g\in D_0} u(g)\lambda_g.$$

□

Example 10.1.4 Consider the Wolfram rule 90 with the local function

$$f_0(e_{-1}, e_0, e_1) = e_{-1}(1 - e_1) + (1 - e_{-1})e_1.$$

If we think of $\{0, 1\}$ as the field $\mathbb{F}_2$, we may also write

$$f_0(e_{-1}, e_0, e_1) = e_{-1} + e_1.$$

The following observation is the key tool that allows to deal with linear cellular automata in a convenient way: the action the of global function on an element in E^Γ can be represented as the multiplication of the corresponding element in $E[\Gamma]$ by a fixed element in $E[\Gamma]_c$. In Theorem 10.1.5 and in the following remarks and propositions up to Corollary 10.1.9 the set of elementary states E could be any commutative ring.

Theorem 10.1.5 *Let* (Γ, D_0, E, f_0), $E = \mathbb{F}_p$, *be a linear cellular automaton,*

$$f_0(u) = \sum_{g\in D_0} \lambda_g\, u(g) \quad \textit{with} \quad \lambda_g \in E.$$

Define

$$p = \sum_{g\in D_0} \lambda_g g^{-1} \in E[\Gamma]_c.$$

Then, for $u \in E^\Gamma$,

$$\Lambda(f(u)) = \Lambda(u)p.$$

Proof We know that $f_0(u|_{\sigma_g(D_0)}) = \sum_{g'\in D_0} \lambda_{g'} u(g\,g')$. Thus,

$$\begin{aligned}
\Lambda(u)p &= \left(\sum_{g\in\Gamma} u(g)g\right)\left(\sum_{g'\in D_0} \lambda_{g'} g'^{-1}\right) = \sum_{g'\in D_0}\sum_{g\in\Gamma} u(g)\lambda_{g'} g\, g'^{-1} \\
&= \sum_{g'\in D_0}\sum_{h\in\Gamma} u(hg')\lambda_{g'} h = \sum_{h\in\Gamma}\sum_{g'\in D_0} \lambda_{g'} u(hg')h \\
&= \sum_{h\in\Gamma} f_0(u|_{hD_0})h = \Lambda(f(u)).
\end{aligned}$$

□

Remark 10.1.6

(1) As $E[\Gamma]_c$ forms a ring, and the local functions can be identified with elements in $E[\Gamma]_c$, this theorem implies that also the set of all linear cellular automata on E^Γ possesses a ring structure.

(2) In most cases we will consider Abelian grids, especially $\Gamma = \mathbb{Z}^d$. Then p and $\Lambda(u)$ commute, and we multiply p from the left.

We discuss the connection to formal Laurent polynomials in the case $\Gamma = \mathbb{Z}^d$.

Remark 10.1.7 Define $E[X, X^{-1}]$ as the set of all formal Laurent series with coefficients in E, and $E[X, X^{-1}]_c$ as the subset of all formal Laurent polynomials. If $\Gamma = \mathbb{Z}^d$, then $E[\Gamma]_c$ can be identified in a natural way with the Laurent polynomials over E,

$$E[\Gamma]_c \equiv E[X, X^{-1}]_c.$$

This can be readily seen if we use the multi-index notation. Let $g = (g_1, \ldots, g_n) \in \mathbb{Z}^d, X \in \mathbb{R}^d$. Denote by $|g| = \sum_{i=1}^n |g_i|$, and $X^g = \prod_{i=1}^n X_i^{g_i}$. We use the fact that the map $\theta : \mathbb{Z}^d \mapsto \{X^z : z \in \mathbb{Z}^d\}, z \mapsto X^z$ is an isomorphism from the additive group $(\mathbb{Z}^d, +)$ into the multiplicative group carried by $\{X^z : z \in \mathbb{Z}^d\}$, where X denotes an independent variable in $\mathbb{R}^d$: the map is bijective and

$$\theta(z_1 + z_2) \mapsto (X^{z_1+z_2}) = X^{z_1} X^{z_2} = \theta(z_1)\theta(z_2).$$

If we define $\Theta : E[\mathbb{Z}^d]_c \to E[X, X^{-1}]_c$ by

$$\Theta(\Lambda(u)) := \sum_{z\in\mathbb{Z}^d} u(z)\theta(z) = \sum_{z\in\mathbb{Z}^d} u(z)X^z \in E[X, X^{-1}]_c \qquad \text{for } u \in (E^\Gamma)_c$$

then $\Theta \circ \Lambda : (E^\Gamma)_c \mapsto E[X, X^{-1}]_c$ is bijective. Similarly, for $\Gamma = \mathbb{Z}^d$, we can identify $E[\Gamma]$ with the set of all formal Laurent series. The natural definition of sum and product on $E[X, X^{-1}]_c$ is

$$\sum_{z\in\mathbb{Z}^d} u(z)X^z + \sum_{z'\in\mathbb{Z}^d} v(z')X^{z'} = \sum_{z\in\mathbb{Z}^d} (u(z) + v(z'))X^z$$

$$\left(\sum_{z\in\mathbb{Z}^d} u(z)X^z\right)\left(\sum_{z'\in\mathbb{Z}^d} v(z')X^{z'}\right) = \sum_{h\in\mathbb{Z}^d}\sum_{z\in\mathbb{Z}^d} u(z)u(h-z')X^h.$$

From this definition, we find at once the following proposition.

Proposition 10.1.8 *$\Theta : E[\mathbb{Z}^d]_c \to E[X, X^{-1}]_c$ is a ring homomorphism.*

We have seen that for a linear cellular automaton there is $p \in E[\Gamma]_c$ such that $\Lambda \circ f(u) = \Lambda(u)p$. If we identify p with the corresponding Laurent polynomial $p(X) = \Theta(p) \in E[X, X^{-1}]_c$, then the action of the cellular automaton can be identified with a multiplication by a polynomial,

$$\Theta \circ \Lambda \circ f(u)(X) = p(X) \;\; \Theta \circ \Lambda(u)(X).$$

Corollary 10.1.9 *Let $(\mathbb{Z}^d, D_0, E, f_0)$, $E = \mathbb{F}_p$, be a linear cellular automaton with global function f. Then there is a Laurent polynomial $p \in E[X, X^{-1}]_c$ and a bijective function $h : E^\Gamma \to E[X, X^{-1}]$ such that*

$$h(f(u)) = p(x)h(u).$$

10.2 Surjectivity, Injectivity and Bijectivity

For general cellular automata, surjectivity, injectivity and bijectivity are decidable problems in the case $\Gamma = \mathbb{Z}$, but in general not in higher dimensions $\Gamma = \mathbb{Z}^d$, $d > 1$ (see Sect. 9.3). Sato [152] observed that the linear structure imposes restrictions that are strong enough to allow for algorithms solving these problems also in $\mathbb{Z}^d$—at least if the ring E is well behaved, e.g. an integral domain or $\mathbb{Z}_q$, where q is a power of a prime number. Later, Dow [45] weakened the assumption of linearity to that of additivity: if E is an additive semi-group and f is additive, $f(u+v) = f(u) + f(v)$, then algorithms can be found that decide about surjectivity and injectivity.

We focus here again on the case $E = \mathbb{F}_p$, p prime, and characterize bijective cellular automata for two grids, the lattice Z^d and the free group over d symbols. It turns out that the reversible automata coincide with the shift automata combined with a permutation of the local states.

Proposition 10.2.1 *A linear cellular automaton $(\mathbb{Z}^d, D_0, \mathbb{F}_p, f_0)$ is bijective if and only if there is $a \in F_p^\times = F_p \setminus \{0\}$ and $g \in \Gamma$ such that the local function can be represented as*

$$f_0(u) = au(g).$$

Proof From the Curtis-Lyndon-Hedlund theorem we conclude that the inverse of a cellular automaton is a cellular automaton (Theorem 9.2.24). The inverse of a linear function is again linear,

$$\begin{aligned} f^{-1}(au+bv) &= f^{-1}(af \circ f^{-1}(u) + bf \circ f^{-1}(v)) \\ &= f^{-1} \circ f(af^{-1}(u) + bf^{-1}(v)) = af^{-1}(u) + bf^{-1}(v), \end{aligned}$$

Lemma 10.1.3 shows that also f^{-1} is a linear cellular automaton.

As $\Gamma = \mathbb{Z}^d$, we use the notation of Laurent polynomials. Let $\tilde{p}(X)$ and $\tilde{q}(X)$ denote Laurent polynomials representing the local function of the cellular automaton resp. its inverse. Then,

$$\tilde{p}(X)\tilde{q}(X) = 1X^0.$$

It is a well known fact (see Proposition A.2.22) that in this case

$$\tilde{p}(X) = aX^\alpha, \quad \tilde{q}(X) = a^{-1}X^{-\alpha}$$

where $a \in \mathbb{F}_q^\times$, $\alpha \in \mathbb{Z}^d$. □

Proposition 10.2.2 *Let $E = \mathbb{F}_p$ and let $\Gamma(G)$ be the free group over d symbols. A linear cellular automaton $(\Gamma, D_0, \mathbb{F}_q, f_0)$ is bijective if and only if there is $a \in \mathbb{F}_q^\times$ and $g \in G$ such that the local function has a representation*

$$f_0(u) = au(g).$$

Proof Let $p, q \in E[\Gamma]_c$ denote the representations of a linear cellular automaton and its inverse,

$$p = \sum_{i=0}^{n} a_i g_i, \quad q = \sum_{j=0}^{m} b_j h_j, \quad pq = \sum_{i=0}^{n}\sum_{j=0}^{m} a_i b_j g_i h_j = 1\, e.$$

Without restriction the elements g_i are distinct, the elements h_j are distinct, and all elements are reduced words in Γ (i.e. nowhere is a generator directly multiplied with its inverse).

Let $H = \{h \in \Gamma \,:\, \exists i, j : h = g_i h_j\}$ be the set of all sites (group elements) that appear in the product of p and q, and, for $h \in H$,

$$I(h) = \{(i,j) \,:\, i \in \{0,\ldots,n\},\ j \in \{0,\ldots,m\},\ g_i h_j = h\}.$$

We know $e \in H$ as $pq = 1\, e$. Equating coefficients yields

$$\sum_{(i,j)\in I(e)} a_i b_j = 1, \qquad \forall h \in H \setminus \{e\} : \sum_{(i,j)\in I(h)} a_i b_j = 0.$$

Let $\nu = \max\{d_c(h, e) \mid h \in H\}$. If $\nu = 0$, we are done, as $H = \{e\}$ in this case, implying $g_i h_j = e$ for all pairs i, j, which forces g_i and h_j to be independent of the index.

Let $\nu > 0$, and $h \in H$ an element with maximal word length in H. Then, we have $\sum_{(i,j)\in I(h)} a_i b_j = 0$. As E is a field, there are at least two different pairs $(i_0, j_0), (i_1, j_1) \in I(h)$.

We distinguish two cases, depending on the length of the words g_{i_0}, g_{i_1} and h_{j_0}, h_{j_1}. Before we start to analyze the two cases, we note that $i_0 = i_1$ leads to a contradiction ($g_{i_0} h_{j_0} = h = g_{i_0} h_{j_1}$, i.e., $j_0 = j_1$ and $(i_0, j_0) = (i_1, j_1)$). Similarly, $j_0 = j_1$ is not possible as it implies $i_0 = i_1$. Hence, without restriction, we may assume that $(i_0, j_0) = (0, 0)$ and $(i_1, j_1) = (1, 1)$.

We know that the words h, g_i and h_i are irreducible, and the concatenation and reduction of the products $g_0 h_0$ reps. $g_1 h_1$ yield h. The front part of g_i agrees with that of h, and the rear part of h_i is identical with the rear part of h ($i = 0, 1$). The rear part of g_0 (g_1) annihilates with the front part of h_0 (h_1). Thus, we may represent the group elements as

$$h = ABC, \quad g_0 = AX, \quad h_0 = X^{-1}BC, \quad g_1 = ABY, \quad h_1 = Y^{-1}C.$$

We know furthermore that AB, BC, AX, $X^{-1}B$, BY, and $Y^{-1}C$ cannot be reduced, and

$$d_c(ABC, e) = \nu, \qquad d_c(g_0h_1, e) = d_c(AXY^{-1}C, e) \leq \nu,$$
$$d_c(g_1h_0, e) = d_c(ABYX^{-1}BC, e) \leq \nu.$$

From $d_c(ABC, e) = \nu \geq d_c(ABYX^{-1}BC, e)$, ABY as well as $X^{-1}BC$ are irreducible, we conclude that YX^{-1} is reducible. Thus, $Y = Y_0Z_0$, $X = X_0Z_0$, where $Y_0X_0^{-1}$ is either irreducible again, or $Y_0 = e$, or $X_0 = e$; therewith, $ABYX^{-1}BC = ABY_0X_0^{-1}BC$.

Assume that $Y_0X_0^{-1}$ is irreducible. The fact that BY and $X^{-1}B$ are irreducible implies that also BY_0 and $X_0^{-1}B$ are irreducible. Therefore, the length of the word $ABY_0X_0^{-1}BC$ is larger than ν. Thus, either $Y_0 = e$ or $X_0 = e$.

Case 1: $Y_0 = e$.
As $g_1h_0 = ABYX^{-1}BC = ABX_0^{-1}BC$ and $X^{-1}B = Z_0^{-1}X_0^{-1}B$ irreducible, $X_0 = X_1B$. Thus, $X = X_1BZ_0$ and $g_1h_0 = AX_1^{-1}BC$. We turn to $g_0h_1 = AXY^{-1}C = AX_1BZ_0Z_0^{-1}C = AX_1BC$. Now, $g_0 = AX = AX_1BZ_0$ as well as BC (the final sequence of h_0) are irreducible, and hence also AX_1BC. As $d(g_0, h_1, e) \leq \nu$, necessarily $X_1 = e$. From this, we find

$$g_0 = AX = ABZ_0 = ABY = g_1$$

in contradiction to the assumption that $g_0 \neq g_1$.

Case 2: $X_0 = e$.
We use a similar reasoning like in Case 1. The starting point is again $g_1h_0 = ABYX^{-1}BC = ABY_0BC$ and $ABY = ABY_0Z_0$ irreducible, $Y_0 = Y_1B^{-1}$. This implies $Y = Y_1B^{-1}Z_0$, and $g_1h_0 = AY_1BC$. Therewith, $g_0h_1 = AXY^{-1}C = AZ_0Z_0^{-1}BY_1^{-1}C = ABY_1^{-1}C$. As neither Y_1B^{-1} nor $Y^{-1}C = Z_0^{-1}BY_1^{-1}C$ (and therefore especially $Y_1^{-1}C$) is reducible, but the length of the word $ABY_1^{-1}C$ is less or equal the length of the word ABC, we find $Y_1 = e$. Then, $h_0 = h_1$ in contradiction to the assumption.

□

Remark 10.2.3 For non-exponential growing groups we know that injectivity is equivalent with bijectivity. The theorem about bijective cellular automata is, at the same time, also a theorem about injective cellular automata for non-exponentially growing groups.

Now we look into surjectivity. We do this for $\Gamma = \mathbb{Z}^d$. We first construct the preimage of a state with compact support, and later proceed to the preimage of a general state.

Proposition 10.2.4 *Let* (Γ, E, D_0, f_0) *be a linear cellular automaton with global function* f. *Let* $\tilde{f}$ *denote the restriction of* f *to the set of states that are mapped to configurations with finite support,*

$$\tilde{f} : D(\tilde{f}) \subset E^\Gamma \to (E^\Gamma)_c, \quad u \mapsto f(u),$$

where $D(\tilde{f}) = \{u \in E^\Gamma : f(u) \in (E^\Gamma)_c\}$ *and* $\Gamma = \mathbb{Z}^d$. *If* f_0 *is not identically zero, then* $\tilde{f}$ *is surjective.*

Proof We focus in the case $\Gamma = \mathbb{Z}^2$. It is straightforward but technical to generalize this idea into a valid proof for $\Gamma = \mathbb{Z}^d$.

As E is a field, the local function of any linear, nonzero cellular automaton is permuting if $E = \mathbb{Z}$ (see Definition 6.2.1). We know that these automata are surjective (see Proposition 6.2.11). These ideas can be directly carried over to the situation at hand; instead of the utmost left or right position, like in the one-dimensional case, a cell in $g_0 = (i_0, j_0) \in D_0$ with non-trivial weight λ_g that is extreme is selected; "extreme" can be characterized in two dimensions in the following way: first determine the maximal first component of sites in D_0 that carry a non-trivial weight λ_g. Among all non-trivial sites with a maximal first component, select that with the maximal second component.

Now we construct one preimage v of a configuration u with finite support. The support of the configuration is covered in $\{(i,j) \in \mathbb{Z}^2 : i \geq z_1, j \geq z_2\}$ if z_1, z_2 are chosen appropriately (see Fig. 10.1). We define the local states of the preimage to be zero, if the first component is smaller than $z_1 + i_0$, or the second component is

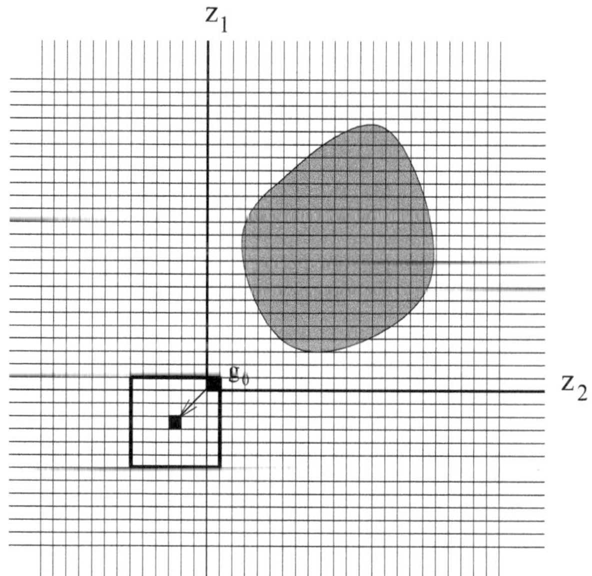

Fig. 10.1 Sketch of the proof of surjectivity: The configuration u has the support in the *gray shaped region*. The preimage v can be defined to be zero in to the *left* of z_1 and below z_2. The *black square* is the local neighborhood of the central point (*black box*). In order to determine the appropriate state in (z_1, z_2), we use the permutivity of the cellular automaton w.r.t. g_0

smaller than $z_2 + j_0$. Then we fill in the local states of a quadrant $[z_1 + i_0 - 1, \infty) \times [z_2 + j_0 - 1, \infty)$, say.

The first nontrivial point is (z_1, z_2); this grid site is the first, where $u(z_1, z_2)$ may be nonzero. As f_0 is permuting with respect to (i_0, j_0), we are able to find $v(z_1 + i_0, z_2 + j_0)$, such that $f(v)(z_1, z_2) = u(z_1, z_2)$. Now we move on to the next cell on the right. Again, we are able to find $v(z_0+i_0+1, z_2+j_0)$, such that $f(v)(z_1+1, z_2) = u(z_1 + 1, z_2)$. In this way, we recursively handle the line with second component z_2. afterwards we proceed to the line with second component $z_2 + 1$ in a similar way. Like a sewing machine, this algorithm constructs line by line the state v, and in this way the surjectivity of $\tilde{f}$ is proven. □

Example 10.2.5 There is some ambiguity in the construction above. We have defined that on the left and below the preimage is zero. We could also have chosen that the preimage should be zero at the right and above. In general it is possible to construct different preimages.

Consider the Wolfram automaton with $E = \mathbb{F}_2$, and $f_0(e_{-1}, e_0, e_1) = e_0 + e_1$, and let $u_0 = \delta_0$. We find two preimages of u_0 in E^Γ,

$$v_1(i) = \begin{cases} 0 & \text{for} \quad i \leq 0 \\ 1 & \text{otherwise} \end{cases}, \qquad v_2(i) = \begin{cases} 1 & \text{for} \quad i \leq 0 \\ 0 & \text{otherwise} \end{cases}.$$

We find non-uniqueness (as could be expected in the light of the theorem about bijective cellular automata), and $v_i \notin (E^\Gamma)_c$.

Remark 10.2.6 Using topological arguments, we can show that the cellular automaton is also surjective as a function of E^Γ into E^Γ: The element $u \in E^\Gamma$ can be approximated (with respect to the Cantor topology) by elements $u_n \in (E^\Gamma)_c$. The sequence of preimages of the elements u_n possess a converging subsequence. The limit of this subsequence is the preimage of u.

Corollary 10.2.7 *Any linear automaton on $\mathbb{Z}^d$ (with $E = \mathbb{F}_q$, q prime) that is nonzero is surjective.*

Remark 10.2.8 If E is not a field, but only a finite, commutative ring, then a linear cellular automaton $\Gamma = \mathbb{Z}^d$ is surjective if and only if one of the weights λ_g that represent the local function is a unit (see [152]).

10.3 Fractal Sets and Linear Cellular Automata

The time-space diagram of some linear cellular automata show nice self-similar structures. Willson [177–179] investigated this property in a series of papers, starting in 1978. Later, Haeseler et al. [81–83] deepened this approach and allowed for more general cellular automata. Their basic assumptions are still close to those of Willson. Gütschow [80] relaxed these properties to less restrictive algebraic structures. Triggered by examples of this type, Nekrashevych [140] developed a theory of self-similar groups, an abstract generalization of the present setting. However, limit sets can also be defined based only on topological methods [132]. This approach allows to assign limit sets to any cellular automaton on an Abelian grid, independent on any algebraic structure. Here, we will focus to the work of Willson, van Haeseler, and coworkers.

10.3.1 Introductory Example and the Fermat Property

Let us start with the Wolfram rule 90 that adds the left and the right neighbor in $\mathbb{F}_2$,

$$f_0(u_{-1}, u_0, u_1) = u_{-1} + u_1.$$

The trajectory starting with configuration δ_0 (zero everywhere except a '1' at site 0) yields the Pascal's triangle modulo 2 (see Sect. 1.4),

$$\begin{array}{ccccccccccccc}
 & & & & & & 1 & & & & & & \\
 & & & & & 1 & & 1 & & & & & \\
 & & & & 1 & & 2 & & 1 & & & & \\
 & & & 1 & & 3 & & 3 & & 1 & & & \\
 & & 1 & & 4 & & 6 & & 4 & & 1 & & \\
 & 1 & & 5 & & 10 & & 10 & & 5 & & 1 & \\
1 & & 6 & & 15 & & 20 & & 15 & & 6 & & 1
\end{array}$$

$$\begin{array}{ccccccccccccc}
 & & & & & & 1 & & & & & & \\
 & & & & & 1 & & 1 & & & & & \\
 & & & & 1 & & 0 & & 1 & & & & \\
 & & & 1 & & 1 & & 1 & & 1 & & & \\
 & & 1 & & 0 & & 0 & & 0 & & 1 & & \\
 & 1 & & 1 & & 0 & & 0 & & 1 & & 1 & \\
1 & & 0 & & 1 & & 0 & & 1 & & 0 & & 1
\end{array}$$

For the supports of states at time points 2, 4, 8, ... we find

$$\text{supp}(f^2(\delta_0)) = \{-2, 2\}, \quad \text{supp}(f^4(\delta_0)) = \{-4, 4\}, \quad \text{supp}(f^8(\delta_0)) = \{-8, 8\}, \cdots$$

and we conjecture that after 2^n iterations the we have $\text{supp}(f^{2^n}(\delta_0)) = \{-2^n, 2^n\}$. Can we prove this conjecture? As a starting point we use the polynomial representation of f_0,

$$p(X) = X^{-1} + X.$$

Next we find in $\mathbb{F}_2$

$$(p(X))^2 = (X^{-1} + X)(X^{-1} + X) = X^{-2} + 2 + X^2 = X^{-2} + X^2,$$

and successively

$$(p(X))^{2^n} = X^{-2^n} + X^{2^n}.$$

As we started from the state δ_0, we find

$$\Lambda(f^{2^n}(\delta_0)) = X^{-2^n} + X^{2^n},$$

i.e., $\text{supp}(f^{2^n}(\delta_0)) = \{-2^n, 2^n\}$. This observation suggests that automata associated with polynomials possessing a similar structure may be interesting.

Definition 10.3.1 A linear cellular automaton has the Fermat property (with index k), if the Laurent polynomial associated with the local function satisfies the equation

$$(p(X))^k = p(X^k)$$

for $k > 1$.

How many polynomials with this property exist? The following result helps to identify polynomials with Fermat property.

Proposition 10.3.2 *Let $a, b \in \mathbb{F}_q$, q prime. Then,*

$$(a + b)^q \text{mod } q = a^q + b^q \text{mod } q.$$

Proof Let q be a prime number. Then, for $i = 1, \ldots, q - 1$,

$$q \Big| \binom{q}{i}.$$

As q is prime and $0 < i < q$, the prime number q is not a prime factor of $i!$, i.e., $q \not| i!$. Therefore,

$$\binom{q}{i} \text{mod } q = \frac{q(q-1)\cdots(q-i+1)}{i!} \text{mod } q = 0.$$

Hence,

$$(a+b)^q \bmod q = \left(a^q + \sum_{i=1}^{q-1} \binom{q}{i} a^i b^{q-1} + b^q\right) \bmod q$$
$$= a^q + b^q \bmod q.$$

□

Remark 10.3.3 As $(a+b)^q \bmod q = a^q + b^q \bmod q$, the mapping $\mathbb{F}_q \to \mathbb{F}_q, x \mapsto x^q$, is a homomorphism of the additive group $(\mathbb{F}_q, +)$. Also for a formal polynomial with coefficients in $\mathbb{F}_q$, we find $(aX^{z_1} + bX^{z_2})^q \bmod q = a^q X^{qz_1} + b^q X^{qz_2} \bmod q$, with the same argument as used in the lemma above. Therefore, any Laurent polynomial for $E = \mathbb{F}_q$ with coefficients in $\{0, 1\}$ possesses the property

$$\left(\sum_i X^i\right)^q = \sum_i X^{qi}.$$

The Fermat property holds for a large class of polynomials. In $\mathbb{F}_2$, every polynomial has this property.

We state an elementary consequence of the Fermat property, which will play a central role in the next sections. Let $\delta_{z_0} \in E^\Gamma$ defined by $\delta_{z_0}(z_0) = 1$ and $\delta_{z_0}(z) = 0$ for $z \neq z_0$. If we know the state of $f^a(\delta_0)$, it is possible to determine the state $f^{k\,a}(\delta_0)$ without computing all states for the time steps between a and $k\,a$.

Proposition 10.3.4 *Consider a linear cellular automaton with the Fermat property with index k. Let $\Gamma = \mathbb{Z}^d$, and $\delta_0 \in E^\Gamma$ as defined above. Let $a \in \mathbb{N}$, $z \in \Gamma$. Then,*

$$f^{k\,a}(\delta_0)(k\,z) = f^a(\delta_0)(z), \qquad f^{k\,a}(\delta_0)(k\,z+j) = 0 \quad for\, j \in \{1,\ldots,k-1\}^d.$$

Proof Let $p(X) = \sum_{|\alpha|\le m} a_\alpha X^\alpha$, where $\alpha \in \mathbb{Z}^d$ denotes a multi-index, be the polynomial associated with f_0. Then, $p^k(X) = p(X^k)$. Furthermore, $\delta_0(z)$ has the representation $1 \in E[X, X^{-1}]$, i.e. corresponds to the constant '1' polynomial. Thus, $f^{k\,a}(\delta_0)$ is represented by $p^{k\,a}(X)$. If we compare the representations of $f^a(\delta_0)$ and $f^{ka}(\delta_0)$,

$$(p(X))^a = \left(\sum_{|\alpha|\le m} a_\alpha X^\alpha\right)^a, \qquad (p(X))^{ka} = \left(\sum_{|\alpha|\le m}^{m} a_\alpha X^{k\alpha}\right)^a$$

we find that $f^{k\,a}(\delta_0)(kz) = f^a(\delta_0)(z)$, and $f^{ka}(\delta_0)(kz+j) = 0$ if $j \in \{1,\ldots,k-1\}^d$.

□

This result indicates a scaling behavior: if we iterate k^2 times the initial state δ_0, and restrict to the subgrid $k\mathbb{Z}$, we recover the state that we obtain by iterating δ_0 only k steps. We exploit this scaling behavior in the next section.

10.3.2 Limit Sets of Linear Cellular Automata

Consider a linear cellular automaton that has the Fermat property with index k. The observation above indicates that rescaling the grid after k^a iterations by k^{-a} should yield a result that is approximately constant. In order to give a precise meaning of "approximately constant", Willson proposed in [177–179] to embed the grid $\mathbb{Z}^d$ in $\mathbb{R}^d$, essentially using a construction that is called "graphical representation" (see also [81–83]). Each grid site with a nonzero local state is represented by a square. The union of all squares codes the support of the state.

The main result of the present section—apart from the strict definition of a limit set (the Kuratowski limit)—is the fact, that this set does not depend on the initial state, but is essentially an intrinsic property of the cellular automaton. This observation allows to relate structures of these limit sets directly to the underlying automata. We follow [177].

Definition 10.3.5

(a) Let $\Omega \subset \mathbb{R}^d$, and $\mathcal{H}(\Omega)$ denote the set of all non-empty closed subsets of Ω.
(b) Consider a linear cellular automaton on $\mathbb{Z}^d$ with global function f. Define

$$G : (E^\Gamma)_c \to \mathcal{H}(\mathbb{R}^{d+1}) \cup \{\emptyset\}, \quad u \mapsto G(u) = \cup_{i=0}^{\infty} \cup_{f^i(u)(z)\neq 0} I_{z,i}$$

where the inclusion extends over all $z \in \mathbb{Z}^d$ and $i \in \mathbb{N}_0$, and $I_{z,i}$ denotes the closed unit interval in $\mathbb{R}^{d+1}$ which is symmetrically centered around (z, i).
(c) Let the linear automaton posses the Fermat property with index k.
Define the function $F_a : (E^\Gamma) \to \mathcal{H}(\mathbb{R}^{d+1}) \cup \{\emptyset\}$, $a \in \mathbb{N}$, by

$$F_a(u_0) = k^{-a}\left(G(u_0) \cap \{(x, y) \,:\, 0 \leq y \leq k^a\}\right).$$

For sake of shortness, we will write $\mathcal{H}$ instead of $\mathcal{H}(\mathbb{R}^d)$. The map G indicates the trajectory of a state $u \in E^\Gamma$ in $\mathbb{R}^{d+1}$ by placing unit squares around $(z, i) \in \mathbb{Z}^d \times \mathbb{Z}$, if $f^i(u)(z) \neq 0$ (see also Fig. 10.2). The map F_a rescales the pattern obtained by G according to the scaling property (Fermat property) of the cellular automaton. The first main result of this section is the fact that the set function $a \mapsto F_a(u)$ has a limit for $a \to \infty$. In order to state this result, we define the Kuratowski limit [151] of a sequence of sets.

Fig. 10.2 Appearance of the Sierpinski gasket as limit set of Wolfram's automaton $f(u_{-1}, u_0, u_1) = u_{-1} + u_1$

Definition 10.3.6 Let $A_i \subset \mathbb{R}^d$, $i \in \mathbb{N}$, be a sequence of sets in $\mathbb{R}^d$. Define the inner limit as

$$\liminf A_i = \{x \in \mathbb{R}^d \, : \, \forall i \in \mathbb{N} \; \exists x_i \in A_i \text{ with } x = \lim_{i\to\infty} x_i\}$$

and the outer limit as

$$\limsup A_i = \{x \in \mathbb{R}^d \, : \, \exists (n_i)_{i\in\mathbb{N}} \subset \mathbb{N}, n_i \to \infty \; \exists x_i \in A_{n_i} \text{ with } x = \lim_{i\to\infty} x_i\}.$$

If these two sets are the same then we call $\liminf A_i = \limsup A_i = \lim A_i$ the Kuratowski limit (set) of the sequence.

There are two simple consequences of this definition: first, $\liminf_{i\to\infty} A_i \subset \limsup_{i\to\infty} A_i$, and second, $\liminf_{i\to\infty} A_i$ as well as $\limsup_{i\to\infty} A_i$ are closed. In our situation, where the sets are subsets of one compact set (that depends on the size of D_0), the Kuratowski limit and the Hausdorff limit agree.

Definition 10.3.7 The ε-neighborhood of a set A is defined as $A_\varepsilon = \cup_{x \in A} B_\varepsilon(x)$ where $B_\varepsilon(x)$ denotes the open ε-ball around x. The Hausdorff distance of two non-empty sets is defined as

$$d_H(A, B) = \inf\{\delta > 0 \,:\, A \subset B_\delta \text{ and } B \subset A_\delta\}.$$

The Hausdorff distance is not a metric on the power set of $\mathbb{R}^d$ (even if we exclude the empty set). E.g., in $\mathbb{R}^d$, the distance of an non-empty, open set and of its closure to any other set coincide. The Hausdorff distance is at best a pseudometric. However, the distance is a metric on the set of all nonempty, compact subsets of $\mathbb{R}^d$ (see e.g. [50]).

Proposition 10.3.8 *Let $A_k \subset \mathbb{R}^d$ closed; if there is a compact set $B \subset \mathbb{R}^d$ such that $A_k \subset B$, then the convergence of $(A_k)_{k \in \mathbb{N}}$ in the Hausdorff distance and the Kuratowski limit coincide.*

Proof Hausdorff convergence implies convergence in the sense of Kuratowski:

Let $\varepsilon > 0$. Convergence in the sense of Hausdorff implies that eventually $A_k \subset A_\varepsilon$. Thus, the outer limit $\limsup A_k$ as well as the inner limit $\liminf A_k$ are subsets of A_ε. As A is closed and $\varepsilon > 0$ arbitrary, we find $\liminf A_k \subset \limsup A_k \subset A$. Furthermore, if $A_k \to A$ with respect to the Hausdorff distance, there is for each point $a \in A$ a sequence $a_k \in A_k$ such that $a_k \to a$ (choose a_k such that $d(a_k, a) = \min\{d(a', a) \,:\, a' \in A_k\}$). Thus, $\liminf A_k \supset A$ and all in all we have $\limsup A_k = \liminf A_k = A$.

Convergence in the sense of Kuratowski implies Hausdorff convergence:

We assume that $\limsup A_k = \liminf A_k = A$.
We first show that for all $\varepsilon > 0$ there is $N \in \mathbb{N}$ such that for all $k > N$ the inclusion $A_k \subset A_\varepsilon$ holds. If this is not the case, then we have for each $\varepsilon > 0$ and each $N \in \mathbb{N}$ a point $a_k \in A_k \setminus A_\varepsilon$ for some $k > N$. Therefore we can construct a (sub-)sequence $a_k \in A_{l(k)}$ with $l(k) \to \infty$ strictly increasing, such that $d(a_k, a) > \varepsilon$ for all $a \in A$. As $a_k \in B$, and B compact, we have a converging subsequence of a_k with limit point a_0, such that $d(a_0, a) > \varepsilon$ for all $a \in A$. However, due to the definition of $\liminf A_k$, we also conclude $a_0 \in A$, which is a contradiction.
With similar reasoning it is possible to show that for all $\varepsilon > 0$ there is $N \in \mathbb{N}$ such that for all $k > N$ the inclusion $A \subset (A_k)\varepsilon$ is valid. Hence, for all $\varepsilon > 0$, we eventually find $d_H(A_k, A) < \varepsilon$, and thus $A_k \to A$ in the sense of Hausdorff. □

After these preliminaries we formulate the central theorem of the present section.

Theorem 10.3.9 *For a linear cellular automaton that has the Fermat property with index k, the limit*

$$\lim_{a \to \infty} F_a(u)$$

exists for all $u \in (E^\Gamma)_c$. If u is not identically zero then this limit does not depend on u.

The proof of this theorem is presented as a sequence of propositions. In the following propositions, the assumptions are always those of this theorem. First we show that $\liminf F_a(\delta_0) \subset \liminf F_a(u_0)$, and $\limsup F_a(\delta_0) \subset \limsup F_a(u_0)$ for any $u_0 \in (E^{\mathbb{Z}})_c$. The proof of this statement is based on the following lemma relating $f^i(u_0)$ to $f^i(\delta_0)$: Recall that $\delta_{z_0}(z) = 1$ if $z = z_0$ and 0 else. Due to the linearity of the cellular automaton, $f^i(u_0)$ can be represented as a linear combination of $f^i(\delta_{z_0})$, $z_0 \in \Gamma$. In particular, if $f^i(u_0)(z)$ is nonzero, we find a $z_0 \in \Gamma$ with $u_0(z_0) \neq 0$ such that $f^i(\delta_{z_0})(z) \neq 0$.

Lemma 10.3.10 *Let $\Gamma = \mathbb{Z}^d$, $u_0 \in (E^\Gamma)_c$, and δ_0 as defined above. Then, for all $i \in \mathbb{N}$ and $z \in \Gamma$ it is true, that*

$$f^i(u_0)(z) \neq 0 \quad \Rightarrow \quad \exists z' \in \Gamma,\ u_0(z') \neq 0 : f^i(\delta_0)(z - z') \neq 0.$$

Proof As $u_0(z) = \sum_{z' \in \mathbb{Z}^d} u_0(z')\delta_0(z - z')$, we have the representation

$$u_0 = \sum_{z' \in \mathbb{Z}^d} u_0(z')\sigma_{-z'}(\delta_0).$$

The linearity of f yields

$$0 \neq f^i(u_0)(z) = \sum_{z' \in \mathbb{Z}^d} u_0(z')(\sigma_{-z'} \circ f^i)(\delta_0)(z) = \sum_{z' \in \mathbb{Z}^d} u_0(z')f^i(\delta_0)(z - z').$$

There is at least one $z' \in \Gamma$ with $u_0(z') \neq 0$ such that $f^i(\delta_0)(z - z') \neq 0$. □

Lemma 10.3.11 *For every $u_0 \in (E^{\mathbb{Z}^d})_c$ holds*

$$\liminf F_a(\delta_0) \supset \liminf F_a(u_0) \quad \textit{and} \quad \limsup F_a(\delta_0) \supset \limsup F_a(u_0).$$

Proof If $(x, t) \in \liminf F_i(u_0)$ (respectively $(x, t) \in \limsup F_i(u_0)$), there is a sequence of natural numbers $(n_i)_{i \in \mathbb{N}}$ with $n_i \leq k^i$ (either for all $i \in \mathbb{N}$ in case of the lim inf, or for a subsequence $(i_l)_{l \in \mathbb{N}}$ of natural numbers with $i_l \to \infty$ in case of lim sup) and $f^{n_i}(u_0)(z_i) \neq 0$, such that

$$(x, t) = \lim_{i \to \infty} k^{-i}(z_l, n_l).$$

We know that there is a sequence z_i' with $u_0(z_i') \neq 0$ such that $f^{n_i}(\delta_0)(z_i - z_l') \neq 0$. As $u_0 \in (E^{\mathbb{Z}^d})_c$, the sequence z_i' is bounded. Thus,

$$\lim_{i \to \infty} k^{-i}(z_i - z_l', n_i) = (x, t).$$

and $\liminf F_i(u_0) \subseteq \liminf F_i(\delta_0)$.

The same argument yields $\limsup F_i(u_0) \subseteq \limsup F_i(\delta_0)$ if we use the sequence (i_l). □

The next step is to prove the converse inclusions, $\liminf F_a(\delta_0) \subset \liminf F_a(u_0)$, and $\limsup F_a(\delta_0) \subset \limsup F_a(u_0)$ for any non-trivial $u_0 \in (E^{\mathbb{Z}d})_c$. This proof is more involved as we need to exclude that the linear superposition deletes occupied sites in $F_a(u_0)$ that is present in $F_a(\delta_0)$. The following lemma parallels Lemma 10.3.10.

Lemma 10.3.12 *Let $\Gamma = \mathbb{Z}^d$, and $u_0 \in (E^\Gamma)_c$, $u_0(0) \neq 0$. Let δ_0 as defined above. We find a fixed $n \in \mathbb{N}$, such that for all $a \in \mathbb{N}$, $a > n$ is true that*

$$\forall i \in \mathbb{N}, f^i(\delta_0)(z_0) \neq 0 : \; f^{i\,k^a}(u_0)(k^a z_0) \neq 0.$$

Proof Recall that $k > 1$ is the index of the Fermat property. As $u_0 \in (E^\Gamma)_c$, we find $n \in \mathbb{N}$ such that

$$k^n > k \max\{|z| \,:\, u_0(z) \neq 0\}.$$

Let us fix a number $a \in \mathbb{N}$ with $a \geq n$. The state $f^{k^a i}(\delta_0)$ is represented by the polynomial $p^i(X^{k^a})$. Any site in state $f^{k^a i}(\delta_0)$ that is nonzero has at least distance k^a from any other site that is nonzero. Like usual, we denote the support of a state u by $\text{supp}(u) = \{z \,:\, u(z) \neq 0\}$. Then,

$$\text{supp}((\sigma_{-z'} \circ f^{ik^a})(\delta_0)) \cap \text{supp}((\sigma_{-z''} \circ f^{ik^a})(\delta_0)) = \emptyset$$

for any $z', z'' \in \text{supp}(u_0)$, $z' \neq z''$. Each site in $\text{supp}(u_0)$ contributes to the state $f^{ik^a}(u_0)$ independently. It is not possible that the linear superposition $f^{k^a\,i}(u_0)(z) = \sum_{z' \in \text{supp}(u_0)} u_0(z') f^{k^a\,i}(\delta_0)(z - z')$ becomes zero in a site z for that one term $u_0(z') f^{k^a\,i}(\delta_0)(z - z')$ is nonzero.

As $f^i(\delta_0)(z_0) \neq 0$, we know that $f^{i\,k^a}(\delta_0)(k^a z_0) = f^i(\delta_0)(z_0) \neq 0$ (Proposition 10.3.4). Since $u_0(0) \neq 0$,

$$\begin{aligned}
& f^{k^a\,i}(u_0)(k^a z_0) \\
&= \sum_{z' \in \text{supp}(u_0)} u_0(z')(\sigma_{-z'} \circ f^{k^a\,i})(\delta_0)(k^a z_0) \\
&= \sum_{z' \in \text{supp}(u_0)} u_0(z') f^{k^a\,i}(\delta_0)(-z' + k^a z_0) \\
&= u_0(0) f^{k^a\,i}(\delta_0)(k^a z_0) + \sum_{z' \in \text{supp}(u_0) \setminus \{0\}} u_0(z') f^{k^a\,i}(\delta_0)(-z' + k^a z_0) \neq 0
\end{aligned}$$

□

Proposition 10.3.13 *The inner limit is translational invariant, i.e. for $z \in \Gamma = \mathbb{Z}^d$ holds*

$$\forall z \in \Gamma : \liminf_{a\to\infty} F_a(u) = \liminf_{a\to\infty} F_a(\sigma_z \circ u).$$

Proof

$$\begin{aligned}
\liminf_{a\to\infty} F_a(\sigma_{z_0} \circ u) &= \liminf_{a\to\infty} k^{-a} \left(\cup_{i=0}^{\infty} \cup_{f^i(\sigma_{z_0}\circ u)(z)\neq 0} I_{z,i} \cap \{(x,y) \,:\, 0 \le y \le k^a\}\right)\\
&= \liminf_{a\to\infty} k^{-a} \left(\cup_{i=0}^{\infty} \cup_{\sigma_{z_0}\circ f^i(u)(z)\neq 0} I_{z,i} \cap \{(x,y) \,:\, 0 \le y \le k^a\}\right)\\
&= \liminf_{a\to\infty} k^{-a} \left(\cup_{i=0}^{\infty} \cup_{f^i(u)(z)\neq 0} I_{\sigma_{z_0}(z),i} \cap \{(x,y) \,:\, 0 \le y \le k^a\}\right)\\
&= \liminf_{a\to\infty} k^{-a} \left((z_0, 0) + \cup_{i=0}^{\infty} \cup_{f^i(u)(z)\neq 0} (I_{z,i}) \cap \{(x,y) \,:\, 0 \le y \le k^a\}\right)\\
&= \liminf_{a\to\infty} k^{-a} \left(\cup_{i=0}^{\infty} \cup_{f^i(u)(z)\neq 0} (I_{z,i}) \cap \{(x,y) \,:\, 0 \le y \le k^a\}\right)\\
&= \liminf_{a\to\infty} F_a(u).
\end{aligned}$$

□

Lemma 10.3.14 *We find for any nontrivial $u_0 \in (E^{\mathbb{Z}^d})_c$ that*

$$\liminf F_a(\delta_0) \subset \liminf F_a(u_0) \quad \textit{and} \quad \limsup F_a(\delta_0) \subset \limsup F_a(u_0).$$

Proof As $\liminf F_a(u_0)$ is shift invariant and u_0 non-trivial, we assume without restriction that $u_0(0) \neq 0$. If $(x,t) \in \liminf F_i(\delta_0)$, there is a sequence of natural numbers $(n_i)_{i\in\mathbb{N}}$ with $n_i \le k^i$ and $f^{n_i}(\delta_0)(z_i) \neq 0$, such that

$$(x,t) = \lim_{i\to\infty} k^{-i}(z_i, n_i)$$

We know due to Lemma 10.3.12 that there is a fixed $a \in \mathbb{N}$, such that $f^{k^a n_i}(u_0)(k^a z_i) \neq 0$. As $n_i k^a \le k^{a+i}$, it is $k^{-a-i}(k^a z_i, k^a n_i) \in F_{a+i}$ and

$$\lim_{i\to\infty} k^{-i-a}(k^a z_i, k^a n_i) = (x,t) \in \liminf F_i(u_0).$$

As the argument extends to sequences as well as to subsequences, we find $\limsup F_i(\delta_0) \subseteq \limsup F_i(u_0)$. □

By now, we know for any non-trivial $u_0 \in (E^\Gamma)_c$ that $\limsup F_a(u_0) = \limsup F_a(\delta_0)$ and $\liminf F_a(u_0) = \liminf F_a(\delta_0)$. Now we come to the last proposition, that finalizes the proof of Theorem 10.3.9.

Proposition 10.3.15 *Let δ_0 be the state in $(E^\Gamma)_c$ such that $\delta_0(z) = 0$ for $z \neq 0$, and $\delta_0(0) = 1$. Then,*

$$\limsup_{a\to\infty} F_a(\delta_0) = \liminf_{a\to\infty} F_a(\delta_0).$$

Proof From the definition of the Kuratowski limit, we have at once the inclusion

$$\limsup_{a\to\infty} F_a(\delta_0) \supset \liminf_{a\to\infty} F_a(\delta_0).$$

We show the conversed inclusion.

Step 1: $k^{-a}(z,n) \in F_a(\delta_0)$ *for $a \in \mathbb{N}$ fixed, then* $k^{-a}(z,n) \in \liminf F_i(\delta_0)$.
As $k^{-a}(z,n) \in F_a(\delta_0)$, we find $f^n(\delta_0)(z) \neq 0$. According to Proposition 10.3.4, we find $f^{k^i n}(\delta_0)(k^i z) \neq 0$, and thus

$$\liminf F_{a+i}(\delta_0) \ni \lim k^{-i-a}(k^i z, k^i n) = (z,n).$$

Step 2: $\limsup F_a(\delta_0) \subseteq \liminf F_a(\delta_0)$.
Let $(x,y) \in \limsup F_a(u_0)$. Then, there is a sequence $(z_i,n_i) \in F_{a_i}(\delta_0)$ such that $k^{-a_i}(z_i,n_i) \to (x,y)$. According to step 1, we have $k^{-a_i}(z_i,n_i) \in \liminf F_a(\delta_0)$ for all $i \in \mathbb{N}$. As $\liminf F_a(\delta_0)$ is a closed set, also all limit points of $(k^{-a_i}(z_i,n_i))_{i\in\mathbb{N}}$ are in $\liminf F_a(\delta_0)$, and hence also $(x,y) \in \liminf F_a(\delta_0)$.

□

Definition 10.3.16 For a linear cellular automaton with Fermat property define the limit set

$$\mathcal{F} = \lim_{a\to\infty} F_a(\delta_0).$$

Why is $\mathcal{F}$ interesting? Often enough, we are interested in the long term behavior of a trajectory. Mostly, this long term behavior is characterized by the ω-limit set of a given trajectory. If we consider our standard example, the Wolfram automaton given by $f_0(u_{-1},u_0,u_1) = u_{-1} + u_1$ in $\mathbb{F}_2$, the ω-limit set becomes rather nontrivial: Start with δ_0, we find non-periodicity on any finite subset of $\mathbb{Z}$. The set $\mathcal{F}$, however, nicely displays the long term structure of the state starting in any nontrivial state with finite support.

In the next sections we especially focus on an insight into the structure of limit sets, described by the dimension and the self-similarity. It will turn out that limit sets are often self-similar fractals.

10.3.3 Iterated Function Systems

The limit set $\mathcal{F}$ constructed above may have a rather complex structure. To understand this structure better, we approach this set from another direction and define an iterated function system, which is a standard device to construct self-similar fractals. However, it will turn out that this formulation is too restricted. Later on, we will generalize this setup slightly and only then we are able to understand linear cellular automata with Fermat-property in depth.

Definition 10.3.17

(a) A map $f : \mathbb{R}^d \to \mathbb{R}^d$ is called a similarity if there is a number $r \in (0, 1)$ such that

$$|f(x) - f(y)| = r\,|x - y|.$$

The number r is called the contraction ratio of f.

(b) An iterated function system on $\mathbb{R}^d$ is a nonempty, finite list of similarities $(f_1, \ldots, f_m)$ with $f_i \neq f_j$ for $i \neq j$. $\mathcal{H}(\mathbb{R}^d) \subset \mathcal{P}(\mathbb{R}^d)$ denotes the set of all compact, non-empty subsets of $\mathbb{R}^d$. The global function associated with the iterated function system is defined as

$$F : \mathcal{H}(\mathbb{R}^d) \to \mathcal{H}(\mathbb{R}^d), \quad A \mapsto \cup_{i=1}^m f_i(A).$$

(c) A set $X \in \mathcal{H}$ is called self-similar if it is a fixed point of the global function of an iterated function system.

Remark 10.3.18 For an iterative function system $(f_1, \ldots, f_m)$, the functions f_i are Lipschitz continuous. The image of a compact set under the global function is a finite union of compact sets and is thus compact. The function F is well defined.

Theorem 10.3.19 *The global function of an iterated function system has exactly one fixed point.*

Proof Consider the iterative function system $(f_1, \ldots, f_m)$ with contraction ratios $(r_1, \ldots, r_m)$ and global function $F(.)$. We equip $\mathcal{H}$ with the Hausdorff metric $d_H(.,.)$, and prove that $F(.)$ is Lipschitz-continuous with Lipschitz constant $r = \max\{r_1, \ldots, r_m\}$. Let $A, B \in \mathcal{H}$. We show that $F(A)$ is contained in $(F(B))_{r\,d_H(A,B)}$, the ε-neighborhood of $F(B)$ with $\varepsilon = r\,d_H(A, B)$. Let $\hat{a} \in F(A)$. Then, there is $a \in A$ and $i \in \{1, \ldots, n\}$ such that $\hat{a} = f_i(a)$. Thus, there is $b \in B$ with $d(a, b) \leq d_H(A, B)$. As A, B, $F(A)$ and $F(B)$ are compact,

$$d(f_i(a), f_i(b)) = r_i d(a, b) \leq r d(a, b) \quad \Rightarrow \quad F(A) \subset (F(B))_{rd_H(A,B)}.$$

From symmetry reasons, we have also $F(B) \subset (F(B))_{rd_H(A,B)}$. Hence,

$$d_H(F(A), F(B)) \leq r d_H(A, B).$$

As $r \in (0, 1)$, the function F is a contraction of $\mathcal{H}$. The set $\mathcal{H}$, together with the Hausdorff metric is a complete, compact metric space [50, Chap. 2.4]. The Banach fixed point theorem implies a unique fixed point of $F(.)$ in $\mathcal{H}$. □

Example 10.3.20

(a) Let $f_1 : \mathbb{R} \to \mathbb{R}, x \mapsto x/2$, and $f_2 : \mathbb{R} \to \mathbb{R}, x \mapsto 1/2 + x/2$. The invariant set is the interval [0, 1].

(b) Let $f_1 : \mathbb{R} \to \mathbb{R}, x \mapsto x/3$, and $f_2 : \mathbb{R} \to \mathbb{R}, x \mapsto 2/3 + x/3$. If we start with the interval [0, 1], the first iteration with the global function yields

$$F([0,1]) = [0,1/3] \cup [2/3,1]$$
$$F([0,1/3] \cup [2/3,1]) = [0,1/9] \cup [2/9,3/9] \cup [2/3,2/3+1/9] \cup [2/3+2/9,1]$$

We easily recognize that the iterated set approximates the standard 1/3 Cantor set, and this set is indeed the unique fixed point of this iterated function system (use the representation $\mathfrak{C}_{1/3} = \{\sum_{i=1}^{\infty} a_i 3^{-i} \,:\, a_i \in \{0,2\}\}$).

There are various definitions for (fractal) dimensions. In "nice" situations they all agree. However, we find sets for which different definitions yield different values. We start with the dimension that can be most easily computed: the similarity dimension. In the following definition, we also state the Hausdorff dimension, the perhaps best known notion of dimension. Later on, we will also define the Kolmogorov or box count dimension.

Definition 10.3.21

(a) Consider an iterated function system $(f_1,\ldots,f_n)$ with contraction ratios $(r_1,\ldots,r_n)$. Let $s \in \mathbb{R}$ given by

$$1 = g(s) = \sum_{i=1}^{n} r_i^s.$$

Then, s is called the similarity dimension of the iterated function system, resp. of the fixed point A of the global function F. We denote this dimension by $\dim_S(A)$.

(b) Let $A \subset \mathbb{R}^d$ denote a compact set. A finite or countable family of sets in $\mathbb{R}^d$ with a union that covers A is called a countable cover. If each of the sets in a countable cover has a diameter less or equal ε, the family is called ε-cover. Define

$$m_\delta^\varepsilon(A) = \inf\{\sum_{C \in C_\varepsilon} \operatorname{diam}(C)^d \,:\, C \text{ is } \varepsilon\text{-cover of } A\}.$$

If $\lim_{\varepsilon\to 0} m_\delta^\varepsilon(A)$ exists, we define

$$m_\delta(A) = \lim_{\varepsilon\to 0} m_\delta^\varepsilon(A).$$

A value $D \in [0, d]$ such that $m_\delta(A) = \infty$ for $\delta < D$, and $m_\delta(A) = 0$ for $\delta > D$ is called Hausdorff dimension. We denote this dimension by $\dim_H(A)$.

Of course, it is necessary to show that each compact set has a Hausdorff dimension. A proof can be found e.g. in the book of Edgar [50]. The similarity dimension can be often easily computed. Let us consider an example.

Example 10.3.22 The Cantor set is the fixed point of the iterative function system $f_1 : \mathbb{R} \to \mathbb{R}$, $x \mapsto x/3$, and $f_2 : \mathbb{R} \to \mathbb{R}$, $x \mapsto 2/3 + x/3$. The similarity dimension s is given by $1/3^s + 1/3^s = 1$, this is,

$$\dim_S(\mathfrak{C}_{1/3}) = \log_3(2).$$

We do not discuss dimensions in depth, but show a handy result [177].

Proposition 10.3.23 *Let $A \subset \mathbb{R}^d$ denote a compact set.*

(a) *Let $a \in \mathbb{N}$, $a \geq 2$, and assume that there are vectors $x_1, \ldots, x_m \in \mathbb{R}^d$ such that $A \subset \cup_{i=1}^m (A + x_i)/a$. Then,*

$$dim_H(A) \leq \log_a(m).$$

(b) *Let $a \in \mathbb{N}$, $a \geq 2$, and assume that there are vectors $x_1, \ldots, x_m \in \mathbb{Z}^d$ which are pairwise disjoint modulo a, such that $A \supset \cup_{i=1}^m (A + x_i)/a$. Then,*

$$dim_H(A) \geq \log_a(m).$$

Proof **(a)** We show that $\delta > \log_a(m)$ implies $m_\delta(A) = 0$. We start off with any ε-cover C_ε such that $\mathrm{diam}(C) < \varepsilon$ for $C \in C_\varepsilon$ and some $\varepsilon > 0$. Let $N = \sum_{C\in C_\varepsilon} \mathrm{diam}(C)^\delta$. Since $A \subset \cup_{i=1}^m (A + x_i)/a$, the set

$$\{(x_i + C)/a \; : \; i = 1, \ldots, m, C \subset C_\varepsilon\}$$

also is a ε/a-cover, and we obtain

$$m_\delta^{\epsilon/a}(A) \leq \sum_{C\in C_\varepsilon} \sum_{j=1}^{m} \mathrm{diam}((x_j + C)/a)^\delta = N\, m a^{-\delta},$$

Iterating this procedure, we find for $k \in \mathbb{N}$

$$m_\delta^{\epsilon/a^k}(A) \leq N \left(m a^{-\delta}\right)^k.$$

If $\delta > \log_a(m)$, then $ma^{-\delta} < 1$ and $k \to \infty$ yields $m_\delta(A) = 0$. Thus, $\dim_H(A) \leq \log_a(m)$.

(b) *Step 1:* Fix $\delta > 0$ such that $\delta < \log_a(m)$, i.e. $a^\delta/m < 1$. We show that $m_\delta(A) > 0$. This time our aim is more challenging, as it requires to establish a lower bound for the infimum $m_\delta^\varepsilon(A)$.
Consider an ε-cover C_ε of A, with $N = \sum_{C \in C_\varepsilon} \operatorname{diam}(C)^\delta$. Ultimately, we will derive a lower bound for N (independent of ε). First of all, as A is compact, we have a finite subset of C that still covers A. It is sufficient to consider finite ε-covers. Select $K > 0$, and determine a number B, such that any set X of diameter less or equal K meets $X + z$, for at most B different vectors $z \in \mathbb{Z}^d$. We use K and B during the estimation (step 3).
Step 2: We utilize the assumption $A \supset \cup_{i=1}^m (A + x_i)/a$ to construct more coverings. We find

$$A \supset \cup_{j_0=1}^m (x_{j_0} + A)/a \supset \cup_{j_0=1}^m \big(x_{j_0} + \cup_{j_1=1}^m (x_{j_1} + A)/a\big)/a$$
$$= \cup_{j_1=0}^m \cup_{j_1=1}^m (A + a\,x_{j_1} + a^0 x_{j_2})/a^2 \supset \cdots$$
$$\cdots \supset \quad \cup_{j_0=1}^m \cdots \cup_{j_p=1}^m (A + a^p x_{j_p} + \cdots + a^0 x_{j_0})/a^p$$

Therefore,

$$a^p A \supset \cup_{j_0=1}^m \cdots \cup_{j_p=1}^m (A + a^p x_{j_p} + \cdots + x_{j_0}).$$

We know that the vectors $x_i \in \mathbb{Z}^d$ are distinct modulo a. This fact implies that $a^p x_{j_p} + \cdots + x_{j_0}$ are all distinct: if we assume

$$a^p x_{j_p} + \cdots + x_{j_0} = a^p x_{\tilde{j}_p} + \cdots + x_{\tilde{j}_0}$$

then $\sum_{i=0}^l a^i (x_{j_i} - x_{\tilde{j}_i}) = 0$. Thus, $x_{j_0} = x_{\tilde{j}_0} \bmod a$, and hence, as for $j_0 \neq \tilde{j}_0$ the two vectors are pairwise different modulo a, necessarily $j_0 = \tilde{j}_0$. Dividing the resulting equation by a, we find $\sum_{i=1}^l a^{i-1}(x_{j_i} - x_{\tilde{j}_i}) = 0$. The same argument yields $j_1 = \tilde{j}_1$. Iterating the argument shows that $a^p x_{j_p} + \cdots + x_{j_0} = a^p x_{\tilde{j}_p} + \cdots + x_{\tilde{j}_0}$ implies $x_{j_i} = x_{\tilde{j}_i}$, such that we obtain m^{p+1} different vectors $a^p x_{j_p} + \cdots + x_{j_0}$.
Step 3: Select $q \in \mathbb{N}$ such that $(a^\delta/m)^{q+1} \leq 1/B$. This is possible as $a^\delta/m < 1$. We show that $N \geq (K/a^q)^\delta$. Thereto, we consider two cases: either at least one set $\tilde{C}$ in the ε-cover C has a diameter $\geq K/a^q$ (case 1), or for all elements C in the ε-covering have $\operatorname{diam}(C) < K/a^q$ (case 2).
Step 3a/case 1: $\exists \hat{C} \in C_\varepsilon : \operatorname{diam}(\hat{C}) \geq K/a^q$.
An immediate consequence is $N \geq \operatorname{diam}(\hat{C})^d = (K/a^q)^\delta$.
Step 3b/case 2: $\forall C \in C_\varepsilon : \operatorname{diam}(C) < K/a^q$
Choose $p = q$. As $\operatorname{diam}(a^p C) \leq K$ for all sets $C \in C_\varepsilon$, we conclude from the definition of B that each of the sets $a^q C$ meets at most B of the sets

$A+a^px_{j_p}+\cdots+x_{j_0}$. Let $\Psi(j_0,\ldots,j_p)=\{C\in C_\varepsilon\;:\;a^qC \text{ hits } A+a^px_{j_p}+\cdots+x_{j_0}\}$. As a^qC_ε covers a^qA and $a^qA\supset\cup_{j_0=1}^m\cdots\cup_{j_p=1}^m(A+a^px_{j_p}+\cdots+x_{j_0})$, the set $a^q\Psi(j_0,\ldots,j_p)$ covers $A+a^px_{j_p}+\cdots+x_{j_0}$. Furthermore, a given $C\in C_\varepsilon$ appears in at most B different sets $\Psi(j_0,\ldots,j_p)$. Therewith, we conclude

$$\sum_{(j_0,\ldots,j_p)}\sum_{C\in\Psi(j_0,\ldots,j_p)}\text{diam}(a^qC)^\delta=a^{qd}\sum_{(j_0,\ldots,j_p)}\sum_{C\in\Psi(j_0,\ldots,j_p)}\text{diam}(C)^\delta$$
$$\leq a^{qd}B\sum_{C\in C_\varepsilon}\text{diam}(C)^\delta=Ba^{q\delta}N$$

where the first sum extends over all $(j_0,\ldots,j_p)\in\{1,\ldots,m\}^{p+1}$. Due to our choice $p=q$, there are m^{q+1} different elements $(j_0,\ldots,j_p)\in\{1,\ldots,m\}^{p+1}$. There is at least one vector of indices $(\hat{j}_0,\ldots,\hat{j}_p)$ such that

$$\sum_{C\in\Psi(\hat{j}_0,\ldots,\hat{j}_p)}\text{diam}(a^qC)^\delta\leq Ba^{q\delta}N/m^{q+1}\leq N$$

where we used the defining condition $(a^\delta/m)^{q+1}\leq 1/B$ in the last inequality. Hence $N'=\sum_{\hat{C}\in\Psi(\hat{j}_0,\ldots,\hat{j}_p)}\text{diam}(a^q\hat{C})^\delta$ is a lower bound for N; it is sufficient to prove $N'\geq(K/a^q)^\delta$ in order to prove the desired bound $N\geq(K/a^q)^\delta$.
As $A+a^qx_{\hat{j}_q}+\cdots+x_{\hat{j}_0}$ is just a translate of A, we conclude that $a^q\Psi(\hat{j}_0,\ldots,\hat{j}_p)$ forms an $a^q\varepsilon$-cover of A (up to a translation). This new cover either meets case 1 (there is one set with a diameter larger or equal K/a^q), or we are again in case 2. If we are in case 1, we are done: for a $a^q\varepsilon$-cover we found that $N'=\sum\text{diam}(.)^\delta\geq(K/a^q)^\delta$. If we are again in case 2, we repeat the construction. As the initial covering C_ε has been finite, and the diameters in each loop are increased by a factor $a^q>1$, we eventually hit case 1.
In case 1 as well as in case 2, we find

$$m_\delta^\varepsilon(A)=\inf\{\sum_{C\in C_\varepsilon}\text{diam}(C)^d\;:\;C_\varepsilon\text{ is }\varepsilon\text{-cover of }A\}\geq(K/a^q)^\delta$$

and thus $m_\delta(A)>(K/a^q)^\delta>0$. This implies $\dim_H(A)<\log_a(m)$. □

Theorem 10.3.24 *Let $A\subset\mathbb{R}^d$ denote a compact set. Let $a\in\mathbb{N}$, $a\geq 2$, and assume that there are vectors $x_1,\ldots,x_m\in\mathbb{Z}^d$ which are pairwise different modulo a. Assume furthermore $A=\cup_{i=1}^m(A+x_i)/a$. Then,*

$$dim_H(A)=\log_a(m).$$

If we define the iterated function system $(f_1,\ldots,f_m)$ by $f_i(x)=(x+x_i)/a$, the set A is the fixed point of the global function associated with the iterative function system. Moreover, the Hausdorff dimension of A and the similarity dimension of the iterative function system agree.

Proof From the previous proposition, we immediately conclude $\dim_H(A) = \log_a(m)$. The similarity dimension for the iterated function system is given by the root s of the equation $1 = \sum_{i=1}^{m}(1/a)^s = ma^{-s}$. Thus, $a^s = m$ and $s = \log_a(m)$.

The only thing that remains to show is that the fixed point X of F coincides with A. This is equivalent to $F(A) = A$, which in turn is equivalent with $A = \cup_{i=1}^{m}(A + x_i)/a$. □

Example 10.3.25 The Cantor set is the fixed point of the iterative function system (see Example 10.3.22) $f_1 : \mathbb{R} \to \mathbb{R}, x \mapsto x/3$, and $f_2 : \mathbb{R} \to \mathbb{R}, x \mapsto (2 + x)/3$. We are in the situation of the theorem above, and find

$$\dim_S(\mathfrak{C}_{1/3}) = \dim_H(\mathfrak{C}_{1/3}) = \log_3(2).$$

Now we turn to our standard example of a cellular automaton with Fermat property: the automaton $f(u_{-1}, u_1, u_1) = u_{-1} + u_1$ in $\mathbb{F}_2$.

Lemma 10.3.26 *The limit set $\mathcal{F}$ of the cellular automaton $f(u_{-1}, u_0, u_1) = u_{-1}+u_1$ in $\mathbb{F}_2$ is the fixed point of the iterative function system $f_i : \mathbb{R}^2 \to \mathbb{R}^2$,*

$$f_1(x) = x/2, \quad f_2(x) = [\begin{pmatrix} -1 \\ 1 \end{pmatrix} + x]/2, \quad f_3(x) = [\begin{pmatrix} 1 \\ 1 \end{pmatrix} + x]/2.$$

Proof We show that $\mathcal{F}$ is invariant under the global function F of the iterative function system (f_1, f_2, f_3). The cellular automaton has the Fermat-property with index $k = 2$.

Step 1: Mirror and shift symmetries.

The state δ_0 possesses the mirror symmetry $\delta_0(z) = \delta_0(-z)$. The trajectories of the present cellular automaton preserve this symmetry such that also the limit set $\mathcal{F}$ is invariant under the reflection $(x, t) \mapsto (-x, t)$.

Apart from this spatial symmetry, we also show some invariance under a certain spatiotemporal shift. Let $n \in \{0, \dots, k^i - 1\}$. We prove the two implications:

$$f^n(\delta_0)(z) = 1 \quad \Rightarrow \quad f^{n+k^i}(\delta_0)(z \pm k^i) = 1$$

and, if $|z| \leq n$,

$$f^{n+k^i}(\delta_0)(z \pm k^i) = 1 \quad \Rightarrow \quad f^n(\delta_0)(z) = 1.$$

We know that $f^{k^i}(\delta_0)(z) = 1$ if and only if $z = \pm k^i$). As the distance between these two sites is k^{i+1}, and the information only spreads one site per time step, the configuration emerging from $z = \pm k^i$ at time k^i cannot interact for the next $k^i - 1$ time steps, that is, until time step $k^i + k^i = k^{i+1}$ (note that $k = 2$ in the present case). Therefore, in time steps $[k^i, k^{i+1})$ the configurations looks like two copies of the configuration in time interval $[0, k^i)$, shifted by $\pm k^i$. This observation implies for $|z| \le n$

$$f^{n+k^i}(\delta_0)(z) = f^{n+k^i}(\delta_0)(z \pm k^i).$$

The observation that $f^n(\delta_0)(z) = 1$ implies $|z| \le n$ establishes the desired conditions.

Step 2: $F(\mathcal{F}) \subset \mathcal{F}$.

Let $x = (y, t) \in \mathcal{F}$. There is a sequence (z_i, n_i) with $n_i \le k^i, f^{n_i}(\delta_0)(z_i) = 1$, and $\lim_{i\to\infty} k^{-i}(z_i, n_i) = (y, t)$. As $n_i \le k^i \le k^{i+1}$, we conclude that

$$\mathcal{F} \ni \lim_{i\to\infty} k^{-i-1}(z_i, n_i) = x/k = f_1(x).$$

Furthermore, we also know that $f^{n_i+k^i}(z_i \pm k^i) = 1$, and hence

$$\mathcal{F} \ni \lim_{i\to\infty} k^{-(i+1)}(z_i \pm k^i, n_i + k^i) = (y \pm 1, t + 1)/k = ((\pm 1, 1)^T + x)/2.$$

Also $f_2(\mathcal{F}), f_3(\mathcal{F}) \subset \mathcal{F}$ is true.

Step 3: $F(\mathcal{F}) \supset \mathcal{F}$.

The argument parallels that of step 2. Let $x = (y, t) \in \mathcal{F}$, such that there is a sequence (z_i, n_i) with $n_i \le k^i, f^{n_i}(\delta_0)(z_i) = 1$, and

$$\lim_{i\to\infty} k^{-i}(z_i, n_i) = (y, t) = x.$$

We distinguish three cases: either $t \in [0, 1/2)$, or $(z, t) \in [0, 1] \times [1/2, 1]$, or $(z, t) \in [-1, 0) \times [1/2, 1]$.

Case 1: $t \in [0, 1/2)$. As $t < 1/2 = 1/k$, we know that eventually $n_i < k^{i-1}$ and hence $k^{i-1}(z_i, n_i) \in F_{i-1}$. Thus,

$$\mathcal{F} \ni \lim_{i\to\infty} k^{i-1}(z_i, n_i) = k(y, t) = f_1^{-1}(x)$$

and $\mathcal{F} \cap \{t < 1/2\} \subset f_0(\mathcal{F})$. The arguments for cases 2 and 3 are essentially the same, using $f_2^{-1}(.)$ and $f_3^{-1}(.)$ instead of $f_1^{-1}(.)$. □

Using Theorem 10.3.24 above, we immediately find the dimension of the limit set.

Corollary 10.3.27 *The dimension of the limit set* $\mathcal{F}$ *of the cellular automaton* $f(u_{-1}, u_0, u_1) = u_{-1} + u_1$ *in* $\mathbb{F}_2$ *reads*

$$dim_S(\mathcal{F}) = dim_H(\mathcal{F}) = \log_2(3).$$

There are other cellular automata like $f(u_{-1}, u_1, u_1) = u_{-1} + u_0$ in $\mathbb{F}_2$ that can be handled in this way. In other cases the tools developed so far are not sufficient to determine an iterative function system describing the limit set $\mathcal{F}$. Therefore, in the next section we change the construction and introduce matrix substitution systems.

10.3.4 Matrix Substitution Systems

In this section, we define matrix substitution systems, and prove some basic properties. We later use matrix substitution systems as an alternative way to construct the limit set of a cellular automaton. It will turn out that the advantage to approach limit sets of cellular automata via matrix substitution systems instead of iterated function systems are two-fold: (1) there is a rather general and straight way to understand the self-similarity and (2) it is possible to introduce a further definition for the dimension, the growth rate dimension, that is intuitive and straightforward to compute for matrix substitution systems.

Definition 10.3.28 Consider a finite alphabet E and $\Gamma = \mathbb{Z}^d$. Let $0 \in E$ a distinct element, called resting or zero element. Define $D_0 = \{0, \dots, m-1\}^d$, and a map

$$M_0 : E \to E^{D_0}$$

that maps zero to $M(0) = \overline{0}$ (where the state $\overline{0} \in E^{D_0}$ denotes the configuration consisting of zeros only). Let $u_0 \in (E^\Gamma)_c$ a state that is not identically zero. A matrix substitution system is a tuple (Γ, E, m, M_0, u_0).

A matrix substitution system forms (like a cellular automaton) a dynamical system. The state space is $(E^\Gamma)_c$, and the initial state is defined by u_0. In each step, an element $u(z) \in E$ at location $z \in \Gamma$ is replaced by the matrix $M(u(z)) \in E^U$. This is, the tupel (Γ, E, m, M_0) induces a global function

$$M : (E^\Gamma)_c \to (E^\Gamma)_c$$

by

$$M(u)(m\,r + s) = M_0(u(r))(s), \qquad r \in \mathbb{Z}^d, \quad s \in \{0, \dots, m-1\}^d.$$

In most examples, we take $E = \mathbb{F}_q$ and $u_0 = \delta_0$.

Example 10.3.29 Let $\Gamma = \mathbb{Z}^2$, $m = 2$, $E = \mathbb{F}_2$, and $u_0 = \delta_0$. Define

$$M_0(0) = \begin{pmatrix} 0\,0 \\ 0\,0 \end{pmatrix}, \qquad M_0(1) = \begin{pmatrix} 1\,1 \\ 1\,0 \end{pmatrix}.$$

If we suppress zero, the first iterations read

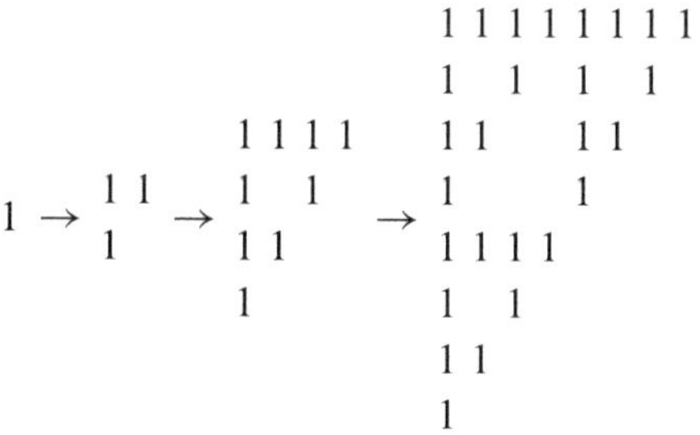

A structure evolves that bears some similarity with the Sierpinski gasked.

Remark 10.3.30

(a) Note that matrix substitution systems are not special cases of cellular automata, as the locality of the function $M : (E^\Gamma)_c \to (E^\Gamma)_c$ is not given: the entry $M(u)(m\,r)$ only depends on the state of grid point r. If r is large, this dependence may span arbitrary long distances. Nevertheless, the map M is continuous on $(E^\Gamma)_c$ with respect to the Cantor topology. However, it is not shift invariant.

(b) Consider (Γ, E, m, M_0, u_0), and assume that $M(u_0)(z) = u_0(z)$ if $u_0(z) \neq 0$. Then, induction shows that $M^i(u_0) = u_0$ if $u_0(z) \neq 0$. Let us consider "the second generation points", i.e., the set $\Omega_1 \subset \Gamma$ defined by

$$\Omega_1 = \{z \in \Gamma \,:\, z = mz_0 + z', u_0(z_0) \neq 0, z' \in \{0, \ldots, m-1\}^d\}.$$

Then, $M^i(u_0)(z) = M^{i-1}(u_0)(z)$ for all $z \in \Omega_1$ and $i \geq 1$. We can iterate this observation: Let

$$\Omega_j = \{z \in \Gamma \,:\, z = m^j z_0 + z', z_0 \in \Omega_{j-1}, z' \in \{0, \ldots, m-1\}^d\}.$$

Then, $M^i(u_0)(z) = M^{i-1}(u_0)(z)$ for all $z \in \Omega_j$ and $i \geq j$. The state of the matrix substitution system becomes immediately constant on all cells that the iteration process reaches.

We define a graphical representation for matrix substitution systems as we did for cellular automata. Recall that $\mathcal{H}(\mathbb{R}^d)$ denotes the set of all closed subsets of $\mathbb{R}^d$.

Definition 10.3.31 Consider a matrix substitution system $(\mathbb{Z}^d, E, m, M_0)$, and define

$$G : (E^\Gamma)_c \to \mathcal{H}(\mathbb{R}^n) \cup \{\emptyset\}, \quad u \mapsto G(u) = \cup_{u(z)\neq 0} I_z$$

where the inclusion extends over all $z \in \mathbb{Z}^d$, and $I_z = z + [0, 1]^d$ denotes the closed unit square in $\mathbb{R}^{d+1}$ which has z in the "lower, left corner".

Define the rescaled graphical representations $\hat{F}_i(u_0) \subset \mathcal{H} \cup \{\emptyset\}$ by

$$\hat{F}_i(u_0) = m^{-i} G(M^i(u_0)).$$

Remark 10.3.32 Note that the unit intervals I_z used here are not symmetrically centered around z (this has been the case in the graphical representation used for cellular automata), but z is a corner of the interval. At the end of the day, the reason is technical: in this construction, the graphical representation is contained in the positive cone of $\mathbb{R}^d$, while the corresponding construction for cellular automata used a half-space $\mathbb{R}^d$ (positive times).

Limit Sets Before we come to the next theorem and show that the rescaled graphical representation of a matrix substitution system has a limit, we state a result about the convergence of nested sequences of compact sets w.r.t. the Hausdorff metric. We use this obvious proposition in the proof of convergence of the sets $\hat{F}_i$.

Proposition 10.3.33 *Let $A_n \subset \mathbb{R}^d$ denote a nested sequence of non-empty, compact sets, $A_i \subset A_{i-1}$. This sequence is a Cauchy sequence w.r.t. the Hausdorf metric, and converges to $\cap_{i\in\mathbb{N}} A_i$.*

Theorem 10.3.34 *Consider the sequence of rescaled graphical representations $\hat{F}_i(u_0)$ of a matrix substitution system $(\mathbb{Z}^d, E, m, M_0, e)$. If $\hat{F}_i(u_0) \neq \emptyset$ for all $i \in \mathbb{N}$, then $\{\hat{F}_i(u_0)\}_{i\in\mathbb{N}}$ form a Cauchy sequence in the metric space $(\mathcal{H}(\mathbb{R}^d), d_H)$.*

Proof As $0 \in E$ is mapped to the matrix with zero entries only, we find that $G(M^i(z)) = 0$ implies that $G(M^{i+1})(mz + j) = 0, j \in \{0, \ldots, m-1\}^d$. Therefore,

$$\hat{F}_{i+1} \subset \hat{F}_i$$

and $\{\hat{F}_i\}_{i\in\mathbb{N}}$ is a Cauchy sequence. □

10.3.4.1 Growth Rate Dimension

In order to measure the fractal dimension of the limit set, the growth rate dimension (sometimes also called Kolmogorov dimension, or box count dimension) is rather simple to handle. Note that sometimes the Kolmogorov dimension, or box count dimension of a set is defined differently. The idea is to count the number $N(\epsilon)$ of boxes with side length ε that cover the set (have a non-empty intersection with the set), and to estimate the growth of this number as $\varepsilon \to 0$. In order to use the special setup of the present construction, we note that the boxes in $\hat{F}_i(u_0)$ have side length m^{-i}, so we have a natural scaling of ε with i.

Definition 10.3.35 Consider a matrix substitution system $(\mathbb{Z}^d, E, m, M_0, u_0)$ with rescaled graphical iterations $\hat{F}_i(u_0) \neq \emptyset$.

(a) Let $\hat{\mathcal{F}}(u_0)$ be the limit of the sequence $\{\hat{F}_i(u_0)\}$

(b) The growth rate dimension of $\hat{\mathcal{F}}(u_0)$ is defined as

$$\dim_G(\hat{\mathcal{F}}(u_0)) := \lim_{i\to\infty} \log_m(|\{z \in \mathbb{Z}^d : M^i(u_0)(z) \neq 0\}|)/i.$$

Example 10.3.36

(a) In order to get some feeling for this definition, consider the (rather trivial) matrix substitution system $\Gamma = \mathbb{Z}^d$, $m \in \mathbb{N}$, $E = \mathbb{F}_2$, and $u_0 = \delta_0$. As before, let $D_0 = \{0, \ldots, m-1\}^d$, and define

$$M(0) = \overline{0} \in E^{D_0}, \quad M(1) = \overline{1} \in E^{D_0}.$$

This is, 1 is mapped to the $m\times m\times\cdots\times m$ matrix with only zero entries, and 1 to that with only '1' entries. Therefore, we expect $\hat{\mathcal{F}} = [0, 1]^d$, and the dimension should be d. Indeed, the number of '1' after the first iteration of δ_0 is m^d; in the next iteration, each of these '1' are again replaced by m^d symbols '1', and so on. Thus,

$$|\{z \in \mathbb{Z}^d : G(M^i(\delta_0)(z) \neq 0\}| = m^{i\,d}$$

and

$$\dim_G(\hat{\mathcal{F}}(\delta_0)) := \lim_{i\to\infty} \log_m(|\{z \in \mathbb{Z}^d : M^i(\delta_0)(z) \neq 0\}|)/i = \lim_{i\to\infty} d\,i/i = d.$$

(b) Now consider the Example 10.3.29, the Sierpinsky gasket. Here, we have $m = 2$, and a "1" is replaced by three symbols "1". Thus,

$$|\{z \in \mathbb{Z}^d : M^i(\delta_0)(z) \neq 0\}| = 3^i$$

and

$$\dim_G(\hat{\mathcal{F}}(\delta_0)) := \lim_{i\to\infty} \log_2(3^i)/i = \log_2(3).$$

We recover the well known dimension of the Sierpinsky gasket (see also Corollary 10.3.27).

Remark 10.3.37 In our case, the growth rate dimension and the Hausdorff dimension of $\hat{\mathcal{F}}$ will always agree. We do not prove this fact here, but a proof can be found in von Haeseler [83], and Wilkens [179]; at the end of the day, the argument is based on the so-called open set condition, see also Edgar [50]. This observation justifies to write *dim*$_G(\hat{\mathcal{F}})$, as the Hausdorff dimension (and hence also the growth rate dimension) is a property of the limit set $\hat{\mathcal{F}}$, and not a property of the matrix substitution system.

In the example above, the alphabet E only consists of two signs, zero and one. This fact allows to directly compute the number of non-void boxes after i iterations. In general, the situation is slightly more involved. We introduce an operator that counts the number of appearances of a given sign in a configuration $u \in (E^\Gamma)_c$. Thus, with $E^\times = E \setminus \{0\}$ (the alphabet without the resting element) we define

$$b : (E^\Gamma)_c \to \mathbb{N}_0^{E^\times}, \qquad u \mapsto b(u) \quad \text{with} \quad (b(u))_e = \sum_{g\in\Gamma} \chi_e(u(g))$$

where $\chi_e(e')$ is the characteristic function, i.e. $\chi_e(e) = 1$, and $\chi_e(e') = 0$ for $e \neq e'$. If we number the elements in $E^\times$, we may write $b(u)$ as a vector, where the j'th component corresponds to the number of appearances of the j'th sign in u. The number of nonzero sites can be expressed by

$$|\{z \in \mathbb{Z}^d \,:\, G(M^i(u_0))(z) \neq 0\}|) = \mathbf{e}^T b(M^i(u_0))$$

where $\mathbf{e} = (1, \ldots, 1)^T$.

What do we gain from the definition of the pattern counting operator $b(\cdot)$? It is possible to introduce a next-generation matrix T that computes $b(M^{i+1}(u))$ from $b(M^i(u))$, only taking into account the vector $b(M^i(u))$ without the need to actually know the configuration $M^i(u)$ explicitly: Any sign $e \in E^\times$ is replaced by $M_0(e)$. Thus, the number of signs e' that appear by replacing e is the number of appearances of e' in $M_0(e)$. Let $T \in \mathbb{N}_0^{E^\times \times E^\times}$ given by

$$T_{e,e'} = \text{number of appearances of sign } e \text{ in } M_0(e').$$

We call T the next generation operator, and find

$$(b(M^{i+1}(u)))_e = \sum_{e'\in E^\times} T_{e,e'}(b(M^i(u)))_{e'}.$$

This equation can be re-written as $b(M^{i+1}(u)) = Tb(M^i(u))$. We iterate with the operator T, or, if we number the elements in $E^\times$, with the matrix T.

For the seed u_0 of the matrix substitution system (Γ, E, m, M_0, u_0), we know that the initial counting vector $b(u_0)$ is non-negative and nonzero. Then,

$$|\{z \in \mathbb{Z}^d \,:\, G(M^i(u_0))(z) \neq 0\}|) = \mathbf{e}^T T^i b(u_0).$$

T is non-negative. If T is also primitive, the theorem of Perron-Frobenius [63] can be used to characterize the asymptotic behavior: Order the absolute values of the eigenvalues. Denote the largest absolute value $\rho(T)$ (also called the spectral radius), and μ the next smaller value. Then, there is $c > 0$ such that

$$|\{z \in \mathbb{Z}^d \,:\, G(M^i(u_0))(z) \neq 0\}| = \mathbf{e}^T T^i b(u_0) = c\,\rho(T)^i + O(\mu^i)$$

Therefore,

$$\begin{aligned}\dim_G(\hat{\mathcal{F}}) &= \lim_{i\to\infty} \log_m(|\{z \in \mathbb{Z}^d : G(M^i(u_0))(z) \neq 0\}|)/i \\ &= \lim_{i\to\infty} \log_m(c\,\rho(T)^i + O(\mu^i))/i \\ &= \lim_{i\to\infty} [i \log_m(\rho(T)) + \log_m(c) + \log_m(1 + O((\mu/\rho(T))^i))]/i\end{aligned}$$

As $|\mu| < \rho(T)$, we find $1 + O((\mu/\rho(T))^i)) \to 1$, such that

$$\dim_G(\hat{\mathcal{F}}(u_0)) = \lim_{i\to\infty} i \log_m(\rho(T))/i = \log_m(\rho(T)).$$

This is, we have shown the following useful result.

Theorem 10.3.38 *Consider the matrix substitution system (Γ, E, m, M_0, u_0). Let T denote the next generation operator $T \in \mathbb{N}_0^{E^\times \times E^\times}$, defined by*

$$T_{e,e'} = \text{number of appearances of sign } e \text{ in } M_0(e').$$

If T is primitive, then the growth rate dimension of the limit set $\hat{\mathcal{F}}$ is given by

$$dim_G(\hat{\mathcal{F}}(u_0)) = \log_m(\rho(T))$$

independently of u_0 (provided that u_0 is not identically zero).

10.3.4.2 Hierarchical Iterative Function Systems

This section connects matrix substitution systems $(\mathbb{Z}^d, E, m, M_0, e)$ and iterative function systems. The definition of an iterative function system is extended to $\Pi_{i=0}^{m-1}\mathbb{R}^d$. These extended iteration systems are called hierarchical iterative function systems. Each component of the state space is related to one sign in E. In this way it is possible to relate the rescaled graphical representation of the matrix substitution systems with hierarchical iterative function systems.

Definition 10.3.39 Let $m, n, d \in \mathbb{N}$. Let W_i, $i = 0, \ldots, n$, nonempty subsets of $\{0, \ldots, m-1\}^{d+1} \times \{1, \ldots, n\}$, such that for a given vector $z \in \{0, \ldots, m-1\}^d$ and a given set W_i, we have at most one entry $(z, j) \in W_i$, i.e.,

$$|\{(z, j) \in W_i : j \in 0, \ldots, m-1\}| \leq 1.$$

The integers m, n, d and the sets W_i, $i = 0, \ldots, m-1$, define a hierarchical iterative function system in the following way:

For $z \in \{0, \ldots, m-1\}^d$ denote by g_z the functions $g_z : \mathbb{R}^d \to \mathbb{R}^d, x \mapsto (z+x)/m$. Define

$$F : \Pi_{i=1}^m \mathcal{H}([0,1]^d) \to \Pi_{i=1}^m \mathcal{H}([0,1]^d), \quad A = (A_1, \ldots, A_n) \mapsto B = (B_1, \ldots, B_n)$$

with

$$B_i = \bigcup_{(z,j) \in W_i} g_z(A_j).$$

We call F the global function of the hierarchical iterative function system.

The hierarchical iterative function systems splits the interval $[0, 1]^d$ into m^d equal-sized intervals. Each of these small intervals either stays empty, or one of the sets $A_0, \ldots, A_{n-1}$ is scaled and copied into it (see Fig. 10.3).

Proposition 10.3.40 *The global function of a hierarchical iterated function system has exactly one fixed point.*

The proof of Proposition 10.3.40 parallels that of Proposition 10.3.19, where we define the distance for two sets $A = (A_1, \ldots, A_n), B = (B_1, \ldots, B_n) \in \Pi_{i=1}^m \mathcal{H}(\mathbb{R}^d)$ by $d_H(A, B) = \max\{d_H(A_1, B_1), \ldots, d_H(A_n, B_n)\}$.

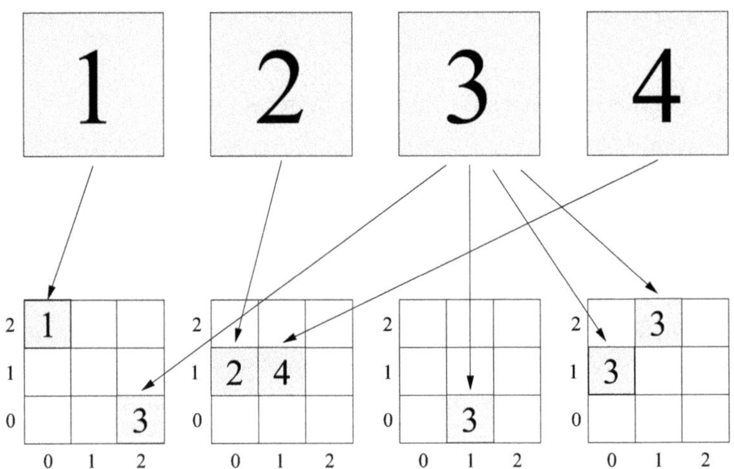

Fig. 10.3 Scheme of a hierarchical iterative function systems with $d = 2$, $m = 3$, $n = 4$, and $W_1 = \{(2, 0, 3), (0, 2, 1)\}$, $W_2 = \{(0, 1, 2), (1, 1, 4)\}$, $W_3 = \{(1, 0, 3)\}$, and $W_4 = \{(0, 1, 3), (1, 2, 3)\}$

We clarify the connection between matrix substitution systems and hierarchical iterative matrix systems. The key observation is the next lemma. Before we state this lemma, we consider an example. The example illustrates the notation and the idea of the lemma.

Example 10.3.41 Consider the matrix substitution system (Γ, E, m, M_0, u_0) where $\Gamma = \mathbb{Z}^2$, $m = 2$, $E = \{0, 1, 2\}$, and $u_0(0, 0) = 1$, $u_0(0, 1) = 2$, and $u_0(z) = 0$ elsewhere.

The map M_0 is defined as

$$M_0(0) = \begin{pmatrix} 0 & 0 \\ 0 & 0 \end{pmatrix}, \qquad M_0(1) = \begin{pmatrix} 1 & 2 \\ 0 & 2 \end{pmatrix}, \qquad M_0(2) = \begin{pmatrix} 2 & 1 \\ 0 & 1 \end{pmatrix}.$$

We find (suppressing zero in the representation)

$$1\;2 \mapsto \begin{array}{cccc} 1 & 2 & 2 & 1 \\ & 2 & & 1 \end{array} \mapsto \begin{array}{cccccccc} 1 & 2 & 2 & 1 & 2 & 1 & 1 & 2 \\ & 2 & & 1 & & 1 & & 2 \\ & & 2 & 1 & & & 1 & 2 \\ & & & 1 & & & & 2 \end{array}$$

Now we play a slightly different game. We work with three iterations at a time, where the initial states $u_i \in (E^\Gamma)_c$, $i = 0, 1, 2$, are given by

$$u_i(0) = i \qquad u_i(z) = 0 \text{ for } z \neq 0.$$

The k'th iteration step generates states u_i^k. These states can be considered as matrices of size $2^k - 1$, since u_i^k are zero outside the region $\{0, \ldots, 2^k - 1\}^2$. In order to determine u_1^k from u_i^{k-1}, note that

$$M_0(1) = \begin{pmatrix} 1 & 2 \\ 0 & 2 \end{pmatrix}.$$

We interpret this matrix as a layout grid for u_1^k in that u_1^k is considered to consist of four matrices of size 2^{k-1}: The matrix in the left, upper side is u_1^{k-1}, the matrix in the right, upper location is u_2^{k-1}, the matrix in the lower, left location reads u_0^{k-1} and the matrix in the lower, right location is u_2^{k-1}. We write more formally

$$u_i^k = \left(\begin{array}{c|c} u^{k-1}_{(M_0(i))_{1,1}} & u^{k-1}_{(M_0(i))_{1,2}} \\ \hline u^{k-1}_{(M_0(i))_{2,1}} & u^{k-1}_{(M_0(i))_{2,2}} \end{array} \right).$$

Let us see what we obtain in this way (we only show the non-trivial part of the state, i.e. the $2^k \times 2^k$ matrices):

$$
\begin{array}{ccc}
u_0^k & u_1^k & u_2^k \\[1em]
(0) & (1) & (2) \\[1em]
\left(\begin{array}{c|c} 0 & 0 \\ \hline 0 & 0 \end{array}\right) &
\left(\begin{array}{c|c} 1 & 2 \\ \hline 0 & 2 \end{array}\right) &
\left(\begin{array}{c|c} 2 & 1 \\ \hline 0 & 1 \end{array}\right) \\[1em]
\left(\begin{array}{cc|cc} 0 & 0 & 0 & 0 \\ 0 & 0 & 0 & 0 \\ \hline 0 & 0 & 0 & 0 \\ 0 & 0 & 0 & 0 \end{array}\right) &
\left(\begin{array}{cc|cc} 1 & 2 & 2 & 1 \\ 0 & 2 & 0 & 1 \\ \hline 0 & 0 & 2 & 1 \\ 0 & 0 & 0 & 1 \end{array}\right) &
\left(\begin{array}{cc|cc} 2 & 1 & 1 & 2 \\ 0 & 1 & 0 & 2 \\ \hline 0 & 0 & 1 & 2 \\ 0 & 0 & 0 & 2 \end{array}\right)
\end{array}
$$

The interesting observation now is, that the configuration $M^k(u_0)$ computed above in the usual way coincides with the concatenation of u_1^k and u_2^k,

$$M^k(u_0) = u_1^k | u_2^k.$$

The next lemma shows, that this finding is no coincidence but reveals a general principle: there are two different ways to construct the states $M^k(u_i)$. Either in the original way introduced in Definition 10.3.31; each local state is replaced by the appropriate matrix $M_0(\cdot)$. Or, in a recursive way; the matrices $M^{k-1}(u_i)$ are used as building-blocks that are assembled according to the blueprint $M_0(\cdot)$.

Lemma 10.3.42 *Consider a matrix substitution system* (Γ, E, m, M_0, u_0) *where* $\Gamma = \mathbb{Z}^d$. *Let* $E = \{0, \ldots, n-1\}$, *and define* $w_i^0 = i\,\delta_0 \in (E^\Gamma)_c$, *i.e.,*

$$w_i^0(0) = i, \qquad w_i(z) = 0 \quad \textit{for } z \neq 0, \qquad i = 0, \ldots, n-1.$$

Define furthermore the maps

$$H_k : \Pi_{i=0}^{m-1}(E^\Gamma)_c \to \Pi_{i=0}^{n-1}(E^\Gamma)_c, \qquad (u_0, \ldots, u_{n-1}) \mapsto (v_0, \ldots, v_{n-1}),$$

where

$$v_i(m^k z + z') = u_{M_0(i)(z)}(z'), \quad z \in \{0, \ldots, m-1\}^d, \quad z' \in \{0, \ldots, m^k - 1\}^d.$$

Let $w^0 = (w_0^0, \ldots, w_{m-1}^0)$, *and* $w^k = H_k(w^{k-1})$, $k = 1, 2, \ldots$. *Then,*

$$M^j(w_i^0)|_{z' \in \{0, \ldots, m^j - 1\}^d} = w_i^j.$$

Proof The claim is obvious for the first iteration. Now assume that the claim is true for all iterations up to $j-1$. Then,

$$M^j(w_i^0) = M^{j-1}(M(w_i^0)).$$

Now we know that $M(w_i^0)$ coincides with $M_0(i)$ for $z \in \{0,\dots,m-1\}^d$, and contains zero elsewhere. In the steps $2,\dots,j$ is $M_0(i)(z)$ replaced by

$$M^{j-1}(M_0(i)(z))|_{z' \in \{0,\dots,m^{j-1}-1\}^d} = w_i^{j-1}.$$

Hence,

$$M^j(w_i^0)|_{z' \in \{0,\dots,m^j-1\}^d} = H_j(w^{j-1}) = w_i^j.$$

□

Remark 10.3.43

(a) A direct consequence is the formula

$$M^j(u_0)(m^j z + z') = H_j \circ H_{j-1} \circ \cdots H_2 \circ H_1(w^0_{u_0(z)})(z')$$

where $z \in \{0,\dots,m-1\}^d$ and $z' \in \{0,\dots,m^j-1\}^d$.

(b) Note that the maps H_k require very little information about the states $H_k(w_0,\dots,w_{m-1})$. If we are only interested in the information "$M^j(w_i^0)(z) = 0$" or "$M^j(w_i^0)(z) \neq 0$", we may start with $\tilde{w}_i^0$, where $\tilde{w}_0^0$ is identically zero (like w_0^0), and the states $\tilde{w}_i^0$ are one in the origin of the grid and zero elsewhere (compare with $w_i^0 = i\,\delta_0$ that are in the origin nonzero, but zero elsewhere). Then, $H_j \circ \cdots \circ H_1(\tilde{w}_0,\dots,\tilde{w}_{m-1})$ have entry zero where (w_i^k) have entry zero, and entry one in all grid points where (w_i^k) assume nonzero local states. The information which of the signs in E is at a certain location is lost—this information, however, is not necessary to decide about zero or nonzero sites. This observation implies the following construction resp. corollary.

Construction 10.3.44 Let a matrix substitution system (Γ, E, m, M_0, u_0), $E = \{0,\dots,n-1\}$, $\Gamma = \mathbb{Z}^d$ with graphical representation $\hat{F}_i(u_0)$ be given. The symbol zero is ignored (coded by $\emptyset$) in the graphical representation. The sets

$$W_i = \{(z,j) \in \{0,\dots,m-1\}^d \times \{1,\dots,n\} \,:\, M_0(j)(z) = i\}.$$

define a hierarchical function iteration system with global function $\tilde{F}$ on $\mathcal{H}([0,1]^d)$.

With this setting, we find the following corollary.

Corollary 10.3.45

(1) Start the hierarchical function iteration system with $A_0 = \Pi_{i=0}^{m-1}[0,1]^d$, *and the matrix substitution system with* $u_j = j\,\delta_0$ *for* $j \in \{0,\ldots,m\}$, *i.e.,* $u_j(0) = j$ *and* $u_j(z) = 0$ *elsewhere. Then, for all* $k \in \mathbb{N}_0$

$$(\tilde{F}^k(A))_j = \hat{F}_k(u_0).$$

(2) If the matrix substitution system is started with $v \in (E^\Gamma)_c$, *we obtain*

$$\hat{F}^k(v) = \bigcup_{z\in\Gamma} z\,m^k + (\tilde{F}^k(A))_{v(z)}.$$

The second point implies, loosely speaking, that we first write $v(z)$ in the appropriate unit intervals at $z \in \mathbb{R}^d$, and then replace each symbol $v(z)$ with its corresponding representation $(\hat{F}^k(A))_{u_0(z)}$.

10.3.5 Cellular Automata and Matrix Substitution Systems

We go back to cellular automata and construct a different way to compute the sets F_n, which we used to determine the limit set $\mathcal{F}$ of a cellular automaton. Justified by the next proposition, we assume that the Laurent polynomial that represents a linear cellular automaton is already a polynomial.

Proposition 10.3.46 *Let* $(\mathbb{Z}^d, D_0, \mathbb{F}_q, f_0)$ *denote a linear cellular automaton with Fermat property and global function* $f(.)$. *Denote the limit set by* $\mathcal{F}$. *The limit set* $\tilde{\mathcal{F}}$ *of the shifted cellular automaton with global function* $\tilde{f}(u) = \sigma_z \circ f(u)$, $z \in \mathbb{Z}^d$, *can be constructed from* $\mathcal{F}$ *by an affine map* $g : \mathbb{R}^{d+1} \to \mathbb{R}^{d+1}$, $(x,t) \mapsto (x - z\,t, t)$, *i.e.,* $\tilde{\mathcal{F}} = g_z(\mathcal{F})$.

Proof Let $\mathcal{F}$ be the limit set of the cellular automaton with global function f, and $\tilde{\mathcal{F}}$ that corresponding to $\sigma_x \circ f$. We show that $g(\mathcal{F}) \subset \tilde{\mathcal{F}}$: If $(x,t) \in \mathcal{F}$, there is a sequence (z_i, n_i) with $n_i \leq k^i$, $f^{n_i}(z_i) \neq 0$, and $(x,t) = \lim_{i\to\infty} k^{-i}(z_i, n_i)$. Thus, $(\sigma_z \circ f)^{n_i}(z_i - n_i z) = (\sigma_z^{n_i} \circ f^{n_i})(z_i - n_i z) \neq 0$, and

$$\lim_{i\to\infty} k^{-i}(z_i - n_i z, n_i) = (x - zt, t) = g(x,t) \in \tilde{\mathcal{F}}.$$

A parallel argument shows that $g(\mathcal{F}) \supset \tilde{\mathcal{F}}$. □

As an affine transformation does not change self-similarity and the (Hausdorff)-dimension of a set, we do not care for this transformation (see also Fig. 10.5 below). If the original local function f_0 of a cellular automaton is represented by the Laurent polynomial $p(x)$, the shifted cellular automaton corresponds to a local function represented by $x^i p(x)$. If we choose i appropriately, we convert a Laurent polynomial into a polynomial. Note, that the Fermat-property with index k is not destroyed,

$$(x^i p(x))^k = x^{ik} p^k(x) = (x^k)^i p(x^k).$$

We assume now the situation that $\Gamma = \mathbb{Z}^d$, and that the local function corresponds to a polynomial $p(x)$ of degree n that possesses the Fermat-property with index k. The neighborhood D_0 of the cellular automaton reads $D_0 = \{-n, \ldots, 0\}^d \subset \mathbb{Z}^d$.

10.3.5.1 Construction of a Matrix Substitution System

We construct a matrix substitution system that generates the time-space diagram of the cellular automaton. We will do this in a constructive way, such that the construction of the matrix substitution system, and the proof that this system indeed generates the time-space diagram of the trajectory starting with state δ_0, are done at the same time.

To make things easier to understand, we do each step twice: for the special automaton given by $\Gamma = \mathbb{Z}$, $E = \mathbb{F}_2$, and $p(x) = 1 + x + x^2$, and for a general linear automaton that possesses the Fermat property with index k on $\Gamma = \mathbb{Z}^d$.

Step 1: *Example.* The key observation and starting point is Lemma 10.3.4. This lemma implies that $f^{2i}(\delta_0)$ can directly be determined from $f^i(\delta_0)$ without computing all time steps inbetween by inserting a zero between any two neighboring cells. E.g., we know that $f^1(\delta_0) = \ldots 0011100$. Then, $f^2(\delta_0) = \ldots 0010101000 \ldots$ (see also Fig. 10.4a and b).

General setting. We introduce the expanding-operator,

$$\text{expand}_k : E^\Gamma \to E^\Gamma, \quad u \mapsto \text{expand}_k(u)$$

with $\text{expand}_k(u)(z') = u(z)$ if $z' = kz$, and $\text{expand}_k(u)(z') = 0$ else. Lemma 10.3.4 guarantees that

$$f^{ak}(\delta_0) = \text{expand}_k(f^a(\delta_0)).$$

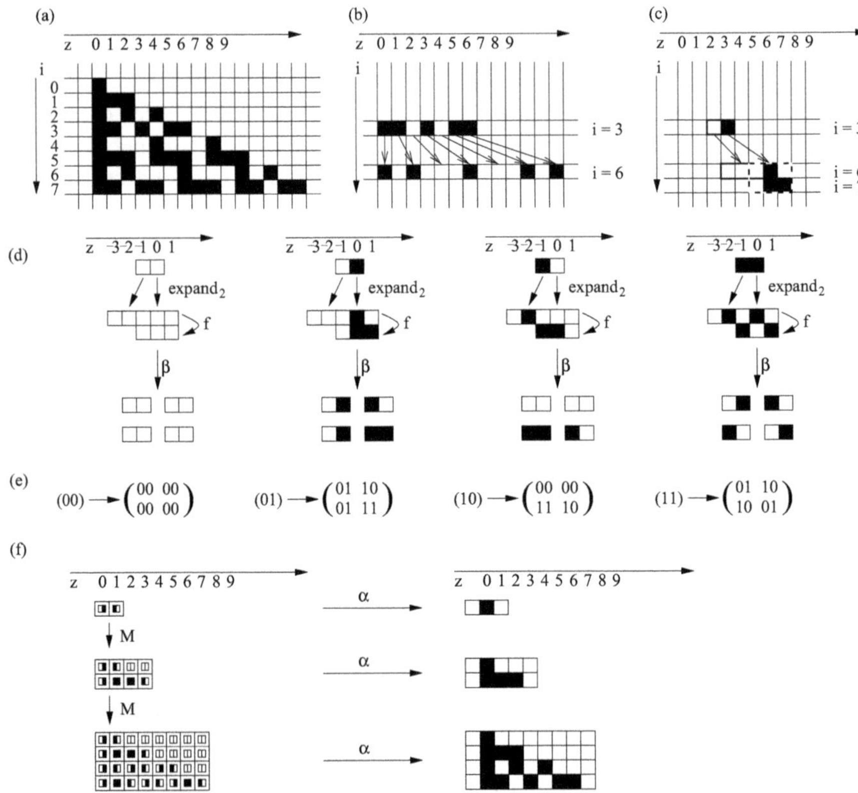

Fig. 10.4 Construction of matrix substitution system. (**a**) Time course of the automaton $p(x) = 1+x+x^2$. (**b**) State for time $i = 6$ can be directly computed from state at time $i = 3$ by expanding. (**c**) To predict the pattern in the *dashed bordered square* ($i = 6, 7$ and $z = 6, 7$), it is necessary to know the states in line $i = 6$, and $z = 3, \ldots, 7$. Therefore, it is sufficient to know the states in line $i = 3$, $z = 2, 3$. (**d**) Pattern in line $2i$ by pattern in line i. (**e**) Resulting matrix substitution system. (**f**) First steps of the matrix substitution system and comparison with the space-time pattern of the cellular automaton

Step 2: *Example.* We extend the idea of step 1. We are able to compute $f^{2i}(\delta_0)$ from $f^i(\delta_0)$ without going through all time steps $i+1, \ldots, 2i-1$. Now we determine the sites in which $f^{2i}(\delta_0)$ can be known if we only know $f^i(\delta_0)$ at a finite interval $\{z-y, \ldots, z\}$. Using Step 1 resp. Lemma 10.3.4, we are able to predict the local state of sites $2(z-y), \cdots 2z$. Furthermore, the local state is zero in sites $2z-2y-1$ and $2z+1$ (again because of Lemma 10.3.4). All in all, we know the local states of sites $2(z-y)-1, \ldots, 2z+1$ (see Fig. 10.4c).

We cannot predict the next line in the time-space diagram by Lemma 10.3.4, as $2i + 1$ is odd. However, as we know state $f^{2i}(\delta_0)$ on an interval, we are able to directly compute $f^{2i+1}(\delta_0)$ on a (smaller) interval, namely for the sites $2(z - y) - 1 + 2, \ldots, 2z + 1$.

General setting. In the general case, we again assume that we know $f^i(\delta_0)$ on sites $z - z'$ where $z' \in \{0, \ldots, y\}^d$, $y \in \mathbb{N}$. Let $\mathbf{e} = (1, \ldots, 1) \in \mathbb{Z}^d$. The expand$_k$()-operator allows us to know the state of $f^{2i}(\delta)$ on the region given by $zk + \mathbf{e}(k-1) - z'$ with $z' \in \{0, \ldots, yk + 2(k - 1)\}^d$. In each time step, a strip of size d on the "left, lower" side of our region is influenced by unknown states from outside. We are able to determine

$$f^{2i+j}(\delta_0)(zk + \mathbf{e}(k-1) - z') \qquad \text{for } z' \in \{0, \ldots, yk + 2(k-1) - jn\}^d$$

for $j \geq 0$ if $yk + 2(k - 1) - jn \geq 0$.

Step 3: *Example.* At the end of the day, we aim at a matrix substitution system. We have seen in the last step, that a finite pattern is able to control a whole region in the space-time diagram. If we interpret this finding cleverly, we can turn this observation into a matrix substitution system. The dimensions of the matrices will be, of course, 2×2. In order to define this system, we do not consider the states of single cells z, but consider the pattern given by the states of sites $z - 1$ and z. We introduce a new alphabet $\hat{E} = E^2$, and a map $\beta : E^{\mathbb{Z}} \to \hat{E}^{\mathbb{Z}}$, $u \mapsto v$ with

$$v(z) = \beta(u)(z) = (u(z-1), u(z)).$$

Note that $\beta(E^{\mathbb{Z}})$ is a proper subset of $\hat{E}^{\mathbb{Z}}$ that is topologically closed, and closed under the shift operator. We also find that the shift operator commutes with β, this is $\sigma_z \circ \beta = \beta \circ \sigma_z$ (where the two shift operators in this equation act on different spaces). A given state in $\beta(E^{\mathbb{Z}})$ determines uniquely a state in $E^{\mathbb{Z}}$. Let $\alpha : \beta(E^{\mathbb{Z}}) \subset \hat{E}^{\mathbb{Z}} \to E^{\mathbb{Z}}$ be defined by $\alpha \circ \beta = id$. On $\beta(E^{\mathbb{Z}})$, there is a lift of our cellular automaton,

$$\hat{f} : \beta(E^{\mathbb{Z}}) \subset \hat{E}^{\mathbb{Z}} \to \beta(E^{\mathbb{Z}}), \qquad v \mapsto \beta \circ f \circ \alpha(v).$$

It is possible to extend $\hat{f}$ to $\hat{E}^{\mathbb{Z}}$ in such a way that $\hat{f}$ is the global function of a cellular automaton; as we will not use this fact, we do not go into the details of this construction.

We rewrite the finding of Step 2 using this notation. In order to know a state at cell $\beta(u)(z)$, we need to know u on $z - 1$ and z. This is, if we know $\beta(f^i(\delta_0))(z)$, we know (again, look at Fig. 10.4c) that $f^j(\delta_0)(z')$ for (z', j) contained in the set

$$\{(2z-3, j), \cdots, (2z+1, j) : j = 2i\} \cup \{(2z-1, j), \ldots, (2z+1, j) : j = 2i+1\}.$$

With this information we also know $\beta(f^j(\delta_0))(z')$ for

$$(z,j) \in \{(2z-2,j),\ldots,(2z+1,j)) \,:\, j = 2i\} \,\cup\, \{(2z,j),(2z+1,j) \,:\, j = 2i+1\}.$$

In particular, we know $\beta(f^j(\delta_0))(z')$ in the square $(z',j) \in (2z,2i) + \{0,1\} \times \{0,1\}$.

General setting. We repeat the construction in the general setup. $D_0 = \{-n+1,\cdots,0\}^d$, and $\hat{E} = E^{D_0}$. The function $\beta : E^\Gamma \to \hat{E}^\Gamma$ is given by

$$\beta(u)(z) = u|_{z+D_0}.$$

Note that β commutes with the shift operator, $\sigma_z \circ \beta = \beta \circ \sigma_z$ (where the shift operators on the left hand side and the right hand side of this equation act on different spaces, on E^Γ and on $\hat{E}^\Gamma$). We make use of this fact below, in the proof of Lemma 10.3.49. Next we assume that we know $\beta(f^i(\delta_0))(z)$, this is, we know $f^i(\delta_0)(z-z')$ for $z' \in \{0,\ldots,n-1\}^d$. This implies (see Step 2), that we know $f^{ki+j}(\delta_0)(z')$ especially on

$$\{zk+\mathbf{e}(k-1)-z') \,:\, j \in \{0,\ldots,k-1\},\;\; z' \in \{0,\ldots,(n-1)k+2(k-1)-(k-1)n\}^d\}$$

(take $y = n-1$ and consider $j \leq k-1$; recall $\mathbf{e} = (1,\ldots,1)^T$). As $(n-1)k+2(k-1)-(k-1)n = (k-1)+(n-1)$, we can determine

$$\beta(f^{ki+j}(\delta_0))(kz+z') \qquad \text{for } (z',j) \in \{0,\ldots,k-1\}^{d+1}.$$

Step 4: Now we define the matrix substitution system.

Example. We have four patterns,

$$\hat{E} = \{(0,0),(0,1),(1,0),(1,1)\}.$$

According to Step 3, each of these patterns located at (z,i) in the space-time interval determines a 2×2 pattern located in $(2z,2i)+\{0,1\}^2$. Let us first determine these matrices in $\hat{E}^{2\times 2}$ (see Fig. 10.4d and e). We take a pattern $v \in \hat{E}$, and embed this pattern into $E^{\mathbb{Z}}$, by defining

$$v \mapsto \tilde{u}_v \quad \text{where } \tilde{u}_v(-1) = v(0), \quad \tilde{u}_v(0) = v(1).$$

All other entries of u can be chosen at random. We then apply the $\text{expand}_2(.)$ operator, and compute one step with the global function of the cellular automaton. We then read off the resulting patterns by inspecting $f^i(\text{expand}_2(u))(z)$ for $i \in \{0,1\}$,

$z \in \{-1, 0, 1\}$. This procedure yields

$$(00) \mapsto \begin{pmatrix} 00\ 00 \\ 00\ 00 \end{pmatrix}, \quad (01) \mapsto \begin{pmatrix} 01\ 10 \\ 01\ 11 \end{pmatrix},$$

$$(10) \mapsto \begin{pmatrix} 00\ 00 \\ 11\ 10 \end{pmatrix}, \quad (11) \mapsto \begin{pmatrix} 01\ 10 \\ 10\ 01 \end{pmatrix}.$$

General setting. Also here, we embed a pattern $v \in \hat{E}$ into E^Γ by defining

$$\tilde{u}_v(z) = v(z) \text{ for } z \in D_0.$$

All sites in $\Gamma \setminus D_0$ can be defined at discretion. Using this embedding, we derive at the following construction.

Construction 10.3.47 Consider a cellular automaton with $\Gamma = \mathbb{Z}^d$, $E = \mathbb{F}_q$, and a local function that corresponds to a polynomial p of degree n. Assume that the cellular automaton possesses the Fermat property with index k.

Let $\tilde{D}_0 = \{-n+1, \ldots, 0\}^d$, and $\hat{E} = E^{\tilde{D}_0}$. Let $\overline{0} \in \hat{E}$ the configuration that is zero on all sites in $\tilde{D}_0$. Define the function $M_0 : \hat{E} \to \hat{E}^{\{0,\ldots,k-1\}^{d+1}}$ by

$$M_0(v)(z,i) = \beta(f^i(\text{expand}_k(\tilde{u}_v)))(z), \quad (z,i) \in \{0, \ldots, k-1\}^{d+1}$$

and define $\tilde{u}_v$ (in dependence of v) as described above. The initial state u_0 is given by

$$u_0(z,i) = \begin{cases} \beta(\delta_0)(z) & \text{for } i = 0 \\ \overline{0} & \text{else} \end{cases}, \qquad (z,i) \in \mathbb{Z}^{d+1}$$

where, as usual, $\delta_0 \in E^\Gamma$ is defined by $\delta_0(0) = 1$, and $\delta_0(z) = 0$ for $z \neq 0$. We call $(\mathbb{Z}^{d+1}, \hat{E}, k, M_0, u_0)$ the associated matrix substitution system.

Remark 10.3.48

(a) We obtain $M_0(\overline{0})(z,i) = \overline{0}$ for all $(z,i) \in \{0, \ldots, k-1\}^{d+1}$, as the cellular automaton is linear. The associated matrix substitution system is well defined.

(b) Consider the region $\{(z,i) : \text{ with } i < 0 \text{ or } z < 0\}$. On this region, we find $u_0(z,i) = \overline{0}$. For all $j \in \mathbb{N}_0$, $M^j(u_0)(z,i) = \overline{0}$ in this region. The only nonzero states appear in the positive cone of $\mathbb{Z}^{d+1}$.

In the following propositions we review the relation between the time-space diagram of a cellular automaton and its associated matrix substitution system. The first proposition shows the condition for the conclusion in Remark 10.3.30(b): the matrix substitution system becomes constant on an increasing region. This finding is an ingredient for the proof of the following proposition that establishes the equivalence of the time-space diagram and the substitution system. Last we show that the limit sets of both systems coincide.

Lemma 10.3.49 *Consider the matrix substitution system defined in Construction 10.3.47. In particular* $u_0 \in \hat{E}^{\mathbb{Z}^d}$ *is defined there. Then,*

$$M(u_0)(z,0) = u_0(z,0), \quad z \in \mathbb{Z}^d.$$

Proof By definition, $u_0(.,0) = \beta(\delta_0)$. Now we inspect $M(u_0)(z,i)$ for $i = 0$ and $z \in \mathbb{Z}^d$. We write $z = kz_0 + z'$, where $z' \in \{0,\ldots,k-1\}^d$, and find

$$\begin{aligned} M(u_0)(kz_0 + z',0) &= M_0(u_0(z_0))(z',0) = \beta(f^0(\text{expand}_k(\tilde{u}_{u_0(z_0)})))(z') \\ &= \beta(\text{expand}_k(\tilde{u}_{u_0(z_0)}))(z') \end{aligned}$$

where $\tilde{u}_v$ has been defined in step 4 of the construction above.

Case 1: Assume $u_0(z_0,0) = \beta(\delta_0)(z_0) \neq \overline{0}$. If we re-embed the symbol $\beta(\delta_0)(z_0) \in \hat{E}$ in E^Γ, we simply find

$$\tilde{u}_{u_0(z_0,0)}(z') = \delta_0(z_0 + z') = \sigma_{z_0}(\delta_0)(z') \quad \Rightarrow \quad \text{expand}_k(\sigma_{z_0}(\delta_0)) = \sigma_{kz_0}(\delta_0).$$

This last equation leads to

$$\begin{aligned} M(u_0)(z,0) &= M(u_0)(kz_0 + z',0) = \beta(\sigma_{kz_0}(\delta_0))(z') \\ &= \beta(\delta_0)(kz_0 + z') = u_0(kz_0 + z',0) = u_0(z,0). \end{aligned}$$

Case 2: Assume $u_0(z_0,0) = \beta(\delta_0)(z_0) = \overline{0}$. Then, $M(u_0)(z,0) = M_0(\overline{0})(z',0) = \overline{0}$. We show that $u_0(z_0,0) = \overline{0}$ also implies $u_0(z,0) = 0$, such that the desired equation is true.
Either $(z,0)$ is outside of the non-negative cone of $\mathbb{Z}^{d+1}$. In this case, $u_0(z,0) = 0$ (see remark above). Or, $(z,0)$ is located in the non-negative cone. In this case, also z_0 is non-negative, and $z = kz_0 + z' \geq z_0$ (if we take the inequality component-wise). As $u_0(z,0) = \beta(\delta_0)$, $u_0(z_0,0) = \overline{0}$ together with $z \geq z_0$ implies $u_0(z,0) = 0$. □

Remark 10.3.50 The state u_0 is nonzero on the sites $\{0,\ldots,n\}^d \times \{0\}$. Lemma 10.3.49 (together with Remark 10.3.30) implies that the matrix substitution system becomes constant on $\{(z,i) \in \{0,\ldots,n\,k^j - 1\}^d \times \{k^j - 1\}$ after j iterations. We will use the fact in the proof of the next proposition.

Proposition 10.3.51 *Consider the matrix substitution system defined in Construction 10.3.47. In particular* $u_0 \in \hat{E}^{\mathbb{Z}^d}$ *is defined there. For* $j \in \mathbb{N}_0$ *holds*

$$M^j(u_0)(z,i) = \beta(f^i(\delta_0))(z), \qquad i \in \{0,\ldots,k^j-1\}, \quad z \in \mathbb{Z}^d.$$

Proof First of all, state u_0 is nonzero only on the sites $\{0,\ldots,n-1\}^d \times \{0\}$. Outside of $\{0,\ldots,nk^j-1\}^d \times \{0,\ldots,k^j-1\}$, the state of the matrix substitution system is zero. If we turn to the space-time pattern of the cellular automaton with initial state δ_0, we find

$$f^i(\delta_0)(z) = 0 \quad \text{if } (z,i) \not\in \{0,\ldots,n(k^j-1)\}^d \times \{0,\ldots,k^j-1\}.$$

Therefore,

$$\beta(f^i(\delta_0))(z) = \overline{0} \quad \text{if } (z,i) \not\in \{0,\ldots,n(k^j-1)+d-1\}^d \times \{0,\ldots,k^j-1\}.$$

As $n(k^j-1)+n-1 = nk^j-1$, the regions where $M^j(u_0)(z,i) = \overline{0}$ resp. $\beta(f^i(\delta_0))(z) = \overline{0})$ agree. We concentrate now on the region $\{0,\ldots,nk^j-1\}^d \times \{0,\ldots,k^j-1\}$.

In order to prove the desired equality, we use induction on j. For $j=0$, the statement is immediately clear, as $M^0(u_0) = u_0$ and $u_0 = \beta(\delta_0)$ according to the definition.

Assume that the statement is true for j. We show that the statement is also true for $j+1$. We checked already in Remark 10.3.50 that

$$M^j(u_0)(z,i) = M^{j+1}(u_0)(z,i) \qquad \text{for} \quad i \le k^j-1, \quad z \in \{nk^j-1\}^d.$$

Therefore, the desired equation is true for $i = 0,\ldots,k^j-1$. We show that the equation is also true for $i = k^j,\ldots,k^{j+1}-1$. Let $(z,i) = (kz_0+z', ki_0+i')$ where $(z',i') \in \{0,\ldots,k-1\}^{d+1}$. Let $v = M^{j-1}(u_0)(z_0,i_0)$. Then,

$$M^j(u_0)(z,i) = M_0(v)(z',i') = \beta(f^{i'}(\text{expand}_k(\tilde{u}_v)))(z').$$

Due to the assumption of the induction we know that $v = \beta(f^{i_0}(\delta_0))(z_0)$, and therewith also $\text{expand}_k(\tilde{u}_v)(kz+z') = f^{ki_0}(\delta_0)(kz+\tilde{z}')$, where the allowed values for $\tilde{z}'$ are discussed in Step 3 above. Thus,

$$M_0(v)(z',i') = \beta(f^{k\,i_0+i'}(\delta_0))(k\,z_0+z')$$

(see formula at end of step 3 and definition of $M_0(.)$). This observation establishes the desired relation,

$$M^j(u_0)(z,i) = M_0(v)(z',i') = \beta(f^{k\,i_0+i'}(\delta_0))(k\,z_0+z') = \beta(f^i(\delta_0))(z).$$

□

All in all, we have shown the following corollary, that allows to compute the dimension of limit sets of cellular automata in an efficient way.

Corollary 10.3.52 *The limit set $\mathcal{F}$ of a cellular automaton with Fermat property and the limit set of the corresponding matrix substitution system $\hat{\mathcal{F}}$ agree,*

$$\mathcal{F} = \hat{\mathcal{F}}.$$

Proof Let $t_i = (k^{-i}, \ldots, k^{-i}) \in \mathbb{R}^{d+1}$ be a shift in space and time; this time shift is required as the graphical representation of matrix substitution system uses non-centered intervals, while that for cellular automata is defined via centered intervals.

Step 3: $\mathcal{F} \subset \hat{\mathcal{F}}$.
Let $(x, t) \in \mathcal{F}$. Then there is a sequence $(z_i, n_i)k^{-i} \to (x, t)$ with $f^{n_i}(z_i) \neq 0 \Rightarrow \beta(f^{n_i})(\hat{z_i}) \neq 0$ and $t_i + (z_i, n_i)k^{-i} \in \hat{F}_i$. This observation implies $(x, t) \in \hat{\mathcal{F}}$.

Step 2: $\hat{\mathcal{F}} \subset \mathcal{F}$.
If $(x, t) \in \hat{F}$, there is a sequence $t_i + (z_i, n_i)k^{-i} \in \hat{F}_i$ with $\beta(f^{n_i})(z_i) \neq 0$. As before, we denote by $D_0 = \{-d + 1, \ldots, 0\}^d$ where d is the degree of the polynomial that represents the local function of the cellular automaton. As $\beta(f^{n_i})(z_i) \neq 0$, there exists $z_i' \in z_i + D_0$ with $f^{n_i}(z_i') \neq 0$ and thus $(z_i', n_i)k^{-i} \to (x, t)$ with $f^{n_i}(z_i') \neq 0$. Hence, $(x, t) \in \mathcal{F}$.

□

Example 10.3.53 Consider our example from above, with the associated matrix substitution system

$$(00) \mapsto \begin{pmatrix} 00 & 00 \\ 00 & 00 \end{pmatrix}, \quad (01) \mapsto \begin{pmatrix} 01 & 10 \\ 01 & 11 \end{pmatrix},$$

$$(10) \mapsto \begin{pmatrix} 00 & 00 \\ 11 & 10 \end{pmatrix}, \quad (11) \mapsto \begin{pmatrix} 01 & 10 \\ 10 & 01 \end{pmatrix}.$$

We compute the next generation operator by counting the number of the pattern that are generated by a certain pattern,

$$(01) \mapsto 2 \times (01), \quad 1 \times (11), \quad 1 \times (10)$$

$$(11) \mapsto 2 \times (01), \quad 0 \times (11), \quad 2 \times (10)$$

$$(10) \mapsto 0 \times (01), \quad 1 \times (11), \quad 1 \times (10).$$

The vector $b = (u, v, w)^T \in \mathbb{N}_0^3$ indicates that we have u times the pattern (01), v times the pattern (11), and w times the pattern (10), we find

$$T = \begin{pmatrix} 2 & 1 & 1 \\ 2 & 0 & 2 \\ 0 & 1 & 1 \end{pmatrix}$$

This matrix is irreducible. The characteristic polynomial reads

$$p(\lambda) = -\lambda(2-\lambda)(1-\lambda) + 4 - 4(2-\lambda) = 5(\lambda - 1) - (\lambda - 1)^3.$$

The eigenvalues are $\lambda = 1$, and $\lambda = 1 \pm \sqrt{5}$. The spectral radius is $\rho(T) = 1 + \sqrt{5}$, and thus (note that the cellular automaton has the Fermat property with index 2) the limit set has the dimension

$$\dim_H(\tilde{F}) = \log_2(1 + \sqrt{5}).$$

If we inspect the limit set in Fig. 10.5, it is at least difficult to imagine an iterated function system that generates this fractal set. For more simple sets, like the Sierpinski gasket, it is possible to guess this system. In this case, however, we know

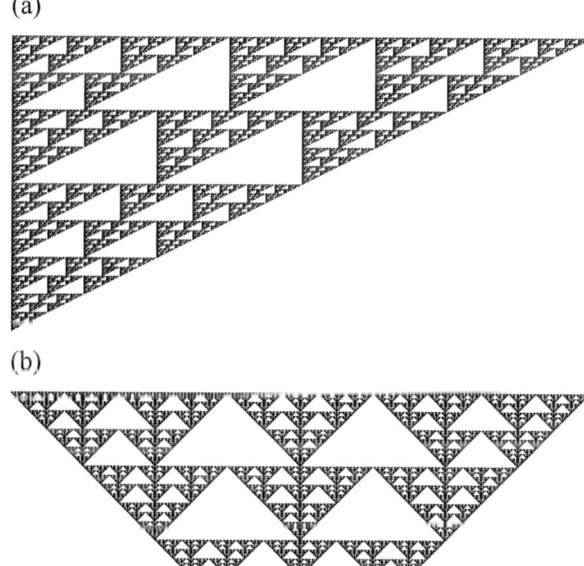

Fig. 10.5 (**a**) Sketch of the limit set for example 10.3.53, $p(x) = 1 + x + x^2$. (**b**) Limit set of the shifted automaton, $p_1(x) = p(x)x^{-1} = x^{-1} + 1 + x$

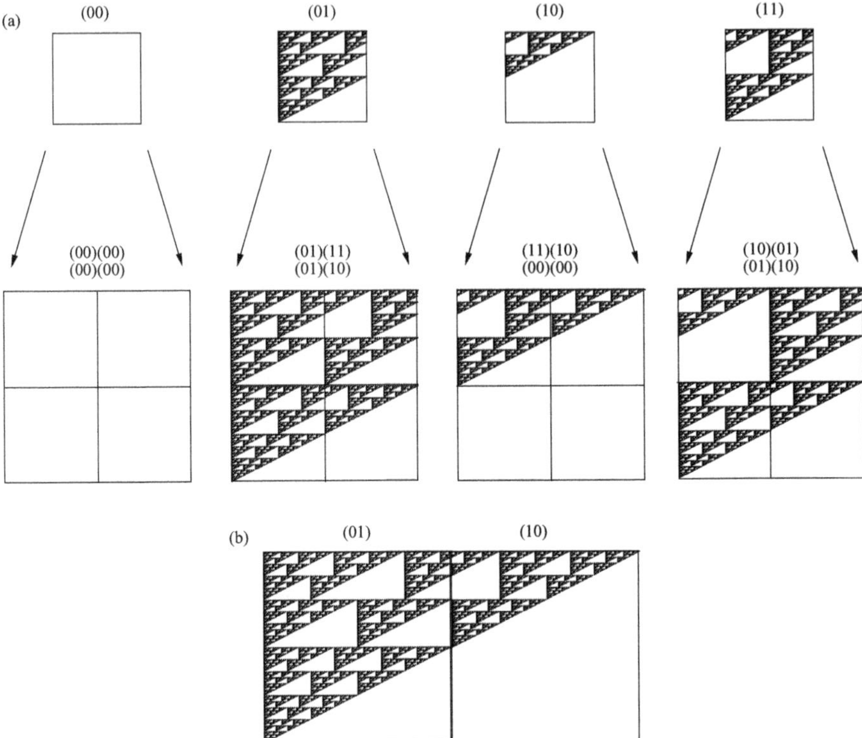

Fig. 10.6 (**a**) Sketch of the limit set for the hierarchical iterative function system set associated with Example 10.3.53, $p(x) = 1 + x + x^2$. (**b**) Limit set of the automaton as the concatenation of the limit sets for *two symbols*

that the matrix substitution system is associated with a hierarchical iterative function system, which has a unique fixed point. It is possible to concatenate the elements of this fixed point to obtain the limit set $\tilde{F}$ (see Fig. 10.6). Therefore we know that $\tilde{F}$ possesses self-similar properties and is justly called "self-similar fractal set".

Chapter 11
Particle Motion

Classical models for spatio-temporal processes are formulated in terms of partial differential equations. Since Newton and Leibniz, it turned out that this is a most powerful formalism. Cellular automata provide an alternative mathematical structure. However, a justification is required why cellular automata could be preferred over partial differential equations. Cellular automata are for sure more simple to define (even to formulate a derivative needs a certain effort, let alone to prove the existence of solutions for partial differential equations), to communicate, and to simulate (efficient numerical schemes for partial differential equations are in general non-trivial), but the tools to analyze the behavior of cellular automata are still less developed than those for continuous models.

There are two different bridges between cellular automata and partial differential equations: either a partial differential equation is discretized in one way or another, or a continuum limit of a cellular automaton is performed. We review examples for both approaches, starting with a formal limit based on finite differences (Sect. 11.1), then proceed to the ultradiscrete limit (Sect. 11.2), a discretization scheme that is more appropriate for the situation at hand, and end up with the continuum limit of a model describing particle movement in a microscopic way (Sect. 11.3).

11.1 Particle Motion: Formal Approach

The heat equation is one of the most common continuous models for particle movement. First, we briefly recall how to motivate the heat equation in one dimension, and then we present a naive transformation of the heat equation into a cellular automaton. For more detailed results about partial differential equations, see any textbook, e.g., the book of Renardy and Rogers [147].

K.-P. Hadeler, J. Müller, *Cellular Automata: Analysis and Applications*,
Springer Monographs in Mathematics, DOI 10.1007/978-3-319-53043-7_11

11.1.1 Modelling Diffusion by Continuous Models

Let us consider a particle density $u(x, t)$, where $x \in \mathbb{R}$ denotes space, $t \in \mathbb{R}_+$ time. The particle mass between two locations a and b is given by

$$\int_a^b u(x, t)\, dx.$$

Since we do not consider annihilation or birth of particles, the mass within this interval only changes via particle flux over the boundaries a and b. Let $j(x, t)$ denote the flux, i.e., the net number of particles crossing x from left to right per time interval (particles moving from right to left are counted negatively, those going from left to right positively; $j(t, x)$ indicates the sum). Hence,

$$\frac{d}{dt}\int_a^b u(x, t)\, dt = \int_a^b u_t(x, t)\, dx = j(a, t) - j(b, t) = -\int_a^b j_x(x, t)\, dx.$$

Since $[a, b]$ is an arbitrary interval, we conclude that

$$u_t = -j_x.$$

This is the mass conservation law.

In order to obtain a closed equation in u, a second relation between u and j is required. Therefore, the first Fickian law is introduced,

$$j = -Du_x.$$

This equation indicates that particles move towards the direction defined by the negative gradient: particles tend to roll downhill. Please note that the conservation law is based on first principles. The first law of Fick is a pure modeling assumption; it may be reasonable, but there are alternative possibilities to model this connection (see, e.g., the correlated random walk below).

Taking the conservation law and the first Fickian law together, we find

$$u_t = Du_{xx}.$$

This is the heat equation, also called diffusion equation or the second Fickian law. In two dimensions, the heat equation reads $u_t = D(u_{xx} + u_{yy})$. An explicit solution for the initial value problem on the real axis (one dimension again) with initial value $u(x, 0) = \psi_0(x)$ is given by

$$\psi(x, t) = \frac{1}{2\sqrt{\pi\, D t}}\int_{\mathbb{R}} e^{-(x-y)^2/(4D^2 t)}\psi_0(y)\, dy.$$

The equation respects positivity, that is, $\psi_0 \geq 0$ implies $u(t, \cdot) \geq 0$. In accordance to the derivation of the equation via the Fickian laws, also mass is preserved, i.e., if $\psi_0 \in L^1(\mathbb{R})$, then

$$\int_{-\infty}^{\infty} u(t,x)dx = \int_{\infty}^{\infty} \psi_0(x)dx.$$

We note two remarkable property of stationary solutions:

1. The scaling property. The heat equation is invariant under the parabolic scaling,

$$x \mapsto ax, \quad t \mapsto a^2 t.$$

   This scaling behavior is one of the main characteristics of a class of partial differential equations behaving similarly to the heat equation, the parabolic equations.
2. The mean value property. A stationary solution $u(x)$ of the one-dimensional heat equation satisfies

$$0 = Du_{xx}$$

   i.e., is a linear function in space. Hence,

$$u(x) = \frac{1}{2}(u(x-h) + u(x+h)).$$

This property carries over to the stationary solutions in higher dimensions [98], e.g., for two dimensions we have

$$u(\vec{x}) = \frac{1}{2\pi r} \int_{|\vec{y}-\vec{x}|=r} u(\vec{y})\, do.$$

An immediate consequence is the Hopf maximum principle for stationary solutions: no maximum or minimum can be located in the interior of a region. If we have a proper maximum (minimum) in the interior, we find a circle such that the values on this circle are smaller (larger) than that in its center, contradicting the mean value property. Particularly, on a torus there must be no maximum or minimum at all, and the only stationary solution is constant. And indeed, it is possible to show that any solution of the heat equation on a torus asymptotically tends to a constant solution.

11.1.2 Naive Cellular Automata Models for Diffusion

A large class of numerical methods to solve partial differential equations are based on the fact that a derivative can be well approximated by a finite difference,

$$u_t(x,y,t) \approx \frac{1}{\delta t}(u(x,y,t+\delta t) - u(x,y,t)),$$

$$u_{xx}(x,y,t) \approx \frac{1}{\delta x^2}\Big(u(x+\delta x,y,t) - 2u(x,y,t) + u(x-\delta x,y,t)\Big),$$

$$u_{yy}(x,y,t) \approx \frac{1}{\delta x^2}\Big(u(x,y+\delta x,t) - 2u(x,y,t) + u(x,y-\delta x,t)\Big)$$

(where δx and δt denote small spatial resp. time steps). If we use these approximations for the heat equation, we obtain

$$u(x,y,t+\delta t) \approx u(x,y,t) + \frac{D\,\delta t}{\delta x^2}\Big(u(x+\delta x,y,t) + u(x-\delta x,y,t)$$
$$+u(y,y+\delta x,t) + u(y,y-\delta x,t) - 4u(x,y,t)\Big).$$

Formally, this equation is close to the global function of a cellular automaton: We define D_0 as the Neumann neighborhood, and $f_0 : \mathbb{R}^{D_0} \to \mathbb{R}$ by

$$f_0(v|_{D_0}) = v(0,0) + \frac{D\,\delta t}{\delta x^2}\Big(v(-1,0) + v(1,0) + v(0,-1) + v(0,1) - 4\,v(0,0)\Big).$$

Then,

$$v^{n+1}(i,j) = f_0(v^n|_{(i,j)+D_0})$$

yields an approximation of $u(i\,\delta x, j\,\delta x, n\,\delta t)$. The notation is that of a cellular automaton, only that the local state space is not a finite set, but $\mathbb{R}$. This observation is the starting point for a naive definition of a cellular automaton resembling the heat equation (see, e.g., [153, 166, 174, 184]). The local state space $\mathbb{R}$ is replaced by some finite set $E \subset \mathbb{R}$, for example $E = \{0,\ldots,N\}$. In general, E^{Z^2} is not invariant under the dynamics as it is given by now. However, a simple way out is to define a projection $\Pi_E : \mathbb{R} \to E$. For $x \in \mathbb{R}$, we determine the minimal distance of x to points in E. If there is exactly one point $e \in E$ at this distance, we define $\Pi_E(x) = e$; if there are two points $e_1, e_2 \in E$ with this distance, we take the smaller one, $\Pi_E(x) = \min\{e_1, e_2\}$. Then, $(\mathbb{Z}^2, U, E, \tilde{f}_0)$ with $\tilde{f}_0 = \Pi_E \circ f_0$ is a well defined cellular automaton. For example, the IEEE representation of real numbers is a finite set. Using this set for E yields an implementation of the original numerical scheme by a standard programming language. It is well known that these numerical schemes

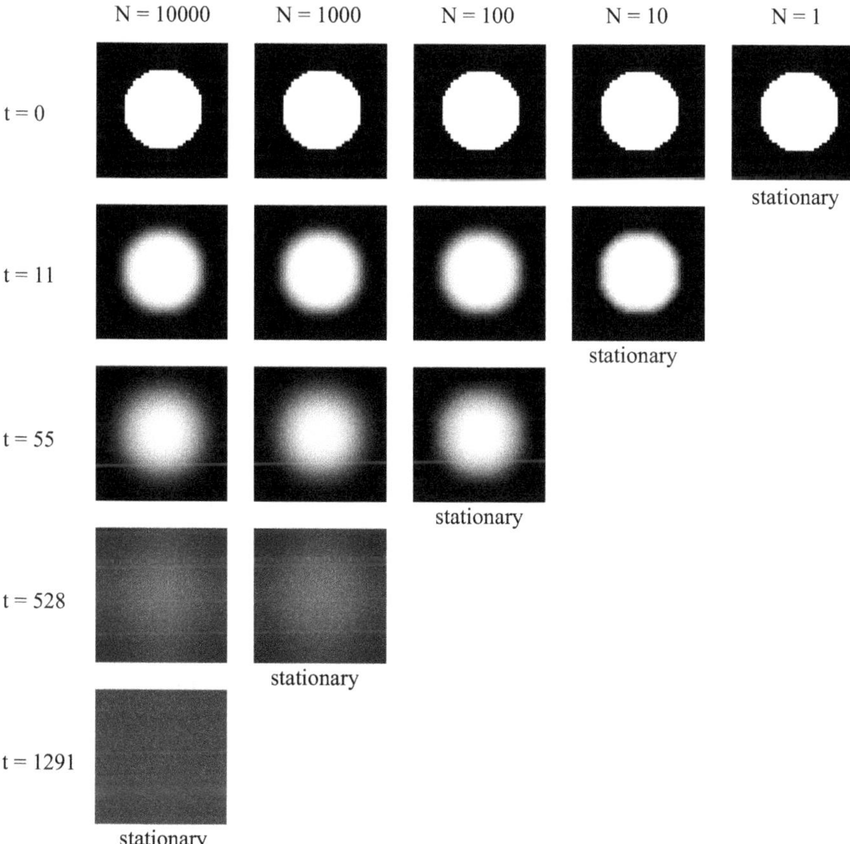

Fig. 11.1 Simulation of the diffusion cellular automaton with $D\,\delta t/\delta x^2 = 0.2$, gridsize 50× 50 cells, and local state space $E = \{0,\ldots,N\}$ for different N. The simulations run until a stationary state is reached. The *gray value* is scaled such that *black* always corresponds to zero, and *white* to N

and their practical implementations yield a reasonable approximation of the solution of the partial differential equation. A cellular automaton based on these ideas should be a reasonable model of the physical process.

The set of IEEE representations of numbers is finite but large—the philosophy of cellular automata targets at few local states. In the literature, one often finds $E = \{0,\ldots,N\}$. Simulations for $N = 2000$ yield a good result, see Fig. 11.1. The visual inspection of the solution shows that it tends to a fairly constant particle density. However, if we use fewer states and reduce N to values of $N - 50$ or even $N = 1$, we find spurious stationary states. We investigate this effect more in detail, concentrating on the one-dimensional case. Let us assume $1-2D\,\delta t/\delta x^2 > 0$. Then, the local function reads

$$\left\lfloor u_i^n + \frac{D\,\delta t}{\delta x^2}(u_{i-1}^n - 2u_i^n + u_{i+1}^n) + 0.5 \right\rfloor.$$

A state u is stationary, if

$$\begin{aligned} u_i &= \left\lfloor u_i + \frac{D\,\delta t}{\delta x^2}(u_{i-1} - 2u_i + u_{i+1}) + 0.5 \right\rfloor \\ \Leftrightarrow & \frac{D\,\delta t}{\delta x^2}(u_{i-1} - 2u_i + u_{i+1}) \in [-0.5, 0.5) \\ \Leftrightarrow & \; u_i \in \frac{1}{2}(u_{i-1} + u_{i+1}) + \frac{\delta x^2}{D\,\delta t}(-0.5, 0.5]. \end{aligned}$$

This inclusion tells us, that a state is stationary if it satisfies an approximate version of the mean value property. If N is small, this condition is not restrictive. Let us consider $N = 1$, and $\delta t\, D/\delta x^2 = 0.2$. Then,

$$(u_{i-1} + u_{i+1})/2 \in \{0, 0.5, 1\} \quad \text{and} \quad \frac{\delta x^2}{D\,\delta t}(-0.5, 0.5] = (-2.5, 2.5].$$

That is, for $N = 1$, any state is stationary. Only for $N > 3$ we find non-stationary states. The absolute roughness of a stationary state does not depend on N—a stationary state u for $N = 1$ is also a stationary state for $N = 1000$; even the state v defined by $v_i = u_i + 500$ is stationary for $N = 1000$. What is different for $N = 1$ and $N = 1000$ is the relative variation of u_i/N. This relative variation decreases, as we can see in Fig. 11.1.

Since the idea of a cellular automaton is to use a *small* set for the local states, this rather formal route via the direct discretization is in general not successful.

11.2 From PDE to Cellular Automata: Ultradiscrete Limit

In the present section, we replace the rather formal approach based on the finite difference scheme by a more refined method. We will, step by step, discretize space, time, and state. The discretization of space and time is done by standard numerical schemes. The most interesting aspect is how to reduce the continuous state at a grid point to a set of elementary states. One possibility based on a discretization of the state space is described in [107]. The approach considered here, mainly developed by Nishinari and Takahashi [141], is the so-called ultradiscretization method. Before the heat equation and in particular the Burgers equation are discussed, first the basic idea of the ultradiscrete limit is outlined on an informal level: Starting point is a partial differential equation

$$\psi_t = A\psi, \qquad \psi(x,0) = \psi_0(x),$$

where A denotes a differential operator (like $A = \partial_x^2$) and $\psi(x,t)$ a solution. Time and space are discretized. The solution $\psi(x,t)$ is approximated at given points (x_i, t_n), $\psi(x_i, t_n) \approx \zeta_i^n$. For this approximation, an explicit numerical scheme is used. In this way, a discrete difference equation is obtained

$$\zeta_i^{n+1} = F(\zeta_{i-m}^n, \dots, \zeta_{i+m}^n)$$

for $m \in \mathbb{N}$ fixed. So far, the method resembles the formal approach described in Sect. 11.1. Recall that the method developed there has some intrinsic problems. A workaround to this situation is achieved by ultradiscretization. The new idea is to apply the transformation

$$\zeta_i^n = e^{u_i^n/\varepsilon}.$$

Depending on the equation, also parameters may be scaled. An equation of the form

$$u_i^n = \varepsilon \log(F(e^{u_{i-m}^n/\varepsilon}, \dots, e^{u_{i+m}^n/\varepsilon}; \varepsilon))$$

is obtained. In many cases, the limit $\varepsilon \to 0$ yields a well defined result. As a simple example, consider a linear function F with positive coefficients a_i, $F(\zeta_{i-m}^n, \dots, \zeta_{i+m}^n) = \sum_{j=-m}^m a_j \zeta_{i+j}^n$. Then,

$$\lim_{\varepsilon\to 0} \varepsilon \log\left(\sum_{j=-m}^{m} a_j\, e^{u_{i+j}^n/\varepsilon}\right) = \max\{u_{i-m}^n, \cdots, u_{i+m}^n\}.$$

This procedure is called the ultradiscrete limit. Basically, the continuous state variable ζ_i^k is replaced by u_i^n. If $u_i^n < 0$, then $\zeta_i^n = 0$, if $u_i^n = 0$ then $\zeta_i^n = 1$, and if $u_i^n > 0$ then ζ_i^n is arbitrary large. The continuous state ζ_i^n is basically replaced by the information: ζ_i^n is very small, ζ_i^n is of order $O(1)$, or ζ_i^n is very large. The new state u does not continuously scan the state space of ζ, but focuses on extreme values. It is astonishing, that such an extreme discretization method yields a dynamical system that still can be related to the original equation. There is no quantitative relation, for example a limit for $|E| \to \infty$ that recovers the partial differential equation. However, often enough it is possible to carry over essential properties or special solutions of the PDE. If one restricts the initial conditions u_i^0 to a finite set (of integers), one obtains a discrete recursive equation and often, but not always, the global function of a cellular automaton. In the following, we discuss two examples.

11.2.1 Heat Equation

One of the simplest examples of a linear parabolic partial equation is the one-dimensional heat equation

$$\frac{\partial}{\partial t}\psi(x,t) = D\frac{\partial^2}{\partial x^2}\psi(x,t), \qquad \psi(x,0) = \psi_0(x).$$

Recall the standard explicit discretization of the heat equation (which is a possible, but numerically not the most efficient scheme), where derivatives are replaced by difference quotients,

$$\frac{\partial}{\partial t}\psi(x,t) \mapsto \frac{\psi(x,t+\Delta t) - \psi(x,t)}{\Delta t}$$
$$\frac{\partial^2}{\partial x^2}\psi(x,t) \mapsto \frac{\psi(x+\Delta x,t) - 2\psi(x,t) + \psi(x-\Delta x,t)}{\Delta x^2}.$$

Choose a time step Δt and spatial step Δx. Let $x_i = i\Delta x$, $t_n = n\Delta t$. The difference equation becomes

$$\frac{\zeta_i^{n+1} - \zeta_i^n}{\Delta t} = D\frac{\zeta_{i-1}^n - 2\zeta_i^n + \zeta_{i+1}^n}{\Delta x^2}$$

or

$$\zeta_i^{n+1} = \zeta_i^n + \delta(\zeta_{i-1}^n - 2\zeta_i^n + \zeta_{i+1}^n)$$

where

$$\delta = D\frac{\Delta t}{\Delta x^2}.$$

We apply the ultradiscrete transformation to the heat equation in its descretized form (see e.g. [141]). Assume $\delta < 1/2$. With

$$\zeta_i^n = e^{u_i^n/\varepsilon}, \qquad 1 - 2\delta = e^{-L/\varepsilon}$$

we obtain

$$\begin{aligned}
u_i^{n+1} &= \varepsilon\log\left(\frac{1}{2}(1-e^{-L/\varepsilon})e^{u_{i-1}^n/\varepsilon} + e^{(u_i^n-L)/\varepsilon} + \frac{1}{2}(1-e^{-L/\varepsilon})e^{u_{i+1}^n/\varepsilon}\right)\\
&= \varepsilon\log\left((1-e^{-L/\varepsilon})e^{u_{i-1}^n/\varepsilon-\log 2} + e^{(u_i^n-L)/\varepsilon} + (1-e^{-L/\varepsilon})e^{u_{i+1}^n/\varepsilon-\log 2}\right)\\
&= \quad \varepsilon\log\left(e^{u_{i-1}^n/\varepsilon-\log 2} + e^{(u_i^n-L)/\varepsilon}/(1-e^{-L/\varepsilon}) + e^{u_{i+1}^n/\varepsilon-\log 2}\right)\\
&\quad + \ \varepsilon\log\left(1-e^{-L/\varepsilon}\right)
\end{aligned}$$

We keep $L > 0$ fixed and let $\varepsilon \to 0$ (such that $\delta \to 1/2$). Then,

$$u_i^{n+1} = \max\{u_{i-1}^n, u_i^k - L, u_{i+1}^n\}.$$

If $E = \{0, .., m\} \subset \mathbb{N}_0$ and $L \in \mathbb{N}_0$, this relation defines a local rule for a cellular automaton. The neighbourhood is given by $D_0 = \{-1, 0, 1\}$. If L is large enough, e.g. $L > m$, the rule reduces to

$$u_i^{n+1} = f_0(u_{i-1}^n, u_i^n, u_{i+1}^n), \qquad f_0(u_{-1}, u_0, u_1) = \max\{u_{-1},\, u_1\}.$$

Let us check whether this automaton and the heat equation have any features in common. At first glance it does not show the typically behavior of the heat equation such as the mean value property. This example clearly indicates that the ultradiscrete limit is a singular limit and not a regular approximation. However, if we recall the transformation, it becomes clear that taking the maximum value *is* in accordance the mean value property for stationary solutions: the different values $0, \ldots, m$ for u_i do not indicate a linear scale, but qualitative differences. These are comparable with the different orders, e.g., $O(e^{\varepsilon^{-1}})$ and $O((e^{\varepsilon^{-1}})^2)$; these terms are not quantitatively but qualitatively different in size. If we average terms of different orders, the average is of the largest order, $O(e^{\varepsilon^{-1}})/2 + O((e^{\varepsilon^{-1}})^2)/2 = O((e^{\varepsilon^{-1}})^2)$.

In case of $E = \{0, \ldots, n\}$ and $L > n$ any trajectory $(u^k)_{k \in \mathbb{N}_0} \subset E^{\mathbb{Z}}$ tends to the constant state $u \in E^{\mathbb{Z}}$ with $u_i = \max\{u_j^0 \,:\, j \in \mathbb{Z}\}$. Also the global attractor is simple to characterize: it consists of all constant states $[\overline{e}] \in E^{\mathbb{Z}}$ with $e \in E$, and all monotone running fronts $u \in E^{\mathbb{Z}}$ with either $u_{i+1} \leq u_i$ for all $i \in \mathbb{Z}$ or $u_{i+1} \geq u_i$ for all $i \in \mathbb{Z}$. These running fronts have a counterpart in the original PDE: the unbounded solution

$$\psi(x, t) = e^{(c/D)(x - ct))} + A$$

connects $\psi = A \in \mathbb{R}$ and $\psi = \infty$, and runs with constant velocity c to the left or to the right (according to the sign of c).

11.2.2 The Burgers Equation

The Burgers equation serves as a one-dimensional toy model of the Navier-Stokes equation. Although it is not directly applicable to fluids, it incorporates some of the most important structures of the Navier-Stokes equation. In particular, it is a prototype of an equation where the solutions may develop shocks. The equation has also been used as a simple model for traffic flow; the solution $\psi(t, x)$ represents the density of cars at time t and location x. At the end of the section we will discuss this interpretation more closely. Let $j(x,t)$ be the flux. Conservation of mass yields the equation

$$\psi_t = -j_x.$$

The essential modelling assumption formulates the dependence of the flux on the particle density,

$$j = \psi^2 - \nu\psi_x.$$

This law should be compared to that of the heat equation, $j(x,t) = -D\psi_x$. The flux of the Burgers equation consists of two terms. The first term, ψ^2, becomes very large if the density becomes large—and plays no role if the density is small. The second term $-\nu\psi_x$ expresses the viscosity of the system. This second term leads to a dissipative character of the equation. The choice of the flux law yields the (viscid) Burgers equation

$$\psi_t + 2\psi\,\psi_x = \nu\psi_{xx}.$$

Let the initial data be $\psi(x,0) = \psi_0(x)$, $x \in \mathbb{R}$. A substitution $x \mapsto -x$ transforms the equation to $\tilde{\psi}_t - 2\tilde{\psi}(x,t)\,\tilde{\psi}_x = \nu\tilde{\psi}_{xx}$. Note that both versions of the Burgers equation are used in the literature.

The so-called inviscid Burgers equation is given by the formal limit $\nu \to 0$,

$$\psi_t + 2\psi\,\psi_x = 0.$$

This is a hyperbolic equation, where diffusive terms do not counteract jumps. We will not present the full theory of the Burgers equation. However, it is necessary to understand some basic properties of the solutions, in particular in case that the initial condition possesses jumps.

11.2.2.1 Cole Hopf Transformation

It is remarkable that it is possible to transform the nonlinear, viscid Burgers equation into the linear heat equation. This transformation, called the Cole-Hopf transformation, opens the possibility to obtain an explicit representation of solutions in terms of initial conditions.

Proposition 11.2.1 *Assume that*

$$\eta(x,t) = e^{-\int_{-\infty}^{x}\psi(\sigma,t)\,d\sigma/\nu}$$

is well defined for the solution if the Burgers equation $\psi_t + 2\psi\psi_x = \nu\psi_{xx}$, $\psi(x,0) = \psi_0(x)$. *In this case,* $\eta(x,t)$ *satisfies*

$$\eta_t(x,t) = \nu\eta_{xx}(x,t), \qquad \eta(x,0) = \eta_0(x) = e^{-\int_{-\infty}^{x}\psi_0(\sigma)\,d\sigma/\nu}$$

and

$$\psi(x,t) = \frac{-1}{\nu}\,\partial_x\,\ln(\eta(x,t)).$$

The transformation $\psi \to \eta$ is called Cole-Hopf transformation.

Proof Let

$$\zeta(x,t) = \int_{-\infty}^{x}\psi(\sigma,t)\,d\sigma$$

then

$$\zeta_t = \int_{-\infty}^{x} -(\psi^2)_x + \nu\psi_{xx}\,dx = -\psi^2(x,t) + \nu\psi_x = -(\zeta_x)^2 + \nu\zeta_{xx}.$$

We are now looking for a function $\phi : \mathbb{R} \to \mathbb{R}$, such that $\eta(x,t) = \phi(\zeta(x,t))$ satisfies the heat equation $\eta_t = \nu\eta_{xx}$. Assuming that ζ is differentiable in time and space, we find

$$\eta_t = \phi'(\zeta)\zeta_t, \quad \eta_x = \phi'(\zeta)\zeta_x, \quad \eta_{xx} = \phi''(\zeta)(\zeta_x)^2 + \phi'(\zeta)\zeta_{xx}.$$

Hence, the equation for η reads

$$\begin{aligned}\eta_t &= \phi'(\zeta)\zeta_t = \phi'(\zeta)(\zeta_x)^2 + \nu\phi'(\zeta)\zeta_{xx}\\ &= \phi'(\zeta)(\zeta_x)^2 + \nu\eta_{xx} + \nu\phi''(\zeta)(\zeta_x)^2\\ &= \nu\eta_{xx} + [\phi'(\zeta) + \nu\phi''(\zeta)](\zeta_x)^2\end{aligned}$$

If we define

$$\phi(x) = e^{-x/\nu}$$

then $\phi' + \phi''\nu = 0$ and

$$\eta_t = \nu\eta_{xx}.$$

Differentiating the defining equation for η yields the relation

$$\psi = \frac{-1}{\nu}\,\frac{\eta_x}{\eta} = \frac{-1}{\nu}\,\partial_x\ \ln(\eta).$$

□

Remark 11.2.2 If we aim at $\psi \geq 0$, then the admissible $\eta(x,t)$ are monotonously decreasing, nonnegative and $\lim_{x\to-\infty}\eta(x,t) = 1$. This is, the initial values ψ_0 should decrease fast enough for $x \to -\infty$ such that the integral $\int_{-\infty}^{x}\psi_0(\tau)\,d\tau$ is well defined. A class of initial condition allows for the Cole Hopf transformation. Especially all initial conditions with bounded support are within the admissible class.

Similar transformations exist for a whole class of partial differential equations of the form $\zeta_t - a\zeta_{xx} + b(\zeta_x)^2 = 0$ (see e.g. [56, Chap. 4.4]).

An alternative approach is the transformation

$$\tilde{\eta}(x,t) = e^{-\int_x^{\infty}\psi(\sigma,t)\,d\sigma/\nu}.$$

Also that transformation connects the Burgers equation to the heat equation.

Corollary 11.2.3 *The solution of the Burgers equation with initial data* $\psi_0 \in L^1(\mathbb{R})$ *is given by*

$$\psi(x,t) = \frac{\int_{\mathbb{R}}\frac{x-y}{t}\,e^{-(x-y)^2/(4\nu t)-\hat{\psi}_0(y)/\nu}\,dy}{2\int_{\mathbb{R}}e^{-(x-y)^2/(4\nu t)-\hat{\psi}_0(y)/\nu}\,dy}.$$

where

$$\hat{\psi}_0(x) = \int_{-\infty}^{x}\psi_0(\sigma)\,d\sigma.$$

Proof As the solution of the heat equation has an explicit representation, we also obtain an explicit representation of the solution of the Burgers equation. Let

$$\hat{\psi}_0(x) = \int_{-\infty}^{x}\psi_0(\sigma)\,d\sigma.$$

We find

$$e^{-\int_{-\infty}^{x}\psi(\sigma,t)/\nu} = \eta(x,t) = \frac{1}{2\sqrt{\pi\,\nu\,t}}\int_{\mathbb{R}}e^{-(x-y)^2/(4t\nu)-\hat{\psi}_0(y)/\nu}\,dy$$

and hence

$$\psi(x,t) = \frac{\partial}{\partial x}\,(-\nu)\,\log\left(\frac{1}{2\sqrt{\pi\,\nu\,t}}\int_{\mathbb{R}} e^{-(x-y)^2/(4\nu t)-\hat{\psi}_0(y)/\nu}\,dy\right)$$

$$= \frac{\int_{\mathbb{R}}(x-y)\,e^{-(x-y)^2/(4\nu\,t)-\hat{\psi}_0(y)/\nu}\,dy}{2t\int_{\mathbb{R}} e^{-(x-y)^2/(4\nu\,t)-\hat{\psi}_0(y)/\nu}\,dy}.$$

□

Because of this formula the Burgers equation is called integrable—there is an explicit integral of the partial differential equation.

11.2.2.2 Viscosity Solutions

In order to better understand the consequences of a jump in the initial condition of the inviscid Burgers equation, we investigate the limit $\nu \to 0$ for the corresponding solution of the viscid equation. These limiting functions are called "viscosity solutions". The viscosity solutions are weak solutions of the inviscid Burgers equation. More detailed explanations can be found e.g. in the books by Evans [56] or Smoller [156].

Proposition 11.2.4 *Consider the initial condition*

$$\psi_0(x) = \begin{cases} 1 & \textit{for} \quad x \in (0,1) \\ 0 & \textit{else} \end{cases}$$

of the Burgers equation. T For $t \in [0,1]$, the viscosity solution is given by (see Fig. 11.2)

$$\psi(x,t) = \begin{cases} 0 & \textit{for } x \le 0 \textit{ or } x > 1+t \\ 1 & \textit{for} \quad 0 < x < 2t \\ x/(2t) & \textit{for} \quad 2t < x \le 1+t \end{cases}$$

and in case of $t > 1$ by

$$\psi(x,t) = \begin{cases} 0 & \textit{for } x \le 0 \textit{ or } x > 2\sqrt{t} \\ x/(2t) & \textit{for} \quad 0 < x \le 2\sqrt{t} \end{cases}.$$

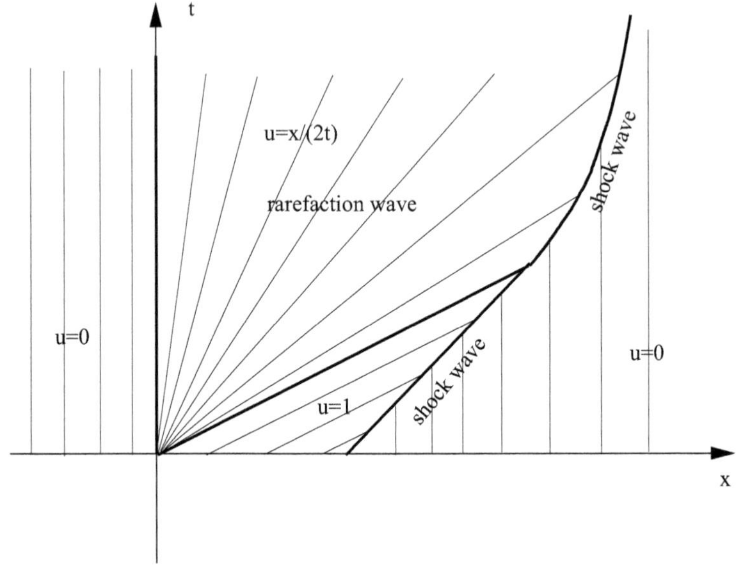

Fig. 11.2 Viscosity solution of the inviscid Burgers equation, see Proposition 11.2.4

Proof According to Corollary 11.2.3 the solution of the viscid Burgers equation reads

$$\begin{aligned} \psi(x,t) &= \frac{\int_{\mathbb{R}} (x-y)\, e^{-(x-y)^2/(4\nu\, t) - \hat{\psi}_0(y)/\nu}\, dy}{2t \int_{\mathbb{R}} e^{-(x-y)^2/(4\nu\, t) - \hat{\psi}_0(y)/\nu}\, dy} \\ &= \int_{\mathbb{R}} \frac{x-y}{2t}\, \alpha(\,y; x, t)\, dx \end{aligned}$$

where

$$\hat{\psi}_0(y) = \int_{-\infty}^{y} \psi_0(x)\, dx = \begin{cases} 0 & \text{for} \quad y < 0 \\ y & \text{for} \;\; y \in (0,1) \\ 1 & \text{for} \quad y > 1 \end{cases} .$$

and

$$\alpha(\,y; x, t) = \frac{e^{-(x-y)^2/(4\nu\, t) - \hat{\psi}_0(y)/\nu}}{\int_{\mathbb{R}} e^{-(x-y)^2/(4\nu\, t) - \hat{\psi}_0(y)/\nu}\, dy}.$$

With

$$\beta(\,y; x, t) = (x-y)^2/(4\,t) + \hat{\psi}_0(y)$$

we may rewrite the integral kernel as

$$\alpha(y;x,t) = e^{-\beta(y;x,t)/\nu} \left(\int_{\mathbb{R}} e^{-\beta(y;x,t)/\nu}\, dy \right)^{-1}.$$

Hence, the integral kernel $\alpha(y;x,t)$ is non-negative on $\mathbb{R}$ and the integral is one, $\int_{\mathbb{R}} \alpha(y;x,t)\, dy = 1$. If we assume that the function $\beta(y;t,x)$ assumes a unique minimum for given t and x at the location $y_0 = y_0(t,x)$, then

$$e^{-(\beta(y_0;t,x)-\beta(y;t,x))/\nu} \to 0 \quad \text{for } \nu \to 0 \text{ and } y \neq y_0.$$

If β is regular enough (for example, if it is approximately quadratic locally at y_0 and uniformly bounded away from its minimum), then

$$\alpha(y;x,t) = \frac{e^{-(\beta(y_0;t,x)-\beta(y;t,x))/\nu}}{\int_{\mathbb{R}} e^{-(\beta(y_0;t,x)-\beta(z;t,x))/\nu}\, dz} \to \delta_{y_0}(x) \quad \text{for } \nu \to 0$$

and hence

$$\psi(x,t) = \frac{x - y_0(t,x)}{2t}.$$

A straightforward computation shows that $\beta(y;t,x)$ assumes its minimum for $t < 1$ at

$$y_0(x,t) = \begin{cases} x & \text{for } x \in (-\infty, 0] \cup (1+t, \infty) \\ 0 & \text{for } \qquad 0 < x < 2t \\ x - 2t & \text{for } 0 < x, \quad 2t < x < 1+t \end{cases}$$

and for $t \geq 1$ at

$$y_0(x,t) = \begin{cases} x & \text{for } x \in (-\infty, 0] \cup (2\sqrt{t}, \infty) \\ 0 & \text{for } \qquad x \in (0, 2\sqrt{t}] \end{cases}.$$

If we replace $y_0(x,t)$ by this expression in the function $\psi(x,t) = (x - y_0(t,x))/(2t)$, then we obtain the result. □

Remark 11.2.5 We find two noteworthy elements in the solution: (1) a jump at the line $x = t + 1$ (for $t \in [0,1]$) resp. $x = 2\sqrt{t}$ (for $t > 1$). This jump is called shock wave. (2) The behavior of the solution between $x = 0$ and the line $x = 2t$ (for $t \in [0,1]$) resp. the line $x = 2\sqrt{t}$ (for $t > 1$). Here, the jump in the initial condition has been smoothed out—a rather atypical behavior for a hyperbolic equation, that mostly transports jumps in the initial condition, but does not smooth a jump. This part of the solution is called rarefaction wave. We again refer to the books of Evans [56] or Smoller [156] for a deeper discussion of shocks and rarefaction waves.

11.2.3 Ultradiscrete Limit and Burgers Equation

11.2.3.1 Discretization

We want to find a discretization for the viscid Burgers equation that preserves essential features. We use the Cole Hopf transformation because we know how to preserve essential features of the heat equation,

$$\eta = e^{-\int_{-\infty}^{x} \psi(x,t)} \quad \Rightarrow \quad \psi(x,t) = \frac{-1}{\nu} \frac{\partial}{\partial x} \ln(\eta(x,t)).$$

The function $\eta(x,t)$ satisfies the heat equation

$$\eta_t = \nu \eta_{xx}.$$

In Sect. 11.1.2 we discussed the discretization of the heat equation. We obtained the difference equation

$$\eta_i^{n+1} = (1-2\delta)\ \eta_i^n +\ \delta\ (\eta_{i-1}^n + \eta_{i+1}^n)$$

where $\delta = \nu\ \Delta x^2/\Delta t$. The relation between ψ and η suggests that the approximation ψ_i^n of $\psi(i\Delta x, n\Delta t)$ can be obtained as

$$\psi_i^n = \frac{-1}{\nu} \frac{\ln(\eta_{i+1}^n) - \ln(\eta_i^n)}{\Delta x} = \frac{\ln(\eta_i^n/\eta_{i+1}^n)}{\nu\ \Delta x} \quad \Leftrightarrow \quad e^{\nu\ \psi_i^n \Delta x} = \frac{\eta_i^n}{\eta_{i+1}^n}.$$

We define

$$\zeta_i^n = c\, e^{\psi_i^n \Delta x} = c\, \frac{\eta_i^n}{\eta_{i+1}^n}$$

introducing artificially the additional positive constant c that will give us an extra degree of freedom. Instead of ψ_i^n we compute ζ_i^n, replacing η_i^n by ζ_i^n in the recursion formula for η_i^n. We obtain

$$\begin{aligned}
\zeta_i^{n+1} &= c \frac{\eta_i^{n+1}}{\eta_{i+1}^{n+1}} = c \frac{(1-2\delta)\ \eta_i^n +\ \delta\ (\eta_{i-1}^n + \eta_{i+1}^n)}{(1-2\delta)\ \eta_{i+1}^n +\ \delta\ (\eta_i^n + \eta_{i+2}^n)} \\
&= c\, \frac{\eta_{i+1}^n}{\eta_{i+2}^n} \frac{(1-2\delta)\ \eta_i^n/\eta_{i+1}^n +\ \delta\ (\eta_{i-1}^n/\eta_{i+1}^n + 1)}{(1-2\delta)\ \eta_{i+1}^n/\eta_{i+2}^n +\ \delta\ (\eta_i^n/\eta_{i+2}^n + 1)}
\end{aligned}$$

$$= c\,\frac{\eta_{i+1}^n}{\eta_{i+2}^n}\;\frac{(1-2\delta)\;\eta_i^n/\eta_{i+1}^n + \;\delta\;(\eta_{i-1}^n/\eta_i^n)(\eta_i^n/\eta_{i+1}^n) + 1)}{(1-2\delta)\;\eta_{i+1}^n/\eta_{i+2}^n + \;\delta\;(\eta_i^n/\eta_{i+1}^n)/(\eta_{i+1}^n\eta_{i+2}^n) + 1)}$$

$$= \zeta_{i+1}^n\;\frac{c(1-2\delta)\;\zeta_i^n + \;\delta\;\zeta_{i-1}^n\zeta_i^k + c^2}{c(1-2\delta)\;\zeta_{i+1}^n + \;\delta\;\zeta_i^n\zeta_{i+1}^n + c^2}$$

$$= \zeta_{i+1}^n\;\frac{(1-2\delta)/c\;\zeta_i^n + \;\delta/c^2\;\zeta_{i-1}^n\zeta_i^n + 1}{(1-2\delta)/c\;\zeta_{i+1}^n + \;\delta/c^2\;\zeta_i^n\zeta_{i+1}^n + 1}$$

This numerical scheme is a discretization for the Burgers equation if take into account that ζ_i^n does not directly approximate the Burgers equation, but the approximate solution of the Burgers equation is given by $\ln(\zeta_i^n/c)/(\nu\Delta x)$.

11.2.3.2 Ultradiscrete Limit

We use the ultradiscretization method to obtain a cellular automaton.

Proposition 11.2.6 *Let L, M be given positive constants, $\varepsilon > 0$, and let c, δ depend on ε such that*

$$c^2/\delta = e^{L/\varepsilon}, \qquad (1-2\delta)/c = e^{-M/\varepsilon}.$$

Let furthermore $u_i^n \geq 0$ defined by $\zeta_i^n = e^{u_i^n/\varepsilon}$. For $\varepsilon \to 0$ we obtain

$$u_i^{n+1} = u_i^n - \min\{u_i^n, M, L - u_{i-1}^n) \\ + \min\{u_{i+1}^n, M, L - u_i^n)$$

and, if $u_i^n \in \{0, \ldots, L\}$ for all $i \in \mathbb{Z}$, also $u_i^{n+1} \in \{0, \ldots, L\}$ for all $i \in \mathbb{Z}$.

Proof We find

$$u_i^{n+1} = u_{i+1}^n + \varepsilon \ln\left(1 + e^{(u_i^n - M)/\varepsilon} + e^{(u_{i-1}^n + u_i^n - L)/\varepsilon}\right) \\ -\varepsilon \ln\left(1 + e^{(u_{i+1}^n - M)/\varepsilon} + e^{(u_i^n + u_{i+1}^n - L)/\varepsilon}\right)$$

and for $\varepsilon \to 0$

$$u_i^{n+1} = u_{i+1}^n + \max\{0, u_i^n - M, u_{i-1} + u_i^n - L) \\ - \max\{0, u_{i+1}^n - M, u_i^n + u_{i+1}^n - L).$$

As $\max(A, A+B) = A + \max(0, B)$ and $-\max(A, B) = \min(-A, -B)$, we obtain

$$u_i^{n+1} = u_{i+1}^n + \left(u_i^n + \max\{-u_i^n, -M, u_{i-1}^n - L)\right)$$

$$
\begin{aligned}
&- \left(u_{i+1}^n + \max\{-u_{i+1}^n, -M, u_i^n - L)\right) \\
&= u_i^n + \max\{-u_i^n, -M, u_{i-1}^n - L) - \max\{-u_{i+1}^n, -M, u_i^n - L) \\
&= u_i^n - \min\{u_i^n, M, L - u_{i-1}^n) + \min\{u_{i+1}^n, M, L - u_i^n)
\end{aligned}
$$

Assume that $u_i^n \in \{0, \ldots, L\}$ for all $i \in \mathbb{Z}$. Then, u_i^{n+1} are integers, and

$$
\begin{aligned}
u_i^{n+1} &\leq u_i^n + \min\{u_{i+1}^n, M, L - u_i^n) = \min\{u_{i+1}^n + u_i^n, M + u_i^n, L) \leq L \\
-u_i^{n+1} &\leq -u_i^n + \min\{u_i^n, M, L - u_{i-1}^n) \\
&= \min\{0, M - u_i^n, L - u_{i-1}^n - u_i^n) \leq 0
\end{aligned}
$$

and hence $u_i^{n+1} \geq 0$. All in all, $u_i^{n+1} \in \{0, \cdots, L\}$ for all $i \in \mathbb{Z}$ □

With the notation introduced above, we obtain the following corollary.

Corollary 11.2.7 *Let* $\Gamma = \mathbb{Z}$, $D_0 = \{-1, 0, 1\}$, $E = \{0, \ldots, L\}$, *and*

$$
f_0(u_{-1}, u, u_1) = u - \min\{u, M, L - u_{-1}) + \min\{u_1, M, L - u).
$$

Then, (Γ, D_0, E, f_0) *is a cellular automaton (the Burgers automaton).*

11.2.3.3 Interpretation in Terms of Particles

In the present section, we assume $M > L$. In order to better understand the local rule of the automaton, we first consider $L = 1$,

$$
\begin{aligned}
u_i^{n+1} &= u_i^n - \min\{u_i^n, L - u_{i-1}^n) + \min\{u_{i+1}^n, L - u_i^n) \\
&= u_i^n - u_i^n(1 - u_{i-1}^n + (1 - u_i^n)u_{i+1}^n \\
&= u_i^n u_{i-1}^n + (1 - u_i^n)u_{i+1}^n.
\end{aligned}
$$

Interpret u_i^n as the indicator function for the presence of a particle. Introduce the rule: A particle moves to the left if the left hand site is empty, otherwise it stays (see also Fig. 11.3).

$$
u_i^{n+1} = \underbrace{(1 - u_i^n)}_{\text{site } i \text{ is empty}} u_{i+1}^n + \underbrace{u_i^n}_{\text{site } i \text{ is occupied}} u_{i-1}^n.
$$

This is precisely the local rule of the Burgers automaton with $L = 1$ (which coincides with Wolframs rule 92). Also the case $L > 1$ can be understood in similar terms. Each site provides space for at most L particles. In order to determine the state in the next time step, first the free capacity of the left neighbor is checked. This gives the maximal number of particles that will move to the left. Then, as many

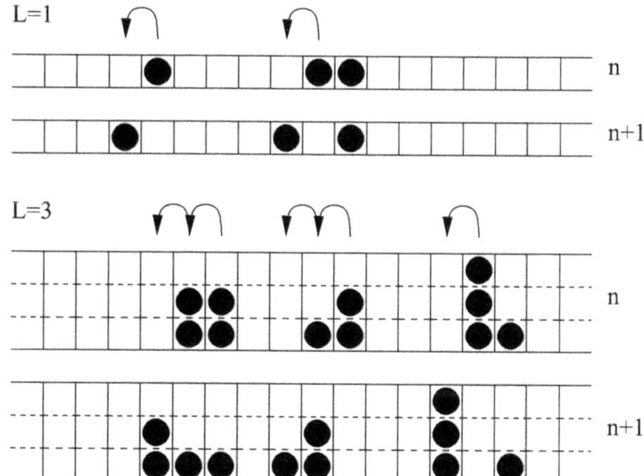

Fig. 11.3 Particle-based interpretation of the Burgers automaton. One time step for $L = 1$ resp. $L = 3$

particles as possible (the minimum of (1) the number of particles present in a site and (2) the free space available in the left neighboring site) jump to the left. If we formulate this recipe as local rule, we find back the Burger's cellular automaton.

Consider an explicit example for $L = 2$:

← direction of movement

···0 0 0 0 1 2 2 2 1 1 1 1 1 1 1···
···0 0 0 1 1 1 2 2 2 1 1 1 1 1 1···
···0 0 1 1 1 1 1 2 2 2 1 1 1 1 1···
···0 1 1 1 1 1 1 1 2 2 2 1 1 1 1···

We find at the l.h.s. of the block of '2' that a rarefaction wave is forming, while a shock wave is located at the rear of this block. This shock wave runs to the right, though the particles run to the left. However, in contrast to the viscosity solution for an initial condition with compact support, here the rarefaction wave does not reach the shock wave; the block 222 just moves with velocity 1 to the right. Let us consider a second example with $L = 3$:

← direction of movement

···0 0 0 0 1 3 3 3 1 1 1 1 1 1 1···
···0 0 0 1 2 1 3 3 2 1 1 1 1 1 1···
···0 0 1 2 1 2 1 3 3 1 1 1 1 1 1···
···0 1 2 1 2 1 2 1 3 2 1 1 1 1 1···
···1 2 1 2 1 2 1 2 1 3 1 1 1 1 1···
···2 1 2 1 2 1 2 1 2 1 2 2 1 1 1···

In this case, the block of '3' vanishes; the shock wave (right hand side of the block) moves with velocity $1/2$ to the right, while the rarefaction wave (left hand side of the block) moves with velocity 1 to the right. The rarefaction wave reaches the shock wave after a finite number of steps.

We easily obtain that the cellular automaton satisfies mass conservation (from our interpretation of the states as particle numbers). We may introduce a discrete flux: Let J_i^k denote the number of particles that jump form site i to site $i-1$ in the time step from k to $k+1$,

$$J_i^k = -\min\{u_i^k, L - u_{i-1}^k);$$

where we choose the minus sign, as the particles move to the left. For a spatial homogeneous global state, $u_i^k = u$, we have $J_i^K = -\min\{u, L-u)$. This is a tent map. Neither the direction nor the shape of the flux shows an obvious similarity to the flux in Burgers equation, which is for a constant state $u^2 - \nu u_x = u^2$.

This observation leads us to consider a transformation of the Burgers equation: Let $\rho(x,t) = u(x,t) + 1$. Then, $(u^2)_x = -((2-\rho)\rho)_x$, the equation for ρ reads

$$\rho_t - ((2-\rho)\rho)_x = \nu\rho_{xx},$$

and the flux is given by $j = -(2-\rho)\rho - \nu\rho_x$. The direction as well as the shape of the flux is now in line with that of the Burgers cellular automaton. This transformed model is well suited as a model for traffic: if there are only few cars on the motorway, the flux is small, and if there are many cars (bumper to bumper), the flux is again small. In between there is a car density maximizing the flux. Shock waves describe the end of a traffic jam, where the density suddenly is increased due to additional cars approaching from behind. At the head of the traffic jam, a rarefaction wave describes the transition to the undisturbed traffic ahead.

We observe that the same flux can be achieved by two different densities of cars. Traffic management systems try to keep the density on the increasing branch. Here, small fluctuations in the car density lead to (positively correlated) fluctuations of the flux; there is no risk of a jam. On the decreasing branch, however, a small increase of the density leads to a decrease in the flux, which in turn increases the density, s.t. a shock wave (traffic jam) appears.

In view of this interpretation, the different elementary states $0, \ldots, L$ do rather not correspond to single particles, but to particle density classes. And this, in turn, is exactly what is required for traffic management systems. Not the precise particle density is of interest, but density classes that can be related to the need for action.

11.3 Microscopic Models for Diffusion

An attempt to improve cellular automata models for particle motion is based on thinking about first principles. For example, in gas dynamics, particles move straight until they are scattered by collisions with other particles. A "good" cellular automaton model mimics this motion, that is, it should be a microscopic model.

The considerations below show how to implement such models and how to bridge the spatial scales between cellular automaton (microscopic) and partial differential equation (mesoscopic).

11.3.1 Straight Movement

We start with a partial differential equation model. The model describes particles in continuous time that run straight on the real line with constant velocity on the real axis and in continuous time. In this very basic model, particles do not interact; left-runners stay left-runners, and right-runners stay right-runners. If we assume that the velocity of a particle is either γ or $-\gamma$, and that the initial density of right (left) running particles is $u_0^{\pm}(x)$, then we find at time t the particle densities $u^{\pm}(x,t) = u_0^{\pm}(x \mp \gamma t)$. That is, the densities satisfy the partial differential equations (just plug them in!)

$$u_t^+(x,t) + \gamma u_x^+(x,t) = 0$$

$$u_t^-(x,t) - \gamma u_x^-(x,t) = 0.$$

In particular,

$$(\partial_t + \gamma\partial_x)(\partial_t - \gamma\partial_x)u^{\pm}(x,t) = u_{tt}^{\pm}(x,t) - \gamma^2 u_{xx}^{\pm}(x,t) = 0.$$

Also the total particle concentration $v(x,t) = u^+(x,t) + u^-(x,t)$ satisfies the wave equation. Recall the characteristic scaling of time and space for the heat equation: $x \mapsto ax$, $t \mapsto a^2 t$. This scaling is not appropriate for this equation. In the present case, the equation is invariant under the hyperbolic scaling, that treats time and space equally: $x \mapsto ax$, and $t \mapsto at$.

Now we extend our model: We assume that two particles which hit in a frontal collision bounce back and change their direction. Are our model equations affected by this modification? No, they are not. If particles collide and turn around, we have before and after the collision one left- and one right-running particle (see Fig. 11.4). In this very simple setting (one dimension, with only one value for the absolute velocity), it does not matter whether the particles do not interact or bounce back in an elastic impact. Later, in models with more degrees of freedom (dimension), collisions will matter.

It is easy to define a cellular automaton that plays this game [155, 161]. We take $\mathbb{Z}$ or $\mathbb{Z}_k$ as a grid, $D_0 = \{-1, 0, 1\}$, $E = \{0,1\} \times \{0,1\}$ (where a "1" in the first component indicates a right running particle, and a "1" in the second component a left running particle), see Fig. 11.5. Accordingly, we define the local function

$$f_0(e_{-1}, e_0, e_1) = ((e_{-1})_1, (e_1)_2).$$

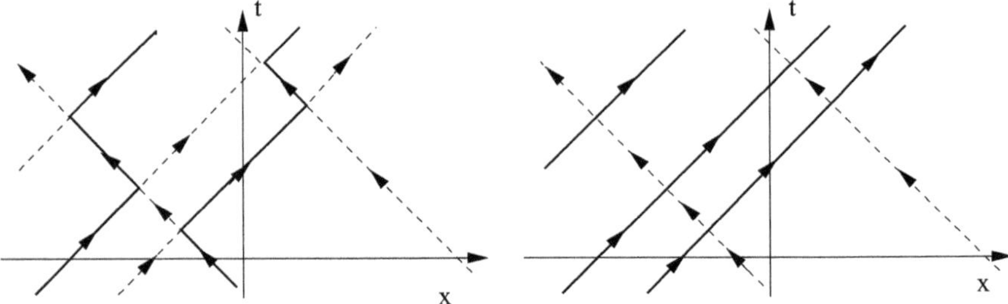

Fig. 11.4 In the two panels, individual particles are indicated by *dashed* resp. *solid lines*. *Left*: particles bounce back. *Right*: particles do not interact. The numbers of left- and right running particles are identical in both figures, only the identity of the particles are not

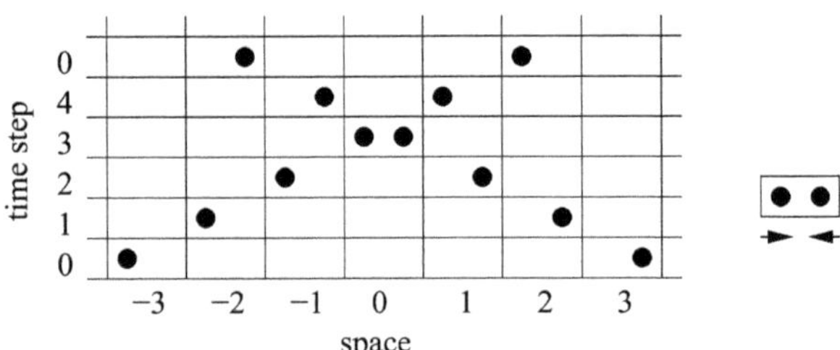

Fig. 11.5 Cellular automaton describing right/left running particles. The *dot* at the *left hand side* within a cell indicates a right running particle, the *dot* at the *right hand side* a left running particle

Select an initial state $u_0 \in E^{\mathbb{Z}}$, and define $m^+(n,i) = (f^n(u_0))_1 \in \{0,1\}$, respectively $m^-(n,i) = (f^n(u_0))_2 \in \{0,1\}$. Then,

$$m^+(n+1,i) = m^+(n,i-1), \quad m^-(n+1,i) = m^-(n,i+1).$$

Next we construct a connection between this cellular automaton and the corresponding partial differential equation introduced above for $u^\pm(x,t)$. Although in the present case, this step is relatively simple (the particles do not interact), we will present this procedure slowly and thoroughly, since we will use it also later in more complex situations.

We first select an initial state according to a random law,

$$P((u_0(i))_1 = 1) = \nu^+(i), \quad P((u_0(i))_2 = 1) = \nu^-(i)$$

for some functions $\nu^\pm : E \to [0,1]$. This construction defines a random measure on E^Γ. In case of a finite grid, $\Gamma = \mathbb{Z}_k$, this random measure is simple to construct. With

$$\zeta(a;\lambda) = \lambda a + (1-\lambda)(1-a), \quad \lambda \in [0,1], \ a \in \{0,1\},$$

the map

$$\mu : E^\Gamma \to [0,1],\quad u \mapsto \mu(u) = \Pi_{i\in\mathbb{Z}_k}\zeta\big((u(i))_1;\nu^+(i)\big)\,\zeta\big((u(i))_2;\nu^-(i)\big)$$

assigns a probability to each initial state. In case of an infinite grid, the random measure is not constructed by assigning a probability to single state $u \in E^\Gamma$, but to cylinder sets $\Sigma_\ell(u) = \{v \in E^\Gamma \,:\, u|_{\Gamma_\ell} = v|_{\Gamma_\ell}\}$,

$$\mu(\Sigma_\ell(u)) = \Pi_{i\in\Gamma_\ell}\zeta\big((u(i))_1;\nu^+(i)\big)\,\zeta\big((u(i))_2;\nu^-(i)\big), \qquad \ell \in \mathbb{N}.$$

The cylinder sets and μ define a probability space $(E^\Gamma, \mathfrak{S}, \mu)$, where $\mathfrak{S}$ denotes the Borel algebra generated by the cylinder sets and μ contains all information about the initial state. If we apply the global function of the cellular automaton constructed above $f : E^\Gamma \to E^\Gamma$, the probability for states change—a state that has a high probability in the initial time step may become a low probability after a few iterations and vice versa. Let μ_n denote the probability measure after n steps, then

$$\mu_n(v) = \mu(\{u \in E^\Gamma \,:\, f^n(u) = v\})$$

resp.

$$\mu_n(\Sigma_\ell(v)) = \mu(\{u \in E^\Gamma \,:\, f^n(u) \in \Sigma_\ell(v)\}).$$

Since the global function of our cellular automaton is bijective, we may write as well

$$\mu_n(v) = \mu(f^{-n}(v)) \quad \text{and} \quad \mu_n(\Sigma_\ell(v)) = \mu(f^{-n}(\Sigma_\ell(v))).$$

Until now we still work with the (deterministic) cellular automaton, we only introduced a random initial value. Now we aim to define smooth functions in time and space (based on the state of the cellular automaton and an appropriate limit) that satisfy the partial differential equation for $u^\pm(x,t)$. Thereto we introduce particle concentrations for the different particle species (left- and right running particles),

$$\overline{m}^+(i,n) = E_{\mu_n}((u_i)_1), \quad \overline{m}^-(i,n) = E_{\mu_n}((u_i)_2).$$

That is, we focus on a grid point $i \in \mathbb{Z}$, and check the probability to find a left- or right running particle at this point and time step n. Since $E_{\mu_n}(u) = E_{\mu_{n-1}}(f^{-1}(u))$, we obtain from the definition of the local function that

$$\overline{m}^\pm(i,n+1) = \overline{m}^+(i \mp 1, n).$$

We introduce the discrete derivatives

$$\Delta_t f(i,n) = f(i,n+1) - f(i,n), \quad \Delta_x f(i,n) = f(i+1,n) - f(i,n)$$

and, for later use, also the second derivative

$$\begin{aligned}\Delta_{xx}f(x,y,n) &= \Delta_x f(x+1,y,n) - \Delta_x f(x,y,n)\\ &= f(x+2,y,n) - 2f(x+1,y,n) + f(x,y,n).\end{aligned}$$

With the help of these derivatives, we formulate the following corollary.

Corollary 11.3.1 *The concentrations of right-(left-) running particles satisfy*

$$\Delta_t\overline{m}^+(i,n) + \Delta_x\overline{m}^+(i-1,n) = 0, \quad \Delta_t\overline{m}^-(i,n) - \Delta_x\overline{m}^-(i,n) = 0.$$

We now assume a certain scaling behavior of the initial condition,

$$\overline{m}_0^{\pm}(i) = w_0^{\pm}(i\,\delta_x)$$

for $\delta_x \in \mathbb{R}_+$ small, and w_0 smooth (twice differentiable). Then, $\overline{m}^{\pm}(i,n) \approx w(x,t)$ with $x = i\,\delta_x$, $t = n\,\delta_t$, and

$$\Delta_t\overline{m}^{\pm}(i,n) = \delta_t\, w_t^{\pm}(x,t) + O(\delta_t^2), \quad \Delta_x\overline{m}^{\pm}(i,n) = \delta_x\, w_x^{\pm}(x,t) + O(\delta_x^2)$$

such that in lowest order

$$w_t^+(x,t) + \gamma w_x^+(x,t) = 0, \quad w_-^-(x,t) - \gamma w_x^-(x,t) = 0$$

with $\delta_x/\delta_t = \gamma$. Note that we used the hyperbolic scaling discussed above, and indeed, we obtain a hyperbolic partial differential equation. This approach, originating in statistical physics, yields a connection between a (deterministic) cellular automaton with random initial conditions and a partial differential equation.

Remark 11.3.2 The basic steps from cellular automaton to partial differential equation are:

Step 1: We start from the microscopic, deterministic equations, in present case:

$$m^{\pm}(n+1,i) = m^{\pm}(n,i\mp 1), \qquad m^{\pm} \in \{0,1\}.$$

These variables are discrete in state, space and time.
Step 2: Then we obtain microscopic equations for the particle densities determined by a random initial condition, in present case:

$$\overline{m}^{\pm}(n+1,i) = \overline{m}^{\pm}(n,i\mp 1), \qquad \overline{m}^{\pm} \in [0,1].$$

These state variables are already continuous, but still discrete in space and time.
Step 3: Last, the initial condition is scaled such that it varies only weakly from grid point to grid point. This assumption allows to embed the grid in continuous

space, and the discrete time in a continuous time $(x, t) = (h\,i, h\,t)$ for h small. In this way, we end up with a partial differential equation.

In case of more complex models, we will introduce one more intermediate step: if the particles interact, they change their state. As the spatial variation of the state is rather weak, the structured particle density locally tends to an equilibrium: The dynamics forms locally an irreducible and aperiodic Markov chain that tends to its invariant measure. Especially, the velocities locally approach some kind of uniform distribution. Using this assumption of local equilibrium distributions, the approximate equations are more simple to derive. This kind of reasoning has been introduced to obtain from the Boltzmann equation the Euler (or, in a more refined way, also to the Navier-Stokes) equation.

When we derived the heat equation, we did not work with particles running in a certain direction, but with the total particle density and the particle flux. We take a second look at our dynamics of left- and right runners in terms of density and flux. Therefore we introduce the velocities $c^{\pm}$, which incorporate in contrast to γ not only the absolute value but also the direction of the velocity.

Proposition 11.3.3 *Let $c_{\pm} = \pm 1$ the velocities for the particles, and*

$$v(i,n) = m^+(i,n) + m^-(i,n), \qquad j(i,n) = c_+\, m^+(i,n) + c_-\, m^-(i,n),$$

$\overline{v}(i,n) = \overline{m}^+(i,n) + \overline{m}^-(i,n)$, *and* $\bar{j}(i,n) = c_+\,\overline{m}^+(i,n) + c_-\,\overline{m}^-(i,n)$. *Then,*

$$\Delta_t \overline{v}(i,n) = -\Delta_x \bar{j}(i,n) + \Delta_x^2 \overline{m}^+(i-1,n)$$
$$\Delta_t \bar{j}(i,n) = -\Delta_x \overline{v}(i,n) + \Delta_x^2 \overline{m}^+(i-1,n)$$

Proof Note that

$$\Delta_x^2 \overline{m}^+(i-1,n) = \Delta_x(\overline{m}^+(i,n) - \overline{m}^+(i-1,n)) = \Delta_x \overline{m}^+(i,n) - \Delta_x \overline{m}^+(i-1,n).$$

Therewith we obtain

$$\begin{aligned}
\Delta_t \overline{v}(i,n) &= \Delta_t \overline{m}^+(i,n) + \Delta_t \overline{m}^-(i,n)\\
&= -\Delta_x \overline{m}^+(i-1,n) + \Delta_x \overline{m}^-(i,n)\\
&= -\Delta_x \bar{j}(i,n) + \Delta_x \overline{m}^+(i,n) - \Delta_x \overline{m}^-(i-1,n)\\
&= -\Delta_x \bar{j}(i,n) + \Delta_x^2 \overline{m}^+(i-1,n),
\end{aligned}$$

and

$$\begin{aligned}
\Delta_t \bar{j}(i,n) &= -\Delta_x \overline{v}(i,n) + \Delta_x \overline{m}^+(i,n) - \Delta_x \overline{m}^+(i-1,n)\\
&= -\Delta_x \overline{v}(i,n) + \Delta_x^2 \overline{m}^+(i-1,n).
\end{aligned}$$

□

Proposition 11.3.4 *If $\overline{v}(i,n) \approx \nu(x,t)$ and $\bar{j}(i,n) \approx \iota(x,t)$ with $x = i\,\delta_x$, $t = n\,\delta_t$, and ν, $\iota \in C^2$, we find in lowest order*

$$\nu_t(x,t) = -\gamma \iota_x(x,t), \quad \iota_t(x,t) = -\gamma \nu_x(x,t),$$

where $\gamma = \delta_x/\delta_t$ and

$$\nu_{tt}(x,t) = \gamma^2 \nu_{xx}(x,t).$$

Proof With the same scaling arguments as above, we have $\Delta_t \overline{v}(i,n) \approx \delta_t\, \nu_t(x,t) + O(\delta_t^2)$, $\Delta_x \overline{v}(i,n) \approx \delta_x\, \nu_x(x,t) + O(\delta_x^2)$, $\Delta_x^2 \nu^+(i,n) \approx O(\delta_x^2)$, and similar expansions for $\Delta_t \bar{j}$, $\Delta_x \bar{j}$. These expansions immediately yield the desired equations in lowest order. □

11.3.2 Lattice Gas Cellular Automata

Lattice gas cellular automata mimic the dynamics of gas molecules: the molecules move straight, and bounce back in elastic collisions. At micro-scale, this structure resembles a billiard game. There has been interest in using balls as a model for computations since Leibniz (1646–1716) who developed in the manuscript *De progressione Dyadica*, 1679, a binary computer based on balls moving in a rack: ball present means 1, ball absent 0. This computer has never been realized during his lifetime, but 1971 a working specimen was build [157]. Fredkin presented the concept of a mechanical computer, which is computational universal, consisting of (ideal) billiard-balls [58]. Later, Margolus [121] published 1984 a cellular automaton version of this billiard model. Following [180], we consider a similar but slightly simpler model, the HPP model (named after Hardy, de Pazzis and Pomeau who proposed this model in 1973 [84]). A thorough introduction can be found in the book by Wolf-Gladrow [180] or in the review paper by Wolfram [182].

11.3.2.1 The HPP Model

The spatial structure of the HPP model is given by $\Gamma = \mathbb{Z}^2$ or by a finite torus $\mathbb{Z}_n \times \mathbb{Z}_{n'}$. Particles are only allowed to move horizontal or vertical, with the velocities

$$c_{+0} = \begin{pmatrix} 1 \\ 0 \end{pmatrix}, \qquad c_{-0} = \begin{pmatrix} -1 \\ 0 \end{pmatrix},$$

$$c_{0+} = \begin{pmatrix} 0 \\ 1 \end{pmatrix}, \qquad c_{0-} = \begin{pmatrix} 0 \\ -1 \end{pmatrix}.$$

At each site, there may be up to four particles, one for each velocity. We choose $E = \{0, 1\}^4$. The four components of $e \in E$ are either indexed by $\{(\pm, 0), (0, \pm)\}$ in line with the definition of $c_{\pm,0}$, $c_{0,\pm}$ (where, e.g., $e_{+0} = 1$ indicates that there is a particle moving to the right, or $e_{0-} = 0$ signals no particle running downwards is present), or we order the components in a fixed sequence,

$$e = (e_{+0}, e_{-0}, e_{0+}, e_{0-}),$$

such that for example $e_2 = 1$ indicates that there is a particle moving to the left. The indices $\{(\pm, 0), (0, \pm)\}$ and $1, \dots, 4$ are interchangeable.

The local rule describes two different events: straight movement of a particle and collisions. Let us first neglect collisions, and assume that particles do not interact at all but only move straight. To describe movement, we choose D_0 as the Neumann neighborhood, and define the local function $f_m : E^{D_0} \to E$ (see Fig. 11.6)

$$(f_m(u))_{+0} = u(-1, 0)_{+0}, \quad (f_m(u))_{-0} = u(1, 0)_{-0},$$
$$(f_m(u))_{0+} = u(0, -1)_{0+}, \text{ and } (f_m(u))_{0-} = u(0, 1)_{0-}.$$

Next, let us define collisions. Here we use the notation that numbers the states (and velocities). Collisions only happen locally, within one site. If there are two balls face to face (e.g. a particle with velocity $c_{+0} = c_1$ and one with velocity $c_{-0} = c_2$), and no other particle, then the velocities are rotated by 90 degree: the state $(1, 1, 0, 0)$ is mapped on $(0, 0, 1, 1)$ and vice versa, all other states stay invariant.

$$f_c : E \to E, \quad f_c(e) = \begin{cases} (1100) & \text{if } \; e = (0011) \\ (0011) & \text{if } \; e = (1100) \\ e & \text{else.} \end{cases}$$

The complete local function for the HPP model is the concatenation of movement and collisions, $f_0 = f_c \circ f_m : E^{D_0} \to E$.

Remark 11.3.5 The global function f of the HPP model is reversible. It is easy to construct f^{-1}: given a state $u \in E^{\Gamma}$, we first apply f_c at each site (since f_c is idempotent, $f_c \circ f_c = id_E$), and then apply the local function that moves particles in the opposite direction they are moved by f_m,

$$(\tilde{f}_m(u))_{+0} = u(1, 0)_{+0}, \quad (\tilde{f}_m(u))_{-0} = u(-1, 0)_{-0},$$
$$(\tilde{f}_m(u))_{0+} = u(0, 1)_{0+}, \text{ and } (\tilde{f}_m(u))_{0-} = u(0, -1)_{0-}.$$

We will use that fact that f is invertible later on.

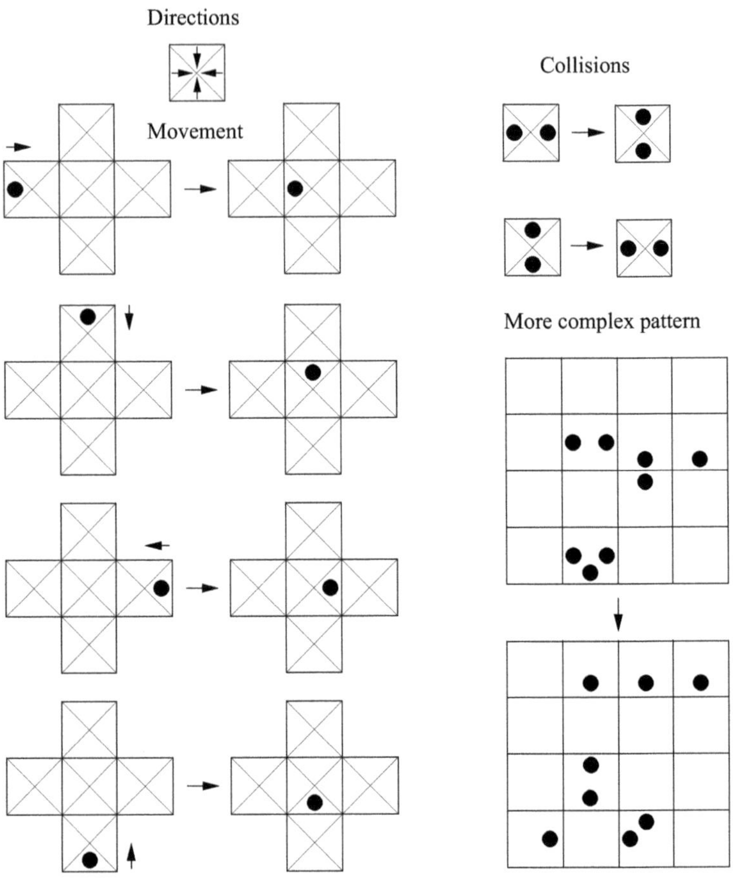

Fig. 11.6 Movement of particles according to the HPP model (*left*), collisions in the HPP model (*right*, *upper panel*), one step in a more complex situation (*right*, *lower panel*)

Proposition 11.3.6 *Consider* $u \in (E^{\Gamma})_c$*, an initial condition with a compact support. The HHP model leaves the total number of particles*

$$N = \sum_{(x,y)\in\mathbb{Z}^2} \sum_{i=1}^{4} (u(x,y))_i$$

and the total particle flow (impulse)

$$J = \sum_{(x,y)\in\mathbb{Z}^2} \sum_{i=1}^{4} c_i \; (u(x,y))_i$$

invariant. Also the quantities

$$M_k^x = \sum_{x \in \mathbb{Z}} \Big(u(x,k)_{+0} - u(x,k)_{-0} \Big)$$

and

$$M_k^y = \sum_{y \in \mathbb{Z}} \Big(u(k,y)_{0+} - u(k,y)_{0-} \Big)$$

for $k \in \mathbb{Z}$ are invariant under the dynamics given by the cellular automaton.

Proof The invariance of the total particle number and the total particle flux is a simple consequence of the local rules. In particular, particles always change their velocity in pairs. Their velocity sum up to zero, before and after the collision. This is also the mechanism leading to the invariance of M_k^x and M_k^y. □

For the total particle number and total particle flux (that is, the total mass and total impulse of the system), this invariance is a desired property. The invariance of $M_k^{x/y}$, however, is rather surprising. Non-local invariants of this type are called "spurious invariants" or "staggered invariants". Methods to uncover these invariants are for example developed by Takesue [162]. If we consider other models for fluid dynamics, e.g. the Boltzmann equations, we do not find these invariants. These invariants are a consequence of the very few directions resp. velocities incorporated by the HPP model. Improved models use more directions with the consequence that they have less or none spurious invariants.

The HPP model is a well defined, deterministic cellular automaton. The analysis, however, is based on the methods developed for statistical physics. The initial condition will be chosen according to a random distribution. In order to simplify the present analysis, the initial condition is defined in such a way that the expectation of the total flux is zero. Let $\nu_0 : \mathbb{Z}^2 \to [0,4]$ with $\sum_{x,y} \nu_0(x,y) < \infty$. ν_0 determines the expected number of particles at site (x,y). Given $(x,y) \in \Gamma$, we construct a state $u \in E^\Gamma$ by

$$P(u(x,y)_i = 1) = \nu_0(x,y)/4 \quad \text{for } (x,y) \in \Gamma, \quad i \in \{1,\ldots,4\}.$$

Recall the notation and definition of μ_n in Sect. 11.3.1. Also here, a family of random measures is defined by

$$\mu_n(\Sigma_\ell(u)) = \Pi_{(x,y)\in\Gamma_\ell} \Pi_{i=1}^4 \zeta((f^n(u)(x,y))_i, \nu_0(x,y)/4).$$

Proposition 11.3.6 implies immediately the following proposition.

Proposition 11.3.7 *Consider an initial state defined by a probability measure μ, and μ_n the induced measure at time step n. Then, the total particle number*

$$\overline{N}(n) = E_{\mu_n}\left(\sum_{(x,y)\in\mathbb{Z}^2}\sum_{a,b\in\{+,-\}}(u(x,y))_{ab}\right)$$

and the total particle flow (impulse)

$$\overline{J}(n) = E_{\mu_n}\left(\sum_{(x,y)\in\mathbb{Z}^2}\sum_{a,b\in\{+,-\}}c_{ab}\ (u(x,y))_{ab}\right)$$

invariant.

Also the quantities

$$\overline{M}_k^x(n) = E_{\mu_n}\left(\sum_{x\in\mathbb{Z}}\left(u(x,k)_{+0} - u(x,k)_{-0}\right)\right)$$

and

$$\overline{M}_k^y(n) = E_{\mu_n}\left(\sum_{y\in\mathbb{Z}}\left(u(k,y)_{+0} - u(k,y)_{-0}\right)\right)$$

for $k \in \mathbb{Z}$ are invariant under the dynamics given by the cellular automaton.

We now construct a PDE that approximates under suitable conditions the dynamics of the cellular automaton. Therefore the particle density and the flux at time $n \in \mathbb{N}$ for an initial state $u \in E^{\mathbb{Z}^2}$ is defined by

$$\begin{aligned}
v(x,y,n;u) &= \sum_{i=1}^{4}(f^n(u)(x,y))_i) \\
j_x(x,y,n;u) &= (f^n(u)(x,y))_{+0} - (f^n(u)(x,y))_{-0} \\
j_y(x,y,n;u) &= (f^n(u)(x,y))_{0+} - (f^n(u)(x,y))_{0-}.
\end{aligned}$$

Additionally, we define $m_{ab}(x,y,n;u) = (f^n(u)(x,y))_{ab}$ for $(a,b) \in \{(\pm,0),(0,\pm)\}$. We do not start with a deterministic state, but with a probability measure μ on E^Γ, and write accordingly

$$\overline{m}_{ab}(x,y,n) = E_{\mu_n}(m_{ab}(x,y,t;u)), \quad a,b \in \{(\pm,0),(0,\pm)\}$$

$$\overline{v}(x, y, n) = E_{\mu_n}\left(\sum_{i=1}^{4} u(x, y)_i\right) = \sum_{(a,b)} \overline{m}_{ab}(x, y, n)$$

$$\overline{j}_x(x, y, n) = E_{\mu_n}\,(u(x, y)_{+0} - u(x, y)_{-0})$$

$$\overline{j}_y(x, y, n) = E_{\mu_n}\,(u(x, y)_{0+} - (u(x, y)_{0-})\,.$$

Note that the dynamics of the cellular automaton is now hidden in the probability measure μ_n.

We use the discrete derivatives, introduced in Sect. 11.3.1.

Proposition 11.3.8

$$\begin{aligned}
\Delta_t\overline{v}(x, y, n) &= -\Delta_x\overline{j}_x(x, y, n) - \Delta_y\overline{j}_y(x, y, n)\\
&\qquad +\Delta_x^2\,\overline{m}_{-0}(x-1, y, n) + \Delta_y^2\,\overline{m}_{0-}(x, y-1, n)\\
\Delta_t\overline{j}_x(x, y, n) &= -\Delta_x\overline{m}_{+0}(x, y, n) - \Delta_x\overline{m}_{-0}(x, y, n) + \Delta_x^2\overline{m}_{+0}(x-1, y, n)\\
\Delta_t\overline{j}_y(x, y, n) &= \Delta_y\overline{m}_{0+}(x, y, n) - \Delta_y\overline{m}_{0-}(x, y, n) + \Delta_y^2 m_{0+}(x, y-1, n)
\end{aligned}$$

Proof We first consider one state $u \in E^\Gamma$. As the local particle number $v(x, y, t; u)$ is not changed by collisions, we find

$$\begin{aligned}
v(x, y, n+1; u) = &\quad m_{+0}(x-1, y, n; u) + m_{-0}(x+1, y, n; u)\\
&+ m_{0+}(x, y-1, n; u) + m_{0-}(x, y+1, n; u).
\end{aligned}$$

Similarly, the flux at a certain position is not affected by collisions: before and after collision the local velocity of the two colliding particles add up to zero.

$$\begin{aligned}
j_x(x, y, n+1; u) &= m_{+0}(x-1, y, n; u) - m_{-0}(x+1, y, n; u)\\
j_y(x, y, n+1; u) &= m_{0+}(x, y-1, n; u) - m_{0-}(x, y+1, n; u).
\end{aligned}$$

Therewith we compute Δ_t for v, j_x and j_y, and then replace the states m_{ab} at the right hand side as far as possible by v, j_x and j_y. We suppress u in the argument.

$$\begin{aligned}
&\Delta_t v(x, y, n)\\
&= \quad m_{+0}(x-1, y, n) + m_{-0}(x+1, y, n) + m_{0+}(x, y-1, n) + m_{0-}(x, y+1, n)\\
&\quad -m_{+0}(x, y, n) - m_{-0}(x, y, n) - m_{0+}(x, y, n) - m_{0-}(x, y, n)\\
&= -(m_{+0}(x, y, n) - m_{-0}(x+1, y, n)) + (m_{+0}(x-1, y, n) - m_{-0}(x, y, n))\\
&\quad -(m_{0+}(x, y, n) - m_{0-}(x, y+1, n)) + (m_{0+}(x, y-1, n) - m_{0-}(x, y, n))\\
&= -j_x(x, y, n) + j_x(x-1, y, n) + \Delta_x m_{-0}(x, y, n) - \Delta_x m_{-0}(x-1, y, n)
\end{aligned}$$

$$
\begin{aligned}
&-j_y(x,y,n)+j_y(x,y-1,n)+\Delta_y m_{0-}(x,y,n)-\Delta_y m_{-0}(x,y-1,n)\\
&=-\Delta_x j_x(x,y,n)-\Delta_y j_y(x,y,n)+\Delta_x^2\, m_{-0}(x-1,y,n)+\Delta_y^2\, m_{0-}(x,y-1,n)
\end{aligned}
$$

and

$$
\begin{aligned}
\Delta_t j(x,y,n) &= m_{+0}(x-1,y,n)-m_{-0}(x+1,y,n)-(m_{+0}(x,y,n)-m_{-0}(x,y,n))\\
&=-\Delta_x m_{+0}(x-1,y,n)-\Delta_x m_{-0}(x,y,n)\\
&=-\Delta_x m_{+0}(x,y,n)-\Delta_x m_{-0}(x,y,n)+\Delta_x^2 m_{+0}(x-1,y,n)
\end{aligned}
$$

The formula for j_y can be obtained by a parallel computation. Since these equations hold true for each single trajectory, they are also true if we average w.r.t. a given probability measure. □

Now we go for a thermodynamic limes. The (formal) computation relies on two different ideas: (1) We assume that locally the distribution tends to an equilibrium distribution. In this equilibrium distribution, at a given site each velocity posses the same probability (the so-called local equilibrium measure). Due to symmetry reasons, in thew HPP model, this equilibrium measure is uniform, $\overline{m}_{ab}(x,y,n)\approx\overline{\nu}(x,y,n)/4$. (2) We rescale space and time, and assume that (under this scaling) all averaged state variables approximate smooth functions.

Remark 11.3.9 Note that

$$\Delta_x f(\delta_x x,\delta_x y,\delta_t t)=f(\delta_x x+\delta_x,\delta_x y,\delta_t t)-f(\delta_x x,\delta_x y,\delta_t t).$$

If we define $X=\delta_x x$, $Y=\delta_x y$, $\tau=\delta_t t$, then we find via Taylor expansion

$$
\begin{aligned}
\Delta_x f(\delta_x x,\delta_x y,\delta_t t) &= f(X+\delta_x,Y,\tau)-f(X,Y,\tau)=\delta_x f_x(X,Y,\tau)+O(\delta_x^2)\\
\Delta_x^2 f(\delta_x x,\delta_x y,\delta_t t) &= O(\delta_x^2).
\end{aligned}
$$

Similar formulas are true for $\Delta_t f$, $\Delta_y f$ and $\Delta_y^2 f$.

For the next corollary, we assume that the solution varies only slowly in space and time, $\overline{\nu}(x,y,n)\approx\rho(\delta_x x,\delta_x y,\delta_t n)$, $\bar{j}_x(x,y,n)\approx\iota_1(\delta_x x,\delta_x y,\delta_t n)$, $\bar{j}_y(x,y,n)\approx\iota_2(\delta_x x,\delta_x y,\delta_t n)$, and also that $\overline{m}_{ab}$ are in the local equilibrium measure. Proposition 11.3.8 implies that

$$
\begin{aligned}
\delta_t\,\partial_t\rho(X,Y,t) &= \delta_x\nabla\iota(X,Y,t)+O(\delta_x^2)\\
\delta_t\,\partial_t\iota(X,Y,t) &= \frac{-2\,\delta_x}{4}\nabla\rho(X,Y,t)\;+O(\delta_x^2)
\end{aligned}
$$

where $\iota=(\iota_1,\iota_2)^T$.

Corollary 11.3.10 *We introduce scale variables δ_x, δ_t. Assume that the expected value of the state variables approximate smooth functions ($\overline{\nu}(x,y,n) \approx \rho(\delta_x x, \delta_x y, \delta_t n)$, $\bar{j}_1(x,y,n) \approx \iota_x(\delta_x x, \delta_x y, \delta_t n)$, $\bar{j}_2(x,y,n) \approx \iota_y(\delta_x x, \delta_x y, \delta_t n)$ and also assume that $\overline{m}_{ab}$ are in the local equilibrium measure. Then, the lowest order approximation of the dynamics for ρ, $\iota = (\iota_1, \iota_2)^T$ reads*

$$\rho_t = -c\,div\;\iota, \qquad \iota_t = -2c\,\nabla\rho.$$

where $c = \delta_x/\delta_t$.

Remark 11.3.11 If we combine the two equations, we find

$$\rho_{tt} = -c\,\mathrm{div}\iota_t = 2c^2\Delta\rho,$$

that is, the wave equation. Note that time and space are scaled with a constant of similar magnitude. This is a hyperbolic scaling; a velocity (space over time) is kept approximately constant.

The wave equation allows to compute the solution also backwards in time—it has a hyperbolic character. A diffusion equation is different in this aspect: it is well posed only forward, but not backward in time. The conservation of impulse and mass, together with the spurious conservation laws do not allow for a strong dissipative character of the HPP model. Visual inspection of simulations indeed show that the solution resembles water waves (Fig. 11.7).

11.3.2.2 Singular Perturbation Theory

We shortly introduce the basic ideas of time scale analysis (singular perturbation theory) at an informal level. For a profound introduction, see e.g. the book of O'Malley [142]. The starting point is an ordinary differential equation of the type

$$\begin{aligned} x' &= f(x,y;\varepsilon) \\ y' &= \varepsilon g(x,y;\varepsilon) \end{aligned}$$

where ε is a small parameter (typically $\varepsilon \approx 0.01$), and f, g are smooth functions of order $O(\varepsilon^0)$. That is, generically the derivative of $x(t)$ is much larger than that of $y(t)$. We call $x(t)$ the fast variable (or process), and $y(t)$ the slow variable. Rather often one observes that $y(t)$ hardly changes, while $x(t)$ tends to an equilibrium. In order to better understand this behaviour, we rescale time

$$\tau = t\varepsilon$$

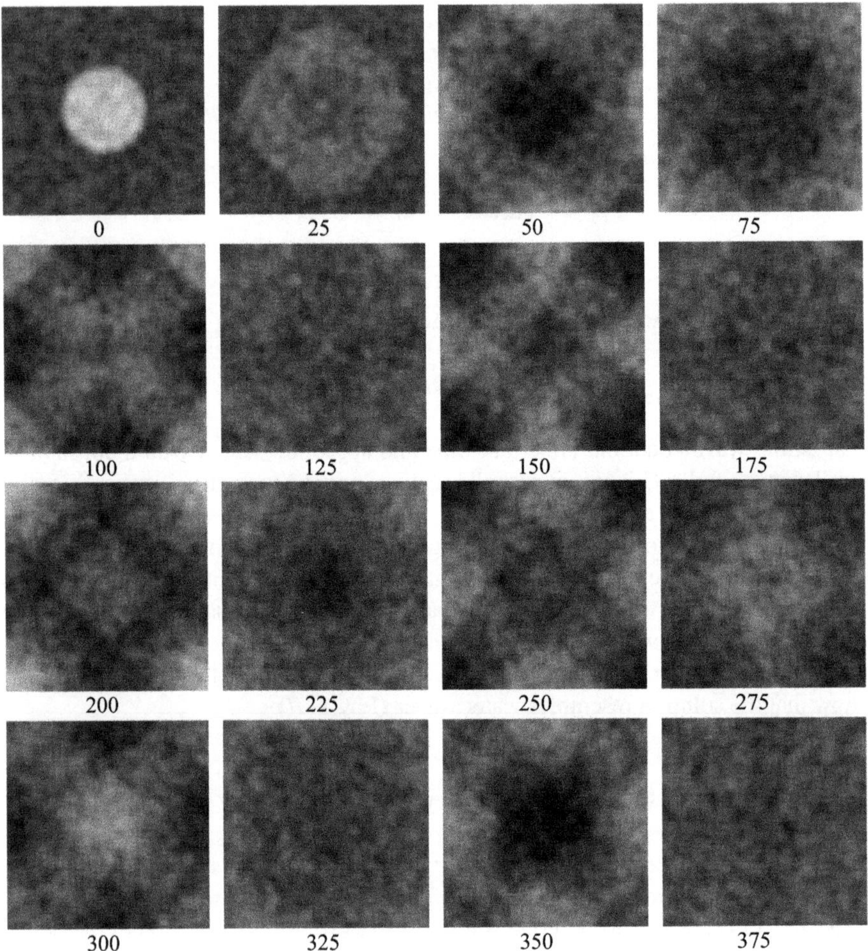

Fig. 11.7 Simulation of the HPP model on a grid of size 100×100. We draw the particle density (averaged over 5×5 cells) every 25 time steps. The dynamics resembles water waves in a rectangle vessel

and find

$$\varepsilon \dot{x} = f(x, y; \varepsilon)$$
$$\dot{y} = g(x, y; \varepsilon)$$

where $\dot{x} = dx/d\tau$. The original time t is appropriate for x, the fast variable, while the time τ is appropriate for y, the slow variable. Therefore, we call the system of ODEs with time t the fast system, and the ODEs with time τ the slow system. The transition from t to τ is a completely regular transformation of time. Both systems are equivalent. Now we go to a singular limit: we let (formally) ε tend to zero, and find

$$\begin{aligned} x' &= f(x, y; 0) & \qquad 0 &= f(x, y; 0) \\ y' &= 0 & \qquad \dot{y} &= g(x, y; 0). \end{aligned}$$

In this singular limit, it is not possible any more to transform one system into the other.

The fast system, $x' = f(x, y; 0)$, $y' = 0$, indicates that $y(t) \equiv y_0$ does not change at all in the very first time interval, the so-called initial layer. Within this time interval, $x(t)$ tends to the ω-limit set. We assume (this is the most simplest case), that this ω-limit set is a stationary point. In general, this stationary point depends on y_0,

$$\lim_{t\to\infty} x(t) = \Theta(y_0).$$

The state $x = \theta(y_0)$ is called quasi-stationary state. We do not get more information from the fast time scale. Now we turn to the slow time scale. We know by now, that $x = \Theta(y_0)$. If y starts to change slowly, then the fast process will keep x on the manifold $\{x = \Theta(y)\}$. Therefore, the curve $(\Theta(y), y)$ is also called the slow manifold. The dynamics of y in the singular limit is given by

$$\dot{y} = g(y, \Theta(y))$$

while x is determined by the algebraic condition $0 = f(x, y)$, which indicates $x = \Theta(y)$.

The Fenichel theorems [99] guarantee that the solution $x(t), y(t)$ for $\varepsilon > 0$ (but small) stays close to the orbits described by the singular systems (see Fig. 11.8).

11.3.2.3 The Diffusive HPP Model

Our analysis above indicated that the HPP model has a rather hyperbolic character. A closer look reveals that the high degree of symmetry in the dynamics together with the simplicity of the grid are essential to create this hyperbolicity. We may change the model in three aspects: either the collisions are extended to more complex cases (three particle collisions), the number of allowed velocities is increased, or the grid is changed to something slightly more complex (e.g. a hexagonal grid). In the present section, we focus on the collisions.

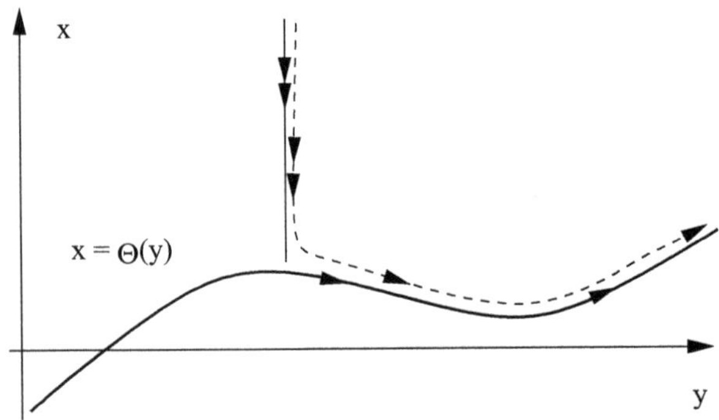

Fig. 11.8 Sketch of the singular perturbation theory. The singular, fast system (*vertical line*) saddles fast on the slow manifold $x = \Theta(y)$. The (singular) slow system evolves along this manifold. The original system (*dashed line*) stays close to the singular trajectories if $\varepsilon > 0$ is small

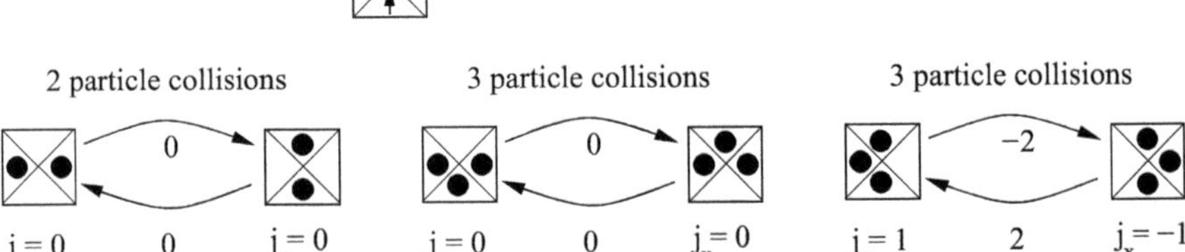

Fig. 11.9 Two- and three particle collisions possible for the diffusive HPP model, together with the value for the net flux j_x within the site (indicated below the pattern), and the change of j_x annotated at the *arrows* indicating the result of collisons

We keep the notation of the last section, and only adapt the local function where it describes collisions. Two-particle collisions are not changed, situations with none or four particles cannot be changed without violating mass conservation locally. It remains to define the tree-particle situation. One setting (among several possible) is depicted in Fig. 11.9,

$e = (e_{+0}, e_{-0}, e_{0+}, e_{0-})$				$f_c(e) = (\tilde{e}_{+0}, \tilde{e}_{-0}, \tilde{e}_{0+}, \tilde{e}_{0-})$			
e_{+0}	e_{-0}	e_{0+}	e_{0-}	$\tilde{e}_{+0}$	$\tilde{e}_{-0}$	$\tilde{e}_{0+}$	$\tilde{e}_{0-}$
1	1	1	0	1	1	0	1
1	1	0	1	1	1	1	0
1	0	1	1	0	1	1	1
0	1	1	1	1	0	1	1

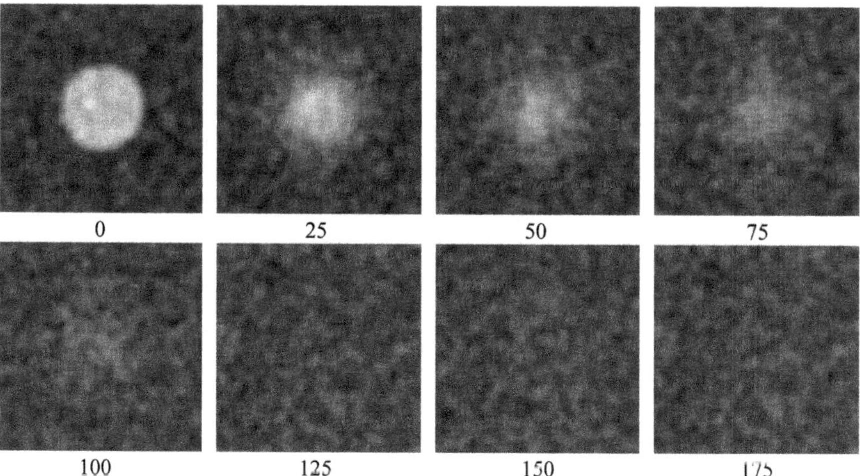

Fig. 11.10 Simulation of the diffusive HPP model on a grid of size 100×100. We draw the particle density (averaged over 5×5 cells) every 25 time steps. Note that the initial condition is identical with that of Fig. 11.7

With this extension, f_c is still idempotent, and the global function is still invertible. However, in particular the conservation of local (and total) flux resp. impulse is not given any more, and the character of the dynamics changes (compare simulations in Fig. 11.7 and 11.10). We keep the notation as introduced for the HPP model. The following proposition parallels Proposition 11.3.8.

Proposition 11.3.12

$$
\begin{aligned}
\Delta_t \overline{v}(x,y,n) &= -\Delta_x \overline{j}_x(x,y,n) - \Delta_y \overline{j}_y(x,y,n) \\
&\quad + \Delta_x^2\, \overline{m}_{-0}(x-1,y,n) + \Delta_y^2\, \overline{m}_{0-}(x,y-1,n) \\
\Delta_t \overline{j}_x(x,y,n) &= -\Delta_x \overline{m}_{+0}(x,y,n) - \Delta_x \overline{m}_{-0}(x,y,n) + \Delta_x^2 \overline{m}_{+0}(x-1,y,n) \\
&\quad -2\Big(\overline{j}_x(x,y,n) - \Delta_x \overline{m}_{+0}(x-1,y,n) + \Delta_x \overline{m}_{-0}(x,y,n)\Big) \\
&\qquad \times\ (\overline{m}_{0+}(x,y,n) - \Delta_y \overline{m}_{0+}(x,y-1,n)) \\
&\qquad \times\ (\overline{m}_{0-}(x,y,n) + \Delta_y \overline{n}^{0-}(x,y,n)) \\
\Delta_t \overline{j}_y(x,y,n) &= -\Delta_y \overline{m}_{0+}(x,y,n) - \Delta_y \overline{m}_{0-}(x,y,n) + \Delta_y^2 \overline{m}_{0+}(x,y-1,n) \\
&\quad 2\Big(\overline{j}_y(x,y,n) - \Delta_y \overline{m}_{0+}(x,y-1,n) + \Delta_y \overline{m}_{0-}(x,y,n)\Big) \\
&\qquad \times\ (\overline{m}_{+0}(x,y,n) - \Delta_x \overline{m}_{+0}(x-1,y,n)) \\
&\qquad \times\ (\overline{m}_{-0}(x,y,n) + \Delta_x \overline{m}_{-0}(x,y,n))
\end{aligned}
$$

Proof We first move the particles one step, and then change their direction according to local collision. As collisions do not change the number of particles at a site, the computation of $\Delta_t\overline{v}(x,y,n)$ does not change at all in comparison with the HPP model. However, the computation of $\Delta_t\bar{j}_x(x,y,n)$ is affected. The collisions incorporate the collisions in the HPP model, and additionally three-particle collisions. The horizontal flux is only affected if two vertical particles and one horizontal particle are located within one site (see Fig. 11.9). As all other collisions are the same as in the HPP model. We find

$$\begin{aligned}
\Delta_t\bar{j}_x(x,y,n) &= \underbrace{-\Delta_x\overline{m}_{+0}(x,y,n) - \Delta_x\overline{m}_{-0}(x,y,n) + \Delta_x^2\overline{m}_{+0}(x-1,y,n)}_{HPP} \\
&\quad -2\overline{m}_{+0}(x-1,y,n)(1-\overline{m}_{-0}(x+1,y,n))\overline{m}_{0+}(x,y-1,n)\overline{m}_{0-}(x,y+1,n) \\
&\quad +2(1-\overline{m}_{+0}(x-1,y,n))\overline{m}_{-0}(x+1,y,n)\overline{m}_{0+}(x,y-1,n)\overline{m}_{0-}(x,y+1,n) \\
&= -\Delta_x\overline{m}_{+0}(x,y,n) - \Delta_x\overline{n}_{-0}(x,y,n) + \Delta_x^2\overline{m}_{+0}(x-1,y,n) \\
&\quad -2\left(\overline{m}_{+0}(x-1,y,n) - \overline{m}_{-0}(x+1,y,n)\right)\overline{m}_{0+}(x,y-1,n)\overline{m}_{0-}(x,y+1,n) \\
&= -\Delta_x\overline{m}_{+0}(x,y,n) - \Delta_x\overline{m}_{-0}(x,y,n) + \Delta_x^2\overline{m}_{+0}(x-1,y,n) \\
&\quad -2\left(\bar{j}_x(x,y,n) - \Delta_x\overline{m}_{+0}(x-1,y,n) + \Delta_x\overline{m}_{-0}(x,y,n)\right) \\
&\qquad (\overline{m}_{0+}(x,y,n) - \Delta_y\overline{m}_{0+}(x,y-1,n))(\overline{m}_{0-}(x,y,n) + \Delta_y\overline{m}_{0-}(x,y,n)).
\end{aligned}$$

The formula for $\bar{j}_y$ follows by a parallel argument. □

Now we use the scaling argument as before and obtain

$$\begin{aligned}
\rho_t &= \frac{-\delta_x}{\delta_t}\operatorname{div}\iota + \text{higher order terms} \\
\partial_t\iota &= \frac{1}{\delta_t}\left(-2\rho^2\,\iota - 2\delta_x\nabla\rho + 4\delta_x\rho\begin{pmatrix}\iota_1\partial_x\rho \\ \iota_2\partial_y\rho\end{pmatrix} + \text{higher order terms}\right)
\end{aligned}$$

where $\iota = (\iota_1, \iota_2)^T$. For the following considerations we neglect the higher order terms, and assume a particular scaling of time and space, the so-called parabolic scaling:

$$\delta_x^2/\delta_t \approx D$$

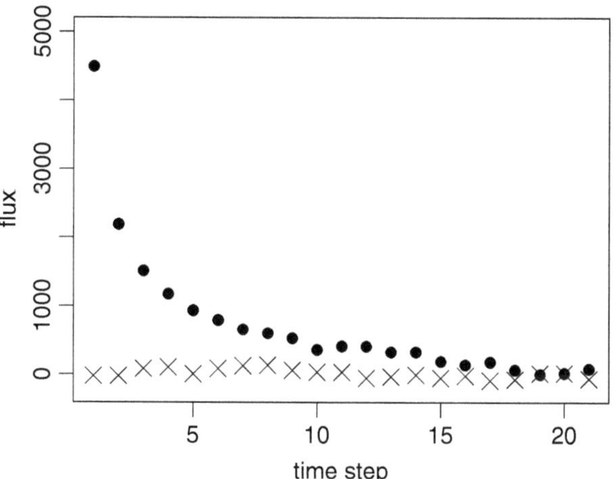

Fig. 11.11 Simulation of the diffusive HPP model. A grid of size 100×100 is initialized with a homogeneous particle density, a nonzero (spatially constant) mean flow in x direction, and a zero flow in y direction. *Bullets* denote the mean flux in x-direction, *crosses* that in y-direction

where D is a constant (recall that we used the hyperbolic scaling in the HPP case). If we start with $O(1)$ terms for particle density and flux, we find in particular that the dynamics of the flux in leading order reads

$$\delta_t\ \partial_t \iota = -2\rho^2\, \iota,$$

that is, the flux decreases exponentially and rapidly. This decrease is only stopped if the other terms (that are of order δ_x) influence the right hand side, i.e., if the flux itself becomes of order δ_x. This observation indicates that any nonzero flux in the initial condition is reduced to order $O(\delta_x)$. In the classical HPP model, a nonzero flux at initial condition leads to a drift that is conserved in time. In the present case, the mean flux vanishes rapidly (see Fig. 11.11). This is the first, and very rapid time scale.

At the latest after this initial layer, the flux is small (of order δ_x). We rescale the flux, introducing

$$\iota(x, y, t) = \delta_x \kappa(x, y, t).$$

If we do so, and if we neglect the higher order terms, we obtain

$$\rho_t = \frac{-\delta_x^2}{\delta_t} \operatorname{div}\kappa = D \operatorname{div}\kappa$$

$$\delta_t\ \partial_t \kappa = \left(-2\rho^2\, \kappa - 2\nabla\rho + 4\delta_x\, \rho \begin{pmatrix} \kappa_1 \partial_x \rho \\ \kappa_2 \partial_y \rho \end{pmatrix}\right)$$

After the first, very rapid time scale we find now a singular perturbed system; the fast time scale of this system (during that the solution saddles on the slow manifold) is the second time scale. During this phase, we still expect a hyperbolic character of the dynamics.

For δ_t small, κ will saddle on the slow manifold, given by (we neglect the term of order $O(\delta_x)$)

$$\kappa = -\rho^{-2}\nabla\rho.$$

Plugging this expression into the equation for $\partial_t\rho$, we find

$$\rho_t = D\nabla\left(\rho^{-2}\nabla\rho\right).$$

The third and slow time scale shows the diffusive behavior; the flux is not determined any more via its time derivative, but via an algebraic condition.

Also the original HPP model incorporates these three time scales, though they did not appear in the analysis explicitly. In the first, very short time interval, the equilibrium distribution of the velocities is reached. In the analysis, we assumed that this already had happened. In the second time period, we find a hyperbolic behavior. This is the behavior we find in the simulations. In the long run, dissipative terms (lower order terms excluded form our first order analysis) take over, and a diffusion-like behavior is displayed by the automaton: the solution approximates in the long run a stationary solution, fluctuating around some mean value. The diffusive HPP model and the original HPP model only show a shift of these three time phases, but basically run though a sequence of similar behavior.

If the initial conditions are scaled appropriately, the cellular automaton behaves as a diffusion equation with fast diffusion [51]. The larger the density, the slower is the diffusion. This observation is in line with physical intuition: If we consider a gas, say, the average free path length between two collisions increases if the concentration of particles decreases, and in this, the diffusion becomes faster. However, the approximation is not valid if the particle density ρ drops below a certain level, as the term ρ^2/δ_t becomes small, though δ_t is small. The assumption of a singular perturbed system is not correct any more. Nevertheless, in many cases the approximation is valid as long as the particle density stays away from zero and we find the dissipative, parabolic character of this cellular automaton back in simulation (see, e.g., Fig. 11.12).

We have been able to understand the difference of the HPP model and the diffusive HPP model, but only via the detour through partial differential equation. Strictly spoken, the grid Γ, the local states E and the local function contain enough information to directly deduce the character of the cellular automaton (the global function), but at the time being there is no general method to directly link the definition of a cellular automaton with its global, long-term behavior.

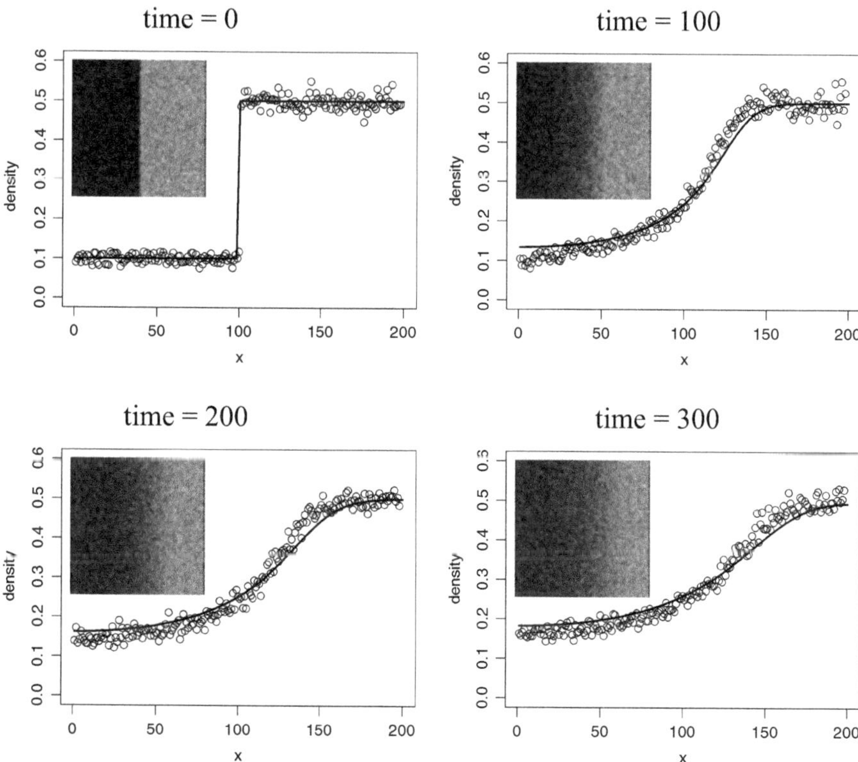

Fig. 11.12 Simulation of the diffusive HPP model. A grid of size 200×200 is initialized with a particle concentration of 0.1 for $x \leq 100$, and 0.5 for $x > 100$ (constant in y direction, each direction with identical probability). Reflecting boundaries have been placed at $x = 200$. The density over x are computed by averaging over the respective column. The *solid line* indicates the solution of $\rho_t = D\nabla\rho^{-2}\nabla\rho$ (finite difference scheme, $D = 1$, $\Delta_x = 0.1$, $\Delta_t = 0.0001$). The inlays show the state of the cellular automaton

Chapter 12
Pattern Formation

12.1 Fractal Mollusc Patterns

Surface patterns are common in biological objects. The stripes of zebras, the spots of the leopard are ubiquitous in the popular scientific literature [137], but also invertebrates and plants show various patterns. In the present context the patterns on the shells of molluscs (especially snails and mussels) are of special interest. These patterns are not produced all at once. Since the shell gradually grows by adding material to the outer edge and since the pattern is not changed later, the pattern is essentially located in a space-time continuum, with the space axis parallel to the edge and the time axis perpendicular to the edge. Although the strange patterns on several shells have been admired for centuries (e.g. the Bedouin tents of *Oliva porphyrea*) and Sierpinski diagrams have been known since the thirties, it took quite some time to establish a connection. The first results in this direction is probably due to Waddington and Cowe [170] who generated patterns by a computer simulation model, and to Ermentrout et al. ([53] and quotations therein) who suggested a neural network (essentially a cellular automaton) and presented some theoretical considerations and computer simulations. Then Meinhardt [125, 126] adapted reaction diffusion equations, and, in order to get a large variety of different patterns (many astonishingly similar to patterns appearing in real mollusks) used systems of reaction diffusion equations of ever increasing complexity. Later Markus [123] went back to cellular automata. Since he has shown that rather simple cellular automata produce more or less all the patterns requested (see, e.g., the fractal pattern generated by some linear cellular automata, Sect. 10.3), one might as well take cellular automata as the underlying generators of these patterns, even more so as nothing is known on the chemical nature of activators and inhibitors, let alone whole reaction chains of these [18]. Unfortunately lack of knowledge does not only apply to the generation of the pattern but also to their significance for the molluscs.

K.-P. Hadeler, J. Müller, *Cellular Automata: Analysis and Applications*,
Springer Monographs in Mathematics, DOI 10.1007/978-3-319-53043-7_12

12.2 Turing Pattern

Skin pattern of fish or mammals form differently than those of molluscs. Molluscs basically add a one-dimensional stripe per time step; in mammals, the pattern evolves at a two-dimensional surface. Local exchange of information leads to the identification of globally well defined spots, where, e.g., the skin becomes black. Turing discovered a mechanism that is able to solve this task [137, 164].

12.2.1 Turing-Pattern in Partial Differential Equations

We study mechanisms of pattern formation based on reaction-diffusion equations. First of all, we consider a linear system (in one dimension), where two substances interact with each other. Pattern formation does mean that the solutions are spatially inhomogeneous. A basic, linear model for two chemical substances which interact and diffuse in space reads

$$\begin{aligned} u_t &= D_1 u_{xx} + au + bv, \qquad & u_x(0,t) = u_x(L,t) = 0 \\ v_t &= D_2 v_{xx} + cu + dv, \qquad & v_x(0,t) = v_x(L,t) = 0. \end{aligned}$$

Since these equations have non-flux boundary conditions, there are spatially constant solutions $u(x,t) = \overline{u}(t)$, $v(x,t) = \overline{v}(t)$ satisfying

$$\frac{d}{dt}\begin{pmatrix}\overline{u}\\ \overline{v}\end{pmatrix} = A_0 \begin{pmatrix}\overline{u}\\ \overline{v}\end{pmatrix} \quad \text{with } A_0 = \begin{pmatrix} a & b\\ c & d\end{pmatrix}.$$

We require that all spatially constant solutions tend to zero. The Hurwitz criterion yields

$$\text{tr}(A_0) = a + d \le 0, \qquad \det(A_0) = ad - bc \ge 0.$$

We aim at parameters that allow for growing spatially non-constant solutions, though the spatially constant solutions always tend to zero. Starting point for the analysis is the observation, that the r.h.s. of the partial differential equation above leaves the space

$$\mathcal{L}_k = \{(\zeta_1,\eta_1)^T \sin(kx) + (\zeta_2,\eta_2)^T \cos(kx)\ :\ \zeta_1,\zeta_2,\eta_1,\eta_2\}$$

invariant. The boundary conditions selects spatial modes k as well as the phase of the trigonometric functions,

$$\mathcal{L}_k = \{(\zeta,\eta)^T \cos(kx)\ :\ \zeta,\eta\}, \qquad kL = n\pi,\ n \in \mathbb{Z}.$$

We identify solutions of the partial differential equation exponentially growing in time, that is, eigenfunctions in $\mathcal{L}_k$ for the r.h.s. of the partial differential equation:

$$A_k \begin{pmatrix} \zeta \\ \eta \end{pmatrix} = \lambda \begin{pmatrix} \zeta \\ \eta \end{pmatrix} \quad \text{with} \quad A_k = \begin{pmatrix} -D_1k^2 + a & b \\ c & -D_2k^2 + d \end{pmatrix}.$$

Again we use the Hurwitz criterion to check for stability. The solution will grow in this eigendirection if either of the two inequalities

$$\begin{aligned} \text{tr}(A_k) &= a + d - D_1k^2 - D_2k^2 = \text{tr}(A_0) - (D_1 + D_2)k^2 > 0 \\ \det(A_k) &= (a - D_1k^2)(d - D_2k^2) - bc \\ &= \det(A_0) - (D_1 d + D_2 a)k^2 + D_1\, D_2 k^4 < 0 \end{aligned}$$

are satisfied. Since $\text{tr}(A_0) < 0$ and $(D_1 + D_2)k^2 \geq 0$, we have $\text{tr}(A_k) < 0$. The first inequality can never be true. Next we turn to the second inequality. If $D_1 = D_2$, we find that

$$(D_1 d + D_2 a) = D_1(a + d) = D_1 \text{tr}(A_0) < 0.$$

Also in the case $a, d < 0$ we obtain

$$(D_1 d + D_2 a) < 0.$$

In booth cases, we conclude

$$\det(A_0) - \underbrace{(D_1 d + D_2 a)}_{<0} k^2 + D_1\, D_2 k^4 > \det(A_0) > 0$$

and the second inequality cannot be true. Necessary conditions for the second inequality to be true are $D_1 \neq D_2$ and $a\,d < 0$ (since $\text{tr}(A_0) < 0$, a and d cannot be simultaneously non-negative). Without restriction, choose $a > 0 > d$, and $D_2 \gg D_1$ such that $(D_1 d + D_2 a) \gg 0$. If we consider $\det(A_k)$ as a polynomial in k, the minimum is located at $k = k_0$ with

$$k_0^2 = \frac{(D_1 d + D_2 a)}{2\, D_1\, D_2}$$

and assumes the value

$$\det(A_{k_0}) = \det(A_0) - \frac{(D_1 d + D_2 a)^2}{2\, D_1\, D_2} + \frac{1}{2} D_1\, D_2\, (D_1 d + D_2 a).$$

If D_2 becomes large enough, this value is negative. There is a parameter set and $k_0 \in \mathbb{R}_+$ such that $\det(A_{k_0}) < 0$. Also if k is located within a sufficient small interval around k_0, the second inequality is satisfied; the region of instability $\{k \in \mathbb{R}_+ : \det(A_k) < 0\}$ is an open, bounded interval. We know that k only assumes the discrete values $n\pi/L$, $n \in \mathbb{N}_0$. If L is small, the first eigenvalue is on the right hand side of the interval of instability. If L increases, eventually this mode moves within this interval. There is a minimal for the region of the partial differential equation necessary for instabilities to appear. If L becomes even larger, the second, third etc. mode move into this interval coming from the right, while the smaller modes are leaving the interval again at its left boundary. The number of local maxima possible increases approximately proportionally to the size of the region L. If $D_1 \neq D_2$, diffusion may destabilize the constant solution, while diffusion always stabilizes the constant solution in case of $D_1 = D_2$.

All in all, we find that one substance (u) activates itself (auto-catalyst) while the second (v) will be degraded. As $ad < 0$ and $\det(A) = ad - bc > 0$, we know that $bc < 0$. There are two different scenarios: either u inhibits v and v enhances u, or vice versa. The typical situation is, that u enhances itself and v, and diffuses slow, while v is inhibiting u and itself, while diffusing fast. This is the so-called **activator-inhibitor-system**, the typical core of the Turing pattern. For an appropriate parameter range, the eigenvalues corresponding to suitable interval of spatial wavelengths $k \in [\underline{k}, \overline{k}]$ become positive. The boundary condition select a finite number of spatial wavelength within that interval. If we perturb the spatial homogeneous state randomly, we will find the pattern with the wavelength corresponding to the maximal eigenvalue. If we e.g. deform the region (enlarge L) it is possible to find patterns with positive, but not maximal eigenvalue: multistability of pattern with different wavelengths are possible.

It is possible to understand the mechanism for pattern formation intuitively: Let us assume that locally there is a maximum of the activator. This maximum will not disperse as the activator has a small diffusion coefficient. On the same time, at the site of this maximum, a lot of inhibitor is created. In contrast to the activator which stays localized, the inhibitor diffuses fast into the surrounding, inhibiting other activator-peaks to evolve (see Fig. 12.1). Only far away from the local peak of activator, the inhibitor is degraded to levels that are this small, that another peak of activator is able to grow. Several peaks, created at a large distance, will move slowly and arrange themselves until a periodic pattern is approached. In two dimensions, the mechanisms is similar, but more complex pattern may evolve caused by regions with a more complex shape (as the skin of a cow [137]).

12.2.2 Excursion: Hopfield Nets

In the next section, we will set up a cellular automaton that captures the most important mechanisms leading to Turing patterns. For the analysis of this automaton, we use ideas about Hopfield nets, which we introduce in the present section.

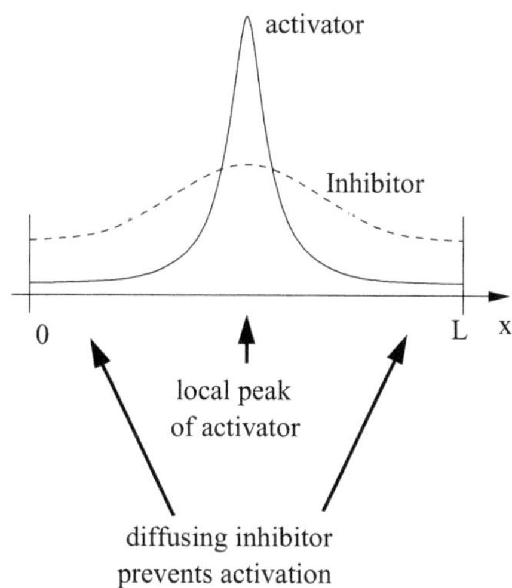

Fig. 12.1 Sketch of Turing-patterns for activator-inhibitor-systems

Definition 12.2.1 Let I be a (finite or countable infinite) index set, and $E = \{0, 1\}$. Let $(C_{ii'})_{i,i' \in I}$ denote a sparse matrix with real entries: for $i_0 \in I$ given, $|\{i \in I : C_{i_0\,i} \neq 0\}| < \infty$. Furthermore, $h \in \mathbb{R}^I$. The global function $f : E^I \to E^I$,

$$(f(u))_i = \begin{cases} 1 & \text{if } \sum_{i' \in I} C_{ii'} u(i') + h_i > 0 \\ 0 & \text{else} \end{cases}$$

is called Hopfield net.

Recall the definition of a Lyapunov function [42].

Definition 12.2.2 Let E finite, I a (finite or countable infinite) index set, and $f : E^I \to E^I$ a function. A function $V : E^I \to \mathbb{R}$ with the properties

(1) V is bounded from below,
(2) $V(f(u)) \leq V(u)$,
(3) $V(f(u)) = V(u)$ if and only if $u = f(u)$,

is called (strong) Lyapunov function. If V only satisfies (1) and (2), it is called weak Lyapunov function.

Note that a strong Lyapunov function forces a trajectory to converge to a continuum of stationary point. In case of a weak Lyapunov function, V is constant on ω-limit sets. The Hopfield net is a generalization of the Ising model, that is well known in statistical physics. The Ising model possesses an energy; the structure of this energy function can be used to construct a Lyapunov function for many Hopfield nets.

Proposition 12.2.3 *Consider a Hopfield net with symmetric coefficients, $C_{ij} = C_{ji}$ and finite index set, $|I| < \infty$. If the matrix $C = (C_{ij})$ is positive definite, the energy*

$$\mathcal{E} : E^I \to \mathbb{R}, \quad u \mapsto -\frac{1}{2} \sum_{i,i' \in I} C_{ii'} u(i') u(i) - \sum_{i \in I} h_i u(i)$$

is a strong Lyapunov function.

Proof We use vector notation, $u = (u_1, \ldots, u_{|I|})$, $v = (v_1, \ldots, v_{|I|})$, $h = (h_1, \ldots, h_{|I|})$. In particular, $\mathcal{E}(u) = \frac{1}{2} u^T C u - h^T u$. Let $v = f(u)$. As C is a symmetric matrix, we find

$$\mathcal{E}(v) - \mathcal{E}(u) = -\frac{1}{2} v^T C v - v^T h + \frac{1}{2} u^T C u + u^T h$$

$$= -\frac{1}{2} \left(v^T C v - 2 u^T C v + u^T C u \right) + u^T C u - u^T C v - (v - u)^T h$$

$$= -\frac{1}{2} (v - u)^T C (v - u) - (v - u)^T (C u + h)$$

We know, as C is positive definite, that $-\frac{1}{2}(v-u)^T C(v-u) \leq 0$. Now, consider one component of $(v-u)(Cu+h) = \sum_{i \in I} (v(i) - u(i))(Cu+h)_i$. If $v(i) = u(i)$, this term does not contribute to $\mathcal{E}$. If $v(i) = 1$ and $u(i) = 0$, we know that $(Cu + h)_i > 0$, that is $-(v(i) - u(i))(Cu + h)_i = (Cu - h)_i < 0$. If $v(i) = 0$ and $u(i) = 1$, we know that $(Cu + h)_i \leq 0$, and $-(v(i) - u(i))(Cu - h)_i = (Cu - h)_i \leq 0$. Therefore, $\mathcal{E}(v) \leq \mathcal{E}(u)$. If we have equality, we know in particular that $(u - v)^T C(u - v) = 0$, which in turn implies that $v = f(u) = u$.

As $|I|$ is finite, also E^I is finite, and the energy possesses a lower bound. Hence, $\mathcal{E}$ is a strong Lyapunov function. □

The interest in Hopfield nets originates in neural networks. They can be considered as simplest models for these networks, where h is an input vector, and C codes the influences between neurons. If we have a finite number of neurons (and C is positive definite), then the neuronal network eventually reaches a stationary state (in dependence on the input vector). This behavior can be used to train the neuronal network to recognize certain patterns, coded by input vectors. The interesting point is that small perturbations of the input pattern lead still to valid results. In this way, neuronal networks can be used for pattern recognition also in case of noisy inputs.

Remark 12.2.4 If C is only semi-definite, the energy is still a Lyapunov function, but not necessarily a strong Lyapunov function. If C is even not positive semi-definite any more, then the energy function may even increase; it is possible that, e.g., a periodic orbit appears. As an example, consider a Hopfield net of two sites with parameters

$$C = \begin{pmatrix} 0 & 1 \\ 1 & 0 \end{pmatrix}, \quad h = \begin{pmatrix} -1/2 \\ -1/2 \end{pmatrix}.$$

Define the Heaviside function $\hat{H}(x)$ by $\hat{H}(x) = 0$ for $x \leq 0$, and $\hat{H}(x) = 1$ else (note that $\hat{H}(0) = 0$, which is different to the standard definition of the Heaviside function, where 0 is mapped to 1). In case the argument is a vector, we apply the function component-wise. If we start with the state $u^0 = (1, 0)^T$, we obtain

$$u^1 = f(u^0) = \hat{H}(Cu^0 + h) = \hat{H}(\begin{pmatrix} 0 - 1/2 \\ 1 - 1/2 \end{pmatrix}) = \begin{pmatrix} 0 \\ 1 \end{pmatrix}.$$

and

$$\hat{H}(Cu^1 + h) = \hat{H}(\begin{pmatrix} 1 - 1/2 \\ 0 - 1/2 \end{pmatrix}) = \begin{pmatrix} 1 \\ 0 \end{pmatrix} = u^0.$$

That is, we obtain a period-two orbit.

Proposition 12.2.5 *If in a Hopfield net only one site is updated in one time step, and the diagonal elements of C are non-negative, $C_{i,i} \geq 0$, then the energy function $\mathcal{E}$ is non-decreasing. Assume that $(Cu - h)_i \neq 0$ for all $u \in E^I$, $i \in I$. Then, the energy $\mathcal{E}$ does not change if and only if the updated site did not change.*

Proof We use the same notation and computation as in Proposition 12.2.3 (recall $v = f(u)$),

$$\mathcal{E}(v) - \mathcal{E}(u) = -\frac{1}{2}(v - u)^T C(v - u) - (v - u)(Cu + h),$$

and now we take into account that only one site is updated, that is, $u - v = (u(i_0) - v(i_0))e_{i_0}$ for some $i_0 \in I$. Hence,

$$\begin{aligned}\mathcal{E}(v) - \mathcal{E}(u) &= -\frac{1}{2}C_{i_0,i_0}\,(u(i_0) - v(i_0))^2 - (v(i_0) - u(i_0))(Cu + h)_{i_0} \\ &\leq -(v(i_0) - u(i_0))(Cu + h)_{i_0}.\end{aligned}$$

With the same argumentation as before, we find that $\mathcal{E}(v) - \mathcal{E}(u) \leq 0$, and $\mathcal{E}(v) - \mathcal{E}(u) = 0$ if and only if $u(i_0) = v(i_0)$. □

Remark 12.2.6 A sequentially updated, finite Hopfield-net with non-negative diagonal in C necessarily tends to a stationary point. If we consider our example from above, and update sites 1 and 2 alternatingly, we find (starting with $u^0 = (1, 0)^T$)

$$u^1 = \begin{pmatrix} \hat{H}(0 \quad 1/2) \\ 0 \end{pmatrix}) = \begin{pmatrix} 0 \\ 0 \end{pmatrix}), \quad u^2 = \begin{pmatrix} \hat{H}(0 - 1/2) \\ 0 \end{pmatrix} = \begin{pmatrix} 0 \\ 0 \end{pmatrix},$$

and

$$u^3 = \begin{pmatrix} 0 \\ \hat{H}(0 - 1/2) \end{pmatrix} = \begin{pmatrix} 0 \\ 0 \end{pmatrix}.$$

That is, we are led to the stationary state $(0, 0)^T$.

There are two possibilities to construct Hopfield nets that do updates of only one site at a time. Either we choose a deterministic sequence of sites, running subsequently trough all elements in the set I (so-called "sweeps"), or we select randomly one site (which is simply possible only in finite networks).

12.2.3 Bar-Yam-Model for Turing Pattern

If we review the basic Turing mechanism as described above, we find centers that activate the close-by neighboring sites, deactivate sites at a larger distance, and have no effects at all on sites far away. In order to model these properties, Bar-Yam [7] proposes the following cellular automaton: We allow for local states 0 (=deactivated) and 1 (=activated) on $\mathbb{Z}_m^2$ (the two-dimensional torus). The neighborhood is given by $D_0 = \{(x, y) : |x|, |y| \leq r_2\}$, where $r_2 > 1$ is a parameter of the model. This local rule is defined as

$$f_0(u) = \hat{H}\left(h + \sum_{(x,y)\in D_0} J(x, y)\, u(x, y)\right)$$

where $\hat{H}$ denotes the function defined above: all positive arguments are mapped to 1, and all other arguments to -1. $h \in \mathbb{R}$ is a parameter describing a kind of "background-inhibition", and $J(x, y)$ is the weighting function

$$J(x, y) = \begin{cases} J_1 & \text{for} \quad |x|, |y| \leq r_1 \\ J_2 & \text{for} \quad r_1 < |x|, |y| \leq r_2 \\ 0 & \text{else.} \end{cases}$$

We choose $J_1 > 0 > J_2$, and obtain the basic inhibition-activation structure we aim at. We call the automaton $(\mathbb{Z}_m^2, D_0, E, f_0)$ the Turing-pattern-automaton (or Bar-Yam automaton).

The cellular automaton described above is a Hopfield net with synchronous update. The connectivity matrix of this Hopfield net reads

$$C_{(x,y),(x',y')} = J(x' - x, y' - y) \quad \Rightarrow \quad C_{(x,y),(x,y)} = J_1 < 0.$$

The diagonal elements of C are negative. If we do not use the cellular automaton in the strict sense (where the state is updated synchronously) but update asynchronously (one site after the other), we know that the state tends to an equilibrium. The cellular automaton in *sensu strictu* may show some oscillations. An asynchronous update can be realized in several ways. Perhaps the most realistic way are random sweeps and stochastic cellular automata as described in [41]; update rules may be decisive, in particular for pattern formation [154]. Here we do not want to go in this direction. We use a sequential order for the update, such that in each time step only one cell is changed. It is straight forward to augment a cellular automaton by a control mechanism in such a way, that we are still in the context of deterministic cellular automata, but only one cell of the Turing-pattern-automaton is updated in a time step.

Stationary states of the automaton with asynchronous update are shown in Fig. 12.2. In particular, we find that the parameter h influences the ratio between 1 and -1 in the stationary pattern. The pictures look rather similar to certain skin pattern, e.g., zebra or cheetah. This does not indicate at all that the mechanism proposed here is indeed the mechanism that creates these pattern. However, the Turing mechanism is rather simple and in accordance to many experiments. It is most likely that a large number of biological systems which form patterns inherit a molecular mechanism that implements in one way or another the idea of Turing.

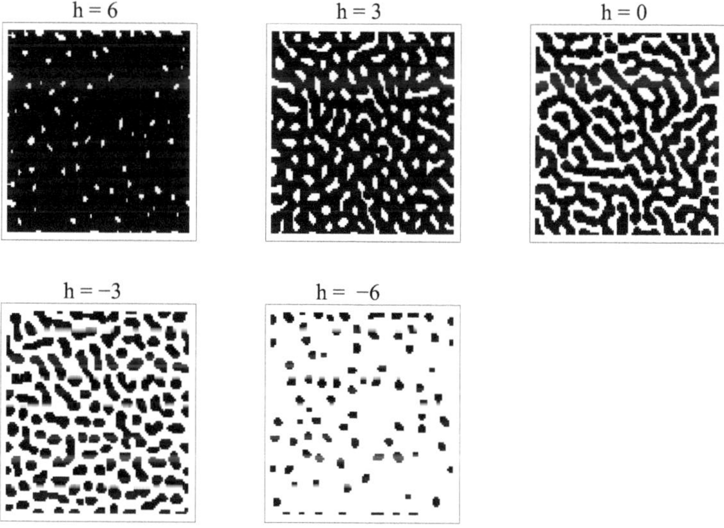

Fig. 12.2 Turing automaton with sequential update; $r_1 = 2$, $r_2 = 6$, $J_1 = 1$, $J_2 = -0.2$; h is varied as indicated. A *black dot* corresponds to 1, a *white dot* to -1

12.3 Greenberg-Hastings Model for Excitable Media

Greenberg and Hastings [74] conceived their model as a simplification or caricature of the Fitzhugh-Nagumo model of nerve excitation which shows periodic behavior if the dendritic input exceeds a threshold. In this model of two coupled differential equations the state is most of the time located on the two stable parts of the slow manifold, the jumps through the fast field are very short and can be neglected almost completely (see, e.g., [37] or [133, Chap. 6]). If one introduces a discrete time (inspecting the solution of the differential equations at equidistant discrete time points) then, once activated, the cell follows, in discrete steps, first the activated branch of the slow manifold, and then the refractory branch (see Fig. 12.3).

We name the resting state '0', and number the states according to their order. Then, we find that there is a number of states within the activated part of the slow manifold, which we call K, and $N - K$ refractory states. In addition we have the resting state, i.e., all in all $N + 1$ states:

$$\underbrace{0}_{resting} \to \underbrace{1 \to 2 \to \cdots K-1 \to K}_{activated} \to \underbrace{K+1 \to K+2 \to \cdots N-1 \to N}_{refractory} \to \underbrace{0}_{resting}.$$

The system runs through these $N + 1$ states in a cyclic fashion.

To make our description complete, we add the following two features:

1. Once a cell is in state '1', then it runs through states $2, \ldots, N$ in a deterministic manner, independent of the neighboring cells; at every time step the state is increased by one, until $N+1 \bmod N+1 = 0$ is reached. The coupling between neighboring cells is so weak that only the resting state can be affected by adjacent cells.
2. If a cell is in the resting state then the transition $0 \to 1$ occurs if at least S cells in the von Neumann neighborhood are in an activated state $1, \ldots, K$.

Definition 12.3.1 Let $\Gamma = \mathbb{Z}^2$ and let D_0 be the von Neumann neighborhood. Let $K, N, S \in \mathbb{N}$, $0 < K < N$, and let $E = \{0, \ldots, N\}$. The state of the system is a map

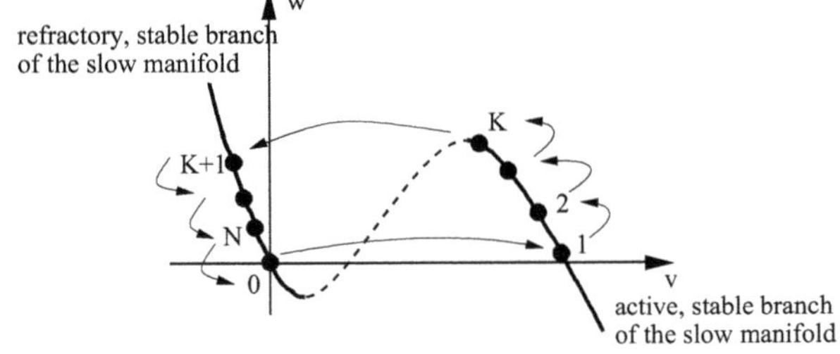

Fig. 12.3 Reduction of the Fitzhugh-Nagumo model to a discrete system. The state follows the activated and then the refractory part of the slow manifold. Using discrete time steps, the system runs through the states that are marked by *black dots*

$u : \Gamma \to E$. The local function is given by

$$f(\varphi) = \begin{cases} \varphi_0 + 1 \bmod N+1 & \text{if } \qquad \varphi_0 \in E \setminus \{0\}. \\ 0 & \text{if } \varphi_0 = 0, \quad \#\{y \in D_0 : 1 \le u(y) \le K\} < S \\ 1 & \text{if } \varphi_0 = 0, \quad \#\{y \in D_0 : 1 \le u(y) \le K\} \ge S. \end{cases}$$

This discrete dynamical system is called a Greenberg-Hastings cellular automaton.

In order to analyze this automaton, several definitions and notations are required, which we introduce in the next section.

12.3.1 Definitions

From now on, we only consider the cases where $S = 1$ and $K \le (N-1)/2$. In order to prove rigorous statements about Greenberg-Hastings automata, we introducc further definitions and hypotheses.

We visualize the elementary state $k \in E$ as the point $e^{2\pi ik/(N+1)}$ on the unit circle in the complex plane $\mathbb{C}$ (see Fig. 12.4 for the next definition).

Definition 12.3.2

(a) The distance between two states $n, m \in E$ is defined as

$$d(n, m) = \min\{|n-m|, N+1-|n-m|\}.$$

(b) The signed distance between two states $n, m \in E$ is defined as follows. If n, m are given and $d(n, m)$ has been defined then $d(n, m)$ corresponds to an arc on the unit circle $\overline{nm}$. This arc may be positively oriented, i.e., counterclockwise, as in Fig. 12.4a or negatively oriented, as in Fig. 12.4b. We define the signed distance as

$$\sigma(n, m) = \begin{cases} d(n, m) & \text{if the arc } \overline{nm} \text{ is orientated counterclockwise,} \\ -d(n, m) & \text{otherwise} \end{cases}$$

(c) Let $\mathcal{P} = (c_1, \ldots, c_M) \in \Gamma^M$ be a finite sequence of cells. This sequence is called a path, if

$$c_{i+1} \in D(c_i), \qquad i = 1, \ldots, M-1.$$

(d) Let $C = (c_1, \ldots, c_M) \in \Gamma^M$ be a path. This path is called cyclic, if $c_1 \in D(c_M)$. Formally, we define $c_{M+1} = c_1$, and thus

$$c_{i+1} \in D(c_i) \qquad \text{for } i = 1, \ldots, M.$$

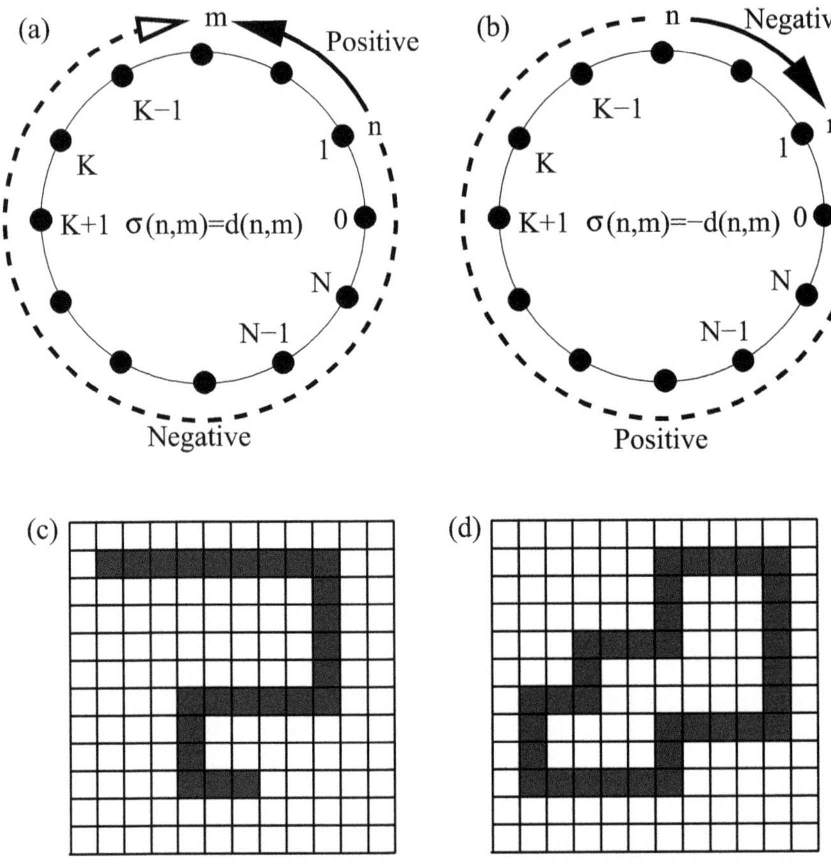

Fig. 12.4 Examples and explanations. (**a**), (**b**) Distance and signed distance $d(n, m)$, $\sigma(n, m)$. (**c**) Example of a path. (**d**) Example of a cycle

(**e**) Suppose there is a state $u \in E^\Gamma$ and a path $\mathcal{P}$ such that the i-th cell carries the elementary state u_i. In other words, if the path is $\mathcal{P} = (c_1, \ldots, c_M)$ then $u|_\mathcal{P}$ is $(u_1, \ldots, u_M)$. In particular, this definition applies to a cyclic path. We call this structure the occupation of the path or of the cycle. If no ambiguity arises, we simply speak of a path or a cyclic path or, in the latter case, of a cycle.

(**f**) A path is called continuous if

$$d(u_i, u_{i+1}) \leq K \text{ for } i = 1, \ldots, M-1.$$

Similarly, a cycle is called continuous if

$$d(u_i, u_{i+1}) \leq K \text{ for } i = 1, \ldots, M-1, M.$$

Observe the convention that for a cycle $c_{M+1} = c_1$ and $u_{M+1} = u_1$.

(g) The winding number of a continuous cycle $C = (c_1, \ldots, c_M)$ with occupation $(u_1, \ldots, u_M)$ is defined as

$$W(u_1, \ldots, u_M) = \frac{1}{N+1} \sum_{i=1}^{M} \sigma(u_i, u_{i+1}).$$

Notice that the underlying (geometric) cycle of cells does not enter the definition of a winding number.

Remark 12.3.3 We give examples and interpretations of these definitions.

a) The states "live" on $\mathbb{Z}_{N+1}$. Hence, we may interpret them as located on a circle. On S^1, there are two distances possible between two points: one may use the length of either the clockwise, or the counterclockwise arc that connects the two points. The definition of $d(\cdot, \cdot)$ chooses the shorter arc.

b) In contrast to $d(.,.)$, the signed distance includes information about the orientation (see Fig. 12.4).

c), d) Examples of a path and a cycle are given in Fig. 12.4.

We now show some elementary properties.

Lemma 12.3.4 *Assume $K \leq (N-1)/2$. If $0 \leq n, m \leq K+1$ or $N-K \leq n, m \leq N$, then*

$$d(n, m) = |n - m|, \qquad \sigma(n, m) = n - m.$$

Proof

Case 1: Let $0 \leq n, m \leq K + 1$: First of all, since $0 \leq n, m \leq K + 1$, we find

$$|n - m| \leq |(K + 1) - 0| = K + 1,$$

and thus $N + 1 - |n - m| \geq N + 1 - (K + 1) = N - K$. Since $2K \leq N - 1$, we have $K \leq N - K - 1$ and $K + 1 \leq N - K$. Hence $|n - m| \leq N + 1 - |n - m|$ and

$$d(n, m) = \min\{|n - m|, N + 1 - |n - m|\} = |n - m|.$$

Furthermore, the arc from n to m is positively orientated if and only if $n < m$ (see Fig. 12.4a, b). Therefore,

$$\sigma(n, m) = m - n.$$

Case 2: Let $N - K \leq n, m \leq N$: The argument follows similar lines.

$$0 \leq |n - m| \leq |(N - K) - N| = K,$$

and thus $N+1-|n-m| \geq N+1-K = N+1-K$. Again, since $K \leq N-1-K$, $|n-m| < N+1-|n-m|$ and

$$d(n,m) = \min\{|n-m|, N+1-|n-m|\} = |n-m|.$$

Since the arc from n to m is positively oriented if and only if $n < m$,

$$\sigma(n,m) = m - n.$$

□

Now we follow the occupation of a fixed geometric cycle while the state u^t of the cellular automation changes in time according to the deterministic evolution. For a general cellular automaton it would be difficult to find any properties which are preserved on some isolated array of cells. Here the situation is different since the evolution of a single cell depends only on the cell itself except when the cell is in the resting state.

Lemma 12.3.5 *Consider any two adjacent cells c_1, c_2 (not necessarily on a given path or cycle). Let u_1^t and u_2^t be their states at some time t and suppose*

$$d(u_1^t, u_2^t) \leq K.$$

Then for all times $t' > t$

$$d(u_1^{t'}, u_2^{t'}) \leq \max\{d(u_1^t, u_2^t), 1\}.$$

Proof It is sufficient to prove the inequality for $t' = t + 1$. The conclusion then follows by induction over t'.

To prove the claim for $t' = t + 1$, we distinguish four cases.

Case 1: $u_1^t, u_2^t \neq 0$:
Then, $u_1^{t+1} = u_1^t + 1 \bmod (N + 1)$, $u_2^{t+1} = u_2^t + 1 \bmod (N + 1)$ and thus

$$d(u_1^{t+1}, u_2^{t+1}) = d(u_1^t, u_2^t).$$

On the other hand, assume that one of the states (without restriction u_1^t) is zero at time t. Since $d(u_1^t, u_2^t) \leq K$, we have

$$u_2^t \in \{N - K + 1, \ldots, N\} \cup \{0\} \cup \{1, \ldots, K\}.$$

which leads to the following three cases.

Case 2: $u_1^t = 0$, $u_2^t = 0$.
Then, $u_1^{t+1}, u_2^{t+1} \in \{0, 1\}$ and thus

$$d(u_1^{t+1}, u_2^{t+1}) \leq 1.$$

Case 3: $u_1^t = 0$, $u_2^t \in \{1, \ldots, K\}$.
Then, $u^1_{t+1} = 1$ and $u_2^{t+1} = u_2^t + 1$,

$$d(u_1^{t+1}, u_2^{t+1}) = d(u_1^t, u_2^t).$$

Case 4: $u_1^t = 0$, $u_2^t \in \{N - K + 1, \ldots, N\}$.
Thus, $u_1^{t+1} \in \{0, 1\}$ and $u_2^{t+1} = u_2^t + 1 \bmod (N + 1)$, i.e.,

$$d(u_1^{t+1}, u_2^{t+1}) \in \{d(u_1^t, u_2^t), d(u_1^t, u_2^t) - 1\}.$$

□

Remark 12.3.6 Lemma 12.3.5 implies that "continuous stays continuous", i.e., a cycle that is continuous at time t stays continuous for all later times $t' \geq t$.

12.3.2 The Winding Number

In this section we prove two important statements about the winding number: the winding number is an integer and the winding number of a continuous cycle is invariant in time. Both results appear unexpected for discrete systems although analogous claims for continuous systems can be easily visualized, think of a closed curve on a torus. As the definition of a winding number does not make explicit use of the underlying geometric cycle, also the proof does not refer to the grid.

Proposition 12.3.7 *Let $u_1, \ldots, u_M \in E$, and define $u_{M+1} = u_1$. Define the function $\tilde{W} : E^M \to \mathbb{R}$,*

$$\tilde{W}(u_1, \ldots, u_M) = \frac{1}{N+1} \sum_{i=1}^{M} \sigma(u_i, u_{i+1}).$$

Then

$$\tilde{W}(u_1, \ldots, u_M) \in \mathbb{Z}.$$

Proof We use induction over M. Note, that we do not assume that this sequence of states originates from a continuous cycle, a cycle is not even required. In this sense, we generalize the definition of the winding number (which we have indicated by the tilde).

Case $M = 2$:

$$\tilde{W}(u_1, u_2) = \frac{1}{N+1}(\sigma(u_1, u_2) + \sigma(u_2, u_1)).$$

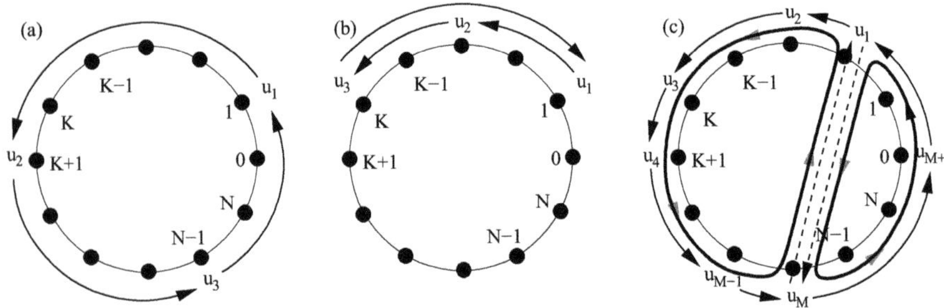

Fig. 12.5 (**a**), (**b**) Winding number (or its generalization) for a three-cycle. (**c**) Sketch of how to split the cyclic sum in two shorter cyclic sums

Whatever u_1, u_2 are, the signed distance $\sigma(u_2, u_1)$ is equal to $-\sigma(u_2, u_1)$. Hence $\tilde{W}(u_1, u_2) = 0$.

Case $M = 3$:

$$\tilde{W}(u_1, u_2, u_3) = \frac{1}{N+1}(\sigma(u_1, u_2) + \sigma(u_2, u_3) + \sigma(u_3, u_1)).$$

We distinguish two cases (see also Fig. 12.5a, b):

(a) $\sigma(u_i, u_{i+1}) > 0$ for $i = 1, 2, 3$ (or all negative).
(b) Two of the three distances $\sigma(u_i, u_{i+1})$ are positive, one is negative (or, two are negative and one positive).

In case (a), we find

$$\sigma(u_1, u_2) + \sigma(u_2, u_3) + \sigma(u_3, u_1) = \pm(N+1)$$

and thus $\tilde{W} \in \{\pm 1\}$. In case (b), we have

$$\sigma(u_1, u_2) + \sigma(u_2, u_3) + \sigma(u_3, u_1) = 0$$

i.e., $\tilde{W} = 0$.

Induction step, $M \to M+1$:

The idea is to split the cyclic sum of length $M+1$ into two cyclic sums with lengths smaller than $M+1$,

$$(N+1)\,\tilde{W}(u_1, \ldots, u_{M+1}) = \sum_{i=1}^{M+1} \sigma(u_i, u_{i+1})$$

$$= \sum_{i=1}^{M} \sigma(u_i, u_{i+1}) + \sigma(u_M, u_1) + \sigma(u_1, u_M) + \sigma(u_M, u_{M+1}) + \sigma(u_{M+1}, u_1)$$

$$= (N+1)\,\tilde{W}(u_1, \ldots, u_M) + (N+1)\,\tilde{W}(u_1, u_M, u_{M+1}).$$

Now we have

$$\tilde{W}(u_1, \ldots, u_{M+1}) = \tilde{W}(u_1, \ldots, u_M) + \tilde{W}(u_1, u_M, u_{M+1}).$$

Since the last two cycles are shorter than $M + 1$, we have $\tilde{W}(u_1, \ldots, u_M) \in \mathbb{Z}$ and $\tilde{W}(u_1, u_M, u_{M+1}) \in \mathbb{Z}$ and hence $\tilde{W}(u_1, \ldots, u_{M+1}) \in \mathbb{Z}$. □

Remark 12.3.8 An immediate consequence is the fact that the winding number of a cycle is an integer for all t,

$$W(u_1^t, \ldots, u_M^t) \in \mathbb{Z}.$$

Another straightforward consequence is the fact, that from $\sum_{i=0}^{M} \sigma(u_i, u_{i+1}) = 0$ it follows that $u_0 = u_M$.

Proposition 12.3.9 *The winding number is invariant under the dynamics of the cellular automaton: let C be a cyclic path and let $(u_1^t, \ldots, u_M^t)$ be a continuous cycle at time t. Then for $t' > t$*

$$W(u_1^{t'}, \ldots, u_M^{t'}) - W(u_1^t, \ldots, u_M^t).$$

In order to prove this proposition we need a lemma. Let us denote the sum of the signed states over a path $\mathcal{P}$ by

$$Q_t(\mathcal{P}) = \sum_{i=1}^{M-1} \sigma(u_i^t, u_{i+1}^t).$$

We investigate this sum for certain patterns on the path.

Lemma 12.3.10 *Let $\mathcal{P} = (c_1, \ldots, c_M)$ with occupation $(u_1^t, \ldots, u_M^t)$ at time t such that $u_1^\tau \neq 0$, $u_M^\tau \neq 0$ while $u_2^t = \cdots = u_{M-1}^t = 0$. Let $Q_t(\mathcal{P}) = \sum_{i=1}^{M-1} \sigma(u_i^t, u_{i+1}^t)$. Then,*

$$Q_{t+1}(\mathcal{P}) = Q_t(\mathcal{P}).$$

Proof If $M = 2$ we find $u_1^{t+1} = u_1^t + 1 \bmod (N+1)$ and $u_2^{t+1} = u_2^t + 1 \bmod (N+1)$. Therefore,

$$\sigma(u_1^{t+1}, u_2^{t+1}) = \sigma(u_1^t, u_2^t).$$

Now assume $M > 2$. Since $u_i^t = 0$ for $i = 2, \ldots, M-1$, we find $u_i^{t+1} \in \{0, 1\}$. Thus, according to Lemma 12.3.4, $\sigma(u_i^{t+1}, u_{i+1}^{t+1}) - u_{i+1}^{t+1} \quad u_i^{t+1}$, and hence

$$\begin{aligned} Q_{t+1}(\mathcal{P}) &= \sigma(u_1^{t+1}, u_2^{t+1}) + \sum_{i=2}^{M-2} \left(u_{i+1}^{t+1} - u_i^{t+1}\right) + \sigma(u_{M-1}^{t+1}, u_M^{t+1}) \\ &= \left(\sigma(u_1^{t+1}, u_2^{t+1}) - u_2^{t+1}\right) - \left(\sigma(u_M^{t+1}, u_{M-1}^{t+1}) - u_{M-1}^{t+1}\right). \end{aligned}$$

Now consider these two terms separately.
Term $\sigma(u_1^{t+1}, u_2^{t+1}) - u_2^{t+1}$:
($\boldsymbol{\alpha}$) We know $u_1^t \neq 0$, $u_2^t = 0$. We know that $\mathcal{P}$ is continuous at time t, i.e., $d(u_1^t, 0) \leq K$ from where

$$u_1^t \in \{1, \ldots, K\} \cup \{N - K + 1, \ldots, N\}$$

and hence

$$u_1^{t+1} \in \{2, \ldots, K + 1\} \cup \{N - K + 2, \ldots, 0\}.$$

Since $K \leq (N - 1)/2$, we find

$$\sigma(u_j^{t+1}, 1) + \sigma(1, 0) = \sigma(u_j^{t+1}, 0), \qquad \sigma(u_j^{t+1}, 0) + \sigma(0, 0) = \sigma(u_j^{t+1}, 0).$$

($\boldsymbol{\beta}$) We also know $u_2^{t+1} \in \{0, 1\}$ and thus

$$-u_2^{t+1} = \sigma(u_2^{t+1}, 0).$$

Using ($\boldsymbol{\alpha}$), ($\boldsymbol{\beta}$) we have

$$\sigma(u_1^{t+1}, u_2^{t+1}) - u_2^{t+1} = \sigma(u_1^{t+1}, u_2^{t+1}) + \sigma(u_2^{t+1}, 0) = \sigma(u_1^{t+1}, 0).$$

Term $\sigma(u_1^{t+1}, u_2^{t+1}) - u_2^{t+1}$:
The very same argument leads to

$$\sigma(u_{M-1}^{t+1}, u_M^{t+1}) - u_{M-1}^{t+1} = \sigma(u_M^{t+1}, 0).$$

Hence,

$$Q_{t+1}(\mathcal{P}) = \sigma(u_1^{t+1}, 0) - \sigma(u_M^{t+1}, 0) = \sigma(u_1^{t+1}, 0) + \sigma(0, u_M^{t+1}).$$

Since u_1^t and u_M^t both are increased by one (modulo $N + 1$), we find

$$Q_{t+1}(\mathcal{P}) = \sigma(u_1^{t+1}, 0) + \sigma(0, u_M^{t+1}) = \sigma(u_1^t, 0) + \sigma(0, u_M^t) = Q_t(\mathcal{P}).$$

□

Proof (of Proposition 12.3.9) First of all, according to Proposition 12.3.5, the cycle C stays a continuous cycle for $t' > t$. We now show that the winding number is invariant in time. For short we call W_t the winding number at time t.

Case 1: At time t, all states are zero.

Then $W_t = 0$. Since the states $(u_1^t, \dots, u_M^t)$ are zero, we have $u_i^{t+1} \in \{0, 1\}$ and

$$W_{t+1} = \frac{1}{N+1} \sum_{i=1}^{M} (u_{i+1}^{t+1} - u_i^{t+1})$$

since $u_{M+1}^{t+1} = u_1^{t+1}$.

Case 2: At time t, at least one state is nonzero.

Let $1 \le j_1 < j_2 < \dots < j_m \le M$ denote the indices of nonzero states,

$$u_{j_l}^t \neq 0 \quad \text{for } l = 1, \dots, m, \qquad u_i^t = 0 \quad \text{if } i \notin \{j_1, \dots, j_m\}.$$

Define furthermore

$$Q_\tau(j_l, j_{l+1}) = \sum_{i=j_l}^{j_{l+1}-1} \sigma(u_i^\tau, u_{i+1}^\tau)$$

where the indices $i > M$ must be interpreted modulo M, i.e., $u_{M+1}^t = u_1^t$, $u_{M+2}^t = u_2^t$, $u_{M+3}^t = u_3^t$ etc., and $j_{m+1} = j_1$. Each $Q_\tau(j_l, j_{l+1})$ starts with a nonzero state and takes care of all following zero states until the next nonzero state. Therefore

$$W_t = \frac{1}{N+1} \sum_{l=1}^{m} Q_t(j_l, j_{l+1}).$$

According to the last proposition, $Q_{t+1}(j_l, j_{l+1}) = Q_t(j_l, j_{l+1})$ and thus

$$W_t = \frac{1}{N+1} \sum_{l=1}^{m} Q_t(j_l, j_{l+1}) = \frac{1}{N+1} \sum_{l=1}^{m} Q_{t+1}(j_l, j_{l+1}) = W_{t+1}.$$

□

Remark 12.3.11 We may interpret the winding number in the following way: a cycle as well as the state space E can be embedded into S^1. Hence, the states on a cycle "live" in $S^1 \times S^1$, i.e., on a torus (see Fig. 12.6). If we plot the states of the cells on the torus and connect these points by curves (shortest connection), we get a closed curve that winds around the torus. The number of windings of this curve is just the winding number (signed by the direction of the windings).

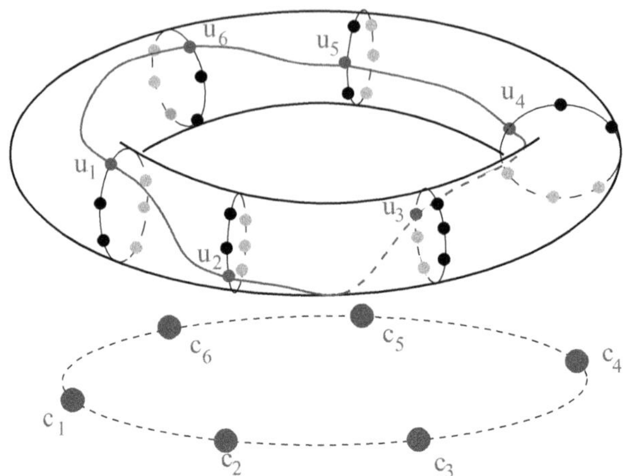

Fig. 12.6 Interpretation of the winding number

12.3.3 *The Potential*

The winding number tells us something about the sum of the differences of states along a closed curve. Insofar, the winding number is similar to an integral of a function along a closed curve. If we consider a vector field $f : \mathbb{R}^2 \to \mathbb{R}^2$ with the property that the integral along any closed curve is zero, we know that there is a potential $V : \mathbb{R}^2 \to \mathbb{R}$ such that f is a gradient field, $f = \nabla V$. Here we have a similar situation in a discrete setting. Assuming that $W_t(C) = 0$ for all continuous cycles, we can define a potential.

Suppose that we start at time $t = 0$ with a state with compact support. Only a finite number of cells have nonzero local state. Then, for all future times, there are only a finite number of cells with nonzero states. In other words, a Greenberg-Hastings automaton preserves the zero state and we can consider the sub-automaton with states with finite support.

Assume that for all $t \geq 0$, every a continuous cycle has winding number zero.

Definition 12.3.12

(a) Let R_t be a "radius" such that all cells $c = (i,j)$ with $|i| + |j| > R_t$ have state zero at time t. Then let $\Omega_t = \{c = (i,j) \mid |i| + |j| > R_t\}$. Thus, Ω_t is the complement of a "diamond".

(b) Suppose a cell c with state u_c can be connected at time t by a *continuous* path $\mathcal{P} = (c_1, \ldots, c_M)$ with Ω_t, i.e., $c_1 = c$, $u_1 = u_c$ and $c_M \in \Omega_t$, $u_M = 0$. Then the potential $h_t(c)$ of cell the c at time t is defined by (u_i is the local state of cell c_i)

$$h_t(c) = \sum_{i=1}^{M-1} \sigma(u_i, u_{i+1}).$$

At this point, at any time the global state has a finite support. This support is always subset of the complement of Ω_t. As D_0 is the Neumann neighborhood, we can choose $R_{t+1} = R_t + 1$. In this way, we easily find Ω_t for all times $t \in \mathbb{N}$, given Ω_0.

If $u \in E^\Gamma$ is a state with finite support at time t then not every cell need to have a potential, as for some cells there may be no continuous path connecting the cell with Ω_t. However, we show, if there is such a continuous path, then the potential is well-defined.

Proposition 12.3.13 *If the winding number for any continuous cycle is zero, then the potential $h_t(c)$ does not depend on the choice of the continuous path from c to Ω_t.*

Proof Assume that there are two paths $\mathcal{P}_a = (c_1^a, \ldots, c_{M^a}^a)$, and $\mathcal{P}_b = (c_1^b, \ldots, c_{M^b}^b)$ with states $(u_1^a, \ldots, u_{M_a}^a)$ and $(u_1^b, \ldots, u_{M_b}^b)$ connecting c with Ω_t such that

$$\sum_{i=1}^{M^a-1} \sigma(u_i^a, u_{i+1}^a) \neq \sum_{i=1}^{M^b-1} \sigma(u_i^b, u_{i+1}^b).$$

Then we can connect the cells c_{M^a} and c_{M^b} by a path $\mathcal{P}_c = (c_1^c, \ldots, c_{M_c}^c) \subset \Omega_t^{M^c}$ completely in Ω_t, i.e., the states along this path are zero. Then the cycle

$$C = (c_1^a, \ldots, c_{M^a}^a, c_1^c, \ldots, c_{M^c}, c_{M^b}, \ldots, c_2^b)$$

is continuous and

$$\begin{aligned} W_t(C) &= \sum_{i=1}^{M^a-1} \sigma(u_i^a, u_{i+1}^a) + \sum_{i=1}^{M^b-1} \sigma(u_{i+1}^b, u_i^b) \\ &= \sum_{i=1}^{M^a-1} \sigma(u_i^a, u_{i+1}^a) - \sum_{i=1}^{M^b-1} \sigma(u_i^b, u_{i+1}^b) \neq 0. \end{aligned}$$

Thus, we have a contradiction to the assumption that the winding number for every continuous cycle is zero. □

Proposition 12.3.14 *Suppose a cell c has a potential at time t, i.e., $h_t(c)$ is defined. Then it has a potential for all $t' \geq t$.*

Proof We show: if there is a continuous path connecting a cell c with Ω_t at time t, then there is a continuous path for any time $t' \geq t$ that connects c with $\Omega_{t'}$. The argument is simple: take the path $\mathcal{P}$ that connects c with Ω_t. Then, the end point of this path in Ω_t can be connected by a continuous continuation to any cell $\tilde{c} \in \Omega_t$. Since $\Omega_{t'} \cup \Omega_t \neq \emptyset$ and a continuous path stays continuous, we find also for times $t' > t$ a continuous path that connects c with $\Omega_{t'}$. □

Remark 12.3.15 We cannot ensure that every cell has a potential at a given time t. There could be even cells which never get a potential. We will see later, that this is not the case, but at the moment we cannot exclude this case.

Now we follow the potential of a single cell in the evolution of the cellular automaton.

Proposition 12.3.16 *Assume that $h_t(c)$ exists. Then*

$$h_{t+1}(c) \in \{h_t(c),\ h_t(c) + 1\}.$$

Furthermore, $h_{t+1}(c) = h_t(c)$ if and only if $u_c^{t+1} = u_c^t = 0$.

Proof Let $\mathcal{P} = (c_1, \ldots, c_M)$ be a continuous path with

$$c_1 \in \Omega_{t+1}, \qquad c_M = c,$$

let $(u_1^t, \ldots, u_M^t)$ be the occupation at time t. The elements of the occupation may be zero or nonzero. Let $i_1 < i_2 < \ldots < i_p$ be those indices such that $u_{i_l}^t \neq 0$, and $u_{i'}^t = 0$ for $i' \notin \{i_1, \ldots, i_l\}$. Then, due to Lemma 12.3.10,

$$Q_{t+1}(i_1, i_p) = Q_t(i_1, i_p).$$

Since the path is continuous and $c_1 \in \Omega_{t+1}$, we know that $u_{i_1}^t \in \{N-K+1, \ldots, N\} \cup \{1, \ldots K\}$.

If $u_{i_1}^t \in \{1, \ldots K\}$, we find $u_{i_1-1}^{t+1} = 1$, $u_{i_1-1}^t = 0$, $u_{i_1-2}^{t+1} = 0$, $u_{i_1-2}^t = 0$, and

$$\sigma(u_{i_1-2}^{t+1}, u_{i_1-1}^{t+1}) = 1 = 1 + \sigma(u_{i_1-2}^t, u_{i_1-1}^t).$$

If $u_{i_1}^t \in \{N - K + 1, \ldots, N\}$, then $\sigma(u_{i-2}^t, u_{i_1-1}^t) < 0$, $d(\cdot,\cdot)$ decreases at the time step while $\sigma(\cdot,\cdot)$ increases, and

$$\sigma(u_{i_1-2}^{t+1}, u_{i_1-1}^{t+1}) = \sigma(0, u_{i_1-1}^t + 1 \bmod (N + 1)) = \sigma(0, u_{i_1-1}^t) + 1.$$

In any case, we obtain

$$\sum_{i=0}^{i_1-1} \sigma(u_i^{t+1}, u_{i+1}^{t+1}) = \sum_{i=0}^{i_1-1} \sigma(u_i^t, u_{i+1}^t) + 1.$$

Hence, if $c_{i_p} = c_M$, then

$$h_{t+1}(i,j) = h_t(i,j) + 1.$$

If $c_{i_p} \neq c_M$, i.e., $u^t_M = 0$, we obtain

$$\sum_{i=i_p}^{M-1} \sigma(u_i^{t+1}, u_{i+1}^{t+1}) \geq \sum_{i=i_p}^{M-1} \sigma(u_i^t, u_{i+1}^t) - 1.$$

If $u_M^{t+1} = 0$, we have

$$\sum_{i=i_p}^{M-1} \sigma(u_i^{t+1}, u_{i+1}^{t+1}) = \sum_{i=i_p}^{M-1} \sigma(u_i^t, u_{i+1}^t) - 1,$$

and otherwise

$$\sum_{i=i_p}^{M-1} \sigma(u_i^{t+1}, u_{i+1}^{t+1}) = \sum_{i=i_p}^{M-1} \sigma(u_i^t, u_{i+1}^t).$$

Taking all together, we conclude that $h_{t+1}(c) \in \{h_t(c),\ h_t(c) + 1\}$, and that the potential does not change if and only if $u^t_M = u^{t+1}_M = 0$. □

Remark 12.3.17 From the proof of the last proposition, we find that $h_{t+1}(i,j) = h_t(i,j)$ if and only if $u^{t+1}_{i,j} = u^t_{i,j} = 0$, and otherwise is increased. We use this fact in the proof of the next proposition.

Proposition 12.3.18 *$h_t(i,j)$ is globally bounded (in time and space).*

Proof

Step 1: Time, at which a cell gains a potential.
Cells in Ω_t have a potential by definition. At time $t = 0$, there are only finitely many cells with nonzero state. Hence there are only finitely many cells which do not have a potential. We also know that if a cell has a potential at time t then it has a potential for $t' > t$. Hence we can assign to each cell (i,j) the time $T(i,j)$ when it obtains a potential; if a cell never obtains a potential, then define $T_{i,j} = \infty$. Then, at time $T = \max\{T_{i,j} : T_{i,j} < \infty\}$ every cell which may get a potential eventually has already obtained it. After time T there are no more changes with respect to having a potential. But even after this time, the potential of some cell could increase.

Step 2: A candidate for a global bound.
Let

$$\Delta' = \max\{h_T(c) : c \in \Gamma,\ \ c \text{ has a potential}\}.$$

Define an integer k by the inequalities $(k-1)(N+1) \leq \Delta' < k(N+1)$, and finally define Δ by $\Delta = k(N+1)$. Then $\Delta' \leq \Delta$, and Δ is a multiple of $N+1$, and Δ is minimal with these properties. We conjecture

$$h_t(i,j) \leq \Delta \qquad \text{for } t \geq T.$$

Step 3: Global bound.

Assume that Δ is not a global bound. At time T, we have $h_T(c) \leq \Delta$ for all cells c which have a potential. Let $t_1 > T$ be the first time such that there is a cell $c \in \Gamma$ which has a potential and $h_{t_1}(c) > \Delta$. According to Proposition 12.3.16, the potential can increase only by 1. Hence $h_{t_1-1}(c) = \Delta$. But $\Delta = k(N+1)$ is a multiple of $N+1$. Hence $u_c^{t_1-1} = 0$ and consequently $u_c^{t_1} = 1$. If such step occurs then there is an excited neighbor c' of c with $u_{c'}^{t_1-1} \in \{1, \ldots, K\}$. The cell c' can be connected to c, the cell c has state 0 and c' has a state in $(0, K]$. Hence also c' has a potential at time $t_1 - 1$ and

$$h_{t_1-1}(c') = h_{t_1-1}(c) + \sigma(0, u_c^{t_1-1}) > h_{t_1-1}(c) = \Delta$$

which contradicts the assumption that t_1 is the first time at which the bound Δ is violated. □

12.3.4 Survival of Configurations

Here our aim is to predict whether a given pattern will survive. Therefore, we have to provide an exact definition of "surviving".

Definition 12.3.19 Consider an initial state with finite support which we call a configuration. We say that this configuration survives if there is a finite domain $\tilde{\Gamma} \subset \Gamma$ such that for all times t there is a time $t' \geq t$ and a cell $c \in \tilde{\Gamma}$ with $u_c^{t'} \neq 0$.

Lemma 12.3.20 *Let u^0 be the initial state. Let m be such that all cells with $u^0(c) \neq 0$ are contained in the set*

$$R_0 = \{(i,j) \in \mathbb{Z}^2 \, : \, |i| + |j| \leq m\}.$$

Let $n \in \mathbb{N}$ and let

$$R_n = \{(i,j) \in \mathbb{Z}^2 \, : \, |i| + |j| \leq m + n\}, \qquad n \in \mathbb{N}.$$

Assume there is a cell $c \in \mathbb{R}_n$ at time t with $u^t(c) = 1$. Then there is a cell $c' \in U(c) \cap R_n$ with $u^t(c') \in \{2, \ldots, K+1\}$.

Proof By definition all cells outside of R_0 are in state $u^0(c) = 0$. Assume that t_0 is the first time when a cell in R_n is activated without having an activated neighbor within R_n, i.e., there is $c_1 \in R_n$ with

$$u^{t_0}(c_1) = 1$$
$$u^{t_0}(c') \notin \{2, \ldots, K+1\} \qquad \text{for } c' \in U(c) \cap R_n.$$

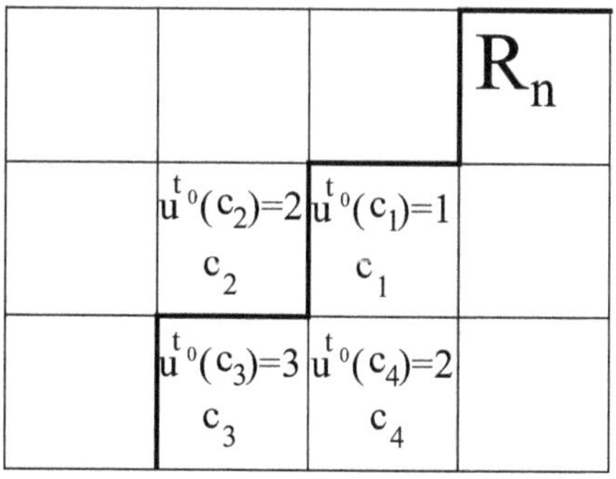

Fig. 12.7 Cells c_1–c_4 in the proof of Lemma 12.3.20

			R_n
	$u^{t_0}(c_2)=2$ c_2	$u^{t_0}(c_1)=1$ c_1	
	$u^{t_0}(c_3)=3$ c_3	$u^{t_0}(c_4)=2$ c_4	

The cell c_1 is necessarily at the boundary of R_n, i.e., $D_0 + c \cap (R_{n+1} \setminus R_n) \neq \emptyset$ and there is a cell $c_2 \in D_0 + c \cap (R_{n+1} \setminus R_n)$ (see Fig. 12.7) with $u^{t_0}(c_2) \in \{2, \ldots, K+1\}$.

Since at time zero we have $u^0(c_1) - u^0(c_2) = 0$, the distance between the states of these two cells is at most one. Thus,

$$u^{t_0}(c_2) = 2.$$

The cell c_2 has been excited by some cell c_3 two time steps before. Where is the cell c_3 located?

1. This cell is not in $R_{n+1} \setminus R_n$, since $D_0 + c_1 \cap (R_{n+1} \setminus R_n) = \emptyset$.
2. If the only cell with $u^{t_0-2}(c) \in \{1, \ldots, K\}$ in $D_0 + c_2$ have been located in $R_{n+2} \setminus R_{n+1}$, then cell $c_2 \in R_{n+1}$ has been activated by a cell outside of R_{n+1} and thus t_0 is not the first time where our lemma failed, in contradiction to the definition of t_0.

Hence, $c_3 \in R_n$. Furthermore, $u^{t_0}(c_3) \in \{3, \ldots, K+3\}$, and the distance of the states of cells c_2 and c_3 is less or equal one. Hence,

$$u^{t_0}(c_3) = 3.$$

Now look at the cell c_4 which is a neighbor to c_1 as well as to c_3. The distance of its state with respect to c_1 and to c_3 cannot exceed 1. Hence $u^{t_0}(c_4) = 2$. This, however, is in contradiction to

$$u^{t_0}(c') \notin \{2, \ldots, K+1\} \qquad \text{for } c' \in D_0 + c \cap R_n.$$

□

Now we show the main result and characterize the conditions for survival pf patterns.

Theorem 12.3.21 *Let $u^0 \in (E^{\mathbb{Z}^2})_c$ be the initial state (the configuration). Let m be such that all cells with $u^0(c) \neq 0$ are contained in the set*

$$R_0 = \{(i,j) \in \mathbb{Z}^2 \ : \ |i| + |j| \leq m\}.$$

Define

$$R_1 = \{(i,j) \in \mathbb{Z}^2 \ : \ |i| + |j| \leq m + 1\}$$

and a time T by

$$T = (N+1)^{|R_1|}.$$

The following two statements hold:

(i) The configuration survives if and only if there is a time t such there is a continuous cycle with nonzero winding number.
(ii) If there is ever a continuous cycle with nonzero winding number then there is such cycle for some $t \leq T$.

Proof As often, one direction of the proof is very simple: If we have a nontrivial winding number then the configuration cannot die out: choose a region G that contains the continuous cycle. Since the winding number is constant in time, and nonzero, at every time there is at least one cell within this cycle which has a non-resting state.

Now we prove the other direction. Also here, we did most of the work by studying the potential. Assume there is never a continuous cycle with nonzero winding number. Then we can define the potential. We know, that the potential is globally bounded and nondecreasing. Thus, the potentials of all cells (that ever gain a potential) eventually becomes constant. Since the only chance for a cell not to increase the potential is to stay in the resting state, all cells that ever get a potential eventually stay in the resting state.

Let $H \subset \Gamma$ be the set of cells that do not eventually stay in the resting state. Since we assume the state to have compact support at time zero (the initial state is a configuration), H is finite. There is necessarily at least one cell $c \in H$ that has a neighbor that has a potential. This neighboring cell eventually stays in its resting state. Therefore, also the cell c eventually stays in the resting state (otherwise c would activate the neighboring, resting cell). This is a contradiction to $c \in H$, i.e. the set H is empty, and all cells eventually go into the resting state.

We know that the cycle with nontrivial winding number has to appear in R_1. The states in R_1 are independent of the states outside of R_1. Thus, the states within R_1 become periodic after at most $(N+1)^{|R_1|}$ time steps. Hence, the cycle with nontrivial winding number must appear until time step $T = (N+1)^{|R_1|}$, or it never will appear. □

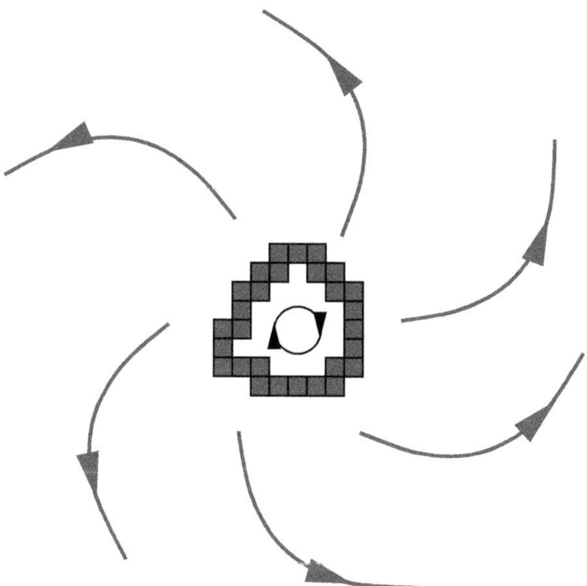

Fig. 12.8 Generation of persisting patterns

The theorem tells us something about the structure of persisting configurations: In the center, there is a cycle of nonzero winding number, where one cell activates the next cell within this cycle. Starting from these cycles, waves of activation sweep over the grid Γ (Fig. 12.8). Also models based on partial differential equations exhibit similar patterns [37].

Chapter 13
Applications in Various Areas

There are many examples where cellular automata contribute to the understanding of scientific phenomena. In the following, we briefly sketch three of these applications to demonstrate the flexibility of cellular automata as a modeling approach. All these models allow for a specific analysis of their dynamics, at least in some heuristic way. The method of this analysis is driven by the application—we are guided by some intuitive insight into the scientific process modeled by the specific cellular automaton. This observation leads to the expectation that—apart from the fundamental, and very valuable classification approaches by Hurley, Gilman, Kůrka, and others—the definition of small, homogeneous "islands" in the vast set of all cellular automata may lead to many classes that allow for analysis and deeper understanding of their dynamics. It may very well be the case, that these "islands" cover most cellular automata which inherit enough structure to be of actual interest.

13.1 Sandpile Automata and Self-Organized Criticality

Earthquakes with a high magnitude (luckily) only happen seldom, while small earthquakes take place all the time. Measurements indicate a clear dependency between the magnitude m and the frequency f of events with this magnitude, $m \approx 1/f^{\alpha}$, where α is a constant. This relation appears in many different systems from noise in music or video broadcasts, to the power spectrum of ocean current velocity, to the change in fitness of biological organisms by evolution and, of course, the avalanches in sand piles. It is considered as a rather universal law of self-organized criticality. As α often is close to one, it is called the $1/f$ law.

Bak et al. [4] introduced a simple model, the Bak-Tang-Wiesenfeld model, that reveals basic mechanisms that lead to the observed behavior. Their predictions have been experimentally confirmed later [59]. Although the analysis of the model was mainly based on numerical experiments, it attracted a lot of attention. The model

K.-P. Hadeler, J. Müller, *Cellular Automata: Analysis and Applications*,
Springer Monographs in Mathematics, DOI 10.1007/978-3-319-53043-7_13

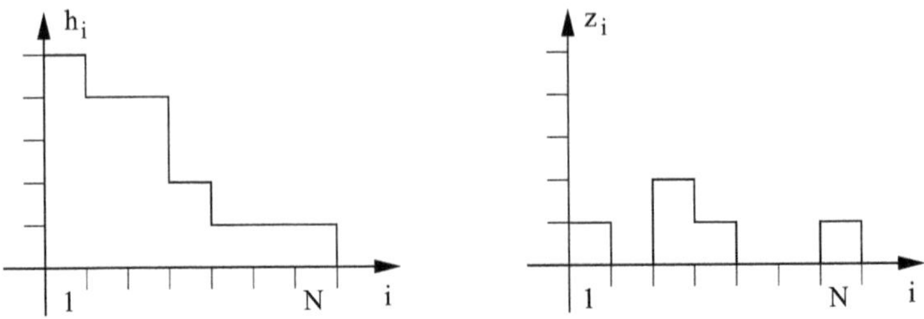

Fig. 13.1 A sandpile, represented by hight h_i (*left panel*), and by slope z_i (*right panel*)

represents a discrete caricature of a sand pile. We only consider the right hand side of a sand-pile, such that the height of the sand-pile decreases from left to right. Let $1, \ldots, N$ denote discrete locations at which the (discrete) heights $h_i \in \mathbb{N}_0$ are given. As indicated, we assume a monotonously decreasing mountain-side, $h_i \geq h_{i+1}$. The dynamics does not depend on the absolute height, but on the slope: an avalanche will occur if the slope exceeds some critical value z_c (this is one central parameter of the model). Therefore we define the dynamics in terms of the slope $z_i = h_i - h_{i+1}$ and not in terms of the height h_i. As a reference, we formally define the height at $N+1$ to be zero, $h_{N+1} = 0$ (see Fig. 13.1).

We denote by h_i^n (z_i^n) the height (the slope) at site i in step n. Let us first look what happens if a sand particle moves down one step in the interior of the region. Then, $h_i^{n+1} = h_i^n - 1$, and $h_i^{n+1} = h_i^n + 1$. That is,

$$z_{i-1}^{n+1} = z_{i-1}^n + 1, \quad z_i^{n+1} = z_i^n - 2, \quad z_{i+1}^{n+1} = z_{i+1}^n + 1.$$

The increment for the cells $(i-1, i, i+1)$ is $(1, -2, 1)$; this finding reminds of the discretized Laplace operator acting on the slope.

We consider a finite, spatial interval $1, \ldots, N$. Therefore, we need to take into account boundary effects (this is no proper cellular automaton as a grid with boundaries cannot be represented by a group). We superimpose no-flux conditions to the left hand boundary. Formally, we define that the slope at $i = 0$ (which is outside of the grid) is zero, $z_0 = 0$. If $z_1^n > z_c$, we obtain

$$z_1^{n+1} = z_1^n - 2, \quad z_2^{n+1} = z_2^n + 1.$$

At the right border, we already introduced formally $h_{N+1} = 0$. This condition corresponds to a homogeneous Dirichlet boundary condition: once a sand particle reaches site $N+1$, it vanishes (think about a sand-pile on a table: if a particle falls from the table it is gone). If $z_N > z_c$, one particle moves from site N to site $N+1$, where it is removed immediately. In terms of the slope, we obtain that

$$z_{N-1}^{n+1} = z_{N-1}^n + 1, \quad z_N^{n+1} = z_N^n - 1.$$

As usual for deterministic cellular automata, the dynamics acts synchronously. The function $\chi_{x>z_c}$ denotes the characteristic function, $\chi_{x>z_c} = 1$ if $x > z_c$ and zero else.

Definition 13.1.1 Let $E = \{0, \ldots, \ell\}$, $\ell > 2$, and $z_c \in \{1, \ldots, \ell - 2\}$ be a fixed constant. Let furthermore $\Gamma = \{1, \ldots, N\}$, and $D_0 = \{-1, 0, 1\}$. Define the local function

$$f_0(z_{-1}, z_0, z_1) = z_0 + \chi_{z_{-1}>z_c} - 2\,\chi_{z_0>z_c} + \chi_{z_1>z_c}$$

and the local functions at the left/right boundary

$$f_0^l(z_1, z_2) = z_1 - 2\,\chi_{z_1>z_c} + \chi_{z_2>z_c}, \quad f_0^r(z_{N-1}, z_N) = z_N + \chi_{z_{N-1}>z_c} - \chi_{z_N>z_c}.$$

The automaton defined in this way is called sand-pile automaton.

The global function of this cellular automaton reads $f : E^\Gamma \to E^\Gamma$, $u \mapsto v$ with

$$\begin{aligned} v(1) &= f_0^l(u(1), u(2)), \\ v(i) &= f_0(u(i-1), u(i), u(i+1)) \quad \text{for } i = 2, \ldots, N-1, \\ v(N) &= f_0^r(u(N-1), u(N)). \end{aligned}$$

Lemma 13.1.2 *The sand pile automaton is well defined.*

Proof We prove that $f(u) \in E^\Gamma$ if $u \in E^\Gamma$. First we show that the local states stay non-negative. If $u(i) > 1$, then $f(u)(i) \geq u(i) - 2 \geq 0$. If $u(i) \in \{0, 1\}$, then $u(i) \leq z_c$. Therefore at site i the state is not decreased, and $f(u)(i) \geq u(i) \geq 0$.

Next we show that $f(u) \leq \ell$. If $u(i) \leq \ell - 2$, then $f(u)(i) \leq \ell$, as a state is at most increased by 2. If $u(i) \in \{\ell - 1, \ell\}$, then $u(i) > z_c$. Therefore, the state either stays constant, or is decreased by 1 or 2. In any case, the state is not increased, and hence $f(u)(i) \leq \ell$. □

(E^Γ, f) is a dynamical system; we introduce the discrete topology on E^Γ. In this way, we can think about ω-limit sets. According to this topology, an ω-limit set is either a stationary point or a periodic orbit. If an initial state is not a stationary state, the dynamics of the cellular automaton mimics an avalanche. As an avalanche tends to become stationary in the long run, we expect also for our cellular automaton that there are no periodic orbits present. The following proposition confirms this expectation.

Proposition 13.1.3 *The ω-limit set of a state always consists of a single stationary state.*

Proof We make use of a series of Lyapunov functions. Let

$$h_i : E^\Gamma \to \mathbb{N}, \qquad u \mapsto \sum_{j=i}^{N} u(j)$$

and

$$L_i : E^\Gamma \to \mathbb{N}, \qquad u \mapsto \sum_{j=1}^{i} h_j(u).$$

$h_i(u)$ can be interpreted as the absolute height of the i'th pile. We obtain

$$\begin{aligned} h_1(f(u)) &= h_1(u) - \chi_{u(1)>z_c} \\ h_i(f(u)) &= h_i(u) + \chi_{u(i-1)>z_c} - \chi_{u(i)>z_c} \qquad \text{for } i = 2, \dots, N. \end{aligned}$$

That is, the height of a pile is increased if the slope to the left neighboring pile is supercritical, and decreased if the slope to the right neighbor is supercritical. Therewith we obtain

$$L_i(f(u)) = L_i(u) - \chi_{u(i)>z_c}.$$

The functions L_i (that can be interpreted as the total amount of sand within the piles $1, \dots, i$) are non-increasing in time, assume only discrete values, and are bounded from below. They are Lyapunov functions. Hence $L_i(f^n(u)) = \sum_{j=1}^{i} h_i(f^n(u))$ become eventually constant (n large enough). As this is true for all $i = 1, \dots, N$, we conclude that $h_i(f^n(u)) = \sum_{j=i}^{N} f^n(u)(j)$ become eventually constant, and so do $f^n(u)(i)$, $i = 1, \dots, N$. □

Now we introduce perturbation operators. By means of these operators, we will in a certain sense investigate the degree of stability of a stationary state.

Definition 13.1.4 Let $P_j : E^\Gamma \to E^\Gamma, j \in \{2, .., N\}$, defined as

$$P_j(u)(i) = \begin{cases} u(i) - 1 & \text{if } \; i = j - 1 \\ u(i) + 1 & \text{if } \quad i = j \\ \quad u(i) & \text{else} \end{cases} \qquad \text{if } u(j-1) > 0, \;\; u(j) < \ell,$$

and $P_j(u) = u$ else. For $j = 1$, we define

$$P_1(u)(i) = \begin{cases} u(i) + 1 & \text{if } \; i = 1 \\ \quad u(i) & \text{else} \end{cases} \qquad \text{if } u(1) < \ell,$$

and $P_1(u) = u$ else.

Remark 13.1.5 If we drop a sand particle at pile j, the state, described by the slopes, is changed by P_j. The total sum of slopes $\sum_{i=1}^{N} u(i)$ is not changed by the operators $P_2, \dots, P_N$. In the literature about sand pile models, these operators are therefore also called a conservative perturbation operators.

The set of stationary states for the sand pile automaton is given by

$$\mathcal{E} = \{u \in E^\Gamma \ : \ u(i) \leq z_c\}.$$

We introduce a semi-order on $\mathcal{E}$.

Definition 13.1.6 Let $u, v \in \mathcal{E}$ be stationary states. Assume that either $u = v$, or that there is a finite sequence of perturbation operators $P_{i_1}, \ldots, P_{i_m}$ such that $v = P_{i_1} \circ \ldots \circ P_{i_m}(u)$. In this case, we define $v \preceq u$. If $v \preceq u$ and $v \neq u$, we say that u is more stable than v.

It is straight to see that $\preceq$ is a semi-order on $\mathcal{E}$.

The next lemma indicates that $[\overline{z_c}] \in E^\Gamma$, the state that is constant z_c, is the only minimal element in $\mathcal{E}$. $[\overline{z_c}]$ is the least stable element.

Lemma 13.1.7 *Let $u \in \mathcal{E}$. There is a finite sequence of stationary states $v_0, \ldots, v_m \in \mathcal{E}$ and of perturbation operators $P_{j_1}, \ldots, P_{j_m}$ such that $v_0 = u$, $v_m = [\overline{z_c}]$, and $v_i = P_{j_i}(v_{i-1})$ for $i = 1, \ldots m$.*

Proof Let $u \in \mathcal{E}$, $u \neq [\overline{z_c}]$. There is at least one site $i_0 \in \Gamma$ with $u(i) < z_c$; let i_0 be the leftmost site with this property, $i_0 = \min\{i \ : \ u(i) < z_c\}$. Consider (see Fig. 13.2)

$$v_0 = u, \quad v_{j+1} = P_{i_0 - j} v^j \text{ for } j = 0, \ldots i_0 - 1.$$

First of all, $v_j \in \mathcal{E}, j = 0, \ldots, i_0$. Furthermore, $v_{i_0}(i_0) = u(i_0)+1$, and $v_{i_0}(i') = u(i')$ for $i \in \Gamma \setminus \{i_0\}$. That is, by this sequence of perturbation operators, the state u is increased at site i_0 and left unchanged elsewhere.

Proceeding in a similar way, we fill up the leftmost site with $u(i) < z_c$ until we reach, step by step, the state $[\overline{z_c}]$. □

Corollary 13.1.8 $[\overline{z_c}]$ *is the unique minimal element in* $(\mathcal{E}, \preceq)$.

On top of the cellular automaton, we introduce a second level dynamics. From time to time—rather seldom—we drop at a randomly chosen site a sand particle. This particle may or may not cause an avalanche. As the particles are dropped only rarely, the cellular automaton is considered to have enough time to reach a stationary

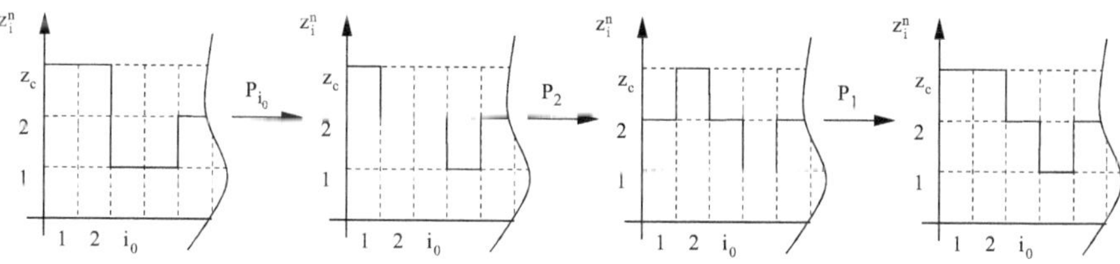

Fig. 13.2 Idea of the proof of Lemma 13.1.7

state; there is a clear separation of time scales between the time points at which sand particles arrive (slow dynamics) and the time span in which avalanches take place (fast dynamics). This idea is formulated in terms of a Markov chain on $\mathcal{E}$.

Definition 13.1.9 Consider an $\mathcal{E}$-valued Markov chain $\{X_n\}_{n\in\mathbb{N}_0}$. Let $I_n \in \Gamma$, $n \in \mathbb{N}_0$, i.i.d. uniformly distributed random variables. Then, X_{n+1} is determined by

$$\omega(P_{I_n}(X_n)) = \{X_{n+1}\}.$$

This Markov chain is called the sand pile Markov chain. The random variable

$$S_n = \min\{\, i \,:\, f^i(P_{I_n}(X_n)) \in \mathcal{E}\}$$

is called the duration of the avalanche (produced in step n).

We reformulate the definition using plain words: Given a state $X_n \in \mathcal{E}$, at a randomly selected site I_n a sand particle is dropped and yields $P_{I_n}(X_n) \in E^\Gamma$. This new state is not necessarily stationary. However, we know that the dynamics of the cellular automata will tend to a stationary state after a finite number of steps—the ω-limit set of $P_{I_n}(X_n)$ consists of one point in $\mathcal{E}$. This point is the next state of the chain X_{n+1}, and S_n the time necessary until this new stationary state is reached.

Lemma 13.1.10 *$X_n \to [\,\overline{z_c}\,]$ for $n \to \infty$ a.s.*

Proof $[\,\overline{z_c}\,]$ is the only absorbing state in $\mathcal{E}$, and according to Lemma 13.1.7, there is a path from any state in $\mathcal{E}$ to $[\,\overline{z_c}\,]$. As the state space of the sand pile Markov chain is finite, X_n eventually jumps into this state with probability one. □

This behavior is meant by "self-organized criticality": The dynamics saddles necessarily (self-organized) on the least stable stationary state. This result allows to compute the long term distribution of the duration of avalanches.

Lemma 13.1.11 *$P(S_n = j) \to 1/N$ for $j \in \Gamma$ and $n \to \infty$, while $P(S_n = j) \to 0$ for $j \notin \Gamma$.*

Proof We know that $X_n \to [\,\overline{z_c}\,]$ a.s.; as the state space is finite, $P(X_n \neq [\,\overline{z_c}\,]) \leq e^{-cn}$ for some positive constant c. Hence, in order to compute the limit of $P(S_n = j)$ for $n \to \infty$, we may assume $X_n = [\,\overline{z_c}\,]$. If we apply P_j to $X_n = [\,\overline{z_c}\,]$, we obtain $X_{n+1} = [\,\overline{z_c}\,]$ (as $[\,\overline{z_c}\,]$ is an absorbing state), and $S_n = N - j + 1$ steps of the global function f are required to reach this state again. As we hit any site in Γ with the same probability, the result follows. □

This finding indicates that our model is too simple to produce the $1/f^\alpha$ characteristics (where we could, of course, say that in our case $\alpha = 0$). Only in higher dimension (two or three) the sand pile model exhibits this power law with $\alpha > 0$. However, already in one dimension the model shows the basic mechanism that underlies self-organized criticality. A more detailed discussion of theoretical and experimental approaches to the phenomenon of self-organized criticality and the $1/f^{\alpha}$-law can be, e.g., found in the review article by Markovic and Gros [122], and references quoted therein.

13.2 Epidemiology

The dynamics of infectious diseases is, of course, of high interest. Cellular automata provide a framework to study the spread of infectious diseases in a spatially structured environment, or, more general, on a given contact network. Most models of this type are of stochastic nature, but also deterministic cellular automata are used [97]. There are many books that give a profound introduction into the modeling of infectious diseases, e.g., the volume of Diekmann and Heesterbeeck [44], the book by van den Driessche et al.[17], and many textbooks about mathematical biology [49, 133, 137].

Before we start with the discussion of epidemic models, we introduce a heuristic concept for the analysis of a cellular automaton: the mean field approximation.

13.2.1 *Mean Field Approximation*

A simple mean field approximation replaces the action of a cellular automaton by a discrete dynamical system for the expected value of an arbitrary cell. The idea is to *assume* that the occupations of the cells are "random" and independent—a probability distribution on E characterizes the states of cells in a given time step completely. One *assumes* that, however complicated the dynamics is in detail, the probabilities for elements in E are more or less constant over space. Next one *assumes* that one can, given these proportions, compute the distribution of patterns in E^{D_0}. Using the local function, it is then possible to compute the expected proportions of elements in E in the next time step. In this way a discrete time dynamical system for the proportions is defined. One expects that the qualitative and even quantitative behavior of this "small" dynamical system reflects the properties of the cellular automaton, at least for infinite or large finite grid domains. Of course, potentially gross errors are committed because, even if the initial state has the assumed independence property, in general the next and all subsequent do not have this property. However, it turns out that in many cases mean field approximations are qualitatively correct and in some cases are even quantitatively reliable. The mean field approximation can be improved by taking into account first order or even higher order correlations, see [152] or [133, Chap. 3.1.10].

As an introductory example we consider Wolfram automata. We have $\Gamma = \mathbb{Z}$, $D_0 = \{-1, 0, 1\}$, $E = \{0, 1\}$. For each $u \in E^{D_0}$ the function $f_0(u)$ assumes either 0 or 1. Let p_n the proportion of '1' in time step n and $q_n = 1 - p_n$. Then, for a given cell the probability to carry '1' is p_n. The probability to find the local state $u = (1, 1, 1)$ is p_n^3, and that for $(0, 0, 1)$ is $q_n^2 p_n$. The probability for $u \in E^{D_0}$ is

$$p_n^{s(u)} q_n^{3-s(u)}$$

with $s(u) = u(-1) + u(0) + u(1)$. Hence the probability for a '1' at the next step is

$$p_{n+1} = M(p_n)$$

where the function $M(p)$ is defined as

$$M(p) = \sum_{u \in E^{D_0}} f_0(u) p^{s(u)} q^{3-s(u)}.$$

$M(p)$ is a polynomial of degree 3 or less. Clearly $M(p) \in [0, 1]$ for $p \in [0, 1]$, that is, M defines a discrete dynamical system on $[0, 1]$. The next question is about the properties of this function M for specific Wolfram automata.

Consider the automaton Wolfram 150 as an example,

$$f_0(u) = (u(-1) + u(0) + u(1)) \bmod 2.$$

In the following scheme we present the 8 local states, the value of the local function and the probability for the local state to occur. Then, by summing up, we compute $M(p)$,

$$\begin{array}{lll} 000 & 0 & q^3 \\ 001 & 1 & q^2 p \\ 010 & 1 & q^2 p \\ 011 & 0 & qp^2 \\ 100 & 1 & q^2 p \\ 101 & 0 & qp^2 \\ 110 & 0 & qp^2 \\ 111 & 1 & p^3 \end{array}$$

We obtain

$$M(p) = 0\left(q^3 + pq^2 + 2qp^2\right) + 1\left(p^3 + 3q^2 p\right) = p^3 + 3(1-p)^2 p.$$

We find immediately that $M(p)$ is strictly increasing, that there are three fixed points $0, 1/2, 1$ with $M'(0) = M'(1) = 3$, $M'(1/2) = 0$.

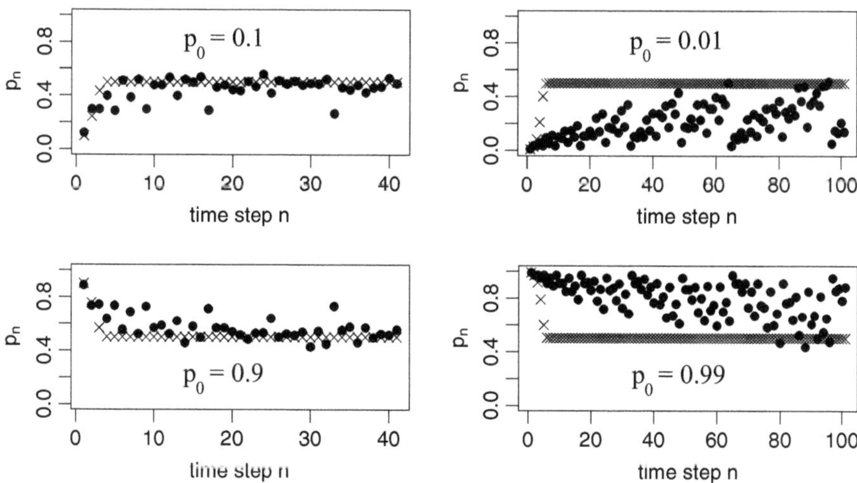

Fig. 13.3 Comparison of mean field approximation (*cross*) and simulated density of 1 (*bullets*) for the Wolfram automaton 150, on a torus of size 500 starting with a random state. The initial probability for 1 is indicated in the four panels

The fixed points 0 and 1 are strongly repelling, the fixed point $1/2$ is strongly attracting. Hence the mean field approximation suggests that if the system does not start with "all 0" or "all 1" then on large grids, after some (short) time, the proportion of '0' and '1' will be each about $1/2$. Figure 13.3 indicate that for randomly chosen initial states this heuristics is confirmed by simulations if p is not too close to 0 or 1.

13.2.2 SIRS Model and Mean Field Approximation

We formulate a model for an infectious diseases in a spatially structured environment. At each site of a Cayley graph an individual is located. This individual may be susceptible, infected or recovered; an infected individual is assumed to also be infectious. A susceptible individual becomes infected if at least one infected individual is located in his/her neighborhood (this is the assumption of a Greenwood type model, in contrast to a Reed-Frost model, where the chance to become infected increases with the number of infected neighbors). An infected individual recovers and becomes immune, and an immune individual becomes susceptible again. A dynamical system with this structure is called SIRS model (Susceptible Infected Recovered-Susceptible).

Definition 13.2.1 A cellular automaton (Γ, D_0, E, f_0) with $D_0 = \Gamma_m$ for some $m \in \mathbb{N}$, $E = \{S, I, R\}$ and local function

$$f_0(u) = \begin{cases} I & \text{if } u(e) = S \text{ and } \exists g \in D_0 : u(g) = I \\ R & \text{if } u(e) = I \\ S & \text{else} \end{cases}$$

is called (epidemic) SIRS automaton.

If we choose $\Gamma = Z^2$ and D_0 as the von Neumann neighborhood, we are in the framework of a Greenberg-Hastings automaton (see Sect. 12.3), where I is the activated, R the refractory, and S the resting state. However, we will not follow the Greenberg-Hastings type of analysis, but apply the mean field technique. This automaton is semi-totalistic; therefore, we need not to consider all different elements in $u \in E^{D_0}$, but only need to know $u(e)$ and $\zeta(u) := \sum_{g\in D_0\setminus\{e\}} \chi_{u(g)=I}$. Or, more specifically, we only need to know $u(e)$ and whether $\zeta(u) > 0$ (in case that $u(e) = S$). Let s_n, i_n, and r_n denote the probabilities to find a site in state S, I and R, respectively. Then, $s_n + i_n + r_n = 1$. It is simple to determine r_{n+1}, as any site in state I becomes R in the next step,

$$r_{n+1} = i_n.$$

Now we aim at i_{n+1}. Given a site $g \in \Gamma$ with local state S: How likely is it, that no site in its neighborhood assumes state I? We have $|D_0| - 1$ sites that can be occupied by I; the probability that none of them actually has state I is $(1 - i_n)^{|D_0|-1}$. Thus, in the mean field approximation,

$$i_{n+1} = s_n \left(1 - (1 - i_n)^{|D_0|-1}\right).$$

The remaining fraction $s_n - i_{n+1} = s_n (1 - i_n)^{|D_0|-1}$ stays susceptible. The recovered of step n are susceptibles in step $n + 1$, and thus,

$$s_{n+1} = s_n (1 - i_n)^{|D_0|-1} + r_n.$$

All in all, the mean field approximation assumes the usual basic form of a deterministic, time-discrete SIRS model:

$$\begin{aligned} s_{n+1} &= s_n (1 - i_n)^{|D_0|-1} + r_n \\ i_{n+1} &= s_n \left(1 - (1 - i_n)^{|D_0|-1}\right) \\ r_{n+1} &= i_n. \end{aligned}$$

Next we determine the stationary points: if $(s_n, i_n, r_n) \equiv (s, i, r)$, then $i = r$ and due to $s + i + r = 1$, $s = 1 - 2i$. A short computation shows that the i-component of the stationary point satisfies

$$\varphi(i) = (1 - 2i)(1 - i)^{\gamma(m)-1} + 3i - 1 = 0.$$

Note that $\varphi(0) = 0$, such that there is always the stationary point $(1, 0, 0)$, where the infection is not present. Furthermore,

$$\varphi'(i) = 3 - 2(1 - i)^{|D_0|-1} - (|D_0| - 1)\,(1 - 2i)\,(1 - i)^{|D_0|-2}$$

and thus $\varphi'(0) = 2 - |D_0|$. If $|D_0| > 2$, then $\varphi'(0) < 0$. Furthermore, $\varphi(1/2) = 1/2 > 0$, and there is at least one more root i^* of φ in $(0, 1/2)$ for $|D_0| > 2$. It is straightforward to see that $\varphi''(i) > 0$ (again in case of $|D_0| > 2$), such that this second root is unique.

Using the discrete-time version of the theorem of Hartman-Grobman, the stability of the trivial stationary point can be investigated by the linearization of the system

$$\begin{aligned} s_{n+1} &= s_n\,(1 - i_n)^{|D_0|-1} + (1 - s_n - i_n) \\ i_{n+1} &= s_n\,\left(1 - (1 - i_n)^{|D_0|-1}\right) \end{aligned}$$

(note that we used $r_n = 1 - s_n - i_n$ to eliminate r_n from our investigation). We obtain the Jacobian of the right hand side at $(s, r) = (1, 0)$

$$J = \begin{pmatrix} 0 & -|D_0| \\ 0 & |D_0| - 1 \end{pmatrix}.$$

If all eigenvalues of this matrix are located within the complex unit circle, the stationary point is locally asymptotically stable. Hence, for $|D_0| \in [0, 2)$, the stationary point is locally asymptotically stable, while for $|D_0| > 2$ one eigenvalue is outside the unit circle, such that we expect a small perturbation to grow exponentially. Indeed, if we consider one infected individual in an island of susceptible individuals, $|D_0| - 1$ is the number of infected produced. Therefore, this number is called the basic reproduction number R_0 for this infection,

$$R_0 = |D_0| - 1.$$

If $R_0 < 1$, the infection dies out, while the linearized mean field model predicts in the case that $R_0 > 1$ an exponential growth of the number of infected in time. And indeed, many models (based on ordinary differential equations, stochastic processes, and other mathematical structures) exhibit this initial exponential growth, in line with many empirical data [17, 44].

13.2.3 Polynomial Growth: Clustering of Contact Networks

In the model considered so far, $R_0 > 1$ implies the instability of the uninfected point, that is, in the linearization of the time-discrete model appears an eigenvalue with absolute value larger one. Hence, in the onset of the endemic, we expect an exponential growth of the number of infecteds; the number of infecteds is also called "prevalence". This exponential increase of the prevalence can be observed for many, but not all infections. In particular, for HIV (and also under some conditions for Ebola), a polynomial growth can be observed [160]. There are several different attempts to explain this polynomial growth [34, 165]. We take up the idea formulated by Szendroi and Csanyi [160]: The starting point of these authors is the observation that individuals only have contacts with few other individuals. That is, the population forms a graph—the contact graph. Within this graph, infected individuals are clustered. This high correlation eventually decreases the net reproduction number. If the net reproduction number is decreased to one, polynomial growth of the prevalence may appear.

A simple setup that allows to investigate this idea is a cellular automaton. We even simplify the SIRS automaton, and consider an SI-automaton: Susceptible individuals become infected if at least one infected individual is located within their neighborhood, and stay infected for ever.

Definition 13.2.2 A cellular automaton (Γ, D_0, E, f_0) with $D_0 = \Gamma_m$ for some $m \in \mathbb{N}$, $E = \{S, I\}$ and local function

$$f_0(u) = \begin{cases} I & \text{if } u(e) = S \text{ and } \exists g \in D_0 : u(g) = I \\ S & \text{else} \end{cases}$$

is called (epidemic) SI automaton or contact automaton.

The SI automaton is again a semi-totalistic cellular automaton. As we have seen in the last section, the basic reproduction number is given by

$$R_0 = |D_0| - 1.$$

Since we assume $D_0 = \Gamma_m$, we have $|D_0| = \gamma(m)$, where γ denotes the growth function of the group. That is,

$$R_0 = \gamma(m) - 1.$$

Consider an individual that became infected in time step n. All persons that ever become infected due to a contact to this individual become infected at time step $n + 1$. Start at time $n = 0$ with one infected individual in site e,

$$u^0(e) = I, \quad u^0(g) = S \text{ else.}$$

Let furthermore $u^i = f(u^{i-1})$ denote the trajectory of our SI-automaton. Define for time step n the number of infected individuals $N(n) = |\{g \in G : u(g) = I\}|$, and formally define $N(-1) = 0$. Furthermore, we introduce the net replacement number for time step $n \geq 1$

$$R(n) = \frac{N(n+1) - N(n)}{N(n) - N(n-1)}.$$

This is the average number of individuals infected by one infector (in time step n). Note that $R(0) = R_0$ is the basic reproduction number.

Remark 13.2.3 $R(n)$ is non-negative and finite for all $n \in \mathbb{N}$. Moreover, $R(n) \leq |D_0|$.

If $R(n) \geq c > 1$, then $N(n) \geq c^n$, and we have an exponential growth in the number of infected persons. Only if $R(n) \to 1$ for $n \to \infty$, the incidence can show an polynomial growth in time.

If we choose in particular $D_0 = \Gamma_1$, then $N(n) = \gamma(n)$ (as indicated above, $\gamma(n)$ denotes the growth function of the underlying group). The growth function may be polynomial or exponential (or intermediate, but this case does not play a role here), depending on the group.

Proposition 13.2.4 $G = \mathbb{Z}^d$, $D_0 = \{z \in \mathbb{Z}^d : \|z\|_\infty \leq 1\}$ *the Moore neighborhood. Then,* $N(n) = (2n+1)^d$, $R_0 = 3^d - 1$, *and*

$$\lim_{n\to\infty} R(n) = 1.$$

Proof We compute $\gamma(n)$, that is, the number of grid points reachable by paths of length at most n. For $d = 1$, this is clearly $2n + 1$. For $d = 2$, the reachable points coincide with the square, centered around the origin of zero, with length $(2n + 1)$, that is $(2n + 1)^2$. Similarly, for general dimension d we find $(2n + 1)^d$.

Hence, $R_0 = 3^d - 1$, and for $n \to \infty$

$$R(n) = \frac{(2n+3)^d - (2n+1)^d}{(2n+1)^d - (2n-1)^d} = \frac{3\,d\,(2n)^{d-1} - d\,(2n)^{d-1} + O(n^{d-2})}{d\,(2n)^{d-1}\,1 - d\,(2n)^{d-1}\,(-1) + O(n^{d-2})} \to 1.$$

□

We find in the case $\mathbb{Z}^d$, that the growth is polynomial, and that the replacement ratio comes down from $R_0 \gg 1$ to one.

Proposition 13.2.5 *Consider the free group over d symbols,* $G = F_d$, *and* $D_0 = \Gamma_1$. *Then,* $N(n) = (d(2d-1)^n - 1)/(d-1)$ *for* $n \geq 1$, $R_0 = 2d$, *and for* $n \geq 1$,

$$R(n) = 2d - 1.$$

Proof At time step zero, we have $N(0) = 1$. In Example 2.5.5 we determined that

$$\gamma(n) = \frac{d(2d-1)^n - 1}{(d-1)}.$$

Hence, $N(n) = \gamma(n)$,

$$R_0 = \gamma(1) - 1 = \frac{d(2d-1) - 1}{(d-1)} - 1 = \frac{2d^2 - d - 1 - d + 1}{(d-1)} = 2d,$$

and for $n > 0$

$$R(n) = \frac{d(2d-1)^{n+1} - d(2d-1)^n}{d(2d-1)^n - d(2d-1)^{n-1}} = 2d - 1.$$

□

Interpretation: In $\mathbb{Z}^2$, there are many different paths from e to a grid point g. The grid is highly clustered. Hence, many attempts to infect someone go to already infected individuals. This effect brings the replacement number $R(n)$ down to one and forces the number of infected individuals to grow polynomially in time. In case of the free group, this is different: There is a unique path from e to any grid points (if we only allow for double-point free paths), and hence a fixed ratio of infection attempts go to susceptible individuals. We have a constant replacement number $R(n) \equiv 2\,d - 1 > 1$ if $d \geq 2$, and the number of infected individuals grows approximately proportionally to $(2\,d - 1)$, in particular exponentially for $d > 1$.

Result: Clustering (multiple ways to infect a individual) is decisive for polynomial growth of the prevalence.

13.3 Evolution

Predator-prey models are classical examples in theoretical and mathematical ecology. Their dynamics is well investigated and well understood. In the present case, the dynamics itself is not the main target of the investigation. The question we seek to answer is connected with populations that live as larvae for several years, and only 1 year as an adult. There are several species with a population dynamics that is not annual, but have a periodic life cycle of several years. Rather often, the population is locally synchronized, such that adults of this species do not appear every year but only every k'th year. For example the cicadas in the genus *Magicicada* follow periodic life cycles of either 13 or 17 years [72, 176]. There are also other species that show long life cycles. It is intriguing that the lengths of these cycles are often prime numbers. Though the period is rather precisely defined, the local populations undergo from time to time shifts; mostly this is either a one- or a four year shift, which is another interesting point [110].

How can cicadas know number theory and select prime numbers for their life cycles? Several mechanisms have been proposed [176]. Among those, the idea that the cicadas try to escape the pressure from a parasite or predator is the most popular. Models based on time discrete dynamical systems [20, 91] and ordinary differential equations [109] are used to investigate this idea, and come to different results (Hoppensteadt [91] and Bulmer [20] confirm this idea, Ryusuke [109] is less convinced that this is the decisive mechanism). Several articles use cellular automata to address this problem in a spatial setting [21, 67, 68, 124]. Before we start to discuss these models, we briefly review the basic idea of mathematical models for evolution.

13.3.1 Evolution

A population has a certain fitness. A measure for the fitness of the population is its fertility. The fertility depends on two different facts: (a) the environmental conditions, and (b) the intrinsic properties of the species. To start with the second point, in a simplifying manner we focus on only one species and one specific property that can be characterized by a real number or an integer. This property is called a trait. For example, the average size of adult individuals can be a trait. Below, we will use the length of the life cycle as a trait. If a population is present in an ecosystem, it will influence the environment. For example, it uses resources and hence less resources are left for other populations that inhabit the same ecosystem. A predominant population (with a distinct trait) is called the resident. If a mutation happens, few individuals with a different trait appear on the scene. The crucial question is now, if these mutants can or cannot spread in presence of the resident. That is, for the given environment we compute the fitness of the mutants. For many models it is true that a mutant with a fitness above a certain threshold (that depends on the resident) outcompetes and replaces the resident. The mutant becomes the new resident—evolution changed the predominant trait. If no mutants are able to invade, the resident is at an evolutionary end point, also called evolutionary stable strategy (ESS). For an introduction into this theory, see the book of Hofbauer and Sigmund [88], or the introduction to adaptive dynamics [43]. In case of two coexisting species, both of them will influence the environment. We are led to the following definition of an ESS.

Definition 13.3.1 Let $H \subset \mathbb{R}^2$ be a set of admissible traits for a system of two species X and Y living in one ecosystem. Let $F_x : H \to \mathbb{R}$, be the fitness of species X, and $F_Y : H \to \mathbb{R}$ that of species Y for $(x, y) \in H$. We call $(x_0, y_0) \in H$ an evolutionary stable strategy (ESS), if

$$\forall x,\ (x, y_0) \in H\ :\ F_x(x, y_0) \leq F_x(x_0, y_0)$$

and

$$\forall y,\ (x_0, y) \in H\ :\ F_y(x_0, y) \leq F_y(x_0, y_0).$$

Remark 13.3.2 This version of an ESS is motivated by game theory: given one species plays a certain strategy what is the best answer of the other species? The ESS (x_0, y_0) is defined in such a way that the fitness of one species is not increased if it changes its strategy *independently* of the other species. There are more ways to introduce an ESS. In particular the approach of adaptive dynamics [43] focuses on the question how an ESS can evolve. There are two more particularities in this definition: (1) The definition is global, not local (often only strategies in a neighborhood of (x_0, y_0) are considered to define an ESS), and (2) we only require weak inequalities (in other versions of this definition it is required that $F_x(x, y_0) < F_x(x_0, y_0)$ for $x \neq x_0$, and similarly $F_x(x_0, y) < F_x(x_0, y_0)$ for $y \neq y_0$).

The definition above is adapted to the model we will consider below: we use a discrete state (such that considerations of arbitrary small perturbations are not meaningful), and we assume that a mutant is only able to invade a resident if its fitness is strictly larger than that of the resident.

13.3.2 Spatial Model

We consider a cellular delay-automaton $(\mathbb{Z}^2, D_0, E, f_0)$, where D_0 is the Moore neighborhood. A cellular delay-automaton is no cellular automaton in the strict sense: not only $u|_{D_0}$, but also a (finite) history of $u|_{D_0}$ is required to define the local rule. If there is an a priori bound for the length of the history needed, we are able to incorporate this history in the local state: Assume we need to know a state not only at time t, but at time points $t, \ldots, t-k$, $k \in \mathbb{N}$ in order to compute the local state a site 0 in the next time step; the local function is given by $f_0(u^t|_{D_0}, \ldots, u^{t-k}|_{D_0})$. This function is no local function of a cellular automaton. We define $\tilde{E} = E^{k+1}$, and consider the cellular automaton $(\Gamma, D_0, \tilde{E}, \tilde{f}_0)$. Note that $\tilde{E}^\Gamma = (E^{k+1})^\Gamma$ can be identified with $(E^\Gamma)^{k+1}$, and likewise $\tilde{E}^{D_0}$ with $(E^{D_0})^{k+1}$. For $v \in \tilde{E}^\Gamma$, v_0 (the first component) represents the state of the delay-automaton at time t, and v_i that at time $t-i$. Consequently, we define

$$(\tilde{f}_0(v))_1 = f_0(v_0, \ldots, v_k), \quad (\tilde{f}_0(v))_i = v_{i-1}(0) \quad \text{for } i = 1, \ldots k.$$

Only the first component of the new state is computed in a non-trivial way, all other components are simple shifts. The dynamics of the cellular automaton $(\Gamma, D_0, \tilde{E}, \tilde{f}_0)$ is equivalent to that of the delay automaton.

As we now know that a cellular delay-automaton can be redefined in terms of a cellular automaton,we return to the notion of a delay-automaton and assume that in time step t, not only u^t, but also u^{t-i} is available for $i = 1, \ldots, k$.

We start to describe the predator-prey system that allows to gain some insight into ESS. A prey as well as a predator are present in every site, and at all time steps. Both are quiescent for most of their live, only for one time step they become active (as predators) respectively available (as prey). Let a_X denote the age of the prey, and a_Y the age of the predator. Then, the tupel (a_X, a_Y) characterizes the age of prey and predator within a given cell. We assume that no prey and no predator has a live span longer than a given constant K. Then,

$$(a_X, a_Y) \in \{1, \ldots, K\}^2.$$

Each individual has a period between 1 and K. These periods can differ from individual to individual. To completely characterize the state of the cell, it is also required to know the period of the live cycle for the prey/predator present in this cell. Let X denote the period for the prey, and Y the period of the predator, then

$$(X, Y) \in \{1, \ldots, K\}^2.$$

All in all, the tupel $(a_X, a_Y, X, Y) \in E = \{1, \ldots, K\}^4$ characterizes completely the state of a cell (that is, the age and the period of predator and prey). For $e \in E$, we denote by e_{a_X}, e_X etc. the corresponding entries in the local state e.

Next we introduce the dynamics of the cellular automaton. All individuals age (one time unit per time step). A prey (predator) individual is a larva if $a_X < X$ ($a_Y < Y$), while it becomes an adult if $a_X = X$ ($a_Y = Y$). Only adults are available as prey, resp. able to predate.

In order to define the local rule, we assign a fitness to prey and predator within a given cell and a given time step. Thereto we first define an instantaneous fitness at time t. The fitness is the average over the instantaneous fitness for a period of time:

Given a state $u^t \in E^\Gamma$ at time t, one ingredient for the instantaneous fitness of prey $\varphi_X(u^t)(x)$ in site x is the number of adult predators in the Moore neighborhood $x + D_0$,

$$s_Y(u^t)(x) = |\{y \in x + D_0 \,:\, (u^t(y))_Y = (u^t(y))_{a_Y}\}|.$$

Several cases are to consider: (a) If the prey is a larva ($u^t(x)_{a_X} < u^t(x)_X$), the prey is not affected by predators. Prey is neither object to predation nor it is able to reproduce. The fitness is zero. (b) If the prey is an adult ($u^t(x)_{a_X} = u^t(x)_X$), and no adult predators are in its Moore neighborhood ($s_Y(u^t)(x) = 0$), the prey performs very well, and its fitness is defined as $p \in \mathbb{N}$. (c) If the prey is an adult ($u^t(x)_{a_X} = u^t(x)_X$), and $s_Y(u^t)(x) > 0$, then the fitness of the prey is defined to be $-s_Y(u^t)(x)$:

$$\varphi_X(u^t)(x) = \begin{cases} 0 & \text{if} \quad (u^t(x))_X < (u^t(x))_{a_X} \\ p & \text{if} \quad (u^t(x))_X = (u^t(x))_{a_X},\ s_Y(u^t)(x) = 0 \\ -s_Y(u^t)(x) & \text{if} \quad (u^t(x))_X = (u^t(x))_{a_X},\ s_Y(u^t)(x) > 0. \end{cases}$$

With similar ideas, the fitness of the predator is defined: Now we count the sites with adult prey in the Moore neighborhood of x,

$$s_X(u^t)(x) = |\{y \in x + D_0 \,:\, u^t(y)_X = u^t(y)_{a_X}\}|$$

and define the instantaneous fitness of predators $\varphi_Y(u^t)(x)$

$$\varphi_Y(u^t)(x) = \begin{cases} 0 & \text{if} \quad u^t(x)_Y < u^t(x)_{a_Y} \\ -p & \text{if} \ u^t(x)_Y = u^t(x)_{a_Y}, \ s_X(u^t)(x) = 0 \\ s_X(u^t)(x) & \text{if} \ u^t(x)_Y = u^t(x)_{a_Y}, \ s_X(u^t)(x) > 0. \end{cases}$$

This instantaneous fitness is not directly used, but—in order to be fair—the average of the instantaneous fitness for one prey (predator) life span over the last $X\,Y$ time intervals is used: The fitness $\Phi_X(x,t)$ of prey (the fitness $\Phi_Y(x,t)$ of predator) at site x and time t is defined by

$$\Phi_X(x,t) = \sum_{\tau=t+1-XY}^{t} \varphi_X(u^\tau)(x)/Y, \quad \Phi_Y(x,t) = \sum_{\tau=t+1-XY}^{t} \varphi_Y(u^\tau)(x)/X.$$

Now we are prepared to define the state of a cell x in the next time step. We do this only verbally. First we go for the prey. There are two possibilities: If the fitness of the prey at site x is smaller than that of the prey at a neighboring site x', the prey is replaced by that in site x'; otherwise, the prey becomes older.

That is, in the first step we compare $\Phi_X(x,t)$ with

$$\Phi'_X(x,t) := \max_{x' \in x+D_0 \setminus \{x\}} \Phi_X(x',t).$$

If $\Phi_X(x,t) < \Phi'_X(x,t)$, the prey component is replaced by a specimen of the prey with larger fitness; we select $x' \in x + D_0$ which has fitness $\Phi_X(x',t) = \Phi'_X(x,t)$, and copy the period $u^{t+1}(x)_X = u^t(x')_X$. The age of the newborn prey with this apparent better fitness is zero, $u^{t+1}(x)_{a_X} = 0$ (if x' is not unique, select one of the sites in $x + D_0$ where the prey fitness is $\Phi'_X(x,t)$ according to some deterministic rule).

If $\Phi_X(x,t) \geq \Phi'_X(x,t)$, then only the age component of the prey is updated (increased modulo $u^t(x)_X + 1$),

$$u^{t+1}(x)_{a_X} = u^t(x)_{a_X} + 1 \bmod u^t(x)_X + 1.$$

The predator component of the cell is updated likewise: if the fitness of the predator is smaller than that of a neighboring predator, the latter replaces the predator (where the age of the predator at this site is taken to zero); if not, the age of the predator is increased.

We simulated this cellular automaton with random initial condition. In order to adapt the history of the initial condition (for delay equations an important aspect of the initial state), we assumed that the types (X and Y component) have been there since ever, without interactions. This procedure yields a suited pseudo-history. Then, we started the cellular automaton until only one type of prey remained. In the numerical experiments, this was always the case; it is by no means obvious that no states exist where several types persist for all times. However, the resulting life cycle

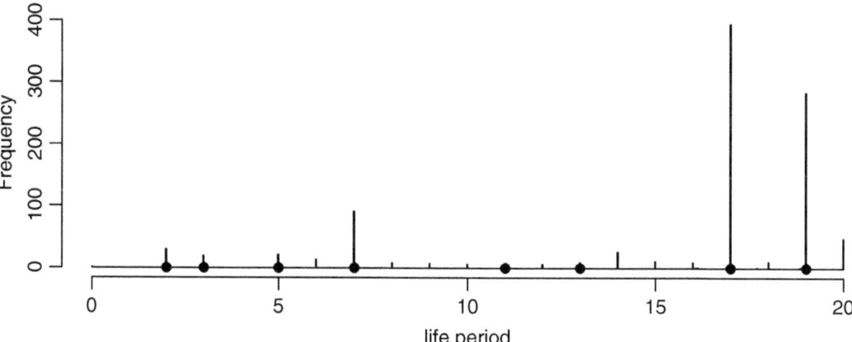

Fig. 13.4 Result of 1000 simulations with random initial conditions. We used a 10×10 grid, $p = 5$, and started with initial periods randomly distributed between 2,...,20. One run has been simulated until only one prey type remained. Frequency of the resulting prey life cycle population is presented. *Black bullets* on the x-axis indicate prime numbers

distribution, generated by 1000 runs, is shown in Fig. 13.4. We find indeed that in the long run prime numbers are predominant.

13.3.3 Heuristic Analysis

In the spirit of the mean field approximation, Goles et al. [67, 68, 124] propose a simple model to obtain some insight why prime periods appear rather often in the cellular automaton described above. However, their approach is not a formal approximation (as the mean field approximation), but rather a toy model. Let us consider a community of predators and prey, where the prey has a life period x, and the predators a life period of y. The period of the combined predator-prey model divides xy. We assume for simplicity that the prey population as well as the predator population are synchronized. We now define a measure for the fitness of prey and predator. Let us first look at the prey: The prey adults appear every x time steps. If there is no predator present, the prey will proliferate. This is a "good time interval" for the prey, and is counted with +1. If the predator is also present, then the prey is subject to predation; this is a "bad time interval", and counts −1. The average number of good years minus the average number of bad years related to one life span of a prey (length x) yields the fitness of the prey, i.e.

$$F_x(x,y) = (\quad |\{\text{prey without predator in time step } t \ : \ t = 1,\dots,xy\}| \\ -|\{\text{prey with predator in time step } t \ : \ t = 1,\dots,xy\}| \quad)/y.$$

We divide by y, since the number of prey generations in the time interval xy is just y. As a "good time interval" for the prey is a "bad time interval" for the predator, (and vice versa), we toggle the signs and divide by x to define the fitness of the predator,

$$F_y(x,y) = \big(\; -|\{ \text{ predator without prey in time step } t \; : \; t = 1,\ldots,xy\}| \\ + |\{\text{predator with prey in time step } t \; : \; t = 1,\ldots,xy\}| \quad \big)/x.$$

Evolution of the prey maximizes F_x, while the evolution of the predator maximizes F_y. Random mutations "test" different values for the "trait" x resp. y. A special state (x_0, y_0) is given, if any mutation of x (given $y = y_0$) does not improve F_x, and similarly any mutation of y (given $x = x_0$) does not improve F_y. This equilibrium bears some similarity with a Nash equilibrium, and is an ESS as introduced in Definition 13.3.1.

In the following, we do not allow for arbitrary periods $(X, Y) \in \mathbb{N}^2$, but we choose a number $L \in \mathbb{N}$, $L > 2$, and require

$$(X,Y) \in H = \{(x,y) \in \mathbb{N}^2 \; : \; 2 \leq y \leq L/2+1, \;\; L/2+2 \leq x \leq L\}.$$

We discuss below (after the next lemma) why this restriction is appropriate from the modeling perspective.

We use for $x, y \in \mathbb{N}$ the notation $\gcd(x,y) = \max\{z \in \mathbb{N} \; : \; x/z \in \mathbb{N}, \; y/z \in \mathbb{N}\}$ (greatest common divisor), $\text{lcm}(x,y) = \min\{z \in \mathbb{N} \; : \; z/x \in \mathbb{N}, \; z/y \in \mathbb{N}\}$ (least common multiple), and $\text{gd}(x) = \max\{z \in \mathbb{N} \; : \; x/z \in \mathbb{N}, \; z < x\}$ (greatest divisor). Note that $\text{gd}(x) = 1$ if x is a prime number, and $\text{gd}(x) \leq x/2$ else.

Lemma 13.3.3 *Choose $L \in \mathbb{N}$, $L > 2$, and define the set of admissible strategies by*

$$H = \{(x,y) \in \mathbb{N}^2 \; : \; 2 \leq y \leq L/2+1, \;\; L/2+2 \leq x \leq L\}.$$

(a) If $(X, Y) \in H$ is an ESS, then X is prime.
(b) If X is prime, and there is $Y \in \mathbb{N}$ such that $(X, Y) \in H$, then (X, Y) is an ESS.

Proof Before we start with the proof of the lemma, we rewrite F_x and F_y. In order to determine $F_x(X, Y)$, we need to know how often the adult prey individuals appear in a time period of XY (this is Y times), and how often adult prey and predators appear at the same time (this is $XY/\text{lcm}(X, Y)$). Therefore, the number of "good time intervals" for the prey reads $Y - XY/\text{lcm}(X, Y)$, and the number of "bad time intervals" $XY/\text{lcm}(X, Y)$. If we take into account that $\text{lcm}(X, Y)\gcd(X, Y) = XY$, we find

$$F_x(X,Y) = (Y - 2XY/\text{lcm}(X,Y))/Y = 1 - 2\gcd(X,Y)/Y$$

Similarly, we obtain

$$F_y(X, Y) = (2XY/\mathrm{lcm}(X, Y) - X)/X = 2\mathrm{gcd}(X, Y)/X - 1.$$

(a) We show: X_0 not prime, then any $(X_0, Y) \in H$ is no ESS.

Assume hat X_0 is not prime, and there is $Y_0 \in \mathbb{N}$ such that $(X_0, Y_0) \in H$ is an ESS. In particular, $F_y(X_0, Y)$ is maximized by the choice of $Y = Y_0$, and hence $\mathrm{gcd}(X_0, Y)$ is maximized. Since $(X_0, Y) \in H$, $Y < X$. The choice for Y that maximizes $F_y(X_0, Y)$ is $Y = Y_0 := \mathrm{gd}(X_0)$. Indeed, since X_0 is not prime, $Y_0 \geq 2$, and $Y_0 \leq X_0/2 \leq L/2 \leq L/2+1$. Hence, $(X_0, Y_0) \in H$ is an admissible strategy.

However, (X_0, Y_0) is no ESS: Since $Y_0 = \mathrm{gd}(X_0)$, we have $a \in \mathbb{N}$ with $X_0 = aY_0$, and $\mathrm{gcd}(X_0 \pm 1, Y_0) = \mathrm{gcd}(a\,Y_0 \pm 1, Y_0) < Y_0 = \mathrm{gcd}(a\,Y_0, Y_0)$. Hence,

$$F_x(X_0 \pm 1, Y_0) = 1 - 2\,\mathrm{gcd}(X_0 \pm 1, Y_0)/Y_0 > 1 - 2\,\mathrm{gcd}(X_0, Y_0)/Y_0 = F_X(X_0, Y_0).$$

Therefore, there is no Y such that $(X_0, Y) \in H$ is an ESS, if X_0 is not prime.

(b) Next we assume that X_0 is prime, and that there is $Y \in \mathbb{N}$ such that $(X_0, Y) \in H$. Since $1 < Y < L/2 + 1 < X_0$, we conclude that $\mathrm{gcd}(X_0, Y) = 1$. Therefore,

$$F_y(X_0, Y) = 1/X_0 - 1$$

does not depend Y. Furthermore, if $X \neq X_0$, and $(X, Y) \in H$, then $\mathrm{gcd}(X, Y) \geq 1$, and hence

$$F_x(X, Y) = 1 - 2\,\mathrm{gcd}(X, Y)/Y \leq 1 - 2/Y = F_x(X_0, Y).$$

All strategies $(X_0, Y) \in H$ with X_0 prime are ESS according to Definition 13.3.1.

□

Remark 13.3.4 The definition of H is motivated by the fact that no strategy (X, jX) for X prime and $1 < j$ is an ESS: It is $\mathrm{gcd}(X, jX) = X$, and $\mathrm{gcd}(k, jX) \leq k < X$ for $k < X$. Thus,

$$F_x(X, jX) < F_x(k, jX) \quad \text{for } k < X.$$

If the predator period is allowed to be much larger than the prey period, the prey is likely to tend to very short life cycles. The general feeling is that this observation is rather an artifact due to the simplicity of the present toy model than a structure reflecting reality. The simulations of the spatial model support this opinion.

Appendix A
Basic Mathematical Tools

A.1 Basic Definitions from Topology

We recall some basic definitions and theorems from topology. They can be found in any standard introductory book about topology, e.g. [117]. Topology starts from a set X. The topology of X is a subset $\mathfrak{T}$ of the power set $\mathfrak{P}(X)$ satisfying three conditions.

Definition A.1.1 Let X be a set and $\mathfrak{T} \subset \mathfrak{P}(X)$ a subset of the power set of X satisfying

(T1) $\emptyset, X \in \mathfrak{T}$
(T2) finite intersections of sets in $\mathfrak{T}$ are in $\mathfrak{T}$:

$$U_i \in \mathfrak{T}, i = 1, \ldots, n \Rightarrow \cap_{i=1}^n U_i \in \mathfrak{T}$$

(T3) arbitrary unions of sets in $\mathfrak{T}$ are in $\mathfrak{T}$:

$$U_i \in \mathfrak{T}, i \in I, I \text{ an index set } \Rightarrow \cup_{i \in I} U_i \in \mathfrak{T}.$$

Then $(X, \mathfrak{T})$ is called a topological space, and $\mathfrak{T}$ its topology. The sets in $\mathfrak{T}$ are called open sets. A set V is called closed if there is an open set U such that $V = X \setminus U$. A set that is open as well as closed is called clopen.

There are two extreme and important examples. If we choose $\mathfrak{T} = \mathfrak{P}(X)$ then any set is open; since for any $U \subset X$ also $X \setminus U$ is open, any set is also closed. This topology is called the discrete topology. The other extreme is to define $\mathfrak{T} = \{\emptyset, X\}$ Then, the topology is the smallest possible. This topology is called indiscrete.

Often it is not feasible to characterize all open sets directly. In this case, only some of the sets are defined, the so-called basis of the topology, together with a recipe how to generate all open sets from the basis.

K.-P. Hadeler, J. Müller, *Cellular Automata: Analysis and Applications*, Springer Monographs in Mathematics, DOI 10.1007/978-3-319-53043-7

Definition A.1.2 Let $(X, \mathfrak{T})$ be a topological space, $B \subset \mathfrak{T}$. The collection B is called basis (or base) of the topology $\mathfrak{T}$, if each set $U \in \mathfrak{T}$ is a (possibly infinite) union of sets in B.

For example, the set of open intervals in $\mathbb{R}$ is the basis of the standard topology of $\mathbb{R}$. However, even a basis is sometimes not this simple to describe. An even smaller set, a subbasis, from that a basis can be constructed, is defined.

Definition A.1.3 Let B a basis of a topology $\mathfrak{T}$. A set $S \subset B$ is called subbasis of the basis B, if any set $U \in B$ can be represented as a finite intersection of sets in S.

Often enough, the aim is not to find a subbasis for a given topology $\mathfrak{T}$, but we are faced with the converse problem: we are given a collection of sets and aim at a topology such that all these sets are open. If we interpret the collection as a subbasis, we are able to construct the desired topology. A typical example is the case of a metric space (X, d). We want the balls $B_\varepsilon(x) = \{y \in X \,:\, d(x, y) < \varepsilon\}$ to be open. The collection $B = \{B_\varepsilon(x) \,:\, x \in X, \varepsilon > 0\}$ is a (sub)basis of the desired topology. We call a space a metric, topological space, if the topology has the collection B as a basis. Conversely, if a metric can be found that generates a given topology in this sense, then we call this topological space metrizable.

Since the family of open sets is closed with respect to finite intersections and infinite unions, the converse statement holds for closed sets: An intersection of infinitely many closed sets is closed, and a union of finitely many closed sets is closed.

The idea of a topology is to distinguish between points in X using open sets. Two points for which no open set exist that only covers one of them are—from the topological point of view—identical. A topological space that is able to distinguish all points is the Hausdorff space.

Definition A.1.4 A topological space $(X, \mathfrak{T})$ is called a Hausdorff space, if for all $x, y \in X$, $x \neq y$, there are open sets $U_1, U_2 \in \mathfrak{T}$ such that

$$x \in U_1, \quad y \in U_2, \quad U_1 \cap U_2 = \emptyset.$$

Related to the idea to separate points is the concept of connected components.

Definition A.1.5 Let $(X, \mathfrak{T})$ be a topological space, and $A \subset X$ a subset of X. U is called connected or connected component, if there are no disjoint open sets U_1, U_2 such that $A \cap U_1$ and $A \cap U_2$ are non-empty and the union of $U_1 \cap A$ and $U_2 \cap A$ yields A.

Also the proximity of points can be measured via the topology. If two different points can be distinguished by the topology (which is always the case e.g. for Hausdorff spaces), they are considered to be close if many or most open sets that contain one of the points also contain the other point. This idea leads to the definition of accumulation points.

Definition A.1.6 An element $x \in X$ is an accumulation point of a subset $Y \subset X$, if and only if all open sets covering x have a non-empty intersection with Y,

$$x \in U \in \mathfrak{T}, x \in U \Rightarrow U \cap Y \neq \emptyset.$$

The set of all accumulation points of Y is denoted by $\overline{Y}$. A point $x \in Y$ is called isolated point in Y, if it is no accumulation point of $Y \setminus \{x\}$, i.e., $x \notin \overline{Y \setminus \{x\}}$. If we consider a sequence, $\{y_i\}_{i \in \mathbb{N}}$, the accumulation points $\mathcal{A}(\{y_i\}_{i \in \mathbb{N}})$ of the sequence are defined to consist of the common accumulation points of all sets $\{y_i \ : \ i > n\}$ for all $n \in \mathbb{N}$,

$$\mathcal{A}(\{y_i\}_{i \in \mathbb{N}}) = \cap_{n=1}^{\infty} \overline{\{y_i \ : \ i > n\}}.$$

The sequence is said to have a limit point $z = \lim_{i \to \infty} y_i$ if the set of all accumulation points consists only of the single point z, $\mathcal{A}(\{y_i\}_{i \in \mathbb{N}}) = \{z\}$.

If the topology is based on a metric, this convergence is the usual convergence w.r.t. this given metric. In a bounded region of $\mathbb{R}^d$, each sequence has a subsequence that has a limit in $\mathbb{R}^d$. Compactness for topological spaces is a property that conserves such feature in an abstract setting.

Definition A.1.7 Let U_α, $\alpha \in I$ with a possibly infinite index set I, be an arbitrary family of open sets that cover a topological space X,

$$\bigcup_{\alpha \in I} U_\alpha = X.$$

If it is always possible to find a finite subset $\tilde{I} \subset I$ such that the sets U_α with $\alpha \in \tilde{I}$ still cover X,

$$\bigcup_{\alpha \in \tilde{I}} U_\alpha = X, \qquad \tilde{I} \subset I, \qquad |\tilde{I}| < \infty,$$

then the topological space is called compact.

We find indeed, that any sequence in a compact space has a converging subsequence.

We now know how to introduce a topology for one set X. Starting with this topological space, we would like to derive in a natural way other topological spaces. First, we shrink X.

Definition A.1.8 Let $(X, \mathfrak{T})$ be a topological space, and $Y \subset X$. The trace topology $\tilde{\mathfrak{T}}$ on Y is defined by

$$\tilde{\mathfrak{T}} = \left\{ \tilde{U} \,\middle|\, \tilde{U} = U \cap Y, \text{ where } U \in \mathfrak{T} \right\}.$$

Next we construct a larger set from X. Assume that we have infinite many copies of this set X_α, $\alpha \in I$, where I denotes an infinite index set and $X_i = X$. The product set is defined as all possible tuples, indexed by I,

$$\times_{\alpha\in I} X_i := \{y = (x_\alpha)_{\alpha\in I} \mid x_\alpha \in X_\alpha = X\}.$$

How can we use the topology on X in order to define a topology on $\times_{\alpha\in I} X_\alpha$? The product topology is the answer to this problem.

Definition A.1.9 Let $(X, \mathfrak{T})$ be a topological space. Let I be a (possibly infinite) index set, and X_α, $\alpha \in I$, identical copies of X. The product topology $\tilde{\mathfrak{T}}$ of the product set $\times_{\alpha\in I} X_\alpha$ is defined as

$$\tilde{\mathfrak{T}} = \left\{ U = \times_{\alpha\in I} U_\alpha \,:\, U_\alpha \in \mathfrak{T},\; U_\alpha = X_\alpha \text{ for all but a finite number of indices} \right\}.$$

This definition somehow avoids the problems connected with infinite products: We force almost all indices to be associated with trivial sets open sets, $U_\alpha = X$. Only a finite number of indices are allowed to contain nontrivial information. These sets are called cylindric sets. This construction is the appropriate one to yield a compact space if we start with compact spaces. This is the Theorem of Tychonov.

Theorem A.1.10 *If X_α are compact topological spaces, $\alpha \in I$ with I a possibly infinite index set, then the product space $\times_{\alpha\in I} X_\alpha$ is again compact.*

The proof can be found, e.g., in [185].

The last point we want to mention is the way how to identify classes of topological spaces. If we find a bijective map ψ between two topological spaces $(X, \mathfrak{T})$ and $(Y, \tilde{\mathfrak{T}})$ such that $\psi(\mathfrak{T}) = \tilde{\mathfrak{T}}$ and $\psi^{-1}(\tilde{\mathfrak{T}}) = \mathfrak{T}$, then, basically, the two topological spaces are identical.

Definition A.1.11 Let $(X, \mathfrak{T})$ and $(Y, \tilde{\mathfrak{T}})$ be two topological spaces. A map $\psi : X \to Y$ such that the preimage of an open set is open is called continuous.

If the map ψ is bijective, and ψ and ψ^{-1} both are continuous, then ψ is called a topological homeomorphism and $(X, \mathfrak{T})$, $(Y, \tilde{\mathfrak{T}})$ homeomorphic topological spaces.

The following proposition is quite useful if we construct homeomorphisms between metric compact topological spaces. It shows that it is sufficient to define a bijection between the two spaces (as sets) and, in addition, to prove that this map is continuous. In this case, the inverse of the map is automatically continuous and thus the map is a homeomorphism.

Proposition A.1.12 *Let $(X, \mathfrak{T})$ and $(Y, \tilde{\mathfrak{T}})$ be two metric, compact, topological spaces. Let $\psi : X \to Y$ be a bijective, continuous function. Then ψ is already a homeomorphism, i.e., also the inverse of ψ is continuous.*

Proof Assume that ψ^{-1} is not continuous at $y \in Y$. Then there is a sequence $\{y_n\}_{n\in\mathbb{N}}$ converging to y while $x_n = \psi^{-1}(y_n)$ does not converge to $x = \psi^{-1}(y)$. As $(X, \mathfrak{T})$ is metric, there is $U \in \mathfrak{T}$ with $x \in U$ and a subsequence $x_{n'} \notin U$.

Since X is compact, there is a converging subsequence of $\{x_{n''}\}_{n''\in\mathbb{N}}$ of $x_{n'}$. As $x_{n'} \notin U$, $x_{n''} \to \hat{x} \notin U$. The function ψ is continuous, hence $y_{n''} = \psi(x_{n''}) \to \hat{y} = \psi(\hat{x})$. However, $y_{n''}$ is subsequence of y_n for which we already know $y_n \to y = \psi(x) \neq \psi(\hat{x}) = \hat{y}$. This is a contradiction. □

At the end of this section, we prove the theorem that all metric (or metrizable) Cantor sets are homeomorphic. The proof of this theorem has been already sketched in Sect. 3.1.2. However, some details are used in Theorem 6.3.14, therefore we present the complete proof here. As a prelude, we show a lemma about a way to decompose clopen sets. Recall the definitions of clopen, perfect, and totally disconnected stated in Sect. 3.1.2.

Lemma A.1.13 *Consider a metric, compact topological space X that is perfect and totally disconnected. Any clopen, non-empty set V in X can be represented as the union of two clopen, non-empty, and disjoint sets V_1, V_2,*

$$V = V_1 \cup V_2, \quad V_1 \neq \emptyset, \quad V_2 \neq \emptyset, \quad V_1 \cap V_2 = \emptyset.$$

Proof Since a perfect set contains no isolated points, any open, non-empty set consists of more than one point. Let $x, y \in V$, $x \neq y$ two different points in the clopen set V. Let δ be the distance between x and y, $\delta = d(x, y)$. Then,

$$U_1 = V \cap \{z \in X : d(x, z) < \delta/2\}$$

is non-empty, open and covers x. There is then a clopen set $V_1 \subset U_1 \subset V$ that covers x (and is thus non-empty). The set $V_2 = V \setminus V_1$ is also clopen, non-empty ($y \in V_2$), has a void intersection with V_1, and $V = V_1 \cup V_2$. □

Theorem A.1.14 *Any metric, compact topological space X that is perfect and totally disconnected is homeomorphic to $\mathfrak{C}_{1/3}$.*

Proof Step 1: First partition of X

For every point $x \in X$ there is an open ball of diameter one. Within this ball, there is a clopen set $V(x)$ covering x. As X is compact, we find a finite subcover $W_0^1, \ldots, W_0^m$. We aim at a cover by disjoint clopen sets of diameter less than one, where the number of these sets is a power of 2. We utilize that the W_i are clopen, and finite unions and sections of clopen sets remain clopen. Let

$$U_0 = W_0^1, \qquad U_i = W_0^i \setminus \left(\bigcup_{j=0}^{i-1} U_j\right), \qquad i = 2, \ldots, m.$$

We dismiss all U_i that are empty. The number of the remaining U_i will in general be no power of two. We assume without restriction $U_1 \neq \emptyset$. Let $n_1 \in \mathbb{N}$ chosen as the minimal number n_1 such that the number of non-empty U_i is less or equal 2^{n_1}. We now split U_1 via Lemma A.1.13 into so many non-empty, disjoint and clopen sets, such that the number of the clopen sets covering X is exactly 2^{n_1}.

A number $i \in \{0, 1, \ldots, 2^{n_1} - 1\}$ can be represented as a finite series $(b_0, \ldots, b_{n_1-1})$, $b_i \in \{0, 1\}$ (via the binary representation of a natural number). We use this binary representation to index the sets of the clopen coverage we did construct, and rename our sets in $I^{(1)}_{(b_0,\ldots,b_{n_1}-1)}$. All in all, we constructed a covering of X, consisting of 2^{n_1} clopen and pairwise disjoint sets $I^{(1)}_{(b_0,\ldots,b_{n_1}-1)}$ with a diameter less or equal one,

$$\mathfrak{I}^{(1)} = \{I^{(1)}_{(b_0,\ldots,b_{n_1}-1)} : (b_0, \ldots, b_{n_1-1}) \in \{0, 1\}^{n_1}\}, \qquad X = \bigcup_{I \in \mathfrak{I}^{(1)}} I.$$

Step 2: Recursive refinement of the partition

Let $\mathfrak{I}^{(l)}$ be a split of X into 2^{n_l} disjoint, clopen sets with diameter less or equal 2^{1-l},

$$\mathfrak{I}^{(l)} = \{I^{(l)}_{(b_0,\ldots,b_{n_l}-1)} : (b_0, \ldots, b_{n_l-1}) \in \{0, 1\}^{n_l}\}.$$

We construct the partition $\mathfrak{I}^{(l+1)}$ into disjoint, clopen sets with diameter less or equal 2^{-l} such that $\#\mathfrak{I}^{(l)} = 2^{n_{l+1}}$, where $n_{l+1} \in \mathbb{N}$ is chosen appropriately.

In order to construct this partition, we first split the set $I^{(l)}_{(0,0,\ldots,0)}$ with the same procedure like we did for the complete set in step 1: cover $I^{(l)}_{(0,0,\ldots,0)}$ by open sets $B_{2^{-l}}(x)$, $x \in I^{(l)}_{(0,0,\ldots,0)}$ and replace for each $x \in I^{(l)}_{(0,0,\ldots,0)}$ the ball by a clopen set that is contained in the ball and that still covers x. Since $I^{(l)}_{(0,0,\ldots,0)}$ is compact, a finite number of these clopen sets are sufficient to cover $I^{(l)}_{(0,0,\ldots,0)}$; from these sets we are able to find a finite covering of $I^{(l)}_{(0,0,\ldots,0)}$ by clopen, non-empty and disjoint sets. Let the number of these sets be denoted by $m_{(0,0,\ldots,0)}$.

In the same way, we split all members of $I^{(l)}_{(b_0,\ldots,b_{n_l}-1)} \in \mathfrak{I}^{(l)}$ into clopen, non-empty and disjoint sets with diameter less or equal to 2^{-l}. The number of these sets needed to split $I^{(l)}_{(b_0,\ldots,b_{n_l}-1)}$ is denoted by $m_{(b_0,\ldots,b_{n_l}-1)}$. We find the smallest power of two that is larger or equal all $m_{(b_0,\cdot,b_{n_l}-1)}$,

$$2^{m_{l+1}-1} < m_{(b_0,\ldots,b_{n_l}-1)} \leq 2^{m_{l+1}} \quad \forall (b_0, \ldots, b_{n_l-1}) \in \{0, 1\}^{n_l}.$$

We may decompose sets such that in the end each member of $\mathfrak{I}^{(l)}$ is split into exactly $2^{m_{l+1}}$ sets. Let $n_{l+1} = n_l + m_{l+1}$. We may number all these sets, and obtain

$$\mathcal{I}^{(l+1)} = \{I^{(l+1)}_{(b_0,\ldots,b_{n_{l+1}}-1)} : (b_0, \ldots, b_{n_{l+1}-1}) \in \{0, 1\}^{n_{l+1}}\}.$$

Then, $\mathcal{I}^{(l+1)}$ consist of sets with the four properties

(1) $I \in \mathcal{I}^{(l+1)}$ is clopen
(2) $I, \tilde{I} \in \mathcal{I}^{(l+1)}, I \neq \tilde{I} \Rightarrow I \cap \tilde{I} = \emptyset$

(3) $X = \cup_{I \in \mathcal{I}^{(l)}} I$
(4) $I \in \mathcal{I}^{(l+1)} \Rightarrow \text{diam}(I) \leq 1/2^l$

Step 3: *Construction of the map $\psi : \Sigma_2 \to X$*
Let $(b_0, b_1, \ldots) \in \Sigma_2$. Since the parts $\mathfrak{I}^{(l)}$ are nested, X is compact, and the diameter of the splits tend to zero for $l \to \infty$, the set

$$\bigcap_{l=0}^{\infty} I^{(l)}_{(b_0,\ldots,b_{n_l-1})} = \{x\}$$

consists of exactly one point. We define $\psi : \Sigma_2 \to X$ by assigning the point x to the sequence $\{b_i\}_{i \in \mathbb{N}_0}$ if and only if $\cap_{l=0}^{\infty} I^{(l)}_{(b_0,\ldots,b_{n_l-1})} = \{x\}$. The map ψ is bijective due to its construction.
Step 4: *ψ is continuous*
Let $b_m, b^* \in \Sigma_2$ and $b_n \to b^*$ for $n \to \infty$ in Σ_2. Then, for each n_l (associated with the a partition $\mathcal{I}^{(l)}$) there is m_l such that the first n_l entries of b_m agree with those of b^* for $m > m_l$. Hence,

$$\psi(b^*), \quad \psi(b_m) \in I^{(l)}_{(b_0,\ldots,b_{n_l-1})} \qquad \text{for } m > m_l$$

and thus

$$d(\psi(b_m), \psi(b^*)) \to 0.$$

Step 5: *ψ^{-1} is continuous*
One may either use the theorem that a bijective map between compact topological spaces that is continuous necessarily has a continuous inverse (Proposition A.1.11). Or, one may prove the continuity of the inverse directly. The argument parallels that of step 4. □

A.2 Basic Algebraic Theory

Some elementary facts about groups, rings and fields are collected in the following sections.

A.2.1 *Group Theory*

The statements about groups in this section can be found in most textbooks, for example in [8, 100].

Definition A.2.1 Let G be a set and $\circ$ a binary operator on G,

$$\circ : G \times G \to G, \quad (g_1, g_2) \mapsto g_1 \circ g_2$$

that satisfies the following conditions:

(1) (associative law) $(g_1 \circ g_2) \circ g_3 = g_1 \circ (g_2 \circ g_3)$.
(2) (unit element) There is $e \in G$ such that $g \circ e = e \circ g = g$ for all $g \in G$.
Then $(G, \circ)$ is called a semigroup. If additionally,
(3) (inverse element) For every $g_1 \in G$ there is a $g_2 \in G$ such that

$$g_1 \circ g_2 = e = g_2 \circ g_1.$$

is fulfilled, then $(G, \circ)$ is called a group.
If the elements commute, i.e., if for all $g_1, g_2 \in G$ holds

$$g_1 \circ g_2 = g_2 \circ g_1,$$

then the group is called commutative or Abelian.

In most texts on group theory the operation sign is omitted, sometimes one uses a multiplication sign $\cdot$, and the unit element is sometimes called '1'. For Abelian groups often an additive notation $g_1 + g_2$ is used, and consequently the unit element is denoted by 0.

Example A.2.2

(a) The integers with addition forms the Abelian group $(\mathbb{Z}, +)$.
(b) If we add integers modulo n, we obtain the finite cyclic group $(\mathbb{Z}_n, +)$.

In a similar spirit as we defined topological product spaces, we define product groups.

Definition A.2.3 Let $(G, \cdot)$ and $(H, *)$ be groups. The product group $(G \times H, \circ)$ is defined via

$$(g_1, h_1) \circ (g_2, h_2) = (g_1 \cdot g, \ h_1 * h_2).$$

It is easy to proof that $(G \times H, \circ)$ is a group. If G and H are Abelian groups, often the notation $G \oplus H$ is used for the product group. For example, $\mathbb{Z}^2 = \mathbb{Z} \oplus \mathbb{Z}$.

Another concept parallels topological homeomorphisms.

Definition A.2.4 A map $\varphi : G \to H$ between two groups $(G, *)$ and $(H, \cdot)$ is called homomorphism if it respects the group operation, $\varphi(g_1 * g_2) = \varphi(g_1) \cdot \varphi(g_2)$. The groups are called isomorphic, if there is a bijective homomorphism.

Now we consider a (not necessarily Abelian) group G with subgroup H.

Definition A.2.5 A subgroup H of a group G is a subset of G that is closed w.r.t. the group action and w.r.t. taking the inverse.

We consider (right) cosets

$$Hg = \{h \circ g \,:\, h \in H\} \qquad \text{for } g \in G.$$

These cosets form a partition of G in sets of equal cardinality.

Proposition A.2.6 *Let $g_1, g_2 \in G$ be two arbitrary elements. If $Hg_1 \cap Hg_2 \neq \emptyset$, then $Hg_1 = Hg_2$. Moreover, $|Hg_1| = |Hg_2|$.*

Proof If $Hg_1 = Hg_2$, then there are $h_1, h_2 \in H$ such that $h_1 g_1 = h_2 g_2$, and hence $g_1 g_2^{-1} = h_1^{-1} h_2 \in H$. Let $g \in Hg_2$, then $g = hg_2$ for some $h \in H$. Thus, $g = hg_2 = hh_2^{-1}h_1h_1^{-1}h_2g_2 = hh_2^{-1}h_1g_1g_2^{-1}g_2 = hh_2^{-1}h_1g_1 \in Hg_1$. Similarly, $g \in Hg_1$ implies $g \in Hg_2$. Thus, $Hg_1 \cap Hg_2 \neq \emptyset$ forces the sets Hg_1 and Hg_2 to be identical.

In order to show $|Hg_1| = |Hg_2|$ we define $\varphi : Hg_1 \to Hg_2$, $hg_1 \mapsto hg_2$. Clearly, φ is surjective. If $h_1g_1, h_2g_1 \in Hg_1$ and $\varphi(h_1g_1) = \varphi(h_2g_1)$, then $h_1g_2 = h_2g_2$ and hence $h_1 = h_2$. φ is also injective. Therefore, $|Hg_1| = |Hg_2|$. □

We define the index of a subgroup.

Definition A.2.7 The index of a subgroup H of G (in symbols: $[G : H]$) is the number of (right) cosets.

Note that we may also work with left cosets gH. However, the index (the number of cosets) is the same for left- and right cosets. The considerations above lead to the theorem of Lagrange.

Theorem A.2.8 *If G is a finite group and H a subgroup, then $[G : H] = |G|/|H|$.*

If a subgroup H has the property that $gH = Hg$ for all $g \in G$, it is called normal subgroup. A normal subgroup allows to construct a new group, the factor group.

Definition A.2.9 Let $(G, \cdot)$ a group, and N a normal subgroup. The factor group G/N is the set of cosets $\{Ng \,:\, g \in G\}$ together with

$$(Ng_1) * (Ng_2) = N(g_1 \cdot g_2).$$

Note that the left- and the right cosets coincide for a normal subgroup. It is easy to check that $(G/N, *)$ is a group, indeed. One possibility to obtain normal subgroups is via homomorphisms. If $\varphi : G_1 \to G_2$ is a group-homomorphism, then $N = \{g \in G_1 \,:\, \varphi(g) = e\}$ is a normal subgroup of G_1 (also called the kernel of φ). Surjective homomorphisms indicate a strong connection between groups.

Theorem A.2.10 *Let φ be a surjective homomorphism from $(G_1, \cdot)$ to $(G_2, \odot)$. Let $N = \{g \in G_1 \,:\, \varphi(g) = e\}$ be the kernel of the homomorphism φ. Then, $(G_2, \odot)$ is isomorphic to the factor group G/N.*

A group can be defined in various ways, e.g., as a permutation group (all permutations of $1, 2, 3, 4, 5$), as a transformation group (all rotations and reflections that carry a regular hexagon into itself). For our purpose defining a group by generators is most convenient. Therefore the concept of subgroups is useful: If $(G, \circ)$ is a group, a subset $H \subset G$ is called subgroup, if $(H, \circ)$ is again a group. Hence, $(\{e\}, \circ)$ is a subgroup for any group.

Definition A.2.11 Let G be a group, $\sigma \subset G$ a subset. Define the collection of all subgroups, $\mathcal{H} = \{H : H \text{ is subgroup of } G\}$, and $\mathcal{H}_\sigma = \{H : H \in \mathcal{H}, \ \sigma \subset H\}$. Then

$$< \sigma >= \cap_{H \in \mathcal{H}_\sigma} H$$

is called the (sub)group generated by σ.

It is straightforward to show that $< \sigma >$ is a subgroup, as an intersection of subgroups. Immediate implications are: $\mathcal{H} \neq \emptyset$ since $\{e\}, G \in \mathcal{H}$; $< \sigma >$ is a subgroup, uniquely determined by the set σ; $< \emptyset >= \{e\}$.

Definition A.2.12 A group G is finitely generated, if there is a finite subset $\sigma \subset G$ such that $G =< \sigma >$.

Obviously, $\mathbb{Z}^d$ is finitely generated. Finitely generated Abelian groups have a particularly simple structure as stated in the structure theorem.

Theorem A.2.13 *If $(G, +)$ is a finitely generated Abelian group, then we find numbers $m, n_1, \dots, n_k \in \mathbb{N}$, such that $(G.+)$ is isomorphic with $\oplus_{i=1}^m \mathbb{Z} \oplus_{\ell=1}^k \mathbb{Z}_{n_\ell}$.*

Another prominent example for a finitely generated group is the free group over n symbols.

Example A.2.14 Construction of the free group over n symbols: Let $\sigma = \{s_1, \dots, s_n\}$ be a finite set, and denote by $\sigma^{-1} = \{s_1^{-1}, \dots, s_n^{-1}\}$ the formal inverse elements. Consider the set of all finite words over $\sigma \cup \sigma^{-1}$ together with the empty word, $(\sigma \cup \sigma^{-1})^*$ (notation as in Sect. 7.1). The concatenation of two words is defined as binary operator $\cdot$ on $\sigma \cup \sigma^{-1}$,

$$\cdot : (\sigma \cup \sigma^{-1})^* \times (\sigma \cup \sigma^{-1})^* \to (\sigma \cup \sigma^{-1})^*, \quad (w_1, w_2) \mapsto w_1 \cdot w_2 = w_1 w_2.$$

Two words w_1 and w_2 are considered to be equivalent if they can be transformed into each other by a finite number of insertions/deletions of pairs $s_i s_i^{-1}$ respectively $s_i^{-1} s_i$. For example,

$$s_1 s_3^{-1} s_2 s_2^{-1} s_3 s_1 \sim s_1 s_1.$$

If $w_1 \sim w_2$ and $v_1 \sim v_2$, then $(w_1 \cdot v_1) \sim (w_2 \cdot v_2)$. Therefore, the product carries over to equivalence classes, $\cdot : (\sigma \cup \sigma^{-1})^* / \sim \times (\sigma \cup \sigma^{-1})^* / \sim \to (\sigma \cup \sigma^{-1})^* / \sim$ is well defined. It is straightforward to show that $((\sigma \cup \sigma^{-1})^* / \sim, \cdot)$ is a group (with the empty word as the neutral element). This group is called the finitely generated group (over the set σ), and often denoted by $\mathbf{F}(\sigma)$.

The free group is "very large", it consists of all words formed from the alphabet of generators. However, we may define smaller groups by additional rules like $\sigma_1^n = 1$ or $\sigma_i\sigma_j = \sigma_j\sigma_i$. These rules make some words collapse into one, and the group gets "smaller". In order to introduce this concept formally, we consider a finitely generated group $G =< \sigma >$. There is a surjective homomorphism from the free group over the set σ into G, $\varphi : \mathbf{F}(\sigma) \to G$. This homomorphism is uniquely defined by $\varphi(s) = s$ for all $s \in \sigma$. Up to an isomorphism G equals $\mathbf{F}(\sigma)/N$, where $N = \{u \in \mathbf{F}(\sigma) : \varphi(u) = e\}$ is the kernel of φ (in particular, N is a normal subgroup). Hence, if we know $\mathbf{F}(\sigma)$ and N, we know G. We again use the idea of finitely generated groups to characterize N. This time, we are not interested in any subgroup, but only in normal subgroups. Let $r \subset \mathbf{F}(\sigma)$, and

$$R(r) = \cap\{N \text{ is normal subgroup of } \mathbf{F}(\sigma) : N \supset r\}.$$

Then, $R(r)$ is a normal subgroup (generated by r). If we find a finite set r, such that $N = R(r)$, then the finitely generated group G is called finitely represented.

Definition A.2.15 Let $G =< \sigma >$ be a finitely generated group. Let N be the normal subgroup of $\mathbf{F}(\sigma)$, such that G is isomorphic with $\mathbf{F}(\sigma)/N$. If there is a finite subset $r \subset \mathbf{F}(\sigma)$ that generates the normal subgroup N, $N = R(r)$, we call G a finitely represented group, and write

$$G =< \sigma : r > .$$

For example, $\mathbb{Z}^2 =< \{\sigma_1, \sigma_2\} | \{\sigma_1, \sigma_2\sigma_1^{-1}\sigma_2^{-1}\} >$. Often, it is more intuitive to drop the inner brackets indicating sets, and to write elements $R \in r$ as equations $R = e$ that can be rearranged, e.g., $\mathbb{Z}^2 =< \sigma_1, \sigma_2 : \sigma_1\sigma_2 = \sigma_2\sigma_1 >$. For the free group itself, the normal subgroup N is trivial, and the set r can be empty. Slightly abusing notation, we often write $\mathbf{F}(\sigma) =< \sigma >$.

Note that not all groups are finitely generated, and not all finitely generated groups can be finitely represented. Moreover, the set r for a finitely represented group is by no means unique, even if we demand $|r|$ to be minimal. We can observe this fact in the next example.

Example A.2.16 Let us consider the symmetry group of the isosceles triangle. Denote by σ the rotation by $2\pi/3$ and by τ the reflection at the y-axis. Then, of course, $\sigma^3 = e$ and $\tau^2 = e$. Furthermore, we find $\tau\sigma = \sigma^2\tau$ (see Fig. A.1). Indeed, we may define

$$D_3 =< \sigma, \tau : \sigma^3 = e, \ \tau^2 = e, \ \tau\sigma = \sigma^2\tau > .$$

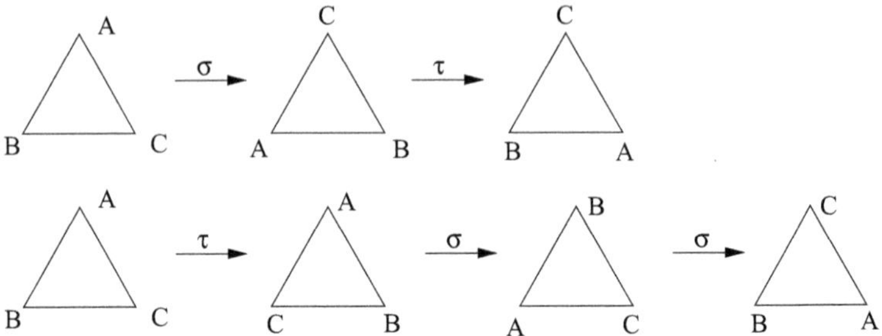

Fig. A.1 $\tau\sigma = \sigma^2\tau$ for D_3

If we multiply $\tau\sigma = \sigma^2\tau$ by $\tau\sigma$, we find $(\tau\sigma)^2 = e$, and hence

$$D_3 = <\sigma, \tau\ :\ \sigma^3 = e,\ \tau^2 = e,\ (\tau\sigma)^2 = e> .$$

A short computation shows (see also Table 2.1) that

$$D_3 = \{1, \sigma, \sigma^2, \tau, \tau\sigma, \tau\sigma^2\}.$$

A.2.2 Ring Theory

We only need very few facts about rings and fields. A ring is an object that resembles $\mathbb{Z}$: multiplication as well as addition is defined. We do require an inverse w.r.t. addition, but not w.r.t. multiplication.

Definition A.2.17

(a) A ring $(R, +, \cdot)$ consists of a non-empty set R, an addition $+ : R \times R \to R$, and a multiplication $\cdot\ : R \times R \to R$ that satisfy

(1) $(R, +)$ is an Abelian group
(2) $(R, \cdot)$ is a semigroup
(3) the two operations act distributively,

$$a \cdot (b + c) = a \cdot b + a \cdot c, \qquad (b + c) \cdot a = b \cdot a + c \cdot a.$$

The neutral element of the addition is called 0.

(b) A commutative ring is a ring with a commutative multiplication, i.e. $a \cdot b = b \cdot a$ for all $a, b \in R$.
(c) A ring that has a multiplicative semigroup with a neutral element is called ring with neutral element. The neutral element is denoted by 1.
(d) A ring is called integral domain, if from $ab = 0$ it follows that either $a = 0$ or $b = 0$.

(d) A finite ring is a ring where the set R is finite.
(e) Given a ring with neutral element, an element $a \in R$ is denoted unit element, if it has an inverse $b \in R$, $a \cdot b = b \cdot a = e$.

Example A.2.18

(a) $\mathbb{Z}$ together with the natural addition and multiplication is a commutative ring with neutral element.
(b) $\mathbb{Z}_n$ together with addition and multiplication modulo n is a finite commutative ring with neutral element.

Definition A.2.19

(a) A subring of a ring $(R, +, \cdot)$ is a subset $R' \subset r$ that is closed w.r.t. addition and multiplication. If the ring R has a neutral element, then this element is contained in any subring R'.
(b) Given two rings $(R, +, \cdot)$, $(S, \oplus, \circ)$ and a map $\phi : R \to S$. ϕ is called homomorphism, if it preserves addition and multiplication and (if R is a ring with neutral element) the neutral element,

$$\phi(a + b) = \phi(a) \oplus \phi(b), \quad \phi(a \cdot b) = \phi(a) \circ \phi(b),$$

and, if $1 \in R$, $s \circ \phi(1) = \phi(1) \circ s = s$. The map ϕ is called isomorphism if it is bijective.

Definition A.2.20 Given a ring R, the set $R[X, X^{-1}]$ contains all formal Laurent polynomials $p(X)$, i.e., for p given there is $n \in N$

$$p(X) = \sum_{i=-n}^{n} r_i X^i$$

with $r_i \in R$.

$R[X, X^{-1}]$ is equipped in a natural way with addition and multiplication: we identify a Laurent polynomial $p(X) = \sum_{i=-n}^{n} r_i X^i$ with a Laurent series by defining $r_i = 0$ for $i > n$. Then,

$$p(X) + q(X) = \sum_{i \in \mathbb{Z}} r_i X^i + \sum_{i \in \mathbb{Z}} s_i X^i = \sum_{i \in \mathbb{Z}} (r_i + s_i) X^i$$

and

$$p(X)\, q(X) = \left(\sum_{i \in \mathbb{Z}} r_i X^i \right) \left(\sum_{i \in \mathbb{Z}} s_i X^i \right) = \sum_{i \in \mathbb{Z}} \left(\sum_{j \in \mathbb{Z}} r_i\, s_{j-i} \right) X^i.$$

It is straightforward to show that $R[X, X^{-1}]$ is again a ring.

This definition can be extended to polynomials in several independent variables $X \in \mathbb{R}^d$ using multi-index notation. Let $\alpha \in \mathbb{Z}^d$, we define $|\alpha| = \sum_{i=1}^d |\alpha_i|$ and $X^\alpha = \prod_{i=1}^d X_i^{\alpha_i}$. Then,

$$R[X, X^{-1}] = \left\{ p(X) \,:\, \exists n \in \mathbb{N} \text{ such that } p(X) = \sum_{|\alpha| \le n} r_\alpha X^\alpha \right\}.$$

Also $R[X, X^{-1}]$ is a ring. We call $R[X, X^{-1}]$ the (d-dimensional) Laurent polynomial ring (over R).

Note that we work with Laurent polynomials, while the standard definition of the polynomial ring $R[X]$ only incorporates polynomials (no negative power of the dummy variable).

Proposition A.2.21 *If R is commutative ring and an integral domain, then the polynomial and Laurent polynomial rings in one or several variables are integral domains.*

Proof First consider the (Laurent) polynomial ring over one independent variable X. Let $p(X) = \sum_{i=-n}^n a_i X^i$, $q(X) = \sum_{j=-n}^n b_i X^i$ and $p(X)q(X) = 0$. We may multiply the expression $p(X)q(X)$ by X^{2n}, and find $\tilde{p}(X) = x^n p(X) \in R[X]$, $\tilde{q}(X) = x^n q(X) \in R[X]$. Thus, we have two polynomials with a product that is zero. If $\text{grad}(\hat{p}) = k$ and $\text{grad}(\hat{q}) = \ell$, then (as R is an integral domain), $\text{grad}(\hat{p}\hat{q}) = k + \ell$. Thus, $\hat{p}(X)\hat{q}(X) = 0$ is only possible if either $\hat{p} = 0$ or $\hat{q} = 0$.

As the polynomial rings over several variables are defined recursively, the result follows per induction. □

Proposition A.2.22 *If R is a commutative ring with one and an integral domain. Let $p, q \in R[X, X^{-1}]$, $X \in \mathbb{R}^d$. Then,*

$$p(X)q(X) = X^r \quad \Rightarrow \quad p(X) = aX^n, \quad q(X) = a^{-1}X^m$$

with $n + m = r$ and some $a \in R$, $a \neq 0_R$.

Proof We use induction over d.
Induction base, $d = 1$.
If $a_u X^{n_1}$ is the term with lowest and $a_o X^{n_2}$ that with largest degree in $p(X)$ and, similarly, $b_u X^{m_1}$ the term with smallest and $b_o X^{m_2}$ the term with largest degree in $q(X)$, then the term with smallest resp. largest degree in $p(X)q(X)$ reads

$$a_u b_u X^{n_1 + m_1}, \quad a_o b_o X^{n_2 + m_2}.$$

As $p(X)q(X) = X^r$, this is only possible if $n_1 + m_1 = n_2 + m_2 = r$. As $n_1 \le n_2$ and $m_1 \le m_2$ we find $n_1 = n_2$, $m_1 = m_2$, and thus the assertion follows.
Induction step.
We may write $R[X, X^{-1}]$ with $X \in \mathbb{R}^d$ as $(R[Y, Y{-}1])[Z, Z^{-1}]$ for $Y \in \mathbb{R}^{d-1}$ and $Z \in \mathbb{R}^1$. The polynomial ring over an integral domain is again an integral domain.

Hence, $p(X) = \sum p_i(Y)Z^i$, and $q(X) = \sum q_i(Y)Z^i$, where $p_i, q_i \in R[Y]$, which is an integral domain. Thus the case $d = 1$ shows that already

$$p = p_0(Y)Z^i, \quad q = q_0(Y)Z^j,$$

where $p_0(Y)q_0(Y) = 1Y^0$. As $Y \in \mathbb{R}^{d-1}$, the proposition follows. □

Proposition A.2.23 *Let $p(X) \in R[X, X^{-1}]$ a Laurent polynomial over a commutative ring R that is an integral domain, then $p(X)$ is a unit if and only if*

$$p(X) = aX^z$$

for some $a \neq 0_R$ and $z \in \mathbb{Z}$.

Proof ⇐: If $p(X) = aX^z$, then $q(X) = a^{-1}X^{-Z} \in R[X, X^{-1}]$ is the inverse element, and thus $p(X)$ is a unit in the Laurent polynomial ring.

⇒: Let $p(X) \in R[X, X^{-1}]$. If $p(X)$ is a unit, then we find $q(X) \in R[X, X^{-1}]$ such that

$$p(X)q(X) = 1_R.$$

As we may write $p(X) = X^{-n}\hat{p}(X)$, $q(X) = x^{-m}\hat{q}(X)$, where $p(X)$ and $q(X)$ are polynomials $\hat{p}(X) = \sum_{i=0}^{k} a_i X^i$, $\hat{q}(X) = \sum_{i=0}^{l} b_i X^i$, we find

$$\hat{p}(X)\hat{q}(X) = X^{n+m}.$$

Thus, with the proposition above, we conclude $\hat{p}(X) = aX^n$ and thus $p(X) = aX^z$ for some $z \in \mathbb{Z}$. □

Remark A.2.24 The similar result holds for Laurent polynomial rings with several independent variables $R[X_1, X_1^{-1}, .., X_n X_n^{-1}]$ as these rings are defined in a recursive way. Thus, if $p(X) \in R[X, X^{-1}, .., X_n X_n^{-1}]$, then $p(X_1, .., X_n)$ is a unit if and only if

$$p(X_1, .., X_n) = a \prod_{i=1}^{n} X_i^{z_i}$$

for some $a \neq 0_R$ and $z_i \in \mathbb{Z}$.

A.2.3 *Fields*

A field resembles $\mathbb{Q}$, $\mathbb{R}$ or $\mathbb{C}$: it is a ring, where—apart of a zero element—all elements have an inverse w.r.t. multiplication.

Definition A.2.25 Let $(F, +, \cdot)$ be a commutative ring, $F^\times = R \setminus \{0\}$. If $(F^\times, \cdot)$ is a group, then $F$is called field.

Apart of the well known examples $\mathbb{Q}$, $\mathbb{R}$, and $\mathbb{C}$, we use finite fields, i.e., fields with $|F| < \infty$. An example of particular interest for us is $\mathbb{Z}_p$ for p prime. These cyclic groups can be equipped with an addition, and then become finite fields.

Proposition A.2.26 *Let $\mathbb{F}_p = \mathbb{Z}_p$, the cyclic group. If we define addition and multiplication in the natural way, then $\mathbb{F}_p$ is a field in case of p prime.*

Proof Clearly, $(\mathbb{Z}_p, +, \cdot)$ is an Abelian ring. What remains to show is that $(\mathbb{Z}_p^\times, \cdot)$ is a group. The associative law and the neutral element are again trivial. It remains to show that for any $z \in Z_p^\times$ there is a $z' \in \mathbb{Z}_p^\times$ such that $z \cdot z' = 1$. Therefore, we first note that $0 \notin z\mathbb{Z}_p^\times$: If there is $z' \in \mathbb{Z}_p^\times$ such that $z \cdot z' = 0$, then (take the multiplication this time in $\mathbb{Z}$)

$$\exists \ell \in \mathbb{Z} : zz' = \ell p.$$

As p is prime, this implies that either z or z' can be divided by p, which in turn contradicts $z, z' \in \{1, \ldots, p-1\}$. Furthermore, there are no different numbers $z', z'' \in \mathbb{Z}_p^\times$ such that $z \cdot z' = z \cdot z''$: If this equality is true, then (take again the multiplication in $\mathbb{Z}$)

$$\exists \ell \in \mathbb{Z} : zz' - zz'' = z(z' - z'') = \ell p.$$

As z is no multiple of p, $z' - z''$ is a multiple of p, such that $z' = z''$ in $\mathbb{Z}_p$, contradicting the assumption $z' \neq z''$ in $\mathbb{Z}_p$.

Hence, $z' \mapsto z \cdot z'$ is a permutation of $\mathbb{Z}_p^\times$. Thus, there is $z' \in \mathbb{Z}_p^\times$, such that $z \cdot z' = 1$. □

It is possible to show that a field F with $|F| = |F_p|$ already is isomorphic to F_p. There is basically only one field with a given prime number of elements.

A.3 Basic Measure Theory

Why would we need measure theory for cellular automata? For instance, we want to make statements about the sizes of basins of attraction. So we must have a measure and a way to determine whether a basin is measurable. An instructive, short introduction into measure theory can be found, e.g., in the book by Rudin [150].

We want to work with intersections and unions of sets which we will call measurable. Hence we need families of sets that are closed under such operations. The most important family is the σ-algebra.

Definition A.3.1 Let X be a non-empty set and $\mathfrak{P}(X)$ its power set. A family of subsets $\mathfrak{B} \subset \mathfrak{P}(X)$ is called a σ-algebra if it has the following properties:

(σ 1) $\emptyset \in \mathfrak{B}$
(σ 2) $A \in \mathfrak{B} \quad \Rightarrow \quad X \setminus A \in \mathfrak{B}$
(σ 3) If $A_i \in \mathfrak{B}$ for $i \in \mathbb{N}$, then $\cup_{i\in\mathbb{N}} A_i \in \mathfrak{B}$.

There is a major difficulty when working with σ-algebras: In general, it is almost impossible to tell whether a given set is a member of a given σ-algebra. On the other hand one knows which sets should be members of the σ-algebra, e.g. all open sets with respect to some given topology. Hence one proceeds as follows. One chooses a certain family of sets in $\mathfrak{P}(X)$ (again the open sets serve as an example) and constructs the σ-algebra *generated* by these sets. After this construction has been performed, one has a σ-algebra at hand and it contains the desired sets. It also contains many other sets, some of which are difficult to be characterized explicitly. The power set $\mathfrak{P}(X)$ is one example of a σ-algebra. It is convenient to have two weaker concepts of families of sets.

Definition A.3.2 A family $\mathfrak{A} \subset \mathfrak{P}(X)$ is called an algebra if it has the following properties:

(A1) $\emptyset \in \mathfrak{A}$
(A2) $A \in \mathfrak{A} \Rightarrow X \setminus A \in \mathfrak{A}$
(A3) $A, B \in \mathfrak{A} \Rightarrow A \cap B \in \mathfrak{A}$.

Let $\mathfrak{A}$ be an algebra. Then the following are statements true:

(A4) $A, B \in \mathfrak{A} \Rightarrow A \cup B \in \mathfrak{A}$
(A5) $A, B \in \mathfrak{A} \Rightarrow A \setminus B \in \mathfrak{A}$

An algebra can also be characterized by the properties (A1)(A2)(A4).

Definition A.3.3 A family $\mathfrak{S} \subset \mathfrak{P}(X)$ is called a semi-algebra if it has the properties

(S1) $\emptyset \in \mathfrak{S}$
(S2) $A, B \in \mathfrak{S} \Rightarrow A \cap B \in \mathfrak{S}$
(S3) $A_i \subset X, i = 1, \ldots, n, A_i \cap A_j = \emptyset$ for $i \neq j \Rightarrow \cup_{i=1}^n A_i \in \mathfrak{S}$.

A semi-algebra can also be characterized by (S1)(S2), and
(S4) $A \in \mathfrak{S} \Rightarrow$ There are $A_i \in \mathfrak{S}, i = 1, \ldots, n, A_i \cap A_j = \emptyset$ for $i \neq j, X \setminus A = \cup_{i=1}^n A_i$.

Clearly, a σ-algebra is an algebra, and an algebra is a semi-algebra.

Suppose we have two σ-algebras $\mathfrak{B}_1$ and $\mathfrak{B}_2$ on the set X. Let $\mathfrak{B}$ be the set of all intersections $A_1 \cap A_2$ where $A_i \in \mathfrak{B}_i, i = 1, 2$. Then $\mathfrak{B}$ is a σ-algebra

Let $\mathfrak{M} \subset \mathfrak{P}(X)$. There are σ-algebras containing $\mathfrak{M}$ as a subset ($\mathfrak{P}(X)$ is one example). In view of the intersection property just mentioned there is a smallest σ-algebra that contains $\mathfrak{M}$. We call it the σ-algebra generated by $\mathfrak{M}$.

In a similar way we can construct the algebra $\mathfrak{A}$ and the semi-algebra $\mathfrak{S}$ generated by the set $\mathfrak{M}$. Trivially we have

$$\mathfrak{M} \subset \mathfrak{S} \subset \mathfrak{A} \subset \mathfrak{B} \subset \mathfrak{P}(X).$$

Definition A.3.4 Let X be a set and $\mathfrak{B} \subset \mathfrak{P}(X)$ a σ-algebra. The pair $(X, \mathfrak{B})$ is called a measurable space. The elements of $\mathfrak{B}$ are called measurable sets.

In the most important example the concepts of measure theory and of topology are linked.

Definition A.3.5 Let $(X, \mathfrak{T})$ be a topological space. Let $\mathfrak{B}(X)$ the σ-algebra generated by the family of open sets $\mathfrak{T}$. Then $\mathfrak{B}(X)$ is called the Borel σ-algebra of $(X, \mathfrak{T})$, and the elements of $\mathfrak{B}(X)$ are called Borel measurable sets.

If X is finite and endowed with the discrete topology then $\mathfrak{B}(X) = \mathfrak{P}(X)$; all sets are Borel measurable.

The definition of the σ-algebra $\mathfrak{B}(\mathfrak{M})$ that is generated by some set $\mathfrak{M}$ has been straightforward, but it may be difficult to characterize the elements of $\mathfrak{B}(\mathfrak{M})$ directly. The next proposition indicates how to characterize the elements of a σ-algebra generated by a semi-algebra.

Proposition A.3.6 *Let $\mathfrak{S}$ be a semi-algebra and $\mathfrak{A}$ the algebra generated by $\mathfrak{S}$. Then*

$$\mathfrak{A} = \{A = \cup_{i=1}^{n} S_i | \, S_i \in \mathfrak{S} : \; S_i \cap S_j = \emptyset \textit{ for } i \neq j, \;\; n \in \mathbb{N}\}.$$

Proof (Idea of proof) Check the axioms and find that $\mathfrak{A}$ is an algebra with $\mathfrak{S} \subset \mathfrak{A}$. From $S_i \in \mathfrak{S}$ it follows that $X \setminus \left(\cap_{i=1}^{n} (X \setminus S_i)\right) = \cup_{i=1}^{n} S_i \in \mathfrak{A}$. Hence $\mathfrak{A}$ is minimal. □

Next we define real functions on semi-algebras, algebras, and σ-algebras that may serve in measuring the "size" of a set.

Definition A.3.7 Let $\mathfrak{S}$ be a semi-algebra. A mapping $\mu : \mathfrak{S} \to \mathbb{R}_+$ is called finitely additive if it has the following two properties:

(1) $\mu(\emptyset) = 0$
(2) $\mu(\cup_{i=1}^{n} A_i) = \sum_{i=1}^{n} \mu(A_i)$ for any $A_i \in \mathfrak{S}$, $i = 1, \ldots, n$ with $A_i \cap A_j = \emptyset$ for $i \neq j$.

The mapping μ is called countably additive or a pre-measure if

(1) $\mu(\emptyset) = 0$
(2) $\mu(\cup_{i=1}^{\infty} A_i) = \sum_{i=1}^{\infty} \mu(A_i)$ for $A_i \in \mathfrak{S}$, $A_i \cap A_j = \emptyset$ for $i \neq j$.

The mapping μ is called a measure if $\mathfrak{B}$ is a σ-algebra, $\mu : \mathfrak{B} \to \mathbb{R}_+$, and

(1) $\mu(\emptyset) = 0$,
(2) $\mu(\cup_{i=1}^{\infty} A_i) = \sum_{i=1}^{\infty} \mu(A_i)$ whenever $A_i \in \mathfrak{B}$ and $A_i \cap A_j = \emptyset$ for $i \neq j$.

The mapping is called a finite measure if it is a measure and, in addition,

(3) $\mu(X) < \infty$.

The mapping is called a probability measure if

(3') $\mu(X) = 1$.

Let $\mathfrak{B} \subset \mathfrak{P}(X)$ be a σ-algebra and μ a finite measure on $\mathfrak{B}$. The object $(X, \mathfrak{B}, \mu)$ is called a measure space. If μ is a probability measure then it is called a probability space.

If μ is a probability measure on the Borel σ-algebra $\mathfrak{B}(X)$ (for some topological space $(X, \mathfrak{T})$) then μ is called a Borel probability measure.

A rather trivial, but (in view of our application) important example is the following. Let X be a finite set with the discrete topology and $p : X \to \mathbb{R}_+$ such that $\sum_{x \in X} p(x) = 1$. Then, $\mu(A) = \sum_{x \in A} p(x)$ is a Borel probability measure with σ-algebra $\mathfrak{B} = \mathfrak{P}(X)$.

Now we construct measures. Usually one has good idea what the measure of a "nice" set should be (e.g., a square in the plane) but one also knows that there may be awkward (say "fractal" sets) that may not be "measurable". The idea is to define a measure for these nice sets (they typically form a semi-algebra) and then to extend this measure to as many sets as possible by way of union and intersection. A given set may be result of different union and intersection operations. On must take care that the extension is well defined. Hence one needs suitable approximation theorems that guarantee uniqueness. Some of these theorems are not straight to prove.

Proposition A.3.8 *Let $\mathfrak{S}$ be a semi-algebra and let $\mu : \mathfrak{S} \to \mathbb{R}_+$ be a pre-measure. There is a unique extension of μ to a pre-measure $\hat{\mu}$ on the algebra $\mathfrak{A}$ generated by $\mathfrak{S}$.*

Proof Step 1: *Definition of the extension.*

Let $A \in \mathfrak{A}$. There are pairwise disjoint sets $S_1, \ldots, S_n \in \mathfrak{S}$ with $A = \cup_{i=1}^n S_i$. Define

$$\dot{\mu}(A) = \sum_{i=1}^{n} \mu(A_i).$$

This extension is well defined: Suppose that also $A = \cup_{i=1}^{n'} S_i'$. Then define sets $T_{i,j} = S_i \cap S_j'$. Since $\cup_{i=1}^n S_i \supset S_j'$, we have $S_j' = S_j' \cup_i S_i = \cup_{i=1}^n T_{i,j}$ and similarly $S_i = \cup_{j=1}^{n'} T_{i,j}$. Then we have

$$\hat{\mu}(A) = \sum_{i=1}^{n} \mu(S_i) = \sum_{i=1}^{n} \sum_{j=1}^{n'} \mu(T_{i,j}) = \sum_{j=1}^{n'} \sum_{i=1}^{n} \mu(T_{i,j}) = \sum_{j=1}^{n'} \mu(\bigcup_{i=1}^{n} T_{i,j}) = \sum_{j=1}^{n'} \mu(S_j').$$

Hence the value $\hat{\mu}(a)$ does not depend on the decomposition of A.

Step 2: $\hat{\mu}$ *is a pre-measure.*
We have $\hat{\mu}(\emptyset) = \mu(\emptyset) = 0$. Suppose $A_i \in \mathfrak{A}$, $i \in \mathbb{N}$, are pairwise disjoint. For each of these A_i we have pairwise disjoint sets $S_{ij} \in \mathfrak{S}$ such that $A_i = \bigcup_{j=1}^{n_i} S_{i,j}$. Then we conclude

$$\hat{\mu}(\bigcup_{i=1}^{\infty} A_i) = \hat{\mu}(\bigcup_{i=1}^{\infty}\bigcup_{j=1}^{n_i} S_{i,j}) = \sum_{i=1}^{\infty}\sum_{j=1}^{n_i} \mu(S_{i,j}) = \sum_{i=1}^{\infty} \hat{\mu}(\bigcup_{j=1}^{n_i} S_{i,j}) = \sum_{i=1}^{\infty} \hat{\mu}(A_i).$$

Step 3: *Uniqueness.*
We know that $\hat{\mu}$ is a pre-measure. Let $\tilde{\mu}$ be another extension of μ. The pre-measures $\hat{\mu}$ and $\tilde{\mu}$ agree on $\mathfrak{S}$. Define the family $\mathfrak{C} = \{A \in \mathfrak{A} : \hat{\mu}(A) = \tilde{\mu}(A)\}$. Trivially $\mathfrak{S} \subset \mathfrak{C}$. The family $\mathfrak{C}$ is an algebra. Since $\mathfrak{A}$ is minimal, $\mathfrak{C} = \mathfrak{A}$. □

If a pre-measure should be extended from an algebra $\mathfrak{A}$ to the σ-algebra generated by $\mathfrak{A}$, one first assigns a real number to every element of the power set $\mathfrak{P}(X)$ based on the pre-measure—these values represent the outer measure (or exterior measure). After that one selects a suitable family $\mathfrak{B}$ of "measurable sets" from the power set and shows that $\mathfrak{B}$ is a σ-algebra, contains $\mathfrak{A}$, and that the restriction of the outer measure to $\mathfrak{B}$ is indeed a measure that extends the given pre-measure. In a final step it is shown that the extension is unique.

Definition A.3.9 Let $\mathfrak{A} \subset \mathfrak{P}(X)$ be an algebra and let $\mu : \mathfrak{A} \to \mathbb{R}_+$ be a pre-measure. The mapping

$$\mu^* : \mathfrak{P} \to \mathbb{R}_+ \cup \{\infty\}, \quad \mu^*(A) = \inf\{\sum_{i=1}^{\infty} \mu(A_i) : A_i \in \mathfrak{A}, \bigcup_{i=1}^{\infty} A_i \supset A\}$$

is called the outer measure on X (for the given measure μ).

The following proposition shows that one can assume the A_i as non-intersecting.

Proposition A.3.10 *Let μ^* be the outer measure for μ. Then*

$$\mu^*(A) = \inf\{\sum_{i=1}^{\infty} \mu(A_i) : A_i \in \mathfrak{A} \textit{ pairwise disjoint}, \bigcup_{i=1}^{\infty} A_i \supset A\}.$$

For sets $A \in \mathfrak{A}$ holds equality, $\mu^*(A) = \mu(A)$. And for $A_i \in \mathfrak{P}(X)$ holds

$$\mu^*(\bigcup_{i=1}^{\infty} A_i) \le \sum_{i=1}^{\infty} \mu^*(A_i). \qquad (*)$$

Proof Step 1: Pairwise disjoint sets suffice.
Let $A_i \in \mathfrak{A}$, $\cup_{i=1}^{\infty} A_i \supset A$. Define

$$\tilde{A}_i = A_i \setminus (\bigcup_{j<i} A_j) = A_i \cap (X \setminus \bigcup_{j<i} A_j) \in \mathfrak{A}.$$

Then $\tilde{A}_i \subset A_i$. The sets $\tilde{A}_i$ are pairwise disjoint, $\cup_i \tilde{A}_i = \cup_i A_i \supset A$, and

$$\mu(A_i) = \mu(A_i \cap \tilde{A}_i) + \mu(A_i \setminus \tilde{A}_i) = \mu(\tilde{A}_i) + \mu(A_i \setminus \tilde{A}_i) \geq \mu(\tilde{A}_i)$$

and hence $\sum \mu(\tilde{A}_i) \leq \sum \mu(A_i)$. Therefore

$$\mu^*(A) = \inf\{\sum_{i=1}^{\infty} \mu(A_i) \ : \ A_i \in \mathfrak{A}, A_i \cap A_j = \emptyset \text{ for } i \neq j, \ \bigcup_{i=1}^{\infty} A_i \supset A\}.$$

Step 2: μ^* is subadditive.
Let $A_i \in \mathfrak{P}(X)$, $i \in \mathbb{N}$. If $\sum_{i=1}^{\infty} \mu^*(A_i) = \infty$ then nothing needs to be proved. Assume $\sum_{i=1}^{\infty} \mu^*(A_i) < \infty$. Choose $\varepsilon > 0$. Choose a fixed $i \in \mathbb{N}$. In view of Step 1 we find pairwise disjoint $B_{i,j} \in \mathfrak{A}$ such that

$$A_i \subset \bigcup_{i=1}^{\infty} B_{i,j}, \qquad \sum_{j=1}^{\infty} \mu(B_{i,j}) \leq \mu^*(A_i) + \frac{\varepsilon}{2^i}.$$

We have

$$\mu^*(\bigcup_{i=1}^{\infty} A_i) \leq \mu^*(\bigcup_{i=1}^{\infty} \bigcup_{j=1}^{\infty} B_{i,j}) \leq \sum_{i=1}^{\infty} \sum_{j=1}^{\infty} \mu(B_{i,j}) \leq \sum_{i=1}^{\infty} \mu^*(A_i) + \varepsilon,$$

and, since ε is arbitrary, we find the inequality $(*)$.
Step 3: $A \in \mathfrak{A} \ \Rightarrow \ \mu^*(A) = \mu(A)$.
Let $A \in \mathfrak{A}$. Choose $n = 1$, $A_1 = A$, and use the definition of the outer measure to get $\mu^*(A) \leq \mu(A)$. Then choose any countable family of pairwise disjoint $A_i \in \mathfrak{A}$ with $\cup_{i=1}^{\infty} A_i \supset A$. Then the sets $\tilde{A}_i = A_i \cap A$ form a family of pairwise disjoint sets in $\mathfrak{A}$ and $A = \cup_{i=1}^{\infty} A_i$. Then

$$\mu(A) = \sum_{i=1}^{\infty} \mu(\tilde{A}_i) = \sum_{i=1}^{\infty} \mu(A_i \cap A) \leq \sum_{i=1}^{\infty} \mu(A_i).$$

Now choose families A_i that yield values arbitrarily close to the infimum in the definition of the outer measure and find $\mu(A) \leq \mu^*(A)$. Hence we have $\mu(A) = \mu^*(A)$. □

Suppose we have an algebra $\mathfrak{A}$ and a pre-measure μ on $\mathfrak{A}$. Using the outer measure μ^* we construct a σ-algebra $\mathfrak{B}$ and a measure that is an extension of the given measure.

Definition A.3.11 Define $\mathfrak{B}$ as the set of all $A \in \mathfrak{P}(X)$ such that for all $B \in \mathfrak{P}(X)$ the following equation holds

$$\mu^*(B) = \mu^*(B \cap A) + \mu^*(B \setminus A). \tag{$*$}$$

Proposition A.3.12 *The set $\mathfrak{B}$ as defined in A.3.11 is a σ-algebra and the restriction of the outer measure to $\mathfrak{B}$ is a measure.*

Proof Step 1: $\emptyset \in \mathfrak{B}$.
Let $B \in \mathfrak{P}(X)$. Then $\mu^*(B) = 0 + \mu^*(B) = \mu^*(\emptyset \cap B) + \mu^*(B \setminus \emptyset)$.
Step 2: $A \in \mathfrak{B}$ *implies* $X \setminus A \in \mathfrak{B}$.
For $B \in \mathfrak{P}(X)$ we have

$$\mu^*(B) = \mu^*(B \cap A) + \mu^*(B \setminus A) = \mu^*(B \setminus (X \setminus A)) + \mu^*(B \cap (X \setminus A))$$

as trivially $B \cap A = B \setminus (X \setminus A)$ and $B \setminus A = B \cap (X \setminus A)$.
Step 3: $A_1, A_2 \in \mathfrak{B}$ *implies* $A_1 \cup A_2 \in \mathfrak{B}$.
Let $\hat{B} \in \mathfrak{P}(X)$. Use the equation $(*)$ with $A = A_1$ and $B = \hat{B} \cap (A_1 \cup A_2)$,

$$\begin{aligned}
\mu^*(\hat{B} \cap (A_1 \cup A_2)) &= \mu^*(\hat{B} \cap (A_1 \cup A_2) \cap A_1) + \mu^*((\hat{B} \cap (A_1 \cup A_2)) \setminus A_1) \\
&= \mu^*(\hat{B} \cap A_1) + \mu^*(\hat{B} \cap (A_1 \cup A_2) \cap (X \setminus A_1)) \\
&= \mu^*(\hat{B} \cap A_1) + \mu^*(\hat{B} \cap A_2 \cap (X \setminus A_1)).
\end{aligned}$$

Use equation $(*)$ with $A = A_2$ and $B = \hat{B} \cap (X \setminus A_1)$ and get

$$\begin{aligned}
\mu^*(\hat{B} \cap (X \setminus A_1)) &= \mu^*(\hat{B} \cap (X \setminus A_1) \cap A_2) + \mu^*((\hat{B} \cap (X \setminus A_1)) \setminus A_2) \\
&= \mu^*(\hat{B} \cap (X \setminus A_1) \cap A_2) + \mu^*(\hat{B} \cap (X \setminus A_1) \cap (X \setminus A_2)).
\end{aligned}$$

Taking these equations together, we have the equation

$$\begin{aligned}
&\mu^*(\hat{B} \cap (A_1 \cup A_2)) + \mu^*(\hat{B} \setminus (A_1 \cup A_2)) \\
&= \mu^*(\hat{B} \cap (A_1 \cup A_2)) + \mu^*(\hat{B} \cap (X \setminus (A_1 \cup A_2))) \\
&= \mu^*(\hat{B} \cap (A_1 \cup A_2)) + \mu^*(\hat{B} \cap (X \setminus A_1) \cap (X \setminus A_2)) \\
&= \left[\mu^*(\hat{B} \cap A_1) + \mu^*(\hat{B} \cap A_2 \cap (X \setminus A_1))\right] \\
&\qquad + \left[\mu^*(\hat{B} \cap (X \setminus A_1)) - \mu^*(\hat{B} \cap (X \setminus A_1) \cap A_2)\right] \\
&= \mu^*(\hat{B} \cap A_1) + \mu^*(\hat{B} \cap (X \setminus A_1)) = \mu^*(\hat{B})
\end{aligned}$$

for every $\hat{B} \in \mathfrak{P}(X)$. Thus, $A_1 \cup A_2 \in \mathfrak{B}$.

Step 4: $A_i \in \mathfrak{B}$, $i \in \mathbb{N}$, *implies* $\cup_{i=1}^{\infty} A_i \in \mathfrak{B}$.

Suppose $A = \cup_{i=1}^{\infty} A_i$ with $A_i \in \mathfrak{B}$. We represent A as a union of pairwise disjoint sets in $\mathfrak{B}$. Define

$$\tilde{A}_1 = A_1, \quad \tilde{A}_i = A_i \setminus (\underset{j<i}{\cup} A_j) \quad \text{for } i > 1.$$

The sets $\tilde{A}_i \in \mathfrak{A}^*$ are pairwise disjoint and in $\mathfrak{B}$. Moreover, $A = \cup_{i=1}^{\infty} \tilde{A}_i$. We show $A \in \mathfrak{B}$.

Let $B \in \mathfrak{B}$. Then we find recursively for any $n \in \mathbb{N}$

$$\begin{aligned}
\mu^*(B) &= \mu^*(B \cap \tilde{A}_1) + \mu^*(B \setminus \tilde{A}_1) \\
&= \mu^*(B \cap \tilde{A}_1) + \mu^*((B \setminus \tilde{A}_1) \cap \tilde{A}_2) + \mu^*((B \setminus \tilde{A}_1) \setminus \tilde{A}_2) \\
&= \mu^*(B \cap \tilde{A}_1) + \mu^*(B \cap (X \setminus \tilde{A}_1) \cap \tilde{A}_2) + \mu^*(B \cap (X \setminus \tilde{A}_1) \cap (X \setminus \tilde{A}_2)) \\
&= \mu^*(B \cap \tilde{A}_1) + \mu^*(B \cap \tilde{A}_2) + \mu^*(B \setminus (\tilde{A}_1 \cup \tilde{A}_2)) \\
&\dots \\
&= \sum_{i=1}^{n} \mu^*(B \cap \tilde{A}_i) + \mu^*(B \setminus (\overset{n}{\underset{i=1}{\cup}} \tilde{A}_i)) \\
&\geq \sum_{i=1}^{n} \mu^*(B \cap \tilde{A}_i) + \mu^*(B \setminus (\overset{\infty}{\underset{i=1}{\cup}} \tilde{A}_i)).
\end{aligned}$$

Since n is arbitrary, we have

$$\mu^*(B) \geq \sum_{i=1}^{\infty} \mu^*(B \cap \tilde{A}_i) + \mu^*(B \setminus (\overset{\infty}{\underset{i=1}{\cup}} \tilde{A}_i)).$$

On the other hand we have

$$\begin{aligned}
\mu^*(B) &= \mu^*((B \cap A) \cup (B \setminus A)) \\
&\leq \mu^*(B \cap A) + \mu^*(B \setminus A) = \mu^*(\overset{\infty}{\underset{i=1}{\cup}} (B \cap \tilde{A}_i)) + \mu^*(B \setminus A) \\
&\leq \sum_{i=1}^{\infty} \mu^*(B \cap \tilde{A}_i) + \mu^*(B \setminus A).
\end{aligned}$$

Hence we have found the equality

$$\mu^*(B) = \sum_{i=1}^{\infty} \mu^*(B \cap \tilde{A}_i) + \mu^*(B \setminus (\overset{\infty}{\underset{i=1}{\cup}} \tilde{A}_i)). \qquad (**)$$

In particular,

$$\mu^*(B) = \mu^*(B \cap (\bigcup_{i=1}^{\infty} \tilde{A}_i)) + \mu^*(B \setminus (\bigcup_{i=1}^{\infty} \tilde{A}_i)).$$

Hence $A \in \mathfrak{B}$ and $\mathfrak{B}$ is a σ-algebra.

Step 5: μ^* is a measure on $\mathfrak{B}$.

We have $\mu^*(\emptyset) = \mu(\emptyset) = 0$. Let $A_i \in \mathfrak{B}$ be pairwise disjoint sets. Use the equality $(**)$ with $A = B = \cup_{j=1}^{\infty} A_j$

$$\mu^*(\bigcup_{i=1}^{\infty} A_i) = \mu^*((\bigcup_{i=1}^{\infty} A_i) \cap (\bigcup_{j=1}^{\infty} A_j)) + \mu((\bigcup_{i=1}^{\infty} A_i) \setminus (\bigcup_{i=1}^{\infty} A_i))$$
$$= \sum_{i=1}^{\infty} \mu^*(A_i \cap (\bigcup_{j=1}^{\infty} A_j)) + 0 = \sum_{i=1}^{\infty} \mu^*(A_i).$$

□

Since $\mathfrak{B}$ is a σ-algebra and contains $\mathfrak{A}$, $\mathfrak{B}$ contains already the algebra generated by $\mathfrak{A}$. The pre-measure can be extended to the σ-algebra generated by $\mathfrak{A}$. It remains to be shown that this extension is unique.

Proposition A.3.13 *Let μ_1, μ_2 be measures on the σ-algebra $\mathfrak{B}$ generated by the algebra $\mathfrak{A}$. Suppose the measures agree on $\mathfrak{A}$. Then they agree on $\mathfrak{B}$.*

Proof We sketch the idea of the proof. Let $Z = \{B \in \mathfrak{B} : \mu_1(B) = \mu_2(B)\}$. Then $\mathfrak{A} \subseteq Z \subseteq \mathfrak{B}$. It is not too difficult to show that Z is already a σ-algebra. In consequence, $Z = \mathfrak{B}$. □

In general we do not know $\mathfrak{B}$ (i.e., it is difficult to know if a given set is in $\mathfrak{B}$) but we know the generating algebra $\mathfrak{A}$ rather well. For this reason one tries to work with $\mathfrak{A}$ instead of $\mathfrak{B}$. The following proposition yields a most useful approximation argument (see [105, Theorem 4.4, p. 84]).

Proposition A.3.14 *Let $\mathfrak{A}$ be an algebra and $\mathfrak{B}$ the σ-algebra generated by $\mathfrak{A}$. Let μ be a probability measure on $\mathfrak{B}$. For every $B \in \mathfrak{B}$ and every $\varepsilon > 0$ there is an $A \in \mathfrak{A}$ such that*

$$\mu(A \Delta B) = \mu((A \setminus B) \cup (B \setminus A)) < \varepsilon.$$

Proof Let $B \in \mathfrak{B}$. Then

$$\mu(B) = \mu^*(B) = \inf\{\sum_{i=1}^{\infty} \mu(F_i) : F_i \in \mathfrak{A} \text{ pairwise disjoint}, \bigcup_{i=1}^{\infty} F_i \supset B\}.$$

For a given $\varepsilon > 0$, there is a family of pairwise disjoint sets $F_i \in \mathfrak{A}$ such that $\cup_{i=1}^{\infty} F_i \supset B$, and

$$\sum_{i=1}^{\infty} \mu(F_i) \leq \mu(B) + \varepsilon/2.$$

The sum converges since we have a probability measure. Now choose n such that $\sum_{i=n+1}^{\infty} \mu(F_i) < \varepsilon/2$ and define

$$A = \bigcup_{i=1}^{n} F_i \in \mathfrak{A}.$$

Then $A, B \subset \cup_{i=1}^{\infty} F_i$, and therefore

$$B \setminus A \subset (\bigcup_{i=1}^{\infty} F_i) \setminus A = \bigcup_{i=n+1}^{\infty} F_i \quad \Rightarrow \quad \mu(B \setminus A) \leq \mu(\bigcup_{i=1}^{\infty} F_i) < \varepsilon/2.$$

Further we have

$$\mu(\bigcup_{i=1}^{\infty} F_i) = \mu((\bigcup_{i=1}^{\infty} F_i) \cap B) + \mu((\bigcup_{i=1}^{\infty} F_i) \setminus B) = \mu(B) + \mu((\bigcup_{i=1}^{\infty} F_i) \setminus B),$$

and hence

$$\mu(A \setminus B) \leq \mu((\bigcup_{i=1}^{\infty} F_i) \setminus B) = \mu(\bigcup_{i=1}^{\infty} F_i) - \mu(B) \leq \varepsilon/2.$$

With $\mu(F \Delta B) = \mu(F \setminus B) + \mu(B \setminus F) < \varepsilon$ we get the desired result. □

Now we use topology to define another approximation property.

Definition A.3.15 Let $(X, \mathfrak{T})$ be a topological space and let $\mathfrak{B}$ be the corresponding Borel σ-algebra. A measure space $(X, \mathfrak{B}, \mu)$ resp. the measure μ is called regular (with respect to $\mathfrak{T}$) if for every $A \in \mathfrak{B}$ and every $\varepsilon > 0$ there is a closed set C_ε and an open U_ε such that

$$C_\varepsilon, \ U_\varepsilon \in \mathfrak{B}, \quad C_\varepsilon \subset A \subset U_\varepsilon, \quad \text{and} \quad \mu(U_\varepsilon \setminus C_\varepsilon) < \varepsilon.$$

Note that closed and open sets are measurable for a Borel σ-algebra, such that $\mu(U_\varepsilon \setminus C_\varepsilon)$ is well defined.

Proposition A.3.16 *Let (X, d) be a metric space. Then every finite Borel measure is regular.*

Proof If a Borel measure is given then all open and all closed sets are measurable. Let $\tilde{\mathfrak{B}}$ be the family of sets that satisfy the condition of definition A.3.15,

$$\tilde{\mathfrak{B}} = \{A \in \mathfrak{P}(X) \ : \ \forall \varepsilon > 0 \ \exists \, U_\varepsilon \text{ open}, C_\varepsilon \text{ closed }, C_\varepsilon \subset A \subset U_\varepsilon : \ \mu(U_\varepsilon \setminus C_\varepsilon) < \varepsilon\}.$$

Step 1: $\tilde{\mathfrak{B}}$ *is a σ-algebra.*

We have $X \in \tilde{\mathfrak{B}}$, since X is open and closed; we can choose $C_\varepsilon = U_\varepsilon = X$. Further we see that if $A \in \tilde{\mathfrak{B}}$ then also $X \setminus A \in \tilde{\mathfrak{B}}$: Let A be given and $\varepsilon > 0$. Let U_ε, C_ε be two sets which satisfy the conditions of Definition A.3.15. Then choose $\tilde{C}_\varepsilon = X \setminus U_\varepsilon$, $\tilde{U}_\varepsilon = X \setminus C_\varepsilon$ for $X \setminus A$. We check that $\tilde{U}_\varepsilon$ is open and $\tilde{C}_\varepsilon$ is closed, and

$$\tilde{C}_\varepsilon \subset X \setminus A \subset \tilde{U}_\varepsilon, \quad \mu(\tilde{U}_\varepsilon \setminus \tilde{C}_\varepsilon) = \mu((X \setminus C_\varepsilon) \setminus (X \setminus U_\varepsilon)) = \mu(U_\varepsilon \setminus C_\varepsilon) < \varepsilon.$$

Finally we check that the countable union of sets $A_i \in \tilde{\mathfrak{B}}$, $i = 1, 2, 3, \ldots$, is an element of $\tilde{\mathfrak{B}}$. Let $A = \cup_{i \in \mathbb{N}} A_i$ and $\varepsilon > 0$. For A_i choose the sets $C_{i,\varepsilon} \subset A_i \subset U_{i,\varepsilon}$ such that

$$\mu(U_{i,\varepsilon} \setminus C_{i,\varepsilon}) < \frac{\varepsilon}{3^i}.$$

Then

$$U_\varepsilon = \bigcup_{i=1}^{\infty} U_{i,\varepsilon}$$

is open. Define the set

$$\tilde{C}_\varepsilon = \bigcup_{i=1}^{\infty} C_{i,\varepsilon}.$$

This set needs not to be closed. Choose some $k > 1$ such that

$$\mu(\tilde{C}_\varepsilon \setminus \bigcup_{i=0}^{k} C_{i,\varepsilon}) < \varepsilon/2$$

and choose $C_\varepsilon = \cup_{i=0}^{k} C_{i,\varepsilon}$. Then $C_\varepsilon \subset \tilde{C}_\varepsilon \subset A \subset U_\varepsilon$, and

$$U_\varepsilon \setminus C_\varepsilon = (U_\varepsilon \setminus \tilde{C}_\varepsilon) \cup (\tilde{C}_\varepsilon \setminus C_\varepsilon).$$

Therefore,

$$\mu(U_\varepsilon \setminus C_\varepsilon) = \mu(U_\varepsilon \setminus \tilde{C}_\varepsilon) + \mu(C_\varepsilon \setminus \tilde{C}_\varepsilon) < \sum_{i=1}^{\infty} \frac{\varepsilon}{3^i} + \frac{\varepsilon}{2} = \varepsilon.$$

We have shown that $\tilde{\mathfrak{B}}$ is a σ-algebra.

Step 2: $\tilde{\mathfrak{B}}$ *contains all closed sets.*

Let A be closed and $\varepsilon > 0$. Choose $C_\varepsilon = A$, and $U_i = \{x \in X : \operatorname{dist}(x, A) < 1/i\}$. The sets U_i are nested, and $A = \cap_{i=1}^{\infty} U_i$. Let $V_i = U_i \setminus U_{i+1}$. Then, $V_i \in \mathfrak{B}$ (the original Borel σ-algebra), and $\cup_{i=1}^{\infty} V_i = U_1 \setminus A$. Since μ is a finite measure, it follows that

$$\infty > \mu(X) \geq \mu(U_1) \geq \mu(U_1 \setminus A) = \mu(\bigcup_{i=1}^{\infty} V_i) = \sum_{i=1}^{\infty} \mu(V_i).$$

We conclude that $0 = \lim_{k\to\infty} \sum_{i=k}^{\infty} \mu(V_i) = \lim_{k\to\infty} \mu(U_k \setminus A)$. Now select i_0 such that $\mu(U_{i_0} \setminus A) < \varepsilon$ and choose $U_\varepsilon = U_{i_0}$.

Step 3: The Borel measure is regular

Since $\tilde{\mathfrak{B}}$ is a σ-algebra and contains the closed sets, it contains the open sets. Hence the Borel σ-algebra is a subalgebra of $\tilde{\mathfrak{B}}$, and hence the Borel probability measure is regular. □

We construct measures on the state spaces of cellular automata. Therefore we start with a measure on the alphabet E (which is finite), and then construct product measures on E^Γ.

In order to understand the idea of product measures, start with $([0,1], \mathfrak{B}, \mu_1)$. How to construct a measure μ_2 on $[0,1]^2$? We first define $\mu_2((a,b)) \times (c,d)) = \mu_1((a,b))\,\mu_1((c,d))$. This definition already implies that open squares are measurable. We use these sets a semi-algebra with pre-measure μ_2, and from that we construct a measure space $([0,1]^2, \mathfrak{B}, \mu_1)$ as described above. If the product is infinite $[0,1]^{\mathbb{N}}$, we use a trick that we already met in the construction of the product topology: As a semi-algebra, we only consider cylindric sets, such that almost all components of a set in this semi-algebra are $[0,1]$. There are only finitely many components that are non-trivial, and for those components we can again use simply the product of their measures to define a pre-measure.

Below we introduce the notion of a product measure (if you are confused by the notation of a Cayley graph, then simply replace G by $\mathbb{Z}$ and write $|g|$ instead of $d_c(g,e)$ for $g \in \mathbb{Z}$).

Construction A.3.17 Let $(X, \mathfrak{B}, \mu)$ be a probability space, let G a finitely generated group, and Γ a Cayley graph $\Gamma(G)$. Let furthermore X_g, $g \in G$, identical copies of X and $Y = \prod_{g\in G} X_g$. Further let

$$\mathfrak{S} = \{s \in \mathfrak{P}(Y) \,:\, \exists n \in \mathbb{N} : s = \prod_{d_c(g,e)>n} X \times \prod_{d_c(g,e)\leq n} B_g,\; B_g \in \mathfrak{B}\}$$

be a semi-algebra. Define the pre-measure $\hat{\mu}$ on $\mathfrak{S}$ by

$$\hat{\mu} : \mathfrak{S} \to [0,1], \quad \hat{\mu}\left(\prod_{d_c(g,e)>n} X \times \prod_{d_c(g,e)\leq n} B_g\right) = \prod_{d_c(g,e)\leq n} \mu(B_g).$$

There is a uniquely determined probability measure on the σ-algebra generated by $\mathfrak{S}$. This measure is called the product measure.

Remark A.3.18

(1) The sets

$$\prod_{d_c(g,e)>n} X \times \prod_{d_c(g,e)\leq n} B_i, \quad B_i \in \mathfrak{B}$$

are called cylinder sets. We do not really use infinite products but we use projections onto finitely many coordinates. The other coordinates are "trivial", i.e., equal to X.

(2) If the original measure μ is a Borel measure then also $\hat{\mu}$ is a Borel measure since both the product topology and the product measure only work with finitely many coordinates that are distinct from X.

(3) The Bernoulli measure we use for the state space of cellular automata is itself a probability product measure. We start off with a probability measure on the set E in assigning to each state $e \in E$ a probability $p(e)$, such that $\sum_{e\in E} p(e) = 1$. The Bernoulli measure is then simply the product measure on E^Γ.

In the proof of Proposition 4.2.4, we intersect countable sets of full measure. We require the following lemma there.

Lemma A.3.19 *Let* $(X, \mathfrak{B}, \mu)$ *a probability space, and* $\Omega_i \in \mathfrak{B}$ *a countable family of sets,* $\mu(\Omega_i) = 1$. *Then,* $\mu(\cap_{i=1}^\infty \Omega_i) = 1$.

Proof Let $A_i = X \setminus \Omega_i$. Then, $\mu(A_i) = 0$, and $\cap_{i=1}^\infty \Omega_i \subset X \setminus (\cup_{i=1}^\infty A_i)$. Since

$$\mu(\cup_{i=1}^\infty A_i) \leq \sum_{i=1}^\infty \mu(A_i) = 0$$

we obtain $\mu(\cap_{i=1}^\infty \Omega_i) = 1$. □

References

1. A.E. Adamatzky, *Collision Based Computing* (Springer, Berlin, 2002)
2. J.-P. Allouchea, G. Skordev, Remarks on permutive cellular automata. J. Comput. Syst. Sci. **62**, 174–182 (2003)
3. N. Aubrun, J. Kari, Tiling problems on Baumslag-Solitar groups. Mach. Comput. Univ. **128**, 35–46 (2013)
4. P. Bak, C. Tang, K. Wiesenfeld, Self-organized criticality. Phys. Rev. A **38**, 364–374 (1988)
5. A. Ballier, M. Stein, The domino problem on groups of polynomial growth. ArXiv e-prints 1311.4222 (2013)
6. J. Banks, J. Brooks, G. Cairns, G. Davis, P. Stacy, On Devaney's definition of chaos. Am. Math. Month. **99**, 332–334 (1992)
7. Y. Bar-Yam, *Dynamics of Complex Systems* (Perseus Books, Cambridge, 1997)
8. B. Baumlsag, B. Chandler, *Group Theory* (McGraw Hill, New York, 1961)
9. S. Beatty, N. Altshiller-Court, O. Dunkel, A. Pelletier, F. Irwin, J.L. Riley, P. Fitch, D.M. Yost, Problems for solutions: 3173–3180. Am. Math. Month. **33**, 159 (1926)
10. S. Beatty, A. Ostrowski, J. Hyslop, Solutions to problem 3173. Am. Math. Month. **34**, 159–160 (1927)
11. R. Berger, Undecidability of the domino problem. Memoirs Am. Math. Soc. **66**, 72–73 (1966)
12. E.R. Berlekamp, J.H. Conway, R.K. Guy, What is life? in *Winning Ways for Your Mathematical Plays*, ed. by E. Berlekamp, vol. 2 (Academic, New York, 1982)
13. A. Besicovitch, *Almost Periodic Functions* (Dover, New York, 1954)
14. F. Blanchard, E. Formenti, Cellular automata in the Cantor, Besicovitch, and Weyl topological spaces. Complex Syst. **11**, 107–123 (1999)
15. F. Blanchard, J. Cervelle, E. Formenti, Periodicity and transitivity for cellular automata in Besicovitch topologies. Lect. Not. Comput. Sci. **2747**, 228–238 (2003)
16. F. Blanchard, J. Cervelle, E. Formenti, Some results about the chaotic behavior of cellular automata. Theor. Comput. Sci. **349**, 318–336 (2005)
17. F. Brauer, P. van den Driessche, J. Wu, *Mathematical Epidemiology* (Springer, Berlin, 2008)
18. A. Budd, C. McDougall, K. Green, B.M. Degnan, Control of shell pigmentation by secretory tubules in the abalone mantle. Front. Zool. **11**, 62 (2014)
19. J.R. Büchi, Symposium on decision problems: on a decision method in restricted second order arithmetic, in *Logic, Methodology and Philosophy of Science, Proceeding of the 1960 International Congress*, ed. by P.S. Ernest Nagel, A. Tarski, vol. 44 (Elsevier, Amsterdam, 1966), pp. 1–11
20. M. Bulmer, Periodical insects. Am. Nat. **111**, 1099–1117 (1977)

K.-P. Hadeler, J. Müller, *Cellular Automata: Analysis and Applications*,
Springer Monographs in Mathematics, DOI 10.1007/978-3-319-53043-7

21. P.R. Campos, V.M. de Oliveira, R. Giro, D.S. Galvao, Emergence of prime numbers as the result of evolutionary strategy. Phys. Rev. Lett. **93**, 098107 (2004)
22. S. Capobianco, Surjunctivity for cellular automata in Besicovitch spaces. J. Cell. Autom. **4**, 89–98 (2009)
23. S. Capobianco, On pattern density and sliding block code behavior for the Besicovitch and Weyl pseudo-distances, in *SOFSEM 2010: Theory and Practice of Computer Science*, ed. by J. van Leeuwen, A. Muscholl, D. Peleg, J. Pokorny, B. Rumpe. Lecture Notes in Computer Sciences, vol. 5901 (Springer, Berlin, 2010), pp. 259–270
24. G. Cattaneo, L. Formeni, L. Margara, J. Mazoyer, A shift-invariant metric on $S^{\mathbb{Z}}$ inducing a nontrivial topology, in *Mathematical Foundations of Computer Sciences*, ed. by I. Prívara, P. Rusika. Lecture Notes in Computer Sciences, vol. 1295 (Springer, Berlin, 1997), pp. 179–188
25. G. Cattaneo, M. Finelli, L. Margara, Investigating topological chaos by elementary cellular automata dynamics. Theor. Comput. Sci. **244**, 219–241 (2000)
26. T. Ceccherini-Silberstein, M. Coornaert, A generalization of the Curtis-Hedlund theorem. Theor. Comput. Sci. **400**, 225–229 (2008)
27. T. Ceccherini-Silberstein, M. Coornaert, *Cellular Automata and Groups*. Springer Monographs in Mathematics (Springer, Berlin, 2010)
28. T. Ceccherini-Silberstein, M. Coornaert, F. Fiorenzi, P. Schupp, Groups, graphs, languages, automata, games and second-order monadic logic. Eur. J. Comb. **7**, 1330–1368 (2012)
29. T. Ceccherini-Silberstein, M. Coornaert, F. Fiorenzi, Z. Sŭnić, Cellular automata on regular rooted trees, in *Implementation and Application of Automata*, ed. by N. Moreira, R. Reis. Lecture Notes in Computer Sciences, vol. 7381 (Springer, Berlin, 2012), pp. 101–112
30. T. Ceccherini-Silberstein, M. Coornaert, F. Fiorenzi, Z. Sŭnić, Cellular automata between sofic tree shifts. Theor. Comput. Sci. **506**, 79–101 (2013)
31. S. Choi, C.-K. Chi, S. Park, Chain recurrent sets for flows on non-compact spaces. J. Dyn. Differ. Equ. **14**, 597–611 (2002)
32. E. Codd. *Cellular Automata* (Academic, New York, 1968)
33. B. Codettoni, L. Margara, Transitive cellular automata are sensitive. Am. Math. Month. **103**, 58–62 (1996)
34. S.A. Colgate, A. Stanley, J. Hyman, S.P. Layne, C. Qualls, Risk behavior-based model of the cubic growth of acquired immunodeficiency syndrome in the united states. Proc. Natl. Acad. Sci. USA **86**, 4739–4797 (1989)
35. C. Conley, *Isolated Invariant Sets and the Morse Index*. CBMS Lecture Notes, vol. 38 (American Mathematical Society, Providence, RI, 1978)
36. B. Courcelle, J. Engelfriet, *Graph Structure and Monadic Second-Order Logic* (Cambridge University Press, Cambridge, 2012)
37. M. Cross, H. Greenside, *Pattern Formation and Dynamics in Nonequilibrium Systems* (Cambridge University Press, Cambridge, 2009)
38. K. Culik II, L. Hurd, Computation theoretic aspects of cellular automata. Phys. D **45**, 357–378 (1990)
39. E. Czeizler, On the size of the inverse neighborhoods for one-dimensional reversible cellular automata. Theor. Comput. Sci. **325**, 273–284 (2004)
40. E. Czeizler, J. Kari, A tight linear bound on the synchronization delay of bijective automata. Theor. Comput. Sci. **380**, 23–36 (2007)
41. A. Deutsch, S. Doormann, *Cellular Automaton Modeling of Biological Pattern Formation* (Birkhäuser, Basel, 2004)
42. R.L. Devaney, *Chaotic Dynamical Systems* (Westview Press, New York, 2003)
43. O. Diekmann, A beginner's guide to adaptive dynamics. Banach Center Publ. **63**, 47–86 (2004)
44. O. Diekmann, H. Heesterbeek, T. Britton, *Mathematical Tools for Understanding Infectious Disease Dynamics* (Princeton University Press, Princeton, 2013)
45. R. Dow, Additive cellular automata and global injectivity. Phys. D **110**, 67–91 (1997)

46. T. Downarowicz, A. Iwanik, Quasi-uniform convergence in compact dynamical systems. Stud. Math. **89**, 11–25 (1988)
47. M. Dubois-Violette, A. Rouet, A mathematical classification of the one-dimensional deterministic cellular automata. Commun. Math. Phys. **112**, 627–631 (1987)
48. B. Durand, The surjectivity problem for 2D cellular automata. J. Comput. Syst. Sci. **49**, 718–724 (1994)
49. L. Edelstein-Keshet, *Mathematical Models in Biology* (McGraw-Hill, New York, 1988)
50. G. Edgar, *Measure, Topology, and Fractal Geometry* (Springer, Berlin, 2008)
51. M. Efendiev, J. Müller, Classification of existence and non-existence of running fronts in case of fast diffusion. Adv. Math. Sci. Appl. **19**, 285–293 (2009)
52. H. Enderton, *A Mathematical Introduction to Logic* (Academic, New York, 1972)
53. B. Ermentrout, Neural networks as spatio-temporal pattern-forming systems. Rep. Prog. Phys. **61**, 353–430 (1998)
54. B. Ermentrout, L. Edelstein-Keshet, Cellular automaton approaches in biological modeling. J. Theor. Biol. **160**, 97–133 (1993)
55. G. Ermentrout, J. Campbell, G. Oster, A model for shell patterns based on neural activity. Veliger **28**, 369–388 (1985)
56. L.C. Evans, *Partial Differential Equations* (American Mathematical Society, Providence, RI, 1998)
57. P. Favati, G. Lotti, L. Margara, Additive one-dimensional cellular automata are chaotic according to Devaney's definition of chaos. Theor. Comput. Sci. **174**, 157–170 (1997)
58. E. Fredkin, T. Toffoli, Conservative logic. Int. J. Theor. Phys. **21**, 219–253 (1982)
59. V. Frette, K. Christensen, A. Malthe-Sørenssen, J. Feder, T. Jøssang, P. Meakin, Avalanche dynamics in a pile of rice. Nature **379**, 49–52 (1996)
60. H. Fukś, A. Skelton, Classification of two-dimensional binary cellular automata with respect to surjectivity, in *Proceedings of the International Conference on Computational Science, CSC 2012*, pp. 51–57, 2012
61. A. Gajardo, A. Moreira, E. Goles, Complexity of Langton's ant. Discret. Appl. Math. **117**, 41–50 (2002)
62. D. Gale, J. Propp, S. Sutherland, S. Troubetzkoy, Further travels with my ant. Mathematical entertainments column. Math. Intell. **17**, 48–56 (1995)
63. F.R. Gantmacher, *Matrizentheorie* (Springer, Berlin, 1986)
64. M. Gardner, The fantastic combinations of John Conway's new solitaire game "life". Sci. Am. **223**, 120–123 (1970)
65. R. Gilman, Classes of linear automata. Ergod. Theory Dyn. Syst. **7**, 105–118 (1987)
66. E. Goles, S. Martínez, *Neural and Automata Networks, Dynamic Behavior and Applications* (Kluwer Academic Press, Dordrecht, 1991)
67. E. Goles, O. Schulz, M. Markus, A biological generator of prime numbers. Nonlin. Phenom. Compl. Syst. **3**, 208–213 (2000)
68. E. Goles, O. Schulz, M. Markus, Prime number selection of cycles in a predator-prey model. Complexity **6**, 33–38 (2001)
69. E. Golod, On nil algebras and finitely residual groups. Izv. Akad. Nauk SSSR Ser. Mat. **28**, 273–276 (1964)
70. E. Golod, I. Shafarevich, On the class field tower. Izv. Akad. Nauk SSSR Ser. Mat. **28**, 261–272 (1964)
71. W. Gottschalk, Some general dynamical notions, in *Recent Advances in Topological Dynamics, Proceedings of Conference in Topological Dynamics*, Yale University, Lecture Notes on Mathematics, vol. 318 (Springer, Berlin, 1973), pp. 120–125
72. P. Grant, The priming of periodical cicada life cycles. Trends Ecol. Evol. **20**, 169–174 (2005)
73. J. Gravner, G. David, Cellular automaton growth on $\mathbb{Z}^2$: theorems, examples, and problems. Adv. Appl. Math. **21**, 241–304 (1998)
74. J. Greenberg, S. Hastings, Spatial patterns for discrete models of diffusion in excitable media. SIAM J. Appl. Math. **34**, 515–523 (1978)

75. J. Greenberg, B. Hassard, S. Hastings, Pattern formation and periodic structures in systems modeled by reaction-diffusion equations. Bull. Am. Math. Soc. **84**, 1296–1327 (1978)
76. J. Greenberg, C. Greene, S. Hastings, A combinatorial problem arising in the study of reaction-diffusion equations. SIAM J. Alg. Discret. Math. **1**, 34–42 (1986)
77. D. Griffeath, C. Moore, Life without death is P-complete. Complex Syst. **10**, 437–447 (1996)
78. R. Grigorchuk, Degrees of growth of finitely generated groups and the theory of invariant means. Izv. Akad. Nauk SSSR Ser. Mat. **48**, 939–985 (1984)
79. B. Grünbaum, G. Shephard, *Tilings and Pattern* (Freeman, New York, 1978)
80. J. Gütschow, V. Nesme, R.F. Werner, The fractal structure of cellular automata on abelian groups, in *Automata. 16'th International Workshop on Cellular Automata and Discrete Complex Systems*, ed. by N. Fatés, J. Kari, T. Worsch, pp. 55–74, 2010 DMTCS Proc. AL
81. F.V. Haeseler, H.-O. Peitgen, G. Skordev, Linear cellular automata, substitution, hierarchical iterated function systems and attractors, in *Fractal Geometry and Computer Graphics*, ed. by J. Encarnação, G. Skordev, H.-O. Peitgen, G. Englert (Springer, Berlin, 1992), pp. 3–23
82. F.V. Haeseler, H.-O. Peitgen, G. Skordev, Cellular automata, matrix substitution and fractals. Ann. Math. Artif. Intell. **8**, 345–362 (1993)
83. F.V. Haeseler, H.-O. Peitgen, G. Skordev, Global analysis of self-similar features of cellular automata: selected examples. Phys. D **86**, 64–80 (1995)
84. J. Hardy, Y. Pomeau, O. de Pazzis, Time evolution of a two-dimensional model system. I. Invariant states and time correlation functions. J. Math. Phys. **14**, 1746–1759 (1973)
85. G. Hedlund, Endomorphisms and automorphisms of the shift dynamical systems. Math. Syst. Theory **3**, 320–374 (1969)
86. H. Hermes, *Enumerability, Decidability, Computability* (Springer, Berlin, 1969)
87. H. Hermes, Entscheidungsproblem und Dominospiele, in *Selecta Mathematica II*, ed. by K. Jacobs (Springer, Berlin, 1970), pp. 3–64
88. J. Hofbauer, K. Sigmund, *Evolutionstheorie und Dynamische Systeme* (Parrey Verlag, Hamburg, 1984)
89. R. Honsberger, *Ingenuity in Mathematics* (The Mathematical Association of America, Washington, DC, 1970)
90. J.E. Hopecroft, J.D. Ullman, *Formal Languages and their Relation to Automata* (Addison-Wesley, Reading, MA, 1969)
91. F. Hoppensteadt, J. Keller, Synchronization of periodical cicada emergences. Science **194**, 335–337 (1976)
92. P.K. Hopper, The undecidability of the turing machine immortality problem. J. Symb. Log. **31**, 219–234 (1966)
93. M. Hurley, Attractors in cellular automata. Ergod. Theory Dyn. Syst. **10**, 131–140 (1990)
94. M. Hurley, Attractors in restricted cellular automata. Proc. Am. Math. Soc. **115**, 536–571 (1992)
95. M. Hurley, Noncompact chain recurrence and attraction. Proc. Am. Math. Soc. **115**, 1139–1148 (1992)
96. E. Jen, Linear cellular automata and recurring sequences in finite fields. Commun. Math. Phys. **19**, 13–28 (1988)
97. A. Johansen, A simple model of recurrent epidemics. J. Theor. Biol. **178**, 45–51 (1996)
98. F. John, *Partial Differential Equations* (Springer, Berlin, 1975)
99. C. Jones, Geometric singular perturbation theory, in *Dynamical Systems*, ed. by R. Johnson, Lecture Notes in Mathematics, vol. 1609 (Springer, Berlin/Heidelberg, 1995), pp. 44–118
100. M. Kargapolov, J. Merzljakov, *Fundamentals of the Theory of Groups* (Springer, Berlin, 1979)
101. J. Kari, Reversibility and surjectivity problems of cellular automata. J. Comput. Syst. Sci. **48**, 149–182 (1994)
102. J. Kari, Rice's theorem for the limit sets of cellular automata. Theor. Comput. Sci. **127**, 229–254 (1994)
103. J. Kari, A small aperiodic set of wang tiles. Discret. Math. **160**, 259–264 (1996)

104. J. Kari, On the undecidability of the tiling problem, in *SOFSEM 2008: Theory and Practice of Computer Science*, ed. by V. Geffert, J. Karhumäki, A. Bettoni, B. Preneel, P. Návrat, M. Bieliková (Springer, Berlin, 2008)
105. J. Kingman, S. Tyalor, *Introduction to Measure and Probability* (Cambridge University Press, Cambridge, 1973)
106. B. Kitchens, *Symbolic Dynamics: One-sided, Two-sided and Countable State Markov Shifts* (Springer, Berlin, 1998)
107. D. Kohler, J. Müller, U. Wever, Cellular non-deterministic automata and partial differential equations. Phys. D **311–312**, 1–16 (2015)
108. P. Koiran, M. Cosnard, M. Garzon, Computability with low-dimensional dynamical systems. Theor. Comput. Sci. **132**, 113–128 (1994)
109. R. Kon, Permanence induced by life-cycle resonances: the periodical cicada problem. J. Biol. Dyn. **6**, 855–890 (2012)
110. G. Kristy, Periodical cicadas. Nature **341**, 288–289 (1989)
111. P. Kůrka, Languages, equicontinuity and attractors in cellular automata. Ergod. Theory Dyn. Syst. **17**, 417–433 (1997)
112. P. Kůrka, *Topological and Symbolic Dynamics* (Society of Mathematics, France, Marseilles, 2003)
113. D. Kuske, M. Lohrey, Logical aspects of Cayley-graphs: the monodid case. Int. J. Alg. Comput. **16**, 307–340 (2006)
114. C.G. Langton, Self-reproduction in cellular automata. Phys. D **10**, 135–144 (1984)
115. D. Lind, B. Marcus, *An Introduction to Symbolic Dynamics and Coding* (Cambridge University Press, Cambridge, 1995)
116. S. Lipschutz, *Set Theory* (McGrawHill, New York, 1964)
117. S. Lipschutz, *Theory and Problems of General Topology* (McGrawHill, New York, 1977)
118. A. Machi, F. Mignosi, Garden of Eden configurations for cellular automata on Cayley graphs of groups. SIAM J. Discret. Math. **160**, 44–56 (1993)
119. A. Mann, *How Groups Grow* (Cambridge University Press, Cambridge, 2012)
120. M. Margenstern, The domino problem of the hyperbolic plane is undecidable. Theor. Comput. Sci. **407**, 29–84 (2008)
121. N. Margolus, Physics-like models of computation. Phys. D **10**, 81–95 (1984)
122. D. Markovic, C. Gros, Power laws and self-organized criticality in theory and nature. Phys. Rep. **536**, 41–74 (2014)
123. M. Markus, Modelling morphogenetic processes in excitable media using cellular automata. Biomed. Biochem. Acta **49**, 681–696 (1990)
124. M. Markus, O. Schulz, E. Goles, Prey population cycles are stable in an evolutionary model if and only if their periods are prime. ScienceAsia **28**, 199–203 (2002)
125. H. Meinhardt, *The Algorithmic Beauty of Shells* (Springer, Berlin, 1995)
126. H. Meinhardt, M. Klinger, A model of pattern formation on shells of molluscs. J. Theor. Biol. **126**, 63–89 (1987)
127. J. Milnor, A note on curvature and fundamental group. J. Differ. Geom. **2**, 1–7 (1968)
128. E. Moore, Machine models of self-reproduction, in *Proceedings of Symposia in Applied Mathematics*, vol. 14 (American Mathematical Society, Providence, RI, 1962), pp. 17–33
129. E. Moore, Machine models of self-reproduction, in *Essays on Cellular Automata*, ed. by R. Bellman (American Mathematical Society, Providence, RI, 1977), pp. 17–34
130. D.E. Muller, P.E. Schupp, Context-free languages, group, the theory of ends, second-order logic, tiling problems, cellular automata, and vector addition systems. Bull. Am. Math. Soc. **4**, 331–334 (1981)
131. D.E. Muller, P.E. Schupp, Groups, the theory of ends, and context-free languages. J. Comput. Syst. Sci. **26**, 295–310 (1983)
132. J. Müller, H. Jiang, Graphical limit sets for general cellular automata. Theor. Comput. Sci. **580**, 14–27 (2015)
133. J. Müller, C. Kuttler, *Methods and Models in Mathematical Biology* (Springer, Berlin, 2015)

134. J. Müller, C. Spandl, A Curtis–Hedlund–Lyndon theorem for Besicovitch and Weyl spaces. Theor. Comput. Sci. **410**, 3606–3615 (2009)
135. J. Müller, C. Spandl, Embeddings of dynamical systems into cellular automata. Ergod. Theory Dyn. Syst. **29**, 165–177 (2009)
136. J. Müller, C. Spandl, Embeddings of dynamical systems into cellular automata, erratum. Ergod. Theory Dyn. Syst. **30**, 1271 (2010)
137. J. Murray, *Mathematical Biology* (Springer, Berlin, 1989)
138. J. Myhill, The converse of Moore's Garden-of-Eden theorem. Proc. Am. Math. Soc. **14**, 685–686 (1963)
139. E. Nagel, J. Newman, *Gödel's Proof* (New York University Press, New York, 1958)
140. V. Nekrashevych, *Self-Similar Groups* (American Mathematical Society, Providence, RI, 2005)
141. K. Nishinari, D. Takahashi, Analytical properties of ultradiscrete Burgers equation and rule-184 cellular automaton. J. Phys. A **32**, 5439–5450 (1998)
142. R.E. O'Malley, *Singular Perturbation Methods for Ordinary Differential Equations* (Springer, Berlin, 1991)
143. J.C. Oxtoby, S.M. Ulam, On the existence of a measure invariant under a transformation. Ann. Math. **40**, 560–566 (1939)
144. P. Pansu, Croissance des boules et des géodésiques fermées dans les nilvarietes. Ergod. Theory Dyn. Syst. **3**, 415–445 (1983)
145. A. Popovici, D. Popovici, Cellular automata in image processing, in *Proceedings of the MINTS*, Notre Dame University (2002)
146. J.W.S.B. Rayleigh, *The Theory of Sound 1* (Macmillan, New York, 1894)
147. M. Renardy, R.C. Rogers, *An Introduction to Partial Differential Equations* (Springer, Berlin, 2004)
148. R. Robinson, Undecidability and nonperiodicity of tilings of the plane. Invent. Math. **12**, 177–209 (1971)
149. P.L. Rosin, Training cellular automata for image processing, in *SCIA 2005*. Lecture Notes in Computer Science, vol. 3540 (Springer, Berlin, 2005), pp. 195–204
150. W. Rudin, *Real and Complex Analysis* (McGraw-Hill, New York, 1987)
151. G. Salinetti, R.J.-B. Wets, On the convergence of sequences of convex sets in finite dimensions. SIAM Rev. **21**, 18–33 (1979)
152. T. Sato, Decidability for some linear cellular automata over finite rings. Inf. Process. Lett. **46**, 151–155 (1993)
153. D. Scholz, *Pixelspiele* (Springer, Berlin, 2014)
154. B. Schönfisch, A. de Roos, Synchronous and asynchronous update in cellular automata. BioSystems **51**, 123–143 (1999)
155. M.A. Smith, Representations of geometrical and topological quantities in cellular automata. Phys. D **45**, 271–277 (1990)
156. J. Smoller, *Shock Waves and Reaction-Diffusion Equations* (Springer, Berlin, 1983)
157. E. Stein, Theoria cum Praxi. Leibniz als technischer Erfinder, in *Der universale Leibniz*, ed. by T. Reydon, H. Heit, P. Hoyningen-Huene (Franz Steiner Verlag, Stuttgart, 2009)
158. K. Sutner, Linear cellular automata and de Bruijn automata, in *Cellular Automata*, ed. by M. Delorme, J. Mazoyer, Mathematics and Its Applications, vol. 460 (Springer, Berlin, 1998), pp. 303–320
159. K. Sutner, Computational classification of cellular automata. Int. J. Gen. Syst. **41**, 1–13 (2012)
160. B. Szendroi, G. Csányi, Polynomial epidemics and clustering in contact networks. Proc. R. Soc. B **271**, S364–S366 (2004)
161. G. 't Hooft, On the foundations of superstring theory. Found. Phys. **43**, 46–53 (2013)
162. S. Takesue, Staggered invariants in cellular automata. Complex Syst. **9**, 149–168 (1995)
163. T. Toffoli, N. Margolus, Invertible cellular automata: a review. Phys. D **45**, 229–253 (1990)
164. A. Turing, The chemical basis of morphogenesis. Philos. Trans. R. Soc. Lond. B **237**, 37–72 (1952)

165. A. Vazquez, Polynomial growth in branching processes with diverging reproductive number. Phys. Rev. Lett. **96**, 038702 (2006)
166. G.Y. Vichniac, Simulating physics with cellular automata. Phys. D **10**, 96–116 (1984)
167. R. Vollmar, *Algorithmen in Zellularautomaten* (Teubner, Stuttgart, 1979)
168. J. von Neumann, *Theory of Self-reproducing Automata*, ed. by A.W. Burks (University of Illinois Press, Illinois, 1966)
169. L. Vuillon, Balanced words. Bull. Belg. Math. Soc. Simon Stevin **10**, 787–805 (2003)
170. C. Waddington, R. Cowe, Computer simulation of a molluscan. J. Theor. Biol. **25**, 219–225 (1969)
171. P. Walters, *An Introduction to Ergodic Theory* (Springer, Berlin, 1982)
172. H. Wang, Proving theorems by pattern recognition II. Bell Syst. Tech. J. **40**, 1–42 (1961)
173. K. Weihrauch, *Computable Analysis* (Springer, Berlin, 2000)
174. J.R. Weimar, J.J. Tyson, L.T. Watson, Cellular automaton models for reaction diffusion equations, in *Proceedings of Sixth Distributed Memory Computing Conference*, ed. by Q. Stout, M. Wolfe (IEEE Computer Society, Los Alamitos, 1991), pp. 431–434
175. N. Wiener, A. Rosenblueth, The mathematical formulation of the problem of conduction of impulses in a network of connected excitable elements, specifically in cardiac muscle. Arch. Inst. Cardiol. Mex. **16**, 205–265 (1946)
176. K. Williams, C. Simon, The ecology, behavior, and evolution of periodical cicadas. Annu. Rev. Entomol. **40**, 269–295 (1995)
177. S.J. Willson, Cellular automata can generate fractals. Discret. Appl. Math. **8**, 91–99 (1984)
178. S.J. Willson, Computing fractal dimensions for cellular automata. Phys. D **24**, 190–206 (1984)
179. S.J. Willson, The equality of fractal dimensions for certain cellular automata. Phys. D **24**, 179–189 (1984)
180. D. Wolf-Gladrow, *Lattice-Gas Cellular Automata and Lattice Boltzmann models. An Introduction*. Lecture Notes in Mathematics, vol. 1725 (Springer, Berlin, 2000)
181. S. Wolfram, Statistical mechanics of cellular automata. Rev. Mod. Phys. **55**, 601–644 (1983)
182. S. Wolfram, Cellular automaton fluids 1: basic theory. Stat. Phys. **45**, 471–707 (1986)
183. S. Wolfram, *A New Kind of Science* (Wolfram Media, Boca Raton, 2002)
184. X.-S. Yang, Y. Young. Cellular automata, PDEs, and pattern formation, in *Handbook of Bioinspired Algorithms and Applications*, ed. by S. Olariu, A.Y. Zomaya (Chapman & Hall/CRC Press, Boca Raton, 2005), pp. 271–282
185. K. Yosida, *Functional Analysis* (Springer, Berlin, 1980)
186. S. Yukita, Tesselation automata on free groups. Hiroshima Math. J. **25**, 561–570 (1995)

Index

K.-P. Hadeler, J. Müller, *Cellular Automata: Analysis and Applications*,
Springer Monographs in Mathematics, DOI 10.1007/978-3-319-53043-7

GPSR Compliance
The European Union's (EU) General Product Safety Regulation (GPSR) is a set of rules that requires consumer products to be safe and our obligations to ensure this.

If you have any concerns about our products, you can contact us on

ProductSafety@springernature.com

In case Publisher is established outside the EU, the EU authorized representative is:

Springer Nature Customer Service Center GmbH
Europaplatz 3
69115 Heidelberg, Germany

www.ingramcontent.com/pod-product-compliance
Ingram Content Group UK Ltd.
Pitfield, Milton Keynes, MK11 3LW, UK
UKHW022031190726
13853UKWH00005B/2191